AF815266

Building Sustainable Communities

Md. Nurul Momen
Rajendra Baikady • Cheng Sheng Li
M. Basavaraj
Editors

Building Sustainable Communities

Civil Society Response in South Asia

palgrave
macmillan

Editors
Md. Nurul Momen
Department of Public Administration
University of Rajshahi
Rajshahi, Bangladesh

Cheng Sheng Li
Department of Social Work
Shandong University
Shandong, China

Rajendra Baikady
Paul Baerwald School of Social Work and Social Welfare
Hebrew University of Jerusalem
Jerusalem, Israel

Department of Social Work
University of Johannesburg
Johannesburg, South Africa

M. Basavaraj
Dept of Economic Studies and Planning
Central University of Karnataka
Kadganchi, Karnataka, India

ISBN 978-981-15-2392-2 ISBN 978-981-15-2393-9 (eBook)
https://doi.org/10.1007/978-981-15-2393-9

© The Editor(s) (if applicable) and The Author(s), under exclusive licence to Springer Nature Singapore Pte Ltd. 2020
This work is subject to copyright. All rights are solely and exclusively licensed by the Publisher, whether the whole or part of the material is concerned, specifically the rights of translation, reprinting, reuse of illustrations, recitation, broadcasting, reproduction on microfilms or in any other physical way, and transmission or information storage and retrieval, electronic adaptation, computer software, or by similar or dissimilar methodology now known or hereafter developed.
The use of general descriptive names, registered names, trademarks, service marks, etc. in this publication does not imply, even in the absence of a specific statement, that such names are exempt from the relevant protective laws and regulations and therefore free for general use.
The publisher, the authors and the editors are safe to assume that the advice and information in this book are believed to be true and accurate at the date of publication. Neither the publisher nor the authors or the editors give a warranty, expressed or implied, with respect to the material contained herein or for any errors or omissions that may have been made. The publisher remains neutral with regard to jurisdictional claims in published maps and institutional affiliations.

Cover illustration: eStudio Calamar

This Palgrave Macmillan imprint is published by the registered company Springer Nature Singapore Pte Ltd.
The registered company address is: 152 Beach Road, #21-01/04 Gateway East, Singapore 189721, Singapore

Foreword

The Task of Building Sustainable Communities

There have been repeated warnings through the media, as a follow-up of the intensive research which has been carried out across the world, to change the way we live and to look for sustainable options. The Himalayan glaciers are melting twice as fast as they were at the start of the twenty-first century (HT 21.06.2019). This is threatening water supply for more than 800 million people, mostly in South Asia. The glaciers are already one-fourth smaller than what they were 40 years ago. This could lead to both floods and droughts. What South Asia will earn in development, it will lose in disasters. Progress, development and livelihood in the coming century would largely depend upon the way nations reduce, reuse and recycle the earth's resources. This is a responsibility which the state may not be able to handle alone and therefore the rationale that a civil society should become a partner to the state rather than simply a beneficiary, recipient and a passive spectator. However, the problem is much beyond defining a partnership role for a civil society. Due to the rise of marketization, globalization and the increasing influence of dominant groups in representative democracy, communities have been losing their traditional, cultural and geographical bonding. To reclaim community bonding, civil society has to rise and respond to its challenges.

A multi-pronged effort is required to build 'Sustainable Communities' and this suggests that the task is too convoluted to be left to the government alone. This book uncovers various micro- and macro-level efforts and instances in which the civil society responded to this challenge and became partners in such a mission. In conventional public administration literature, communities are by themselves considered sustainable and on that logic much effort has been expended in public policy to conserve this surviving 'community system' or to plan programmes within the context of community systems.[1] However, this book refers to case studies relating to how unsustainable communities have been able to explore solutions and strategies to build a sustainable community. Since the critical feature of 'sustainability' embedded within a community seems threatened or lost, the remedies and anthropological narratives suggested in this book would present a meaningful text.

Considering their common vulnerabilities that stem from underdevelopment and environmental challenges, there is greater expectation from South Asian civil societies to join efforts in building sustainable communities in genuine partnerships, since sustainable community is the key to development. It should be restored and reclaimed through civil society efforts. The many authors of this book have brought multi-level case studies to strengthen the belief that civil society efforts can salvage the damage within communities. There is an epistemological question which the book triggers on the nature of civil society vis-à-vis communities since in many studies, and sometimes during programme implementation, the two metaphors are mistakenly used interchangeably. While communities are mostly inward driven and are ready to bend backwards in search of their autonomy, sustenance and sustainability, a civil society is a compact of many loosely tied interest groups ready to bargain with the state. In short, while communities are embedded in a natural

[1] Many studies in Public Administration have been conducted on the belief that communities are a sustainable fact of any policy. While the legendary scholar F.W. Riggs's whole framework of 'Ecology of Public Administration' (1961) is about community-driven implementation, there are continuing discourses in the discipline which reiterate its importance, i.e. Kapucu, N. (2016); Wade, R. (1988); Mansuri, Ghazala and Vijayendra Rao (2003); M. Das Gupta, H. Grandvoinnett and M. Romani (2004); UNHCR (2001).

ecological phenomenon for survival and resilience building, civil society can be competitive within as much as outside. Interestingly, the efforts of civil society suggested in the chapters of this book demonstrate that they may catalyse the process and generate sustainability.

This book investigates the role of civil society in its mission towards restoring sustainable communities. Civil society is recognizable within government as the most uncivil (Glasius 2010) segment of an otherwise symmetrically structured and sophisticatedly regulated state. It is perceived to be informal, unorganized, noisy and mostly impulsive. It is seen to value instinctive responses and provoke rebellious passions which may sometimes cross over to the category of anti-state militants or insurgents (Karriem and Benjamin 2016). Rumford (2001) prefers to define an 'uncivil society' as a 'catch-all term for a wide range of disruptive, unwelcome and threatening elements deemed to have emerged in the spaces between the individual and the state'. On the contrary, civil societies obstinately remain fastened to the state carrying an infallible belief that it plays a major role in the retention of democracy, rights of the vulnerable and the constitutional spirit of inclusive governance. This civil society is believed to be an intangible life-force of a Western state, while communities represent a more natural phenomenon for South Asian societies, which have a long history of their land habitations. Whatever a thin line of distinction between the two, a civil society is an enigma, but always hopeful of its might to achieve an Arcadian bliss when it would bloom!

An Age-Old History of Civil Society in South Asia

The authors are also concerned about globalization as defacement of civil society but micro-level case studies by many others have more to tell. These studies re-inspire faith and prowess of not just people who constitute civil societies across South Asia but also the bureaucracy which is left with no option but to match the energy of civil society organizations (CSOs). The book documents an analysis of CSOs where it is accepted that despite cross-cultural variations, CSOs are a fact of current dispensation as they march through sustainable options for growth and advancement. No one has a deeper reach to an understanding of CSOs than the people of South Asia due to their agrarian reference point, which dates back to some of the oldest anecdotes from non-state voices which were documented much before the Westphalian State[2] came into being. Most ancient kingdoms of South Asia were mostly reflections of their civil societies till mega wars broke down communities into small kingdoms and then into nations. Starting from the oldest works, the *Rig Veda*[3] (earlier than 5000 BC) down to the *Samhitas* (1500 BC) and *Puranas* (500 BC), one can observe the spirit of free will, human rights, conservation of species, environmental protection and the nature of social movements including rebellion against the state as an assertion of good and sustainable governance. It is always enlightening to explore civil society in the lands of South Asia, which has a history of more than 6000 BCE, carrying the world's oldest narratives on civil society, passed on to the famous, completely original, Comilla Experiment[4] of Bangladesh and the Orangi Project of Pakistan.

Civil Society has carried a deep influence on governance everywhere in South Asia. The words *Sabha* and *Samiti* have their origin in both the *Rig Veda* and the *Atharva Veda*. The word *Republic* appears 40 times in the *Rig Veda* and 9 times in the *Atharva Veda*. The *Sabha*, a gathering of people, was crucial for governance as it represented the collective consciousness of the people, called *Samjnana*. This gathering of people was expected to speak powerfully—*Samvadaddhvam*, generate a union of minds—*Sammanah*, and hearts—*Samachittam*, and of hopes and aspirations—*akuti*. This concept of a civil society, carried through most of the ancient texts produced in this region, reflects on the nature of civil

[2] The term 'Westphalian State' emerges from the Treaties of Westphalia in 1648. It refers to a state possessing monopoly of force within their mutually recognized territories and a set of rules which govern inter-state and intra-state relations.

[3] See Lok Tilak (2004) and Kazanas (2002).

[4] The Comilla Model (1959) of Bangladesh and the Orangi Project (1980s) of Pakistan were both celebrated community development initiatives launched by Akhtar Hameed Khan, born in Agra (India), for rural and urban communities, respectively. Even if they under-achieved, the programmes pushed several successful micro-credit programmes such as the BRAC(NGO) and Grameen Bank by Dr. Md. Yunus and Fazle Hasan Abed.

society. The Vedas, Valmiki Ramayana and the Mahabharata of the seventh–fifth century BCE, which is taken to and absorbed in a rational analysis found in Kautilya's *Arthashashtra* (Boesche 2002), third century BCE, a treatise on economics and the economy of nations much before Adam Smith's *Wealth of Nations*. This region also produced meaningful narratives on civil society in Buddhist literature, some of which could also be reflected in the oldest available Bengali poetic volume *Charayapada* of the tenth–eleventh century and Islamic epics in the works of Syed Sultan, Abdul Hakim and the secular texts of Alaol. The coming of Kazi Nazrul Islam re-ignited the praxis of civil society through his celebrated works *Bidrohi* (The Rebel) and his newspaper *Dhumketu* (The Comet) between 1915 and 1922. In much the same spirit, Prof. Nurul Momen wrote *Nemesis* and a number of other works in 1944 and sometime later when ordinary masses were making efforts to change their destinies trapped in political and natural calamities. Going downwards to the coastal rim of South Asia, Sri Lanka's civil society literature demonstrated a phenomenal growth and expansion with the establishment of the Ceylon Bible Society in 1812 and the Mahabodhi Society in 1891. The Ceylon Social Reform Society of 1905 has contributed immensely to cultural conservation of Sri Lankan civil society values. Ironically, much has remained unstudied and unresearched and the scholars in South Asia have missed out on their indigenous wealth under the dominant Western publications. Authors admit that it is difficult to delve into the deep mine of this regional literature in the absence of sufficient archival conservation and due to colonial influences. However, it is about time that the focus is shifted inwards and the inner strengths of the region are explored.

How the Western and South Asian Perspectives on Civil Society Differ

There is a sharp contrast in Western and South Asian perspectives on civil society and this is what constitutes the much brutal interrogation into the very concept of civil society and its uncivil peripheral fragments. There is not much difference in the classical perspective of civil society as understood in India, Bangladesh, Sri Lanka, Nepal and Pakistan due to their much common and overlapping regional history. In clear deviation from an understanding where loosely tied individuals come together for the fulfilment of collectively held interest, South Asians mix in its meaning a complex of regionally determined ethics, morality and sensitivity towards local traditions. It would be nice to read what an outsider has to say on this difference. Sebastian Velassery (2004) writes on the classical Indian political tradition of civil society but it seems quite acceptable for the region. Velassery writes (pp. 39–40) that this tradition is 'society-centred as it subordinated the state and government to societal mandates. This is the idea of an active involvement and participation of a group of people as a community in the conduct of various institutions and organizations they belong to in their day-to-day living and this may exist in a system where government at the centre is paternalistic'. He further found that most of these groups were led by hereditary, caste and local community based laws and norms, as a result of which they enjoyed sufficient internal autonomy and were left free from direct king's interference. This suggests that the prevalent understanding of civil society from a South Asian perspective was more organic and interdependent yet clearly role based, which differed from the Western concept of civil society as a tenuous aggregate of independent self-contained individuals held together by self-interest.

The Western–South Asian difference explains how civil societies in the South Asian region were mostly equivalent to village councils and formed little republics. Some of the primordial concepts are still reflected in the northeastern India, which has a civil society identifiable by a particular community residing on a particular hill. They have their own markets, community-specific artefacts, clothes and norms to follow. The Meitis, Paietes and Kukis or Ukhrul Nagas can never form a common civil society. Similarly, the Jats and Gujjars of western and north-western India follow different norms in accordance with their caste-specific civil societies. Such community-based segregations traverse the region. In Bangladesh, India and Pakistan (and maybe in Sri Lanka and Nepal too) the so-called organic civil societies even constituted their own justice dispensation arrangements in the form of local communitarian court systems, that is, *Shalish* in Bangladesh and India, *Khap Panchayats* in India and *Jirgahs* in Pakistan and Afghanistan. Nepal has *arbitration panchayats* and *village development committees* acting as community mediators. Such a community-linked and tradition-embedded civil society is an asset if governments can channelize their energy and mainstream them into national development. On the other hand, if left unattended, it could be a catalyst to regressive, patriarchal, orthodox and fundamentalist led obstruction to holistic and globally competitive development. Bangladesh government has brought some grassroots changes to link district courts to the Shaleesh in a village. Some of the South Asian countries also practise restorative justice systems with the help of their civil society. Sustainability is the key outcome if these local community groups are brought into mainstream governance and fulfil their share of responsibilities.

Addressing Sustainable Communities and Civil Society

There have been inherent intellectual difficulties in the Western world which has prevented an understanding about Asia in its own right and within its own history and legacy of civilization (Cheung 2013, pp. 249–261). Subsequent to Soviet collapse, Fukuyama mistakenly declared the 'End of History' (Fukuyama 1992) and a triumph of Western liberal ideology. He ignored the emergence of smaller but powerful nation states identical to the 'Rise of the Ants'[5] or the prevalence of burly trans-boundary micro-level forces which were constantly challenging this notion.[6] As a result, South Asia missed an important opportunity to centre its development on communities and grow into their culturally well-rooted and user-friendly structures of governance.

Comparative public administration unravels the mysteries of the historical evolution of community-based institutions spread between cultural relativism and universalism. Riggs worked at the convergence point

[5] A metaphor picked up from the world of science symbolic of natural changes Wilson and Holldobler (2005).

[6] See, Marks, Susan, (1997) reflections on legal thesis.

of history, culture and contemporary socio-economic framework to ignite a discourse on ecology-specific values, that is, Asian values and Asian politics and society as given by Dalton and Shin (2006) and Kim (2010). Under these circumstances, sustainability remained a major problem of governance. A few of these problems are listed in the following:

1. Most developmental initiatives were more individualistic than institution based and that brought to them a premature death once the official was transferred.
2. There was a political aspiration or a dominant vote bank behind an initiative rather than a genuine need for social concern or particular need for development.
3. Planning was mostly elitist and delinked to the ground realities and therefore worked on Western borrowed and implanted frameworks.
4. The legal framework lagged behind the changing times and instead of facilitating implementation, it obstructed and delayed it, only to increase corruption and local mafias.
5. There was no machinery for handling emergencies like floods, landslides, hurricanes, tornadoes and fires which collapsed to a dead end even the best worked out initiatives.

The above five basic causes to unsustainable governance diverted attention in the mid-1990s towards stronger norms for accountability, transparency and performance audits through budget and expenditures. Tough procurement procedures were introduced in departments and wherever government funding was disbursed. Many colonial laws were changed to give way to more appropriate and relevant measures to curb delays, obstructions and corruption. The early years of the twenty-first century brought a speedy investigation of laggard norms of governance to open free spaces for local communities to perform and help as partners in governance. The introduction of e-governance in service delivery was a major transformation in public life. South Asian countries adopted the Right to Information Acts, Public Disclosures Laws and also reforms, even if minimal, were brought to the Representation of People's Acts, which improved public elections to state and central legislatures.

It was during this time that the World Development Report (2004) *Making Services work for the poor people* was published to focus upon some of the most basic services to human development. These included education, health, water, sanitation and electricity. Many dimensions to poverty were highlighted and governments were asked to work towards them. Recently, in 2014, the Indian government's major initiative on 'Swachh Bharat Abhiyaan' has already started proving right the wisdom which the experts had put in the 2004 Report. It is highlighted that sanitation is becoming a key to rising ambition to work and do better. The World Development Report 2004 had set a holistic agenda for the region to scale up sectoral reforms by linking them to public sector, budget management and decentralized administrative reforms. In India, the disbursement to civil society groups increased and there was a mushrooming of Self-Help Groups (SHGs), Community-Based Organizations (CBOs) and voluntary groups in the process of materializing, implementing and outreaching the demand groups, beneficiaries and the stakeholders. The situation was so diffused for the sovereign state that scholars started declaring that the state was hollowing out (Rhodes 1994) or retreating (Strange 1996). As Rhodes (1994) further explained that 'hollowing out' referred to a loss of state capacity and a 'shift from a unitary state to a differentiated polity' (1997, p. 19). The splattering by civil society groups appeared to be eroding or eating away the state but Rhodes was also clear that this neither increased state efficiency nor its effectiveness (2017, p. 119). However, the public sphere became a network of competing civil society groups rather than a Weberian pyramid of known functionaries. While analyzing this changing situation replete with examples of rising influence of marketization and subsequent unbundling of public sector undertakings, deregulations and corporatization led many scholars to call the process a form of a 'predatory state'(Olson 1993; Galbraith 2008).

The rise of civil society inadvertently carried the mission to fill up gaps of accountability and state capacity but it turned to neither of the two. There was a felt need for a deadline of performance and the declaration on Sustainable Development Goals established a meaningful direction within the time availability. The Goal No. 11 on 'sustainable cities and communities' is directed to protect habitats and life as the world grows towards unmanageable and dangerous proportions of growth of population, land-scarcity, environmental degradation and challenges of service delivery. It is amply understandable that there is but one last effort to be made by humanity to save this planet, which is to generate sustainable communities. The Hyogo Declaration 2005, and later the Sendai Framework for Disaster Risk Reduction (2015–2030), emphasizes on community resilience building and 'to build back better in recovery, rehabilitation and reconstruction'. This task is possible only with a vibrant, involved and a committed civil society participating in the task of governance.

Conclusions

Communities and the civil society encounter new responsibilities and new challenges in the current phase of development. The marketization of the 1990s is believed to have disrupted or scattered community bondings and the rise of the civil society is one of the most appropriate options to reclaim and restore sustainability, which once came naturally within communities. Building sustainable communities is one great fiat for development and growth in current times. The two concepts are embedded in the history of South Asia, yet by adopting the Western implanted models. this region has weakened its pace of development. Finally, the effort to build sustainable communities through civil society responses is a reminder that this multi-pronged process requires socio-legal-cultural initiatives simultaneously to build resilience and address issues of state capacity.

Centre for the Study of Law and Governance
Special Centre for Disaster Research
Jawaharlal Nehru University, New Delhi, India

Amita Singh

References

Boesche, R. (2002). *The First Great Political Realist: Kautilya and His Arthashastra*. Lanham: Lexington Books.

Cheung, A. B. L. (2013). Can There Be an Asian Model of Public Administration. *Public Administration and Development, 33*(4), 249–261.

Dalton, R., & Shin, D. C. (Eds.). (2006). *Citizens, Democracy and Markets Around the Pacific Rim: Congruence Theory and Political Culture*. Oxford: Oxford University Press.

Das Gupta, M., Grandvoinnett, H., & Romani, M. (2004). State-Community Synergies in Community-Driven Development. *Journal of Development Studies, 40*(3), 27–58.

Fukuyama, F. (1992). *The End of History and the Last Man*. New York: The Free Press.

Galbraith, J. K. (2008). *The Predator State: How Conservatives Abandoned the Free Market and Why Liberals Should Too* (p. 208). Free Press.

Glasius, M. (2010). Uncivil Society. In H. K. Anheier & S. Toepler (Eds.), *International Encyclopedia of Civil Society*. New York, NY: Springer.

Kapucu, N. (2016). Community-Based Research in Generating Usable Knowledge for Public Policy and Administration. *Administration & Society, 48*(6), 683–710.

Karriem, A., & Benjamin, L. M. (2016). How Civil Society Organizations Foster Insurgent Citizenship: Lessons from the Brazilian Landless Movement. *Voluntas: International Journal of Voluntary and Nonprofit Organizations, 27*(1), 19–36. Retrieved from https://doi.org/10.1007/s11266-015-9549-3.

Kazanas, N. (2002). Indigenous Indo-Aryans and the *Rigveda* (PDF). *Journal of Indo-European Studies, 30*, 275–334. Retrieved 30 December 2009.

Kim, S. Y. (2010). Do Asian Values Exist? Empirical Tests of the Four Dimensions of Asian Values. *Journal of East Asian Studies, 10*, 315–344.

Mansuri, G., & Rao, V. (2003). *Evaluating Community-Based and Community-Driven Development: A Critical Review of the Evidence, Development Research Group*. Washington, DC: World Bank.

Marks, S. (1997). The End of History? Reflections on Some International Legal Thesis. *European Journal of International Law*, No. 449. Retrieved from www.ejil.org/journal/Vol8/No3/art5.pdf.

Olson, M. (1993). Dictatorship, Democracy and Development. *The American Political Science Review, 87*(3), 567–576. Retrieved July 1, 2019, from http://www.jstor.org/stable/2938736.

Rhodes, R. A. W. (1994). The Hollowing Out of the State. The Changing Nature of Public Services in Britain. *Political Quarterly, 65*(1), 138–151.
Rhodes, R. A. W. (1997). *Understanding Governance: Policy Networks, Governance, Reflexivity and Accountability* (Public Policy & Management) (p. 19). Philadelphia: Open University.
Rhodes, R. A. W. (2017). *Network Governance and the Differentiated Polity: Selected Essays* (Vol. 1). London: Oxford University Press.
Riggs, F. W. (1961). *The Ecology of Public Administration*. Bombay: Asia Publishing House.
Rumford, C. (2001). Confronting 'Uncivil Society' and the 'Dark Side of Globalization': Are Sociological Concepts Up to the Task? *Sociological Research Online, 6(3)*, 1–4. https://doi.org/10.5153/sro.640.
Strange, S. (1996). *The Retreat of the State: The Diffusion of Power in the World Economy*. Cambridge: Cambridge University Press.
Tilak, L. (2004). *Indus Civilization*. Delhi: Discovery Publishing House.
UNHCR. (2001, February 15). *Reinforcing a Community Development Approach*. EC/51/SC/CRP.6.
Velassery, S. (2004). Constitution of a Rational Society: A Kautilyan Text, Chap. III. In W. A. Barbieri, R. Magliola, & R. Winslow (Eds.), *Civil Society: Who Belongs?* Washington, DC: Council for Research in Values and Philosophy.
Wade, R. (1988). *Village Republics: Economic Conditions for Collective Action in South India*. Cambridge: Cambridge University Press.
Wilson, E. O., & Holldobler, B. (2005). The Rise of the Ants: A Phylogenetic and Ecological Explanation. *PNAS, 102*(21), 7411–7414. https://doi.org/10.1073/pnas.0502264102.
World Bank. (2004). *World Development Report 2004: Making services Work for Poor People - Overview (English)*. World Development Report. Washington, DC: World Bank Group. Retrieved from http://documents.worldbank.org/curated/en/527371468166770790/World-Development-Report-2004-Making-services-work-for-poor-people-Overview.

Acknowledgment

Editing this book started with many challenges and endless learning. The contents of this book were shaped by the discussion and debates between junior and senior scholars from diverse social science subjects. Commitment, dedication and an aspiration for new learning among all contributing authors resulted in this book. This book presents truly interdisciplinary and international content pertaining to civil society organization and social welfare in the South Asian region.

We the editors sincerely thank all our contributing authors for their valuable contributions, and also thanks go to many other colleagues who offered advice, criticism, and encouragement in various settings. We specially mention the support of Prof. Nasreen Aslam Shah (Pakistan), Dr. Nadarajah Pushparajah (Sri Lanka), Bharat Bushan (India), Dr. Jilly Johan (India), Dr. Ashok D Souza (India), AHM Kamrul Ahsan (Bangladesh), Vasudeva Sharma (India), Dr. Anurada Koumodurai (India), Prof. Awal Hossain Mollah (Bangladesh), Dr. Aslam Khan (Ethiopia), Kanagarajan Eswaran (India), Mahbub Alam Prodip (Bangladesh), Golam Rabbani (Bangladesh), Prof. Md. Faruque Hossain (Bangladesh), Dr. Anjali Kulkarni (India) with peer reviewing chapters.

In addition, a big thanks goes to Prof. Matthew McCartney (University of Oxford), Prof. M. Shamsur Rahman (Ex-Vice Chancellor, Jatiyo Kobi Kazi Nazrul Islam University), Prof. Vimla V. Nadkarni (TISS Mumbai), Prof. He Xuesong (East China University of Science and Technology),

Prof. Jing Guo (University of Hawaii at Manoa), Prof. GAO Jianguo (Shandong University), Sandeep Kaur (Palgrave Macmillan, New Delhi), Sagarika Ghosh (Palgrave Macmillan, New Delhi), NAPSIPAG Sri Lanka and the International Association of Schools of Social Work (IASSW) for their constant support and guidance at various stages of preparation of this book.

We gratefully acknowledge the many friends and family who supported us along the journey with their expertise, encouragement and stimulating suggestions.

Md. Nurul Momen
Rajendra Baikady
Cheng Sheng Li
M. Basavaraj

Contents

Notes on Contributors

Akbaruddin Ahmad is Chairman, Policy Research Centre.bd (PRC.bd); Chairman (Admn), NAPSIPAG (Network of Schools and Institutes of Public Administration and Governance of the Asia Pacific Region); Chief Editor, *The Weekly Finance World* and former Vice Chancellor, DarulIhsan University.

Jewel Ahmed is a sector specialist at BRAC, the Bangladesh-based world's largest NGO, with its Gender Justice and Diversity Programme. Before joining BRAC he worked at Shapla Mahila Sangsta, a Bangladesh-based local NGO, with its Justice for Marginalised programme as Associate Coordinator. He has been working in the development sector for the last 6 years. He received his undergraduate and graduate degrees in Public Administration from the University of Rajshahi, Bangladesh with excellent academic results. His research interests include gender, women's rights, NGOs and Civil society, etc.

Md Zahir Ahmed is Junior Research Coordinator at the Policy Research Centre, Dhaka, Bangladesh. He is currently pursuing his doctoral degree at the Northwest Normal University, China. Previously, he has worked with Policy Research Centre (PRC.bd) as Junior Research Coordinator.

He was former Research Assistant, Government of People's Republic of Bangladesh. His research areas are sexual harassment, disaster management, domestic violence, e-governance, social development, and youth empowerment. He holds a BS and MS in Psychology from the Department of Psychology, University of Dhaka. Alongside, he has also studied professional courses under the University of London.

Tahmina Akhtar is a professor in the Institute of Social Work and Research (ISWR), University of Dhaka, Bangladesh. Working since 1991, Akhtar specializes in psychotherapy and counseling. Her research activities have been concentrated on juvenile delinquents, Community Mental Health, and HIV/AIDS.

Touhid Alam is a senior research associate at Policy Research Centre.bd (PRC.bd), a civil society think-tank in Bangladesh and a member of NAPSIPAG (Network of Schools & Institutes of Public Administration and Governance of the Asia Pacific Region).

Isahaque Ali is a fellow at the Social Work Section, School of Social Sciences, Universiti Sains Malaysia, Penang, Malaysia. He was a senior lecturer in the Sociology and Social Work Department, Gono University, Dhaka, Bangladesh. He was a postdoctoral fellow at the Social Work Section, School of Social Sciences, USM during 2015–2016. He is an honorary research fellow with Comparative Research Programme on Poverty (CROP), International Social Science Council (ISSC) and University of Bergen (UiB), Norway, from 19 January 2015 to 31 December 2018.

Prince Annadurai, MSW, MA (Psychology), PhDDr, is an assistant professor in the Department of Social Work for the last 18 years and also Dean of Student Affairs for the past year at Madras Christian College (Chennai, India). He has extensive experience in the fields of social work and psychology and has also lectured in international universities such as Chan Jung Christian University, China, and Thompson River University, Canada.

Azlinda Azman is Professor and Dean, School of Social Sciences, Universiti Sains Malaysia (USM), Penang, Malaysia. She is also the Convenor of the AIDS Action and Research Group (AARG), USM. A

Fulbright Scholar, she holds a PhD in Clinical Social Work from New York University. Her fields of expertise include social work education/curriculum, theory and methods in social work and social work research. Her areas of research interest include community development, family institution and well-being, poverty, HIV/AIDS and drug-related issues. Azlinda is an executive committee member of the Malaysian Association of Social Workers (MASW). Within the short span of her career as a professional social worker, she is now enthusiastically advocating for the development of Standards for Social Work Curriculum and Competency as well as the Social Worker's Bill for the country. She chairs the National Joint Council Committee on Social Work Education. As Convenor of AARG, USM, she regularly conducts workshops and trainings for agencies and NGOs providing direct services to drug-addiction and HIV/AIDS affected populations, while actively doing research in the areas of HIV/AIDS. AARG manages two government-supported Needle Syringe and Exchange Programme (NSEP) sites each in the states of Penang and Perak (Sungkai and Jelapang), respectively. AARG is the pioneer institution operating the NESP programmes since its introduction in 2006 to mitigate the spread of HIV and AIDS. Azlinda is also the Honorary Secretary of the Malaysian AIDS Council (MAC) (2018–2020) and has immensely contributed in advocating for issues related to HIV and AIDS.

Rajendra Baikady is a Social Work educator and researcher. He is the winner of the Golda Meir Postdoctoral Fellowship at the Hebrew University of Jerusalem, Israel (2019–2020) and the Confucius Studies Understanding China Fellowship (Postdoctoral Research) at Shandong University, Peoples Republic of China (2018–2019). Baikady was one among the 20 selected candidates for the Short-Term Research Award (STRA) by the Ministry of Education, Government of Taiwan, and conducted research at the National Chengchi University, Taiwan during June–July 2018. Baikady was awarded the prestigious INLAKS foundation Research Travel Grant 2015 and Indian Council of Social Science Research, Collect Data abroad Scholarship—2015 for conducting research at Shandong University, China. He has published nearly 15 peer-reviewed research papers and 29 book chapters. His most recent books (co-edited) are *Social Welfare Policies and Programmes in South Asia* (Routledge) and *The Palgrave Handbook of Global Social Work Education* (Palgrave Macmillan), *Social Welfare in India and China—A Comparative Perspective* (Palgrave Macmillan). He is co-editor of the journal special issue of *Social Work and Society* (to be published in 2020) and also has ongoing international collaboration with researchers from China, Israel, Japan, Slovenia and South Africa. Presently, he is at the Hebrew University of Jerusalem Israel and conducting research under the supervision of John Gal. In addition, he is also affiliated to the Department of Social Work, University of Johannesburg, South Africa as Senior Research Associate (a special appointment). His ongoing projects are the *Routledge Handbook of Field Work Education in Social Work* and the *Routledge Handbook of Poverty in the Global South*.

M. Basavaraj is Assistant Professor of Economics at the Department of Economic Studies and Planning School of Business Studies, Central University of Karnataka, India since 2012. Prior to this, he worked 5 years as a lecturer in Economics in Undergraduate and Pre-University Colleges and also qualified for the University Grants Commission Junior Research Fellowship and Meritorious scholarship in Master's Degree. He holds a Master of Philosophy and a Doctor of Philosophy in Economics from S.V. University, Tirupati and the Department of Economic Studies and Research, Gulbarga University, Kalaburagi, respectively. He has published two books, many research articles in national and international journals and presented many papers in various conferences; one of the research papers was appreciated at ISBM-2014-Meiji University, Tokyo, Japan. His research mainly focuses on regional imbalance and development, small-scale industries, agriculture, economics, social welfare, etc.

V. J. Byra Reddy holds a PhD in Business Administration from Mangalore University and a postgraduate in Economics from Bangalore University. His areas of teaching interests include managerial economics, advanced micro economics, international business, research methods and public policy. He has more than 20 years of experience teaching postgraduate management courses and four years of industry experience at the entry and middle level management. Presently, he is a professor in the School of Business and Head, School of Public policy at the University of Petroleum and Energy Studies [UPES], Dehradun, India. In his current role, he is steering International and National accreditations at UPES. He is one of the pioneers of Quality in Business Education and currently serves on the board of one of the largest Business Education accrediting bodies based out of the USA—the International Accreditation Council for Business Education [IACBE]. Recognizing his efforts in the area of quality in Business Education, the IACBE awarded him the 'Frank V. Mastrianna Educational Leader of the year award' for the year 2015 at the IACBE Annual Conference during April 2015 at Baltimore, USA. In addition to delivering MDP/EDP and research-based consultancy programmes to professionals in the Industry, he has published more than 25 scholarly papers in national and international journals and edited books. He also has varied international exposure in terms of participating in academic events, including teaching in various countries in Asia, America, Europe and Africa. During his experience in the last 20 years, he has guided more than 150 postgraduate Management students through their dissertations successfully. He has successfully supervised research scholars through their PhD courses.

Shashidhar Channappa is Head, Department of Social work, The Oxford College of Arts, Bangalore. He completed his Bachelor's degree from National College, Jayanagar and postgraduation from Lorven Educational Centre, Chandapura, which is affiliated to Bangalore University. He was awarded PhD from Bangalore University. He has more than 15 years of experience in the field of teaching, training and development. His research and teaching interests include studies of the informal sector, women empowerment, reproductive health, street vendors and manual scavengers to name a few. He has worked as a trainer for BESCOM (Parivarthana Dhare Module) where he trained around 2500 employees. He was also a resource person who conducted a two-day workshop on Field Work Supervision: A Key Factor for Strengthening Fieldwork program funded by Asia Foundation, which was organized at Talawa, Anuradhapura, Sri Lanka. He has received scholarships to present papers in Cambodia, China, Croatia, Japan and Ireland. He received the Best Presentation Award at The Second International Conference on Transforming Social Welfare and Social Work in China during December, 2016.

Dipankar Chatterjee has worked as assistant professor at the School of Agriculture and Rural Development in Ramakrishna Mission Vivekananda University, Ranchi, since 2007. He holds an MSc and a doctorate in Anthropology from Calcutta University, West Bengal. He has published two edited books and various research articles in peer-reviewed journals. He was the recipient of the National Scholarship, 2001 (MHRD), T.C. Roychowdhury, Silver Medalist, 2013 (Indian Anthropological Society) and Senior Research Fellowship, 2005 (Anthropological Survey of India). He has considerable research and teaching experiences in the field of anthropology. He is Assistant Editor of *South Asian Anthropologist* (Serials Publications, New Delhi). His research interests include culture and cognition, rural and tribal development, ecological anthropology and natural resource management and the anthropology of tourism.

Bharti Chhibber is teaching Political Science in University of Delhi for more than 12 years. She has published books and has more than 50 research papers and articles in mainstream journals to her credit. She has presented papers in various national and international seminars and conferences. Chhibber's specialization includes international relations, South Asia, Indian politics, environmental issues and gender discourse.

Girdhari Dahal has been teaching political science at the Tribhuvan University, Nepal for more than 15 years. He has published 7 books and has more than 40 research papers and more than150 general articles in mainstream journals to his credit. He has presented papers in various national and international seminars and conferences. Dahal's specialization includes democratic movement of Nepal and comparative politics of the world.

Firdous Ahmad Dar has had a distinguished academic career, having completed his MA, MPhil, and PhD from the University of Kashmir, India. Dar is assistant professor at Central University of Kashmir in the Department of Politics and Governance with interests in political theory and civil society/NGOs. Dar has written on a range of issues dealing with the South and Central Asia. His research work has been published in a number of journals accredited both nationally and internationally. His areas of interest include civil society, NGOs, India's Central Asia Policy, and South Asia. Along with teaching, Dar has attended a number of seminars, international conferences and workshops held/organized at both national and international levels. Dar can easily be reached on https://scholar.google.co.in/citation?user=-hdSM9QAAAAJ$hl=en.

Nitin Dhaktode is Doctoral Research Scholar in the School of Development Studies, Tata Institute of Social Sciences, Mumbai. His research areas are transparency and accountability in social policies, tool of social audit to curb corruption and empower people belonging to poorer sections, especially dalit and tribal, social audit and democratic accountability practices. He has also worked with Society for Social Audit Accountability and Transparency (SSAAT), Government of Andhra Pradesh (AP) (including Telangana) for 3 years as programme manager, where he has participated and lead/managed social audits of MGNREGA, pension schemes, state housing scheme, etc. Before joining the AP government, he completed his Bachelors in Social Work with specialization in Rural Development and Masters in Social Work from Tata Institute of Social Sciences, Tuljapur and Mumbai, respectively. After an MPhil in Development Studies, Nitin worked with Tata Institute of Social Sciences as Programme Coordinator to handle the project of social audit trainings in all the states of India, where he worked in collaboration with the Ministry of Rural Development, National Institute of Rural Development and Panchayati Raj, Hyderabad and various State Institutes of Rural Development. He has travelled various states to monitor the social audit trainings offered by TISS, Mumbai.

Adi Fahrudin, PhD is Professor at the Department of Social Welfare, University of Mohammadiyah, Jakarta, Indonesia. His research interests are in the fields of micro social work practice; psychosocial of disaster; aging; HIV/AIDS; social work education and training; child, youth and family welfare; and psychosocial aspects of chronic and terminal illness (oncology, mental illness, etc).

R. Lalitha S. Fernando serves as a professor in the Department of Public Administration, Faculty of Management Studies and Commerce, University of Sri Jayewardenepura, Sri Lanka. She is Head of the

Department of Public Administration and serves as Chairman of the Research Committee of the Faculty of Management Studies and Commerce at the university, and is Secretary General of the Network of Asia-Pacific Schools and Institutes of Public Administration and Governance. She was awarded the prestigious Commonwealth Academic (internal) Scholarship to pursue Postgraduate Diploma in Development Studies, leading to a Master in Development Administration and Management, University of Manchester, UK, for the period 1990–1992. She obtained a full-time scholarship to pursue her doctoral study from Graduate School of Public Administration, National Institute of Development Administration (NIDA), Bangkok, Thailand, for the period 1999–2003. She has published a number of papers, book chapters, conference papers, and articles related to public management and governance and environmental management at both national and international levels. She has co-edited the book, *Strategic Disaster Risk Management in Asia* (2015) with Huong Ha and Amir Mahmood, published by Springer.

Ghouse Basha Ahameed Mustafa (Madurai, Tamil Nadu, India) is an aspiring lawyer, a qualified social worker (and researcher) and freelance trainer on workplace harassment and sexual abuse (especially in corporate companies). He holds a Master in Social Work from Madras Christian College (Chennai, India) and at present he is studying Law at Dr. Ambedkar Law University, Chennai. He is the founder of 'Ashokam', a non-profit organisation working towards research and analysis in diverse areas such as women's welfare, and social and policy issues in South India. Basha earned a Bachelor in Physics, during which time he developed a passion for photography and documentary film making. He has accumulated a wide range of exposure in various professions (from physics to social work to media production and now law) and has been able to adapt to different fields.

H. K. S. Gunasekara holds a Bachelor in Social Work from the National Institute of Social Development (NISD) in Sri Lanka. She is a diploma holder in Diplomacy and World Affairs in Bandaranaike International Diplomatic Training Institute (BDITI) in Sri Lanka. She has considerable experience in the field of social work and research.

H. O. C. Gunasekara is a research assistant in the Department of Public Administration, Faculty of Management Studies and Commerce, University of Sri Jayewardenepura, Sri Lanka. She holds a B.Sc., Public Management (Special) Degree, from the Department of Public Administration, University of Sri Jayewardenepura, Sri Lanka (2011–2015). She has published several research papers, book chapters, and conference papers at both national and international levels and is engaged in several ongoing research projects.

A. K. M. Mahmudul Haque is an associate professor in the Department of Political Science at the University of Rajshahi, Bangladesh. He has BSS (Honors) and MSS in Political Science from the University of Dhaka. He holds a PhD from the Institute of Bangladesh Studies, University of Rajshahi, Bangladesh. Haque's research interest includes environmental governance, sustainable development, local government, civil society, etc. During his PhD, Haque was a visiting scholar at the University of Victoria, Canada. He has researched and written in different national and international journals on various aspects of environment, governance and local government.

Mahfuzul Haque is a former Secretary of the Government of Bangladesh who has been teaching as an adjunct faculty in the Department of Development Studies, Dhaka University and other private and public Universities. A civil servant by profession, Haque holds a Masters and PhD in International Relations and his thesis dealt with the insurgency and integration of the ethnic communities in Bangladesh, India and Sri Lanka. During his stint at the Ministry of Environment and Forests for more than a decade, he gained hands-on training and experience on issues related to environment and development, sustainable development, natural resources management, disaster risk reduction, climate change, biodiversity conservation, atmospheric pollution, desertification, etc. He has authored half a dozen of books and numerous peer-reviewed articles concerning environmental governance, climate change, environmental laws, environmental procedures and guidelines, street children, EIA training manuals, ethnic insurgency in southern Asia, etc.

Zulkarnain A. Hatta is a professor of Social Work with 31 years of experience at the Universiti Sains Malaysia, Penang, Malaysia, and works at the Lincoln University College, Malaysia. He has been a board member of APASWE since 2006 and in 2017 was elected as the president. He is also one of the vice-presidents of IASSW, visiting lecturer of the Shukutoku University, visiting professor of Japan College of Social Work, and consultant for the OECD Korea Policy Center. He is Dean, Faculty of Social Sciences, Arts and Humanities, Lincoln University College, Malaysia.

A. K. M. Monirul Islam is Associate Professor in Social Work at the National University, Bangladesh. He is a PhD fellow at the Institute of Social Welfare and Research, University of Dhaka, Bangladesh.

Md. Rabiul Islam, PhD is Professor at the Institute of Social Welfare and Research, University of Dhaka, Bangladesh. His research interests are aging, poverty and social exclusion of ageing and human rights

Rashidul Islam is a research fellow at the Policy Research Centre.bd (PRC.bd), a civil society think-tank in Bangladesh and Professor of Business English, ICMAB (Institute of Cost & Management Accountants of Bangladesh), Dhaka, and former Registrar, North South University (NSU).

Muhammud Shariful Islam is Professor in the Department of Social Work, University of Rajshahi, Bangladesh. Islam obtained an MSS degree with first class in Social Welfare from the Institute of Social Welfare and Research, University of Dhaka, Bangladesh, and was awarded an MPhil in Social Work from the Department of Social Work, University of Rajshahi, Bangladesh, where he has been working as part of faculty for more than 21 years. He teaches courses entitled 'Human Right, Social Justice and Social Work'; 'Social Thought and Political System', 'Urban Resource Management and Social Work', etc. His research activities are concentrated on Human Rights and Child Rights. He has published more than 16 research articles in various refereed journals at home and abroad.

Atul Jaiswal is a doctoral scholar at the School of Rehabilitation Therapy, Queen's University, Canada. He is from India, where he previously worked for 5 years as an Occupational Therapist, disability rehabilitation social worker, and Community-Based Rehabilitation (CBR) professional.

He is a gold medalist in Masters in Social Work in Disabilities Studies and Action from the Tata Institute of Social Sciences, Mumbai. He is a recipient of the Queen Elizabeth II Diamond Jubilee Scholarship and is pursuing his doctoral research on deafblindness in India.

Paramjit Singh Jamir Singh is Senior Lecturer and Coordinator of the Social Work Programme in the School of Social Sciences, Universiti Sains Malaysia, Penang, Malaysia. He was a visiting PhD Scholar at Wayne State University, Michigan, USA, in 2014. His research interests focus on HIV and AIDS, drug addiction, harm reduction, criminal justice, mental health, and social work with families. Prior to joining the university, he worked as a Senior Police Officer in the Crime Investigation Department, Royal Malaysian Police. He was recently appointed by the Minister of Women, Family and Community Development to serve as an advisor to the Magistrate Court Judge, providing recommendations related to the sentencing and welfare of juvenile offenders.

Neena Joseph retired as Professor at the Institute of Management in Government, Regional Centre, Kochi after putting in 30 years of service. She had been conceiving, designing and implementing training programmes for the government in areas of governance including decentralized governance, gender, training, research methodology, disaster management and general management, and has undertaken research programmes for the government. Her publications include a book, international- and national-level research papers and articles on the above subjects. Her recent research work on Customer Satisfaction for the Social Justice Department is being published. She is a freelance trainer and consultant to corporate and government. She had been involved in the drafting of "Formulation of 13th Five Year Plan 2017–2022, Report on the Working Group on Gender". She is part of the policy-making bodies of the Federal School of Science and Technology, Sakhi Women Resource Centre and Cultural Academy for Peace.

Z. R. M. Abdullah Kaiser is an assistant professor in the Department of Public Administration, University of Chittagong, Bangladesh. His previous research interest was focused on the politics, economic growth and the civil service of Bangladesh. He has published three research papers in international journals. His first research paper, 'RMG in Bangladesh: A

Study of the Effects of Pre-election (2014) Political Turmoil', has been published in the renowned SAGE journal *Foreign Trade Review* in 2015. The second research paper 'Bangladesh Civil Service (BCS) Examination: A Critical Review', has been published in the journal *Public Affairs and Governance* in 2015. The third paper 'Trump and the Ascension of Western Realism: A Critical Discussion on the Western Realists' and Western Liberalists' Evaluation of Globalisation', has been published in the renowned SAGE journal *India Quarterly*.

Niaz Ahmed Khan is Professor at the Department of Development Studies, University of Dhaka, and Executive Director of the Centre for Resources and Development Research. His career reflects a rich blend of academic and practicing development management experiences. Khan holds a PhD and postdoctoral qualifications from the University of Wales, University of Oxford and the Asian Institute of Technology.

Anjali Kulkarni works at the Tata Institute of Social Sciences, School of Rural Development, as an assistant professor. She leads the MA in Social Innovation and Entrepreneurship programme and is also a coordinator for Research and Development at the TISS Tuljapur Campus. She holds a PhD in Sociology and Masters in Business Administration, Personnel Management and Social Work. She has more than 25 years of field experience and 16 years' teaching experience. She has completed 12 research projects and has published 2 books and more than 25 papers in books and international and national journals. She has presented papers in more than 30 international and national conferences. Her research interest areas are entrepreneurship, microfinance and financial services, rural development, empowerment, health, etc. She has undertaken more than ten field action projects especially in rural areas. She has been invited on several committees of different universities as an expert on syllabus formation and revision. She has received special training on 'women empowerment' at Haifa, Israel and was selected in the Faculty Exchange Programme under Linnaeus Palme Grant 2015 with Lund University, Sweden.

Binod Kumar is Assistant Professor (Ad hoc) of Social Work in Dr. Bhim Rao Ambedkar College, University of Delhi. He has submitted his PhD thesis entitled 'Disaster, Law and Community Resilience: A Study of Recent Natural Disasters in India' at the Centre for the Study of Law and Governance (CSLG), Jawaharlal Nehru University (JNU), New Delhi. The PhD work locates disaster prevention and mitigation as a subject matter of law apart from developmental planning. Just as knowledge of weather, acquaintance of existing legal framework can also save the life and property of people from disasters. In this backdrop, his research navigates the intersection of disaster, law and community resilience in India. He completed his MPhil at CSLG, JNU with thesis entitled 'Governance of Nonprofit Organizations in India'. He is Master in Social Work from Tata Institute of Social Sciences, Mumbai. Apart from that, he pursued BA LLB (5 years) from Aligarh Muslim University, Aligarh. His training in different disciplines enables him to approach the research problem from a multidisciplinary perspective. He has been awarded Junior Research Fellowship (JRF) by the University Grants Commission (UGC) and has participated in several national and international conferences.

Uttam Kumar is the Head of Programmes at Sense International India. He has 13 years' experience of working with deafblind population in India. He has spearheaded the pan India deafblind rehabilitation work of Sense India and has been instrumental in initiating deafblind interventions in Southeast Asia in collaboration with Sense India. He has conducted various international trainings on deafblindness in Bangladesh and Uganda and has paved the way for the expansion of deafblind work in India and abroad.

Anju Lis Kurian is UGC Postdoctoral Fellow, School of International Relations and Politics, Mahatma Gandhi University, Kottayam, Kerala, India. She has published various articles in different international and national journals. She was awarded The National Young Political Scientist Award 2015 by the Indian Political Science Association.

Lakshmana Govindappa is an assistant professor in the Department of Social Work, School of Social and Behavioral Sciences at Central University of Karnataka, Kalaburagi. He completed his graduation (BA) and postgraduation in Social Work (MSW) from Bangalore University, Bangalore. He holds an MPhil and a PhD from National Institute of Mental Health and Neuro Sciences (NIMHANS), Bangalore. He has worked as Programme Manager and Psychiatric Social worker at NIMHANS and Full Time Guest Faculty in the Department of Social work, Jnanbharathi Campus, Bangalore University. He also worked as field officer at Swathi Mahila Sanga. He has published 18 articles in national and international journals including *Asian Social Work & Policy Review, Hong Kong Journal of Social Work, Social Work Practice in the Addiction, Indian Journal of Social Work* and has written six chapters in books, he also published two monographs (co-author) and has presented 25 papers in national and nternational conferences. He has completed two research projects. He won the gold medal for best paper presentation at a National Conference during his PhD and also received the best paper award in teacher's category at the ISPSW conference.

Koustab Majumdar holds a master's degree in Rural Development and Management from the University of Kalyani (Department of Rural Development), Nadia, West Bengal. He also holds an MPhil in Rural Development from Tata Institute of Social Sciences (School of Rural Development), Mumbai, where he was the recipient of the MPhil Research Fellowship (2015–2017). He has attended several national (five) and international (four) seminars and conferences in various research institutions and universities. He has published four chapters in edited books and two articles in peer-reviewed journals. He has worked as a research associate at Tata Institute of Social Sciences. He is presently working as Research Associate (ICSSR) and also pursuing his doctoral degree in Agriculture, Rural and Tribal Development at the School of Agriculture and Rural Development, Ramakrishna Mission Vivekananda University, Ranchi. His research interests include agrarian studies, labour issues, policy and development, women empowerment, rural economy and livelihood and tribal rights.

Gazi Arafat Uz Zaman Markony is a university teacher and researcher, with experience in development and concerned research sectors. He is a versatile academic, with research and project involvement in the fields of governance, social inclusion, public procurement, food security, health and nutrition. Markony conducts both qualitative and quantitative research and is experienced as facilitator in research and development project training, workshops and seminars. Markony received training in research methodology and statistical analysis. He is also familiar in working with national and international development institutions and organizations. His field of interest includes urban governance and development, food security, public policy, community development, social inclusion and sustainable development. Markony has a BSS (Honors) and MSS in Public Administration from University of Dhaka, Bangladesh.

Md. Awal Hossain Mollah has been teaching Public Administration at the University of Rajshahi, Bangladesh for the last 16 years. He is working on various issues of politics and governance in the context of Bangladesh and South Asia. He has published two chapters in the *Global Encyclopedia of Public Administration, Public Policy and Governance.* He has a number of peer-reviewed journal articles published by Sage, Springer, Routledge, Emerald, and Palgrave including a book and some book chapters.

Md. Nurul Momen is an associate professor in the Department of Public Administration at the University of Rajshahi (Bangladesh). Since December 2002, he has been offering courses for different semesters, conducting academic research, and providing supervision and consultation for academic activities of the students and researchers. Currently, he teaches 'organization theories' and 'politics and administration in South Asia'. He holds a Bachelor of Social Science (BSS) and a Master of Social Science (MSS) in Public Administration from the University of Rajshahi (Bangladesh), and has completed his Master of Philosophy (MPhil) from the University of Bergen in Norway, and obtained his Doctor of Philosophy (PhD) from Sant Anna School of Advanced Studies in Italy. Momen has published many articles in international journals, including chapters in various books dealing in a range of debates in Public Policy and Law, Governance, and Public Sector Reform in South Asia. He has participated in many national and international seminars, workshops and conferences, and has been member of several professional networks, journal editorial boards, research institutes, and civil society organizations. Momen as a project manager has the extensive experience in implementing a 3-year-long World Bank project on enhancing the quality of higher education in Bangladesh. He has travelled to Norway, Italy, UK, Spain,

France, Belgium, Vatican, Turkey, Romania, Switzerland, Hungary, Germany, India, Nepal, Poland, Austria and China in connection with academic and professional activities.

Ndungi Wa Mungai is a lecturer in Social work and has been at CSU since February 2010. He completed his PhD in 2011 at Deakin University, Victoria, titled: *The Health and Well-being of Young Sudanese Men in Melbourne*. Ndungi has also completed a BSc (agr.) (University of Nairobi); M Agr. Studies (University of Melbourne); Bachelor of Social Work (University of Melbourne), Master of Social Work (RMIT University, Melbourne) and Graduate Certificate in Learning & Teaching in Higher Education (Charles Sturt University). Before joining CSU Ndungi worked in a variety of social work setting including research, telephone relationship counselling for men, counsellor/advocate in torture and trauma, caseworker for asylum seekers, housing support worker, foster care field worker and social worker at Centrelink (locum position).

M. Nagaraj Naik is part of the faculty of the Department of Social Work, Jnanabharathi Campus, Bangalore University, Bengaluru, Karnataka, India. He has been teaching for the last 6 years in the specialization of Mental Health, Human Resource Management, Industrial Relations, Labour Laws, Organization Behaviour and Development, and Counselling. He earned a PhD in Social Work and his research was on 'stress and coping skills amongInformation Technologies (IT) employees', along with which he also studied their quality of life and social support.

Naik completed his postgraduation with a Master of Social Work from Bangalore University, Master of Philosophy (MPhil) from the National Institute of Mental Health and Neuro Sciences (NIMHANS), which is considered an institute of national importance. In addition, he has also qualified UGC-NET conducted by the University Grants Commission in 2012. Prior to his PhD, he has worked in BSE/NSE listed Industries such as OPTO Groups (Infrastructure & Manufacturing), Badve Engineering Limited as Assistant Manager HR for more than 5 years, and he has also published several articles in national and international journals and is member of professional bodies like Indian Society of Professional Social Work (ISPSW), Karnataka Association for professional Social Workers (KAPSW), and National Institute of Personnel

Management (NIPM). His area of interest is in mental health, human resource management, school mental health, and life skills education, corporate counseling, labour welfare, etc.

Naveen Naik Karje is a PhD scholar at the Tata Institute of Social Sciences, Tuljapur, under the School of Rural Development, under the guidance of Anjali Kulkarni. He completed his MPhil in 2017 from the School of Rural Development at TISS, Tuljapur. He is an alumnus of the International Institute for Population Sciences (IIPS), Mumbai. Through his MPhil dissertation, he has explored the influence of Self-Help Groups (SHG) on women empowerment. His research interests are: migration, rural development, women empowerment and education.

Ramesha Naika is an Integrated MPhil-PhD Scholar at the Tata Institute of Social Sciences, Tuljapur Campus Maharashtra, India. His research during his first MPhil examined Empowering Farmers through Sustainable Organic Agriculture in Udupi District of Karnataka. In addition to an MPhil, he holds a Master in Development Management (University of Mysore), an MPhil in Applied Research (The Gandhigram Rural Institute), a Postgraduate Diploma in Rural Development (AU). Prior to joining his present research study at the Tata Institute of Social Sciences, Tuljapur Campus, Ramesh worked as a Young Professional in WASH Sector at SCOPE, Dharwad and Arghyam Bangalore.

Parag Namdeo is Head of Advocacy & Networks at Sense International India. He has spearheaded advocacy initiatives of Sense India and played a key role in inclusion of 'deafblindness' in the new disability Act 'The Rights of Persons with Disability 2016'. He has conducted various deafblind sensitization trainings of Government officials across the 23 states of the country. He also provides leadership to the national networks of Sense India—Families of deafblind people (PRAYAAS), Adults with deafblindness (UDAAN), and teachers of persons with deafblindness (ABHI-PRERNA).

Debasish Nandy, MA, MPhil, PhD is Associate Professor and Head of the Department of Political Science at Kazi Nazrul University, West Bengal, India. He is also Coordinator of the Centre for Studies of South & South-East-Asian Societies, at the same university. He has published

33 research articles in various national and international journals. He has contributed 31 book chapters in a number of edited volumes. He is the sole author of three books, has co-authored two and co-edited six. Nandy has conducted two Government of India funded research projects on South Asian Politics. He is the editor of the *International Journal of South & South-East Asian Research*. He has visited 11 countries on different academic assignments. He regularly contributes articles in newspapers. He also appears on Bengali TV channels as a political analyst. His research interests include South Asian Politics and India's Foreign Policy.

Sri Krishan Sudheer Patoju received a PhD from the School of Economics, Andhra University, Visakhapatnam in 2012. His research interests concern agricultural development, production economics, rural entrepreneurship, social entrepreneurship and rural development. He is Assistant Professor, School of Rural Development, Tata Institute of Social Sciences, Tuljapur, Maharashtra. Previously he worked for Andhra University and GITAM University, Visakhapatnam. He is associated with the International Federation of Organic Agricultural Movement (IFOAM), Germany and the Research Institute of Organic Agriculture (FiBL), Switzerland.

Akhil Paul is Director and founder member of Sense International India. He is the former vice president of Deafblind International (DBI), a world association promoting services for deafblind people. He is also the founder member of the Board of the National Trust for Welfare of People with Autism, Cerebral Palsy, Mental Retardation and Multiple disabilities. He is the recipient of HSBC-Indo British Award—2003 for outstanding services to deafblind people in India. He has also received the Young Leader Award in the 'Community Service and Social Upliftment' category at the prestigious Lakshmipat Singhania—IIM Lucknow National Leadership Awards.

Mostafijur Rahman is Assistant Professor of Law at the Department of Law, Prime University, Bangladesh. He holds a BA and LLM from Rajshahi University and is working towards his MPhil thesis titled 'Administration and Management of Khas Land in Bangladesh: Law and Practice'. His research interests are constitutional law, land laws of

Bangladesh, international law, law of the sea, human rights, contemporary global and political issues, etc.

M. C. Rasmin is a media development expert with 15 years of extensive experience, working on post-conflict and transitional settings. In his previous career at the Sri Lanka Development Journalist Forum (SDJF) as CEO/Director, he was also heavily involved in topics such as gender and media, media for democratization, mobile journalism, radio for peace building, and participatory and video storytelling. Since Joining IREX in 2017, he has started working on media development with a focus on policy reforms and digitalization, with special emphasis in promoting mobile and digital journalism and media literacy in Sri Lanka. Rasmin's specialty areas can be summed up as conflict/gender-sensitive storytelling, radio drama, and community broadcasting.

M. Rezaul Islam, PhD is Professor at the Institute of Social Welfare and Research, University of Dhaka, Bangladesh. He received his masters and PhD from the University of Nottingham, England. His research focuses on poverty and inequality; climate change, disasters and displacement; human rights and community development.

Pema Rinzin is an associate lecturer in the Department of Sustainable Development, Royal University of Bhutan. Previously, he has briefly worked with the National Council Secretariat, Parliament of Bhutan. He has also worked with a national consultant on a Gender Mainstreaming Project for UNDP in Bhutan.

Nasreen Aslam Shah is a Meritorious Professor, Director, Centre of Excellence for Women's Studies and Chairperson, Department of Social Work, University of Karachi. Her teaching experience (service length) covers 28 years and her expertise is in women's issues, self-employed women, qualitative and quantitative research, field work, management skills and administration. She is the editor of 14 volumes of *Pakistan Journal of Gender Studies*, Centre of Excellence for Women's Studies and 5 volumes of *Pakistan Journal of Applied Social Sciences*, Department of Social Work, University of Karachi. She has supervised ten PhD dissertations; currently, 22 MPhil and PhD students are enrolled. Her 46 research articles have been published in various reputed national and international journals, and a book is also published on the Internet via scholar press, Germany. She is the author of ten books and has completed 13 research projects on various socio-economic and environmental issues. Along with her academic achievements she continuously works to improve the status of women. For the first time in Pakistan she has established an organization named Self-Employed Women's Fund in 2011 to provide interest-free loans to Self-Employed Women and is also a founder and General Secretary of *Anjuman Taraqqi-e-Niswan*.

Cheng Sheng-Li is a professor in the department of Social Work School of Philosophy and Social Development, Shandong University, Peoples Republic of China. He has been in postgraduate teaching for more than 25 years. Cheng has carried out research projects funded by international agencies such as UNESCO; Washington University, St. Louis, USA; University of British Columbia in Canada; and Taiwan Dongwu University. He was a visiting professor in many international university departments. Cheng has 6 books, 37 research articles (both in English and Chinese) and 5 international projects to his credit. His areas of work are urban poverty and social assistance, social policy and social welfare, social psychology, family, child and youth. His most recent books are *The Palgrave Handbook of Global Social Work Education* co-edited with Sajid S.M, Cheng Rajendra Baikady and Haruhiko Sakaguchi and published by Palgrave Macmillan, USA, *Social Welfare Policies and Programmes* in South Asia co-edited with Channaveer R.M, Rajendra Baikady and Haruhiko Sakaguchi and published by Routledge, New Delhi.

Abhishek Pratap Singh holds a PhD in East Asian Studies from Jawaharlal Nehru University (JNU), New Delhi. He was a visiting fellow with Centre for India Studies, Shenzhen University, China and was also Fellow, South Asian Democratic Forum (SADF), Brussels, Belgium. He was selected for Visiting Sinologists Program, 2018 at Jinan University, Guangzhou and was also nominated for the BRICS Program 2017 at Fudan University, Shanghai. He has contributed his writings on China to established forums and is currently teaching at University of Delhi.

C. Sivapragasam is a professor at the Centre for Applied Research, the Gandhigram Rural Institute (Deemed to be University) Gandhigram, Tamil Nadu. He has more than 35 years of teaching experience and has published extensively in national and international journals and books. Sivapragasam has three books, 35 research articles and 15 book chapters to his credit. Under his supervision, 19 MPhil and 5 PhD degrees have been awarded.

Jeganathan Thatparan, Human Rights Activist—Asia Region and legal expert has served as an executive director and as President of Council of NGOs. He provides technical supports to the National Institute for Social Development, the state ministries in South Asia, and is a senior member of the Lions Club International. For more than 13 years, Thatparan has worked with several organizations with the focus on humanitarian aid, human rights sector, and has proven his experiences are not limited but include South and Central Asia, European countries, the Middle East and African region particularly focusing on women, children and elders' rights. Further, he has initiated several networks, capacity building initiatives, developed contingency plans, engaged with civil societies on alternative reports, developed strategic plans, joint action plans, partnerships and thematic plans specifically focusing on social protection components. Thatparan is also well known as a legal professional in Child Rights. He holds an MA from the University of Jaffna, Sri Lanka and the American College awarded Postgraduate Diploma in Psychology. He has successfully completed his LLB and reading papers for his PhD. Thatparan has been appointed by the National Child Protection Authority as a panel member of an Investigation unit which assesses the criminal jurisdiction on issues related to child abuse.

Tania Afrin Tonny is an assistant professor in the Department of Public Administration at Bangabandhu Sheikh Mujibur Rahman Science and Technology University, Gopalganj, Bangladesh. Previously, she worked at BRAC (Bangladesh Rural Advancement Committee), the world's largest NGO based in Bangladesh, with its Migration Programme as a field coordinator and worked as a Trainer at Ain o Salish Kendra (ASK), which is a leading human rights organization in Bangladesh. She received her undergraduate and graduate degrees in Public Administration from the University of Rajshahi, Bangladesh with excellent academic results. Her research interests include migration, gender, human rights, public policy, civil society, etc.

K. N. Veena holds a PhD in Business Administration, a postgraduate degree in Statistics from Mysore University and a postgraduate degree in Total Quality Management from Sikkim Manipal University. She is Dean at Dayananda Sagar Business Academy, Bangalore. During the 20 years of her academic career, she has worked as Program Manager for MBA programme and Manager for Entrepreneurship Excellence Centres at reputed B-schools in Bangalore and was a core committee member for writing SSR for NAAC, national accreditation and IACBE, an international accreditation. She has been a resource person for Faculty Development programmes, Management Development programmes, and workshops and has worked as consultant for organizations in the area of Statistics and Management. She has conducted workshops on SPSS for research scholars, faculty, executives and students. She has presented research papers at national and international conferences, published papers in reputed journals. She is one of the very few certified Entrepreneurship Educators from NEN, Stanford University and IIM-B and has undergone various courses related to entrepreneurship and has mentored students and entrepreneurs. She has conducted workshops on topics related to entrepreneurship for faculty and students of various institutes. She has spearheaded *Market Research* for small and medium entrepreneurs in the capacity of Chief Enumerator and Statistician. During her tenure, she has guided more than 150 postgraduate Management students during their dissertation and internship and has also assisted research scholars through data analysis and thesis report preparation during their doctoral thesis.

C. Vinodan is Assistant Professor and Chair, Centre for Strategic and Security Studies, School of International Relations and Politics, Mahatma Gandhi University, Kottayam, Kerala, India. His areas of interest are security studies, disarmament and peace studies, international relations, South Asian security, energy, environment and human security. He has received national and international awards and fellowships. He is member of the editorial board of the *Journal of Human Security* (Canada), *South East Asian Journal of Socio-Political Studies* (Indonesia), *Indian Journal of Politics & International Relations* and *South Asian Journal of Diplomacy*. He has published numerous research papers and books.

W. A. D. P. Wanigasundera holds a PhD in Agricultural Extension and Rural Development from the University of Reading, UK. He is a professor attached to the Department of Agricultural Extension, University of Peradeniya, Sri Lanka. He has 25 years' experience in University teaching, research, curriculum development, together with over 15 years of practical experience as a field extension and rural development worker. As Development Communication Expert, he has worked for many state and non-government sector organizations. He has contributed in the development of interactive audiovisual and multimedia communication approaches and the production of award-winning video documentaries and development of computer-based communication aids.
Currently, Wanigasundera teaches courses in the fields of Development Communication, Development Journalism and Media Use, Communication for Natural Resource Management, Preparation and Use of Audio—Visual Media, Technical Writing and Human Resource Management at bachelors' and postgraduate levels.

Mahathir Yahaya is a lecturer at the School of Social Sciences, Universiti Sains Malaysia, Penang, Malaysia. Prior to joining university, he worked as a journalist at *Utusan Karya*. His research interest focuses on social work community, social work in school, and communication social work. He is associate trainer at various training providers. He has been involved in training for more than 8 years with the intention to develop character in people so that they achieve their organizational objectives. He has also been a guest speaker for numerous government agencies, schools and community organizations.

List of Figures

List of Tables

1

Introduction: Civil Society in the Era of Globalization—Emerging Concerns as a Social Welfare Provider

Md. Nurul Momen, Rajendra Baikady, Cheng Sheng-Li, and M. Basavaraj

In view of the growing interest among the academia, government, international organizations, and media in the present globalized society, with chapters covering Nepal, Pakistan, Bangladesh, India, Sri Lanka, and Bhutan, the book makes an attempt to examine and critically analyse the issues with regard to the development and growth of civil society organizations (CSOs) and their contribution to social welfare across South

Md. N. Momen (✉)
Department of Public Administration, University of Rajshahi, Rajshahi, Bangladesh

R. Baikady
Paul Baerwald School of Social Work and Social Welfare, Hebrew University of Jerusalem, Jerusalem, Israel

Department of Social Work, University of Johannesburg, Johannesburg, South Africa

© The Author(s) 2020

Md. N. Momen et al. (eds.), *Building Sustainable Communities*,
https://doi.org/10.1007/978-981-15-2393-9_1

Asian countries. The term "civil society" is the most contested issue, because many different interpretations exist that are not well connected to and/or articulated with the other definitions and concepts. However, there are constant global challenges that hamper the performance of the third sector in the contemporary globalized world. Given the scenario, this book will provide a new opportunity to present the new development and challenges facing civil society organizations in providing social welfare provisions in South Asia. The book also intend to show the nexus between sustainable roles and cope with the challenges for third sector organizations in the many countries in South Asian region. It is worth a mention that nothing much has been written about the contribution of civil society towards the social welfare sector, especially a comparative perspective on CSOs and Social Welfare focusing on South Asian countries. Given the social-political and cultural development of South Asian countries, the role of civil society as a third sector in development has been more interesting and promising in the recent past.

However, to say that this is the time when civil society organizations gained the trust and confidence of the general people. Given the political diversity and complexity in social dynamics, it is interesting to understand and critically analyse the role of civil society in social welfare across South Asia. It is a paradox of the contemporary South Asian politics that most of the states in the region adopted democracy as the form of government, but simultaneously people are losing faith in political parties, which decreases trust in elected officials and politicians. Growing political apathy, declining membership in political parties, and low voter turnout in many recent parliamentary elections indicate the lack of trust among the masses in political regimes.

C. Sheng-Li
Department of Social Work, Shandong University, Shandong, China

M. Basavaraj
Department of Economic Studies and Planning, Central University of Karnataka, Kadaganchi, India

Historical Development of Civil Society

CSOs are the collective action of several individuals who work outside the state for the development and growth of the society. The primary aim of CSOs is to ensure the betterment and upliftment of society through organized work. However, to date there is no single agreed meaning of the concept (Munene and Thakhathi Munene and Reckson Thakhathi 2017). The concept of civil society goes back to many centuries in Western thinking, with its roots in Ancient Greece (Carothers 1999). While references to "civil society" go back to the sixteenth century, talks of "global civil society" emerged only in the 1990s (Cf. R. A.Falk 1992; Lipschutz 1992; Shaw 1994; Drainville 1998). With state provision of social security going down, active development of civic activities by these civil society organizations rose in the 1980s and 1990s (Cf. G. Esping-Andersen 1996). This time public sector provisions for social security was inadequate, and as a result non-governmental organizations (NGOs) and organizations working at the grassroot stepped into social welfare activities with limited private donations and voluntary assistance. According to the Union of International Associations, in 1998, there were about 16,500 active civic bodies whose members are spread across several countries (Union of International Associations 1998).

Against this backdrop, the origin of the concept of civil society can be traced back to Greek political philosophy, where Aristotle talked about politike koinona (political community/society) to refer to a rule-governed society in which the ruler puts the public good before his private interest. However, with the rise to prominence of the theories of individual rights and the idea of a social contrast during the seventeenth and eighteenth centuries, the concept of civil society took a rebirth. Interestingly, the assumption of human equality, drawn from Christianity, was newly added to the modern use of the term civil society. To explain the phenomenon more precisely, Mary Kalador (2003) noted,

> At that time the civil society and state was not differentiated, both were more or less the same. More interestingly civil society was contrasted only with other kinds of societies and not with the state. More particularly during this time a civil society was a peaceful society, a society in which people treated strangers with civility, in contrast to other violent and 'rude' societies.

It has been observed that since the end of the Cold War there has been confusion among policymakers, activists, and researchers from various parts of the world about the global relevance of civil society in strengthening development and democracy. Later on, in the eighteenth century, influenced by political theorists from Thomas Paine to George Hegel, the idea of modern civil society emerged. Both these philosophers developed the notion of civil society as a domain parallel to but separate from the states (Carothers, 1999). In the 1990s, a trend towards democracy opened up all over the globe that brought a new interest in the activities of civil society. This trend was important in covering the increasing gaps in social service provisions introduced by structural adjustment and other reforms in developing countries. In the second half of the eighteenth century, Adam Smith (1723–1790), John Lock (1635–1704), Thomas Hobbes (1588–1679), Karl Marks (1818–1883), and Hegel (1770–1831) contributed notably to the theoretical development of civil society and recognized the civil body as an independent sphere from the government. In the early twentieth century, civil society emerged as a reaction against "totalitarianism," especially socialist totalitarian regimes. However, it is worth mentioning that the last few decades have witnessed the tremendous growth of civil society organizations such as NGOs and labour unions. Perhaps the introduction of globalization and rapid changes in social, economic, and environmental aspects had contributed to the accelerated growth of civil society organizations all over the globe.

On a particular note, the idea of civil society is rooted in the sixteenth- and seventeenth-century European thinking. J. Arato and J.C. Alexander (2001) argue that the incidents that were taking place in the society guided men to have an "ought to be" model of society and influenced people to unite in the form of civil society organizations. Furthermore, the civil society debates and discussions are not only conceived by the political and social scientists, but also it is a current issue about establishing a idealized society by general public. However, Michael Edwards (2009) noted the importance of examining the concept of civil society in terms of both theory and practice. His argument for re-examining the civil society concept was to understand the question "*Is civil society the big idea for the twenty first century or will the idea of civil society confused, corrupted or captured by elites prove another false horizon in the search for a better world?*"

According to him, recognizing civil society as a contested territory in both theory and reality is the first step towards rescuing a potentially powerful set of ideas from conceptual confusion that threatens to submerge that realistic idea. In the contemporary globalized world, civil society is gaining its importance in most countries irrespective of the political regime and type of administration. Rosa Sánchez Salgado (2007) stated that civil society in Europe tend to gain most importance in the political and academic discourse. The European politics and academia perceive participation in the policy process as a means of improving European governance and connecting people to European policies (Rosa Sánchez Salgado 2007).

Civil Society in the Globalization Era and Social Welfare

Today's society is globalized and interconnected between the nations and within the nations. The developing countries, especially in South Asia, face a number of challenges in ensuring health, education, and other basic needs to their citizens. Distracted economic development, social transformation, labour migration, social delinquency, unorganized sector, crime, poverty, intolerance, and reduced harmonious relationships among the nations are the extended contributions of globalization in developing countries. The civil society organizations in many developing countries are working together between the state and the society to ensure social welfare provisions for citizens. In other words, the civil society intervenes where the state neglects its investment on Social Welfare provisions. The CSOs in India had made a remarkable contribution to the overall development of the society and the upliftment of the marginalized sections of the society prior to economic liberalization. Perhaps this stands true for many other developing South Asian countries. CSOs contribute essentially to the welfare of states and society. In Europe, civil society plays a key role in social service delivery to the people, fulfilling a large variety of functions along with social services delivery. They become voice to the voiceless people by giving them an opportunity to explore the unaddressed issues, offering alternative ways of occupational socialization, or facilitating social inclusion (cf. Kramer 1981; Rose-Ackerman and James 1986; Kendall 2003).

The socio-economic condition of many countries, especially the developing world, has been influenced by the process of globalization for over two decades. On the one hand, globalization has benefited many countries with opportunities such as trade liberalization, a global market for products, foreign direct investment, liberalization, privatization, and information flow. It has also encouraged capital and technological transfer among the countries. On the other hand, globalization through material deprivation and the relative marginalization has also affected the development and growth of the countries which are unable to compete with the developed economies of the world. Thus, globalization has become one of the reasons for regional imbalance across the globe. The countries or regions with well-occupied resources are actually challenging and competing with countries with less or no resources. As a result, even today there are more than a billion people who continue to live in poverty with purchasing power of less than a dollar a day. Malnutrition, unemployment, under-weight children, and lack of basic food and health facilities can be seen even now in the poorest countries in the world. Millions of people even today live without access to clean water, sanitation, basic health services, and education. Current research suggests that the third sector's societal roles considerably vary between countries, depending on the welfare state they are embedded in: Starting with a revision of Esping-Andersen's welfare regime typology (1990) and also based on the earlier work of Moore (1966), Salamon and his colleagues developed a typology of four different "non-profit regimes" (Salamon and Anheier 1998; Salamon et al. 2000).

While recording the importance of civil society, Edward (2005) argues that CSOs are the "associational life of society" that develops values such as tolerance and cooperation among people. However, in countries with a large third sector, CSOs mainly fulfil the service function. Whereas CSOs in Countries with a relatively small number tend to engage in "the expression of political, social, or even recreational interests" (Salamon and Anheier 1998, 229). While conceptualizing civil society, many scholars have different understandings and interpretations. As mentioned elsewhere, CSOs are "the good society" based on the social contracts negotiated between the government, business, and citizens. Nzomo (2003), while conceptualizing CSOs, included family and blood associations and excluded political groups from the purview of CSOs. However, later, in 2005, Edward noted that CSOs are a public space that aims for the common good of the people. As noted by many scholars, the number of CSOs and organized private and no-profit activities has increased remarkably. Perhaps this upward growth in CSO activities in the countries of Asia, Africa, and Latin America was evident from the early 1990s until the late 1990s (Salamon and Anheier 1998; Salamon 1994; Fisher 1993; Brown and Korten 1991). During this regime the CSOs were considerably recognized as relief providers and human rights guardians by both the state and the general public (Salamon and Anheier 1998; Fukuyama 1995; OECD 1995).

Tandon (1986) categorized civil society organizations based on their area of intervention: (a) the civil society organizations that work in the field of environmental degradation, deforestation, land alienation, and displacements, among others—these CSOs try to get the attention of policymakers and thereby make the issue open for wider public debate; (b) the second type experiments with various developmental models and find solutions to the socio-economic problems of the society—these kinds of organizations work in the areas of adult education, primary health care, toilets, irrigation system, bio-gas, and ecologically balanced wasteland development, among others; and (c) the third category works for the most deprived sections of the society. These types of civil society work with women, tribal, landless labourers, and informal sector workers, among others. The political empowerment, social emancipation, and economic development of these sections of people are the main aims of these civil societies.

Lokanath Suar (2011) argued that the role of civil society cannot be undermined in present day. Collective, collaborative, creative, and value-driven work is done by different kinds of civil societies. The civil societies are trying to prepare a society that is just and free that requires in any politics, policymaking or even in the social change sector. Apart from that, the civil society is a means to achieve, and a framework for engaging with each other about ends and means. However, there is the need for integrating and mutually supporting the different perspectives within the civil society. If rightly used, the civil society can bring a great deal of social changes and serve as a practical framework for organizing both resistance and alternative solutions to social, economic, and political problems. David Armstrong et al. (2011) stated that the civil society organizations are playing a very important role in the development aspects of the society, and as a result, in recent years, there is an increasing focus on civil society organizations. The civil society organizations in the current society are playing a crucial role in ensuring legitimacy and effectiveness of global/regional governance.

The definitions and explanations of civil society are continuously being reshaped by various new actors due to new situational environments. However, to say that the importance of the civil society organizations was much felt after the State withdrawal from many of its welfare responsibilities in the 1990s. This is the time when civil society and the NGO sector emerged to take up the welfare responsibilities as the state failed to ensure welfare provision to a large number of people who were in need of state support. It was a challenge for the state during the post-liberalization era to manage the welfare of the citizens and address the concerns of the society. This is the time civil society organizations took the responsibility of developmental tasks, ensuring the provision of services to the people. As Sarah Joseph (2002) writes, in the present condition the state is outsourcing the welfare activities and the civil society and social actors are being encouraged to take the responsibility for the development and welfare functions. The society in the present world is facing problems such as poverty, conflict, exclusion, and marginalization across the world. On the other hand, the resources are concentrated with a handful of people mostly belonging to corporate or government sector which in turn is enabling the civil society organizations to focus on governance and development related issues. However, in the present society international organizations and the UN agencies are working along with the CSOs in South Asia to improve the living conditions for the marginalized population and also providing aid, monitoring and overseeing developmental programmes in regions hit hard by socio-economic problems. Civil society organizations across the world perform many welfare activities. Different scholars have noted a variety of welfare functions performed by CSOs. The first function of CSO is advocacy as a very important welfare activity by doing so they influence authorities and try to impact the policymaking process by expressing cultural, spiritual, political, social, environmental, and ethical concerns (Salamon et al. 2004). This top-down approach can take the form of counter-power actions and protests. Awareness campaigning is also another form of advocacy activities performed by many CSOs. Secondly, CSOs perform service provisions; however, this service provision is a bottom-up approach consisting of service rendering to the people, government, and international institutions. Under this provision, the CSOs provide the welfare provisions such as education, recreation, health and family, humanitarian support, development project implementation and advice, and expertise services to businesses, governments, and international institutions.

Thirdly, CSOs' contributions in the field of people's well-being and improving the living conditions are notable. However, the CSOs cannot be always a substitute for the government in providing the welfare functions, perhaps in many countries CSOs work along with the state in providing welfare provisions to the people. There are several cases where CSOs cooperating with state for the delivery of social services rather than competing with state (Kendall 2003). Fourthly, civil society organizations identify the unaddressed issues and formulate new approaches for solving the societal issues by focusing on them; Kramer (1981), Kendall (2003), and Salamon et al. (2000) called it as "innovation" a most important function of the CSOs. Further Salamon et al. (2000) noted further noted finding innovative solutions to the unaddressed issues in the communities and societies as basic fictions of CSOs. Fifthly, CSOs focus on community building, through which they build social capital and generating sense of community based on mutual trust and reciprocity among the community people (Kendall 2003; and Salamon et al. 2000). However, a group of scholars, among them Kramer (1981), Kendall (2003), Salamon et al. (2000), and Frumkin (2002) noted the CSOs societal function as the expressive and the advocacy function, where expressive function includes the aspects related to cultural, spiritual, professional, or policy values, interests, and the beliefs of the people. The advocacy function deals with changing the policies or collecting the public opinion for the collective good.

There is no doubt in the globalized society that Civil Society Organizations are playing an irreplaceable role in the development and welfare of the society and community. However irrespective of its contributions for the societal development the immediate purpose of the existence of CSOs is not to be a substitute to the state in the welfare provisions. As noted by Michaela Neumayr et al. (2009) in case of state's failure to ensure social welfare to its citizens the CSOs intervene and ensure basic facilities to the people, but in any case CSOs are not existing nether to substitute the state or nor they are efficient in relation to state. In any case, CSOs can be seen as a temporary substitute for the welfare provisions when state fails to fulfil its basic missions (Michaela Neumayr et al. 2009). In the contemporary world, CSOs involve in a wide range of issues such as poverty eradication, energy issues, climate change, agricultural issues, water supply and sanitation, education and capacity building, housing and infrastructure, and access to banking services. However, the role of civil society, non-profit organizations in providing social services are important yet often overlooked features in contemporary South Asian society. The development of social welfare policies and programmes and civil society organizations are largely decided by social, economic, political, and cultural context of any country. These contexts either involve or influence the development and relevance of social welfare provisions.

In sum, despite the growing importance of the civil society organizations, the role and the importance are partially understood in the developing world. Especially in the developing South Asian countries, even basic descriptive information about CSOs such as number, size, area of activity, sources of revenue, and the policy framework within which they operate is not available in any systematic way. As noted by Salamon, Sokolowski, and Associates (2004), there is a major gap in the literature on role and responsibilities of civil society sector, and this literature gap is especially found in the developing South Asian countries CSOs contexts. We in this book aim to gather evidence-based writings on the role of civil society/third sector in social welfare across South Asia, and an attempt was also made to see how different political regimes/ideologies and state administrative systems are influencing or shaping social welfare and civil society sector across South Asia.

Organization of the Book

The roles of the civil society in social services are on the rise around the globe. The book aims to explore South Asian civil society organizations as third sector—the non-profit organizations and providers of social service. The book makes an attempt to (1) define social welfare and describe its relationship to social service programmes and individual well-being; (2) understand the social policy development from the problem identification to policy implementation; (3) describe the range of organization of social service agencies that are responsible for providing social welfare programmes in South Asia; (4) explore the various roles that professional and non-professional bodies provide in the delivery of social welfare and their influence in promoting change in policy development; and (5) understand the umbrella concept of child welfare, welfare of people with disability, and elderly welfare in welfare policy.

All the contributing authors of this book confront these issues, and they analysed it in the perspectives of different countries in South Asia, covering chapters from India, Pakistan, Nepal, Bangladesh, Bhutan, and Sri Lanka, apart from the introduction and conclusion by the editors. Together this provides a clearer understanding of civil society organization and social welfare in South Asian region and the book has been divided into seven parts:

Part I: Changing South Asian Society and Developmental Challenges
Part II: The State, Civil Society Organization and Social Welfare
Part III: Civil Society Initiative for Women Welfare
Part IV: Building Sustainable Communities: Civil Society Response
Part V: Contemporary South Asian Civil Society: Issues Challenges and Concerns
Part VI: Child Welfare in South Asia
Part VII: Towards Sustainable Communities—Realizing 2030 agenda

Editors of the book have given an introduction for each section that sets out the context followed by a brief summary of the chapters covered in that section. While attempts were made to secure contributions from other South Asian Countries, especially from Afghanistan and Maldives, it has been not possible to get a quality academic writing addressing the objectives of the book. Therefore, we note this is a modest attempt to consolidate and develop a balanced account of knowledge base on civil society initiatives in social welfare of South Asian countries. The book is intended for those who are interested in social work, social welfare, and policy-related studies or those who are more interested in working in these areas in particular in South Asian context. We hope, the students, educators, and in particular researchers doing comparative and cross-country studies and exchange programmes find this collection interesting and insightful.

References

Alexander, J. C. (2001). *The New Social Theory Reader*. New Delhi: Polity Press.

Armstrong, D., Beuo, V., Gilson, J., & Spine, D. (2011). *Civil Society and International Governance*. Rutledge/GARNET Series. Europe in the World.

Brown, David L. and David C. Korten (1991). "The Role of Voluntary Organizations in Development," IDR Reports. 8(l): 1–60.

Carothers, Thomas. 1999. Aiding Democracy Abroad: The Learning Curve. Washington, DC: Carnegie Endowment for International Peace.

Carothers, T., & Barndt, W. (1999). *Civil Society. Foreign Policy,* (117), 18–29. doi:10.2307/1149558

Cf. G. Esping-Andersen (Ed.). (1996). *Welfare States in Transition: National Adaptations in Global Economies*. London: Sage.

Cf. R. A. Falk. (1992). The Infancy of Global Civil Society. In G. Lundestad & O. A. Westad (Eds.), *Beyond the Cold War: New Dimensions in International Relations* (pp. 219–239). Oslo: Scandinavian University Press.

Drainville, A. C. (1998). The Fetishism of Global Civil Society. In M. P. Smith & L. E. Guarnizo (Eds.), *Transnationalism from Below*. London: Transaction.

Edward, M. (2005). *Civil Society. The Encyclopedia of Informal Education*. Retrieved from www.infed.org/associational/civil_society.htm.

Edwards, Michael. 2009. *Civil Society*. Cambridge: Polity Press.

Esping-Andersen, G. (1990) The Three Worlds of Welfare Capitalism. Cambridge: Polity Press. Esping-Andersen, G. (1993) 'The Comparative

MacroSociology of Welfare States', in L. Moreno (ed.) Social Exchange and Welfare Development, pp. 123– 38. Madrid: CSIC.

Fisher, Julie (1993). The Road from Rio: Sustainable Development and the Non-Governmental Movement in the Third World. Westport, CT: Praeger Press.

Frumkin, P. (2002). *On Being Nonprofit. A Conceptual and Policy Primer.* Harvard University Press.

Fukuyama, Francis (1995). Trust: The Social Virtues and the Creation of Prosperity. New York: The Free Press.

Joseph, S. (2002). Society vs. State? Civil Society, Political Society and Non-party Political Process in India. *Economic and Political Weekly, 37*(4), 299–305.

Kaldor, M. (2003). *Global Civil Society: An Answer to War.* Cambridge: Polity Press.

Kendall, J. (2003). *The Voluntary Sector. Comparative Perspectives in the UK.* London, New York: Routledge.

Kramer, R. M. (1981). *Voluntary Agencies in the Welfare State.* Los Angeles, London: Berkeley.

Lipschutz, R. D. (1992). Reconstructing World Politics: The Emergence of Global Civil Society. *Millennium, 21*, 389-420; Citizens Strengthening Global Civil Society (Washington: CIVICUS World Alliance for Citizen Participation, 1994).

Moore, B. J. (1966). *Social Origins of dictatorship and Democracy: Lords and Peasant in the Making of the Modern World.* Boston: Beacon Press.

Munene, J. W., & Reckson Thakhathi, D. (2017). An Analysis of Capacities of Civil Society Organizations (CSOs) Involved in Promotion of Community Participation in Governance in Kenya. *Journal of Public Affairs, e1668*, 1–6.

Neumayr, M., Meyer, M., Pospı's˘il, M., Schneider, U., & Maly', I. (2009). The Role of Civil Society Organizations in Different Nonprofit Regimes: Evidence from Austria and the Czech Republic. *Comparative Social Research, 26*, 167–196.

Nzomo M. (2003) "Civil Society in the Kenyan Political Transition: 1990–2002." The Kenyan Politics in Transition: 1990–2002.

OECD (1995). Development Cooperation: Efforts and Policies of the Members of the Development Assistance Committee. Chairman's Report, 1994. Paris: OECD.

Rose-Ackerman, Susan (1996). "Altruism, Nonprofits and Economic Theory." *Journal of Economic Literature 34*(2), 701–726.

Rosa Sánchez-Salgado (2007) Giving a European Dimension to Civil Society Organizations, *Journal of Civil Society*, *3*(3), 253–269, doi:10.1080/17448680701775671
Rose-Ackerman, S., & James, E. (1986). *The Nonprofit Enterprise in Market Economics*. Chur, London, Paris, and New York: Harwood Academic Publishers GmbH.
Salamon, L. M. (1994). The Rise of the Nonprofit Sector. *Foreign Affairs, 73*(3), 111–124.
Salamon, L., & Anheier, H. (1998). Social Origins of Civil Society. *VOLUNTAS: International Journal of Voluntary and Nonprofit Organizations, 9*, 213–248.
Salamon, L.M., Sokolowski, W.S. & Associates. (2004). *Global Civil Society. Dimensions of the Nonprofit Sector*. Bloomfield: Kumarian Press.
Salamon, L., Hems, L., & Chinnock, K. (2000). *The Nonprofit Sector: For What and for Whom?* Working Papers of the Johns Hopkins Comparative Nonprofit Sector Project. Baltimore, The Johns Hopkins University Center for Civil Society Studies.
Shaw, M. (1994). Civil Society and Global Politics: Beyond a Social Movements Approach. *Millennium, 23*, 647–667.
Suar, L. (2011, December). Our Civil Society as an Instrument of Social Change. *Orissa Review*.
Tandon, R. (1986). Regulating NGOs in India: New Moves. *Lokayan Bulletin, 4*, 3–4.
Union of International Associations. (1998). *Yearbook of International Organizations 1998/99* (Vol. I, p. 1764). Munich: Saur.

Part I

Changing South Asian Society and Developmental Challenges

Rajendra Baikady, Md. Nurul Momen, and Cheng Sheng-Li

Introduction

Over the past decade, the South Asian region has seen economic growth and countries in the region continue to grow. Though all countries in the South Asian region are still developing, they are quite distinct in terms of economy, population and demography and the developmental challenges that they encounter. South Asia is considered to be an important region in the world; it includes eight countries—Afghanistan, Bangladesh, Bhutan, India, Maldives, Nepal, Pakistan, and Sri Lanka—known for social, political and economic complexities and challenges, such as conflict, violence and extreme and widespread poverty. Problems such as chronic poverty, lack of legal awareness, low wage, and the financial status of the family in many South Asian countries contribute to high level of child labour and exploitation of children below the working age (Fatima et al. 2018). Problems faced by countries in the South Asian region differ country to country. For better social welfare and economic growth, five of these countries need cooperation with other states in the region: Bangladesh and Pakistan, the least developed countries, and Afghanistan, Bhutan and Nepal, being landlocked countries. Despite South Asia being home to more than a fifth of the global population, the region enjoys abundant natural resources and potentially untapped market. The culture and ethnic and linguistic commonalties bind the states and people of South Asia even though they are distinct in their civilization and development.

Countries in the South Asian region face common problems that pose development challenges. In order to strengthen the regional integration and achieve regional development, the South Asian Association for Regional Cooperation (SAARC), a regional intergovernmental organization and geopolitical union of states in South Asia, including the member states Afghanistan, Bangladesh, Bhutan, India, the Maldives, Nepal, Pakistan and Sri Lanka, was established in 1985. However, being the first regional organization of South Asian countries, the SAARC failed in achieving the goal of regional integration and development in the region. Being the poorest region of the world, all South Asian countries are caught in a poverty trap, unemployment and underdevelopment, which is characterized by the presence of mass illiteracy. All South Asian countries are young independent states that emerged only in the twentieth century. However, neither the geographical proximity nor the other similarities between them encouraged or promoted closer economic cooperation and hence regional integration has not been achieved.

South Asian Society: Demography and Characteristics

Nepal and Bhutan are very different countries in the region, and the problem faced by these two poor landlocked countries, located between the largest democratic country in the world and a one-party authoritarian state: India and China, hence, both Nepal and Bhutan adopts different sets of challenges and priorities, compared to rest of the region. The topography of these two countries has posed serious and potential hindrance for the development of agriculture and infrastructure. Further, Nepal is a country with high diversity in its culture, ethnic, linguistic and religious identity. According to the 2011 National Housing Census surveys, there were 126 ethnic communities, 123 languages and dialects in Nepal. Nepal is home to 26.4 million people with diverse religious and ethnic groups (Population and Population Census 2011). Bangladesh is a traditionally rice-based agricultural country, blessed with soil fertility replenished by the rich alluvial deposits that come with the typical annual flooding of several mighty rivers and their numerous tributaries which criss-cross the entire landscape. Poverty, child labour, and unemployment remain major development challenges in Bangladesh as in other South Asian countries. The geography of Pakistan combines mountains and plains. The Indus Valley development project initiated during the British rule gifted a modern and sophisticated irrigation facility to the country, and thus Pakistan's agriculture is highly benefited by a modern irrigation system. Sri Lanka is an island economy in the region that is geographically opposite to the landlocked countries such as Nepal and Bhutan. Sri Lanka is blessed with fertile land due to generous rainfall from a double monsoon. However, only 25per cent people of Sri Lanka are employed in the agriculture sector, mainly producing rice and wheat; however, the country produces more plantation-based agricultural products such as tea, rubber and coconut. It is worth mentioning that plantation-based agriculture contributed to 30 per cent of GDP in the early 1950s.

Among the South Asian countries, India is the fastest growing economy with US$ 2.957 trillion, and contributes to 85 per cent of the region's economy. The Indian economy is the sixth largest in the world in terms of nominal and the third largest in terms of purchasing power adjusted exchange rate (US$ 10.385 trillion). Among the eight South Asian countries, India is the only powerful member of G-20 major economies, and the BRICS from the region. Even though the gross domestic product (GDP) of the region is showing progress (see Table 1), the region needs more Foreign Direct Investment (FDI), environment-friendly industries and business and rising export in order to keep up the economic growth. The countries in the region need to develop environment-friendly economic development models in order to achieve sustainable development. The relationship between the FDI and economic development in many South Asian countries was examined by numerous scholars. Chakerborty and Nunnenkamp (2006) examined the FDI and economic development in India. Falki (2009), Atique et al. (2004), Yousaf et al. (2008), Malik et al. (2010), and Mohamed and Sidiropoulos

Table 1 GDP growth in South Asian countries

	India	Pakistan	Sri Lanka	Afgha-nistan	Bangla-desh	Nepal	Bhutan
GDP (US $)	2689.992	306.897	92.504	20.367	286.275	28.813	2.624
GDP growth	87.683	1.945	5.156	0.132	24.760	3.933	0.219

Source: International Monetary Fund World Economic Outlook (October 2018)

(2010) examined the impact of FDI on economic growth of Pakistan. All these studies found a close relation between increased FDI and increased economic development.

Climate change challenge and its adaptation strategy in the region is a major issue that needs states' attention. The countries in the region need to consider maintaining sustainable climate as impact on climate change has implication for the gross national product. Evidence shows that agricultural production in many parts of the world is already effected by the climate change that resulted in extreme weather events and sea level rises (SLRs), (United Nations 2018; World Bank 2016). Many studies have shown considerable impact of climate change on economic development (see, e.g. Nordhaus 1991; Tol 2002; Nordhaus 2007; Tol 2012). However, Weitzman (2012) and Stern (2016) emphasized the need for a new generation of economic modelling that may give a real picture of the environmental damage. Despite all developmental challenges faced, the countries in the region have adopted pro-growth policies for accelerating economic development. As a result, they opened up markets and replaced the public sector with the private sector to boost the growth and development (Ahmed 2006). It is also predicted that further acceleration of economic process in the region will allow the countries to utilize their potential and sustain high growth momentum (Ahmed 2006; Ahmed and Ghani 2007).

The region had shown fastest growth in the Human Development Index (HDI) since 1990, but wide inequalities in people's well-being is found; however, human development value of the region still shows widest gap between men and women (16.3). On an average, the HDI level has risen among the developing, and least developed countries across the

Table 2 Human Development Raking of South Asian countries

	India	Pakistan	Sri Lanka	Afghanistan	Bangladesh	Nepal
HDI Rank	130	150	76	168	136	149
HDI Value	0.640	0.562	0.770	0.498	0.608	0.574

Source: Human Development Index 2018

globe; thus, people lead a long and healthy life, are more educated and have greater income opportunities. However, the imbalance with respect to people's well-being still persists across developing countries (Table 2).

Between 1960 and 1980—the region had problems related to low productivity in agriculture, lower wages and least productive jobs, unemployment, problems with industrialization and high level of poverty. However, the scenario of the region changes with the introduction of pro-growth polices by the South Asian countries during the 1980s.

The region had achieved remarkable reduction in poverty over the decades, supported by long-term economic growth, structural economic transformations, and changes in occupation patterns across the region that played a major role in reducing poverty. However, there are country-specific factors that have contributed towards poverty reduction in the region, such as the microfinance revolution led by Grameen Bank in Bangladesh, the egalitarian policy adopted by the Sri Lankan government, and the huge amount of remittance sent by overseas migrant workers to Pakistan and Nepal helped these countries to push out of poverty. In Bhutan, the concept of Gross National Happiness (GNH) perceived by the government helped the people to overcome poverty and achieve a higher standard of living.

Challenges Faced by South Asian Society: An Overview

The South Asian region underwent tremendous changes in economic policy and structural transformation in the last 50 years. It is worth mentioning that the process of globalization and neoliberal economic policies introduced across the globe resulted in the transformation of societies in the global south. Economic transformation helped people to get work, poverty alleviation and uplifting their living standards. Furthermore, the introduction of Millennium Development Goals by the United Nations (UN) strengthens the developmental dreams of South Asian countries. However, as a part of the fast-changing economic process, South Asia still encounters problems such as mass poverty, unemployment, child exploitation, a variety of issues and challenges related to women and societal unrest. South Asian region is historically known for the high rate of malnutrition that impacts on stunting and underweight with negative economic consequences among vulnerable communities. The problem of malnutrition in the South Asian region is far higher than that in sub-Saharan Africa, which is poorer in terms of per-capita national income compared to South Asia. As per the report of the International Food Policy Research Institute (2015), South Asia is home to the world's largest malnourished population; countries throughout the region is the largest home for stunted children under the age of five. In addition, the region is also facing the issues of increasing number of people with diabetes, chronic heart disease and micronutrient deficiencies (IFPRI 2015).

On the other hand, the South Asian region recorded the fastest economic growth in the world (World Bank 2016). While India recorded strongest growing economy, many of its neighbouring countries achieved significant poverty reduction (Bangladesh), significant fall in child malnutrition (Nepal) and declining illiteracy (Sri Lanka) and significant increase in agricultural productivity (Pakistan) (Asian Development Bank 2016, UNICEF 2016). On negative performance, there are signs of growing inequality in education and health outcomes in the region. It was found that the human capital index in the region is also low and the gap between rich and poor children in education entrainment is high. The countries in the region need to work towards strengthening educational institutions, as equal access and opportunities are largely dependent upon educational polices and plans of the nation. A well-planned educational policy helps in reducing disparity between the genders, caste, ethnic group, religious group and classes of people.

Population growth is one of the major demographic challenges faced by the region. According to estimations, by 2030 South Asia is likely to host 2.1 billion people and ensuring sustainable rate of population growth will be a contemporary challenge for the countries in the region. The population growth in the region will ensure economic growth only if it is well managed. Furthermore, countries in the region need to introduce social polices ensuring reduced birth rate, reduced infant mortality, small and secure families. More than 17 per cent of the world's population is accommodated in one country (i.e. India) in the region (United Nations 2015). The United Nations population division estimates that by 2028, India's population will overtake China's population. In addition to rapid population growth in general, the old age population in the region is also growing, and the growth is three times higher than the general population in India (Giridhar, Sathyanarayana et al. 2014). However, population growth in the region is the result of many cultural and traditional practices. Some countries in the region have also shown extreme differences in child sex ratio at birth. Countries such as India, Pakistan, Bangladesh and Nepal have explicit cultural preferences for a male child, and this typical preference results in a sex ratio imbalance between girls and boys (Dyson 2012, Gupta 2006, Kaur 2016, Klasen and Wink 2002). Women empowerment, through education and employment opportunities, decision-making power in both public and private spheres, and access to essential health services, is crucial in South Asia towards the progress of sustainable development goals by 2030.

Child labour remains a big problem throughout the region. It is the most common problem among the developing countries mostly because of poor socio-economic environment. While, there are some underlying reasons for the practice of child labour across the globe that has been reported by researchers pointing out that rapid growth of population, low wages, and increasing cost of living are the major contributing factors in the developing countries (Bukari 2008). Because of these conditions, children involve in the economic activities to earn income (Bass 2004). Other scholars argue that absence of knowledge among the societal members about child labour has also supported the practice of child labour in many countries (Musvoto 2007). Ray (2000) found practising child labour is a cultural aspect, as it is rampant among some communities. Scholars have also argued that the globalization process also contributes to the growth of child labour (Kamer 2013). As per the estimations of the Federal Bureau of Statistics of Pakistan, the country has around

40 million children, of which 3.8 million are in the age group of 5 to 14 years and of which 73 per cent happen to be involved in the agricultural sector and manufacturing industries (Kamer 2013). Researchers have argued about the physical and moral impact of child labour and noted that the brutality faced by the children may lead to dehumanization (Shaw 2006). Robert (2008) noted that children's behaviour becomes delinquent when they are treated with cruelty. Ensuring safety and security, a conducive environment, education and recreational facilities for the children in the region is yet another important aspect for development.

As a measure of protecting child rights, the UN abolished child labour in all its forms (Susan 2004). In addition, at present there is increasing awareness about the negative effects of child labour among the community, and there are efforts to integrate child labour and trafficking concerns into 'education for all' in Cambodia, China, India, Indonesia, Pakistan, the Philippines, Sri Lanka and Vietnam (Khan and Lyon 2015). There are numerous legal and community leaders advocating for the betterment of the condition of children in the region. The International Labour Organization (ILO) launched campaigns such as Red Card to Child Labour in (2002) and the Global Conference on Child labour, where representatives from 97 countries agreed upon elimination of child labour in all of its forms (UNICEF 2014; US DoL 2016). A World Day against Child labour (June 12) with national-level activities and deliberations are also observed by the ILO since 2002. Understanding Children Project (UCW), the International Partnership on Child Labour in Agriculture, the Global Task Force on Child Labour and Education for all and Child Labour Platform are some of the initiatives led by the ILO in collaboration with the UN (ILO 2015).

Child protection in the South Asian Region has been given specific attention. The Rawalpindi (Ministerial) Resolution on Children of South Asia (1996), SAARC Convention on Regional Arrangements for the Promotion of Child Welfare in South Asia (2002), SAARC Convention on Preventing and Combating Trafficking in Women and Children for Prostitution (2002), The Colombo (Ministerial) Statement (2009), The SAARC Development Goals (SDGs), South to South Cooperation on Child Rights 2010 are the instances of state intervention in ensuring safe and protective environment for children in the region (see Table 3 for country-level programmes).

Table 3 The most recent social programmes implemented by South Asian countries to address child labour

Bangladesh	Eradication of Hazardous Child Labour, Phase III
	Services for Children at Risk Project
	Urban Social Protection Initiative to Reach the Unreachable and Invisible and Ending Child Labour Reaching Out-of-School Children II (2012–2017)
	Child Sensitive Social Protection Project (2012–2016)
	Enabling Environment for Child Rights Primary Education Stipend Project, Phase III
	Support Urban Slum Children to Access Inclusive Non-Formal Education
	Country Level Engagement and Assistance to Reduce (CLEAR) Child Labour Project
	Expanding the Evidence Base and Reinforcing Policy Research for Scaling-up and Accelerating Action Against Child Labour Shelter Project
	Child Help Line 1098 National Helpline Centre
	Vulnerable Group Development Programme
India	National Child Labour Project (NCLP) Scheme
	Grants-in-Aid Scheme
	Rehabilitation of Bonded Labour Scheme
	Integrated Child Protection Scheme
	Welfare of Working Children in Need of Care and Protection
	Anti-Human Trafficking Activities
	Childline
	TrackChild
	Testing Methodologies to Support Informal Economy Workers and Small Producers to Combat Hazardous Child Labour in Their Own Sectors
	Education for All Scheme (Sarva Shiksha Abhiyan)
	Enhancing Teacher Effectiveness in Bihar Operation
	Midday Meal Programme
	National Rural Employment Guarantee Scheme
	National Rural Livelihoods Mission

(*continued*)

Table 3 (continued)

Nepal	Comprehensive Child Labour Programme (2011–2015)
	Green Flag Movement (2014–2017)
	Child Helpline—1098
	Country-Level Engagement and Assistance to Reduce Child Labour II (CLEAR II) (2014–2018)
	From Protocol to Practice: A Bridge to Global Action on Forced Labour (The Bridge Project) Towards Achieving the Elimination of the Worst Forms of Child Labour as Priority (ACHIEVE) (2013–2016)
	Project for the Prevention and Reduction of Child Labour in Restaurants in the Kathmandu Valley (PRECLOR) (2011–2015)
	Decent Work Country Programme, Nepal (2013–2017)
	Combating Trafficking in Persons (CTIP) Project (2010–2016)
	Support for Schools
	Compulsory Education Pilot Programme
	School Sector Reform Program (2009–2016)
Pakistan	National Centres for Rehabilitation of Child Labour
	Child Support Programme
	Benazir Bhutto Income Support Programme
	Decent Work Country Program (2010–2015)
	Sabawoon Rehabilitation Centre
	Project to Combat the Worst Forms of Child Labour
	Education Programme for Children of Brick Kiln Workers
	Global Action Programme on Child Labour Issues
	Education Voucher Scheme

Sources: US DoL (2016: 134–135, 537, 762, 810)

Despite these efforts, there are still children out of school, and facing serious safety issues throughout the region. According to available evidence, despite policies and programmes in place to ensure children attend school, 5 per cent children in Nepal are still out of school or of the formal education system (CBS 2011); further, these 5 per cent children are from multiple sub-groups of the community, such as gender, caste, religion and poor families (UNICIEF 2010).

Part I of the book focuses on social welfare provisions in different parts of the South Asian countries, and analyses the set of roles of civil society organizations in social welfare in diverse social, economic and political contexts.

Chapter 2 uses a case-study approach, and examines the role of Sense India as a civil society organization for the inclusion of persons with deafblindness in India. The paper illuminates the institutional journey of Sense India, and how it has transformed the social welfare interventions for persons with deafblindness in the country from a welfare as well as rights-based approach. Chapter 3 evaluates the effectiveness of the Samurdhi Programme, based on a comparison between beneficiaries' and non-beneficiaries' perspective by focusing on one village in Sri Lanka. The issues and problems of the Samurdhi Programme were identified, and strategies were proposed to improve the effectiveness of the programme. Chapter 4 examines various adaptation practices in Bangladesh with respect to the Charland dwellers. The chapter aims to look into the community-based adaptation measures being practised by the river-erosion affected people, and also to review the impediments in promoting community-based adaptation measures.

References

Ahmed, A. D. (2006). The Impact of Financial Liberalization Policies: The Case of Botswana. *Journal of African Development, 1*(1), 13–38.

Ahmed, S., & Ghani, E. (2007). South Asia Growth and Regional Integration. Macmillan India Ltd.

Asian Development Bank. (2016). *Asian Development Outlook 2016: Asia's Potential Growth*. Manila: ADB.

Atique, Z., Ahmad, M. H., & Azhae, U. (2004). The Impact of FDI on Economic Growth Under Foreign Trade Regimes: A Case Study of Pakistan. *The Pakistan Development Review, 43*(4), 707–718.

Bass, L. E. (2004). *Child Labour in Sub-Saharan Africa*. Chicago: Lynne Rienner.

Bukari, S. (2008). *Ministry of Child Labour*. Islamabad: Express Tribune.

CBS (Central Bureau of Statistics). (2011). National Population and Housing Census, Nepal.

Chakraborty, C., & Nunnenkamp, P. (2006). Economic Reforms, Foreign Direct Investment and Its Economic Effects in India. *Kiel Working Papers, 1272*, 1–45.

Dyson, T. (2012). Causes and Consequences of Skewed Sex Ratios. *Annual Review of Sociology, 38*, 443–461.

Falki, N. (2009). Impact of Foreign Direct Investment on Economic Growth in Pakistan. *International Review of Business Research Papers, 5*(5), 110–120.

Fatima, F., Fatima, S., & Ali, S. (2018). Prohibition of Child Labour Leading to Delinquent Behavior in the Constitution of Islamic Republic of Pakistan 1973. *Pakistan Administrative Review, 2*(4), 383–396.

Giridhar, G., et al. (2014). Introduction. In G. Giridhar, K. M. Sathyanarayana, S. Kumar, K. S. James, & M. Alam (Eds.), *Population Ageing in India* (pp. xvii–xxvi). Delhi: Cambridge University Press.

Gupta, M. D. (2006). Cultural Versus Biological Factors in Explaining Asia's "Missing Women": Response to Oster. *Population and Development Review, 32*, 328–332.

International Food Policy Research Institute. (2015). *Global Nutrition Report 2015: Actions and Accountability to Advance Nutrition and Sustainable Development*. Washington, DC: IFPRI.

International Labour organisation (ILO). (2015). Major Results of ILO Work on Child Labour. Retrieved from www.ilo.org/ipecinfo/product/download.do?type=document&id=25895.

International Monetary Fund. (2018). *World Economic Outlook: Challenges to Steady Growth*. Washington, DC, October.

Kamer, D. (2013, May 14). *Datum Corporation*. Retrieved November 19, 2014, from Labour Law Support http://www.sja.gos.pk.

Kaur, R. (2016). *Too Many Men, Too Few Women: Social Consequences of Gender Imbalance in India and China*. Hyderabad: Orient BlackSwan.

Khan, S., & Lyon, S. (2015). *Measuring Children's Work in South Asia: Perspectives from National Household Surveys. International Labour Organization*. New Delhi, India: ILO. http://www.ilo.org/wcmsp5/groups/public/%2D%2D-asia/%2D%2D-ro-bangkok/%2D%2D-sro-new_delhi/documents/publication/wcms_359371.pdf.

Klasen, S., & Wink, C. (2002). A Turning Point in Gender Bias in Mortality? An Update on the Number of Missing Women. *Population and Development Review, 28*, 285–312.

Malik, S., Hayat, M. U., & Hayat, M. K. (2010). External Debt and Economic Growth: Empirical Evidence from Pakistan. *International Research Journal of Finance and Economics, 44*, 88–97.

Musvoto, E. (2007). Child Labour in South Africa: A Brief Overview. *Children World Wide Trust*, 77–89.

National Population and Housing Census (National Report). (2011). *Government of Nepal, National Planning Commission Secretariat*. Nepal: Central Bureau of Statistics. Kathmandu.

Nordhaus, W. D. (2007). A Review of the Stern Review on the Economics of Climate Change. *Journal of Economic Literature, 101*(407), 920–937.

Nordhaus, W. D. (1991). To Slow Or Not to Slow: The Economics of the Greenhouse Effect. *The Economic Journal, 101*, 920–937.

Ray, R. (2000). Child Labor, Child Schooling and Their Interaction with Adult Labour: Empirical Evidence for Peru and Pakistan. *The World Bank Economic*, 347–367.

Robert, R.D. (2008). Child Delinquency in America. In D. Bernaod, *Child Delinquency and the Prophaylaxis of Crime in Early 20th Century Latin America* (pp. volume 24, n. 6). Washington, DC: Rev PanamSalud.

Shaw, J. (2006). State Bank of Pakistan, Underemployment in Pakistan, Socialsection. The State of Pakistan's Economy. (pp. 1–6). Karachi: FY06. Daily Report.

Stern, N. (2016). Economics: Current Climate Models Are Grossly Misleading. *Nature, 530*, 407–409. https://doi.org/10.1038/530407a.

Susan, D. (2004). *World Health Organization Report*. Denmark.

Tol, R. S. (2002). Estimates of the Damage Costs of Climate Change – Part 1: Benchmark Estimates. *Environmental and Resource Economics, 21*, 47–73.

Tol, R. S. (2012). On the Uncertainty About the Total Economic Impact of Climate Change. *Environmental and Resource Economics, 53*, 97–116.

UNICEF. (2014). *Child Labour and UNICEF in Action: Children at the Centre*. New York, USA: UNICEF. https://www.unicef.org/protection/files/Child_Labour_and_UNICEF_in_Action.pdf.

UNICEF (The United Nations Children's Fund). (2016). *South Asia 2016 Progress Report*. Kathmandu: UNICEF Regional Office for South Asia.

United Nations. (2015). *World Population Prospects: The 2015 Revision*. Geneva: United Nations Department of Economic and Social Affairs Population Division.

United Nations. (2018). Goal 13: Take Urgent Action to Combat Climate Change and Its Impacts. Retrieved April 22, 2018, from http://www.un.org/sustainabledevelopment/climate-change-2/.

United Nations Children's Fund (UNICEF). (2010). Child Poverty and Disparities in Nepal. Nepal Report 2010.

Weitzman, M. L. (2012). GHG Targets as Insurance Against Catastrophic. *Journal of Public Economic Theory, 14*, 221–244.

World Bank. (2016). *South Asia Economic Focus, Spring 2016: Fading Tailwinds*. Washington, DC: World Bank.

Yousaf, M. M., et al. (2008). Economic Evaluation of Foreign Direct Investment in Pakistan. *Pakistan Economic and Social Review, 46*(1), 37–56.

2

Role of Civil Society Organization in the Inclusion of People with Deafblindness in South Asia: A Case Study on Sense International India

Akhil Paul, Atul Jaiswal, Uttam Kumar, and Parag Namdeo

Introduction

India, which is a vast South Asian country of 1.2 billion people, is also home to 26.8 million persons with disabilities (Census of India 2011). Despite the impressive aggregate growth of India as a developing nation in South Asia in the past few years, its benefits have very often not reached the marginalized sections of the society, including people with disabilities (Goswami and Tandon 2013; World Bank 2009). Vulnerability and conditions of poverty are still prevalent in the country which make people

A. Paul • U. Kumar • P. Namdeo
Sense International India, Ahmedabad, India

A. Jaiswal (✉)
School of Rehabilitation Therapy, Queen's University, Kingston, ON, Canada
e-mail: atul.jaiswal@queensu.ca

© The Author(s) 2020
Md. N. Momen et al. (eds.), *Building Sustainable Communities*,
https://doi.org/10.1007/978-981-15-2393-9_2

with disabilities the most abject under-served populations (World Bank 2009). Persons with disabilities remain among the most vulnerable populations in these societies and often experience discrimination, marginalization, and human rights violations (World Bank & World Health Organization 2011). People with disabilities face a complex array of social problems such as extreme poverty, perpetuating caste system, poor infrastructure, remoteness, and lack of social support and services, particularly in the areas of health, education, and social protection, resulting in socio-economic inequities across the country (World Bank 2009). There is a huge developmental divide between the interstate and intra-state regions, between urban and rural, and between rich and poor. Moreover, significant regional imbalances are also evident and people with disabilities continue to stay on the boundaries of development, struggling for better social and economic support (World Bank 2009).

In India and most of South Asia, the meaning of disability is embedded in the religious/cultural understanding and basic struggle for survival (Mehrotra 2011). The extent to which a disability impacts someone's participation in society depends on their personal attributes (such as age, gender, type, and level of disability) and environmental factors such as policies and social protection services related to disability in the country (Groce 1999; Thomas et al. 2013; World Bank & World Health Organization 2011). Very often, people with disabilities and their families are stuck within the bidirectional circle of poverty and disability (Barnes and Mercer 2003; Trani and Loeb 2012; World Bank & World Health Organization 2011), and people with disabilities in India are no exception to this phenomena.

In a scenario like this, the role of civil society organizations (CSOs) is indispensable, where these organizations are instrumental in ensuring that the rights of the under-served population are fulfilled and needs met. CSOs, previously known as non-governmental organizations (NGOs), play a complementary role and "act as the third leg to development together with government and development partners (donors)" (Kaulem 2007, p. 2; Ramírez 2016). In some instances, a few CSOs take a different stand and create a parallel system of support services for those marginalized populations whose identity is not recognized in the legal system of the country and almost no welfare policies/services exist for them.

Similar is the condition with Sense International India, which was founded in 1997 to cater to the needs of people with deafblindness in India. Until December 2016, deafblindness was not recognized as a distinct disability in India. Hence, persons with deafblindness remain disadvantaged and had almost no access to governmental social welfare services available for people with disabilities in the country.

Despite the fact that India is adopting a progressive disability policy framework, rigorous efforts are still required to create awareness on disability rights, inclusion, and equal opportunity for people with more than one disability condition such as deafblindness. As per the Census of India (2011), persons with multiple disabilities are approximately 2,144,000 in number (Census of India 2011) and deafblindness is one such unique multi-sensory disability condition (Sense International India 2017a). Deafblindness is a combined dual sensory impairment of vision and hearing causing challenges in communication, mobility, and access to information from the outside world (Dammeyer 2015; Paul et al. 2016; Sense International 2017; Sense International India 2014).

Deafblindness is a heterogeneous condition involving a spectrum of vision and hearing loss that ranges from mild loss in hearing and vision to total deafness and blindness (Dammeyer 2014, 2015). Very often, deafblindness is associated with conditions such as physical, intellectual, developmental, and/or behavioural disorders that further impede their functioning and participation (Sense International India 2014). Deafblindness can be congenital or acquired and individuals with deafblindness vary with regard to the cause or medical aetiology of deafblindness, age of onset, severity of loss, and presence of additional medical conditions.

Congenital deafblindness is commonly caused by the rubella virus during the first three months of pregnancy, while acquired deafblindness can result from an accident, illnesses such as meningitis, genetic conditions, or ageing (Dammeyer 2014; Government of India, UK Aid, & World Bank Group n.d.; Sense International 2015). In some cases, Usher syndrome, which is a genetic disorder, leads to progressive loss of sight and hearing (Dammeyer 2014; Sense International 2015). Individuals with deafblindness are impacted with a combination of two major sensory impairments, sight and hearing, which otherwise contribute to 95%

of the learning (Sense International India 2014). This concurrent loss of hearing and vision results in a situation that is more disabling than the sum of individual impairments (Dammeyer 2014, 2015).

Nevertheless, persons with deafblindness are very often denied access to medical, educational, livelihood, and welfare services for people with disabilities. Many children with deafblindness are denied admission and remain isolated without getting any kind of medical or educational interventions supported by governments in India (Jaiswal and Rizal 2017; Paul et al. 2016; Sense International India 2017b). In the absence of recognition under Indian laws, persons with deafblindness faced extreme discrimination and are considered non-existent and no provisions are made for them (Paul et al. 2016; Sense International India 2014).

Evidence from the global literature on deafblindness also suggests that people with deafblindness have varying experiences in terms of their experiences of discrimination, social isolation, depression, communication difficulties, negative attitude of others, and lack of social support (Jaiswal, Aldersey, Wittich, Mirza & Finlayson 2018). Moreover, individuals with deafblindness are often not accepted by single-disability organizations (working for deaf or blind population) and left on their own to find resources for their education, medical care, and rehabilitation (Sense International India 2014; Wittich et al. 2012).

There is an estimated population of 500,000 persons with deafblindness in India, and a very limited number of CSOs are working for this population in the country (Sense International India 2017b). The reasons behind this limitation could be twofold: first, lack of awareness about deafblindness among various stakeholders such as the general public, policy makers, professionals, and CSOs; and second, limited funding and support available from government and development partners (donors) to support deafblind-specific interventions (Jaiswal and Rizal 2017; Paul et al. 2016; Sense International India 2014).

While deafblindness is not rare in India, many individuals with deafblindness, even those living in urban areas, are not identified or diagnosed early and hence do not get timely support services (Paul et al. 2016). The lack of diagnostic services for this population keeps them away from early intervention, rehabilitation, and educational programmes (Paul et al. 2016). Moreover, there are very few deafblind-specific professionals who can diagnose and provide support services to the deafblind population (Jaiswal and Rizal 2017; Paul et al. 2016).

Currently, persons with deafblindness in India are mainly supported through the collective efforts of Sense International India and its nationwide network of 59 partner CSOs in 22 states of India (Sense International India 2017b). Sense International India is the only organization in India that works for developing comprehensive services for deafblind people throughout the country. Through its network of partners, Sense India supports around 77,500 persons with deafblindness living in the poorest areas of the country for the last two decades (Sense International India 2017b). This chapter aims to highlight the deafblind-specific services developed by Sense India in enhancing the inclusion of individuals with deafblindness in society.

Methods

This chapter uses a case-study approach and examines the role of Sense India as a civil society organization in the inclusion of persons with deafblindness in India. This chapter illuminates the institutional journey of Sense India from 1997 and how it has transformed the interventions for the inclusion of persons with deafblindness in the country with a welfare as well as rights-based approach. This chapter uses multiple sources of evidence in building the case and draws upon organizational documents and reports to highlight the challenges faced by the organization in the process of transforming the welfare landscape for deafblind-specific services in India in the past two decades.

The chapter is organized under three sections. The first section provides an introduction to the deafblind-specific services and their relevance for children and adults with deafblindness. The second section highlights the key accomplishments and challenges faced by the organization in developing deafblind-specific services in the country. Some example case studies on how Sense India facilitated the inclusion of beneficiaries with deafblindness in society are provided in the third section.

Deafblind-Specific Services in India

Before the inception of Sense India in 1997, there was only one centre for persons with deafblindness in India catering to only 23 individuals with deafblindness (Sense International India 2017b). However, understanding the dire need to reach out to many more from the estimated 500,000 deafblind population in the country, Sense India developed its innovative approach to initiate and expand services to remote parts of the country by developing a network of grassroots organizations and building their capacity to work with persons with deafblindness in making them self-reliant. This well-planned consistent effort in the last 20 years enabled Sense India to reach out to more than 77,500 children and adults with deafblindness and support them with deafblind-specific services (Sense International India 2017b).

The partnership approach of supporting local organizations resulted in the setting up of 59 projects in 22 states in a span of 20 years and is one of the most cost-effective models of service delivery for persons with deafblindness in the country (Sense International India 2017b). The unique feature of this innovative approach is that this deafblind-specific intervention model does not require creation of new institutions or infrastructure; it rather uses the infrastructure of the partner CSOs to develop services to avoid duplication of resources. This model focuses on enabling the capacity of stakeholders (like persons with deafblindness, their families, teachers, CSO partners, and government organizations) in developing services for deafblind population. This model facilitates learning through sharing of experiences among groups of professionals, families of persons with deafblindness, and organizations working with persons with disabilities.

Starting with a national perspective, Sense India is now focused on addressing the needs at the regional and local levels, responding to the requests of CSOs through Regional Learning Centres (RLCs) and State Learning Centres (SLCs) on deafblindness. These RLCs and SLCs complement the activities done by Sense India at the national level and act as centres of excellence for technical know-how, referrals, expertise information, and models of good practice for the development of deafblind services in their respective and adjoining states (Sense International India n.d.).

The mission of Sense India is "*to work in partnership with others—people with deafblindness, their families, carers and professionals—to ensure that everyone facing challenges because of deafblindness has access to advice, opportunities and support*" (Sense International India 2017c). The organization aims to improve the lives of people with deafblindness, reduce isolation, and promote their interaction and self-esteem. The organization envisions at least one deafblind service provider in each state so that no child with deafblindness travels more than a day to get needs-based support and advice (Sense International India n.d.). Sense India aims to enhance the capacities of its partner organizations through training, mentoring, awareness visits, and sharing of information and communication on deafblindness (Sense International India n.d.). The main activities of the organization are the following:

(a) **Education:** Highly specialized and tailor-made assessment, communication, and educational services to the child with deafblindness through an Individualized Education Plan (IEP), complemented through training of their parents and family members.
(b) **Teacher Training Programme:** Teachers working with persons with deafblindness receive updated skills on how to work with persons with deafblindness in various areas like independent living, communication, developing IEP, managing challenging behaviours, and vocational rehabilitation, among others, to provide better quality of services.
(c) **Care and Rehabilitation**: Assistive services to build skills for daily living, improving communication, social interaction, behavioural management, and training for vocational skills through home-based, community-based, and day-care rehabilitation programmes.
(d) **Advocacy and Lobbying:** Policy-level advocacy initiatives for inclusion of deafblindness in national policies and through networks of adults with deafblindness, their parents, and special educators in the field across the nation. In addition, the organization strengthens the networks of families, teachers, and adults with deafblindness as a common platform to advocate for their rights and entitlements (Sense International India n.d.).

Sense India works to provide information, guidance, support, and training to help deafblind people and their families overcome specific challenges related to communication, mobility, and access to information (Sense International India n.d.). Through awareness generation and advocacy for the rights, opportunities, and services for this population throughout the nation on the one hand and providing deafblind-specific technical assistance (training and financial support) to its partner CSOs and their staff on the other, Sense India brings meaningful changes in the lives of this unique population. Supplementing to this work, the organization spearheads the collaborative work with the government to ensure the inclusion of deafblindness in policies and the capacity building of government Sarva Shiksha Abhiyan (SSA) teachers (Sense International India n.d.).

Key Accomplishments and Challenges Faced by the Organization

Sense India, along with its partner organizations, has been advocating for the rights of people with deafblindness for the last two decades. The two key accomplishments of the cumulative advocacy efforts of Sense India, its 3 national networks, and 59 partner CSOs are: inclusion of "Rubella vaccination" in the national immunization programme to reduce the incidence of congenital rubella infection (one of the main causes of deafblindness); and recognition of deafblindness in the new disability Act "Rights of Persons with Disabilities Act, 2016" (Government of India 2016; Sense International India 2017a). Recently, Sense India successfully advocated in the National Consultation on Accessible Elections on July 3, 2018, for the inclusion of persons with deafblindness in the electoral process. The organization has also succeeded in getting deafblindness officially recognized as a distinct disability in the state-specific disability policies draft for four Indian states—Rajasthan, Chhattisgarh, Bihar, and Meghalaya (Sense International India 2017a). The inclusion of deafblindness in the national and state-specific disability policies raises hope that now children and adults with deafblindness will have some service support from the Government of India.

In addition, Sense India also played the role of catalyst in developing human resources in terms of teachers/educators and has facilitated the development of teacher-training courses on deafblindness by the Rehabilitation Council of India (Sense International India 2016). The Sense India team, especially the third author (UK) of this chapter, is the main resource person in the development of the module on "Including Children with Deafblindness" for Master Trainers in the training toolkit by the Ministry of Human Resource Development—Government of India, UK Aid, and the World Bank Group (Government of India, UK Aid, & World Bank Group n.d.).

SSA is a national government programme to universalize elementary education for all children including children with disabilities in the country (Sarva Shiksha Abhiyan n.d.-a). In response to the demand for quality education for all children including children with deafblindness and multi-sensory impairments, Sense India is working in partnership with SSA. In 2010, Sense India developed two training modules, one each in deafblindness and multiple disabilities, to assist resource teachers, block resource teachers, and coordinators of SSA with basic skills to identify, assess, and strengthen services for children with deafblindness and multi-sensory impairments in local districts (Sarva Shiksha Abhiyan n.d.-b).

Furthermore, Sense India has also strongly advocated for equal right to education of children with deafblindness and has trained more than 3800 teachers of the mainstream education system (SSA teachers) to build their capacity to teach children with deafblindness in the same classroom with their non-disabled peers (Sense International India 2016). Sense India has worked on curriculum adaptations, that is, adapting the regular classroom text books for students with deafblindness. The strategy of curriculum adaptation is complementing the strategy of providing training to the teachers to attain the goals of inclusive education for children with deafblindness. Moreover, understanding the crucial role of early screening and intervention in the lives of children with deafblindness in their first few years, Sense India has played a pivotal role in developing an early intervention model for newborns with deafblindness (in 2011) and has reached out to more than 5500 children with early intervention support throughout the country (Paul et al. 2016).

Sense India believes that research and practice should go hand in hand to fill the void in deafblind-specific services in India. Hence, understanding the need for producing deafblind-specific scientific literature, Sense India set up its research ethics board in 2017 to ensure evidence-based research on deafblindness-related interventions in India (Jaiswal et al. 2018). Also, Sense India has supported the development and strengthening of its three national networks (UDAAN, a network of 194 adults with deafblindness; PRAYAAS, a network of 1167 family members of individuals with deafblindness; and ABHI-PRERNA, a network of 674 educators of individuals with deafblindness) to build their capacity to advocate for the rights of individuals with deafblindness as enshrined in the new disability law in India (Sense International India 2017a). These networks also play a key role in providing insights to facilitate deafblind research and practice self-advocacy.

The work of Sense India is not limited to India; it has been successfully initiated in South Asia. Sense India trained professionals in Bangladesh, Malaysia, Nepal, and Sri Lanka to initiate deafblind-specific interventions in their respective nations. Sense India has been accredited to *United Nations Conference of States Parties to the Convention on the Rights of Persons with Disabilities* at its eighth session and recently received the Economic and Social Council (ECOSOC) Consultative Status from the United Nations (Sense International India n.d.).

Organizational Challenges and How Sense India Is Trying to Overcome Those

The major challenges faced by the organization in developing deafblind interventions in India are as follows:

1. **Lack of awareness about deafblindness.** Deafblindness, being a low-incidence disability, remains an invisible condition in society. There is a sheer absence of awareness of and lack of understanding about

deafblindness not only in general public but also among the medical professionals due to which most of the time, people with deafblindness get either misdiagnosed or underdiagnosed. The lack of awareness about deafblindness hinders timely identification and referral of children at an early age. Correct screening and early identification would enable children with deafblindness to receive appropriate care. In response to this, Sense India, in collaboration with nation-wide partners, has placed strong emphasis on year-round advocacy campaigning to raise awareness about this unique condition among medical professionals, teachers, government officials, college youth, and community members at large. Awareness generation about deafblindness helps community members to understand the unique nature of the deafblind population and how to accept and include them in society.

2. **Lack of trained human resources in the field of deafblindness**. The currently available human resources in the disability field are not able to provide one-to-one support to meet the needs of this population with deafblindness. Moreover, there are only three training centres throughout India with a limited number of personnel trained in deafblindness, despite the demand. Sense India is not only advocating for more training courses on deafblind interventions, but also conducting national, regional, and state- and local-level training to build capacity of professionals to work with children and adults with deafblindness. Trained professionals in the field of deafblind interventions act as strong pillars in delivering comprehensive rehabilitation care to children and adults with deafblindness and promote their inclusion.
3. **Nature of beneficiary population.** The majority of the persons with deafblindness in India belong to the Below Poverty Line (BPL) category due to the vicious effect of poverty and disability, and are the most vulnerable category in terms of lack of support, recognition, and rehabilitation. They very often lack access to essential services including health, education, public transportation, or access to public spaces such as hospitals, making it inconvenient for them to participate economically, socially, and politically. Cross-sectionalities of gender, age, caste, and religion further exacerbate their marginalization, and are a major deterrent to social acceptance and inclusion. Girls and women with deafblindness are further disadvantaged due to the

double-discrimination effect, and hence, Sense India ensures their equal representation and participation in its programmes.

Moreover, individuals with deafblindness are a heterogeneous population with a range of varying needs and varying abilities in terms of their learning. This variation is based on the type, level, and onset of impairment, time of language acquisition, and degree of vision/hearing loss (Ask Larsen and Damen 2014; Dammeyer 2014, 2015). This heterogeneous nature of disability poses challenges for the professionals in understanding their needs and meeting them. However, to overcome this challenge, Sense India imparts deafblind-specific technical skills to professionals to work with this diverse population in promoting their inclusion in society.

4. **Limited resources to sustain the work**. Due to limited awareness about this condition in society and among donors/funding bodies, Sense India is facing challenges in raising resources in terms of grants and in-kind donations to ensure the sustainability of services to people with deafblindness in the country. However, Sense India has a strong public engagement team to raise resources and has been quite successful in expanding the work to remote parts of the country every year with support from local, national, and international donors/funders.

Case Studies on How Sense India Facilitated the Inclusion of Beneficiaries in Society

Case Study 1: Asha's Journey from Isolation to Inclusion

Despite attending a special school for nearly 10 years, Asha (pseudo name) showed little signs of being able to communicate and perform activities of daily living independently. But within just 2 years of being trained by a team of special educators and therapists from Sense India, the 19-year-old is now able to express herself, communicate with her family, and drink and eat on her own. She has even picked up a vocational skill—weaving!

When Asha was a year old, she started getting epileptic fits. That is when doctors found that her sight and hearing were significantly impaired. As she grew older, Asha was unable to develop speech and, hence, communication became a major challenge for her and her family. Her father is a tailor and is the sole bread earner of the family. Her parents enrolled her in a special school near their home.

"Instead of improving, Asha became more withdrawn. She also became hyperactive and we were unable to understand what she was going through. Her fingers couldn't work properly. We took her out of the school," says her mother.

In May 2014, they met Sense India-supported community workers. We started individualized training and therapy for Asha at home because she did not like going out of her home or meeting new people. Slowly, Asha started responding. She used sounds and gestures to express and communicate. She loves bird sounds and has also started vocalizing a few words.

"Asha likes to go out of the house now—she visits her grandmother sometimes and she also attends family functions. She is still shy of new people, but her self-confidence has grown. People are also accepting her and taking efforts to talk to her," says her mother. Asha does most daily tasks on her own, except for bathing and getting dressed, which her mother helps her with. Asha also helps out with simple chores at home.

In 2015, Asha enrolled to learn floor mat weaving as a vocation. She now gifts floor mats she has woven to family and friends!

Case Study 2: Pakhi—Poised to Fly (as Narrated by Her Educator)

Pakhi (pseudo name), who was born in a remote village of the western region of India, had profound deafness with low vision and learning disability. Due to the traditional misbeliefs, Pakhi was considered God's curse in the family and the village.

When Sense India started its work for deafblind children in the district, things began to improve. We came to know about her through our door-to-door survey, part of our community-based rehabilitation programme. When I saw Pakhi for the first time, she was sitting in a corner dressed as a boy. As I looked at her, I wanted to reach out and break the barriers of isolation and communication. Thus began my struggle.

My first goal was to help her communicate and move around independently inside her house and surrounding areas. "She will fall down, she will hurt herself, she doesn't understand" were a few apprehensions her parents had. But together we overcame these challenges and Pakhi started doing her work all by herself.

Pakhi has a learning disability, so it was not easy training her. I had to come up with various teaching and learning methods to teach her. But she was very receptive; her confidence grew as she started understanding and exploring the world around her. Pakhi began responding and displayed significant progress. She began counting with beads and doing simple calculations; she also learned basic household chores and started helping her mother. She also learned sign language quickly. I urged the local school to allow Pakhi to take admission. Soon, she was able to spell her name in English.

Pakhi was in her transition age—from childhood to adulthood. As she grew, there were other needs that needed to be addressed and that added to my difficulties as a male educator. Not wanting to give up, I asked my wife and two of my female colleagues to help me by accompanying me to Pakhi's house. They joined me every time I went and helped Pakhi understand her body cycle and needs. Slowly with the help of my colleagues, her mother became Pakhi's support through this period of confusion and realization.

Although it took time, slowly and firmly the community was swept by the winds of change. Village farmers began accepting and employing Pakhi's parents in the fields and appreciated Pakhi's progress and achievements.

Today, Pakhi has her own petty shop with the support of Sense India's income generation activity plan. She runs a shop with support from her mother and earns around Rs. 8000–9000 per month. She is contributing to the family income and is supporting her family. Today her family and her educator feel proud to be the voice of Pakhi.

Conclusion

Sense International India completed two decades this year in providing comprehensive services to children and adults with deafblindness in the country. Through its network of 59 partners across 22 states, Sense India is supporting over 77,500 persons with deafblindness in the country. The deafblind work of Sense India is not limited to India; it played a pivotal role in creating awareness about deafblindness and mobilizing services for persons with deafblindness in South Asia—Bangladesh, Nepal, Malaysia, and Sri Lanka. Sense India works to foster awareness about deafblindness in society, builds capacity of professionals, CSOs, and government on deafblind services, and provides comprehensive rehabilitation care and services to individuals with deafblindness and their families from the remote parts of the country.

The inclusion of deafblindness in the new disability Act "Rights of Persons with Disabilities Act, 2016," and the inclusion of rubella vaccination in the national immunization programme because of advocacy efforts of Sense India build hope for a better future for the deafblind citizens of India. This case study on the development of nation-wide deafblind work by Sense India is a perfect example of how CSOs play a significant role in transforming the welfare landscape and are instrumental in bringing meaningful change in the lives of those who are marginalized and vulnerable in society.

Acknowledgements The authors are grateful to the partner organizations of Sense International India, their project staff, and family members of the children and adults with deafblindness receiving services from the deafblind projects across the country. The authors would also like to acknowledge the support received from Sense India team—Nafeeza Pavri, Deepak Krishna Sharma, and Sachin Rizal. We are also highly thankful to our Communication team—Sherine Stanley and Naishil Jha—for the two case studies on Asha and Pakhi.

References

Ask Larsen, F., & Damen, S. (2014). Definitions of Deafblindness and Congenital Deafblindness. *Research in Developmental Disabilities, 35*(10), 2568–2576. https://doi.org/10.1016/j.ridd.2014.05.029.

Barnes, C., & Mercer, G. (2003). *Disability: Key Concepts*. Cambridge: Polity.

Census of India. (2011). *Data on Disability*. Office of the Registrar General & Census Commissioner. New Delhi. Retrieved from http://www.disabilityaffairs.gov.in/upload/uploadfiles/files/disabilityinindia2011data.pdf.

Dammeyer, J. (2014). Deafblindness: A Review of the Literature. *Scandinavian Journal of Public Health, 42*(7), 554–562. https://doi.org/10.1177/1403494814544399.

Dammeyer, J. (2015). Deafblindness and Dual Sensory Loss Research: Current Status and Future Directions. *World Journal of Otorhinolaryngology, 5*(2), 37–40. https://doi.org/10.5319/wjo.v5.i2.37.

Goswami, D., & Tandon, R. (2013). Civil Society in Changing India: Emerging Roles, Relationships, and Strategies. *Development in Practice, 23*(5–6), 653–664. https://doi.org/10.1080/09614524.2013.800839.

Government of India. (2016). *The Rights of Persons with Disabilities Act,* 2016. Government of India (GoI). Gazette of India, 28 December. Retrieved from http://www.disabilityaffairs.gov.in/upload/uploadfiles/files/RPWD%20ACT%202016.pdf.

Government of India, UK Aid, & World Bank Group. (n.d.). *Module 4: Including Children with Deafblindness*. World Bank Group: Toolkit for Master Trainers in Preparing Teachers for Inclusive Education for Children with Special Needs. Retrieved from http://ssashagun.nic.in/docs/module4.pdf.

Groce, N. E. (1999). Disability in Cross-Cultural Perspective: Rethinking Disability. *The Lancet, 354*(9180), 756–757. https://doi.org/10.1016/S0140-6736(99)06140-1.

Jaiswal, A. & Rizal, S. (2017). Partnerships in Early Intervention Services for Children with Deafblindness in India. *The Educator. 31*(2), 19–25. Retrieved from http://icevi.org/publications/educator/pdf/may_2017/The-Educator-2017-January-Partnerships-in-Early-Intervention-Vol-XXXI-Issue-2.pdf.

Jaiswal, A., Kumar, U., & Paul, A. (2018, January). Why Deafblindness Research is Necessary in India. *DbI Review*, 47–51.

Jaiswal, A., Aldersey, H. M., Wittich, W., Mirza, M., & Finlayson, M. (2018). Participation Experiences of People with Deafblindness or Dual Sensory Loss: A Scoping Review of Global Deafblind Literature. *PLOS ONE, 13*(9), e0203772. http://doi.org/10.1371/journal.pone.0203772.

Kaulem, J. (2007). *The Role of Civil Society in Social Policy*. UN Commission on Social Development. Retrieved from http://www.icsw.org/images/docs/Regions/esafrica/activities/Kaulem%20Role%20of%20CS%20in%20Soc%20Pol%20NY%209.2.07.doc.

Mehrotra, N. (2011). Disability Rights Movements in India: Politics and Practice. *Economic and Political Weekly, 46*(6), 65–72. Retrieved from http://www.jstor.org/stable/27918121.

Paul, A., Mathew, B., Kumar, U., Rizal, S., & Jaiswal, A. (2016). Early Intervention (EI) for children with deafblindness in India: Barriers and enablers. *Disability and International Development, 1*, 4–11. Retrieved from http://www.zbdw.de/projekt01/media/pdf/2016_1_BIE.pdf.

Ramírez, M. J. (2016). American Network for Intervention in Situations of Social Suffering (RAISSS): A Case Study of a Civil Society Organization Network. *VOLUNTAS: International Journal of Voluntary and Nonprofit Organizations, 27*(1), 209–221. https://doi.org/10.1007/s11266-015-9563-5.

Sarva Shiksha Abhiyan (SSA). (n.d.-a). *ABOUT SSA*. Ministry of Human Resource Development, Government of India. Retrieved from http://www.ssa.nic.in/about_ssa.html.

Sarva Shiksha Abhiyan (SSA). (n.d.-b). *Training Module on Deafblindness*. Ministry of Human Resource Development, Government of India. Retrieved from http://enabled.in/wp/wp-content/uploads/2014/06/Module-2-Deafblindness.pdf.

Sense International. (2015). *Causes of Deafblindness*. Retrieved June 7, 2017, from https://www.senseinternational.org.uk/about-deafblindness/causes-deafblindness.

Sense International. (2017). Supporting Deafblind People to Live, Learn and Thrive Booklet. Retrieved from https://www.senseinternational.org.uk/sites/default/files/SI%20booklet%20low%20res.pdf.

Sense International India. (2014). *Handbook on Deafblindness*. Ahmedabad, India. Retrieved January 7, 2017, from http://www.senseintindia.org/resources/publications/index/10.

Sense International India. (2016). Annual Report. Ahmedabad. Retrieved from http://www.senseintindia.org/resources/annual-report.

Sense International India. (2017a). *Advocacy*. Retrieved from http://www.senseintindia.org/our-work/advocacy.

Sense International India. (2017b). *History*. Retrieved from http://www.senseintindia.org/about/history.

Sense International India. (2017c). *Vision*. Retrieved from http://www.senseintindia.org/about/our-vision.

Sense International India. (n.d.). *Strategy Document.* Ahmedabad. Unpublished Report.

Thomas, T., Rajaraman, D., Shankar, K. K., & Vaz, M. (2013). Social, Economic, and Demographic Factors Affecting Risk of Severe Disability and Employability in India. *International Journal on Disability and Human Development, 12*(1), 45–51. https://doi.org/10.1515/ijdhd-2012-0129.

Trani, J.-F., & Loeb, M. (2012). Poverty and Disability: A Vicious Circle? Evidence from Afghanistan and Zambia. *Journal of International Development, 24*, S19–S52. https://doi.org/10.1002/jid.1709.

Wittich, W., Watanabe, D. H., & Gagne, J. P. (2012). Sensory and Demographic Characteristics of Deafblindness Rehabilitation Clients in Montreal, Canada. *Ophthalmic and Physiological Optics, 32*(3), 242–251. https://doi.org/10.1111/j.1475-1313.2012.00897.x.

World Bank. (2009). People with Disabilities in India: From Commitments to Outcomes. Washington. Retrieved from http://documents.worldbank.org/curated/en/577801468259486686/pdf/502090WP0Peopl1Box0342042B01PUBLIC1.pdf.

World Bank & World Health Organization. (2011). *World Report on Disability.* Geneva. Retrieved from http://www.who.int/disabilities/world_report/2011/en/.

3

Samurdhi Programme as a Poverty Alleviation Strategy: An Evaluation Study Based on Sri Bodhi Gramma Village in Gampaha District in Sri Lanka

R. Lalitha S. Fernando, H. O. C. Gunasekara, and H. K. S. Gunasekara

Introduction and Problem Identification of the Study

Sri Lanka has experimented with various policies for poverty alleviation for several decades, but most of them did not succeed as intended (Fernando 2002b). Initially, the government provided social welfare facilities, subsidies, land to landless supports for agricultural activities, employment promotion programmes, employment creation schemes, vocational and technical education programmes, self-employment programmes, and so on (Damayanthi 2014). In 1989, Sri Lanka stepped to

R. L. S. Fernando (✉) • H. O. C. Gunasekara
Department of Public Administration, University of Sri Jayewardenepura, Nugegoda, Sri Lanka
e-mail: rlsf@sjp.ac.lk

H. K. S. Gunasekara
National Institute of Social Development, Colombo, Sri Lanka

© The Author(s) 2020
Md. N. Momen et al. (eds.), *Building Sustainable Communities*,
https://doi.org/10.1007/978-981-15-2393-9_3

the milestone of poverty alleviation history introducing "Janasaviya programme" as a social safety net. Moreover, in 1995 the government introduced the national level poverty alleviation policy, "Samurdhi Programme" under the Act No. 30 of Samurdhi Authority in 1995 placing a remarkable sign on the path to alleviate the poverty of the country (Department of Samurdhi Development 2017). Even though the governments were changed from time to time, the concept of the Samurdhi programme had continued with more improvements under different names until today.

The Samurdhi programme has 1,400,868 beneficiary families by the end of March 2017 (Department of Samurdhi Development 2017). The government allocates around 0.2 per cent of gross domestic products for safety net transfers by annually (Kesavarajah, n.d. cited by Damayanthi 2014) and it is around 0.1 per cent of total government expenditure (Jayaweera 2010). The government has allocated approximately 4–5 per cent of the national budget for safety net assistance of the Samurdhi programme (www.med.gov.lk cited by Damayanthi 2014).

By the year 1996, 28.8 per cent out of the population lived below the poverty line in Sri Lanka and it was 15.2 per cent in 2007 (indexmundi 2017). Further, the poverty level was declined up to 6.7 per cent by 2015 (Asian Development Bank 2017). Thus, the poverty level in Sri Lanka has been reduced gradually since the inception of the programme (Kelegama 2001).

There are three main components of the Samurdhi programme: (1) income transfer, it provides consumption support and funds as social insurance support to protect the poor during extreme situations such as health and loss of life; (2) community development, by means of investing in economic, social infrastructure and other agriculture, nutrition and small enterprises; (3) serve the poor by way of encouraging savings and providing financial markets for them (Department of External Resources 2017).

Even though, the poverty alleviation scenario is in favourable in that sense, there are different views on it. Previous research contributing to the factor mal-targeting problem is very limited (Bandara 2016; Yusuf 2010; Phillips et al. 2014). Fernando (2009) revealed that the Samurdhi programme has not achieved its intended objectives at a satisfactory level. Thus, the findings of this study will be useful for identifying the loopholes of the programme and improving the strategies for the development of the poverty alleviation policy of Sri Lanka. Several types of research have examined the success of the Samurdhi programme and reported varied results. In this scenario, this study evaluates the effectiveness of the Samurdhi programme, based on a comparison between the beneficiaries' and non-beneficiaries' perspective by focusing on one village in Sri Lanka. Issues and problems of the Samurdhi programme will also be identified and strategies will be proposed to improve the effectiveness of the programme. In this context, the main research question of this study is twofold:

1. Is the Samurdhi programme effective?
2. What are the issues and problems of the Samurdhi programme?

Thus, the study contains the following objectives.

- To identify issues and problems of the Samurdhi programme.
- To evaluate the effectiveness of the Samurdhi programme, based on a comparison between the beneficiaries' and non-beneficiaries' perspective.
- To propose strategies to improve the effectiveness of the Samurdhi programme.

Methodology

The study area is Sri Bodhi Gramma village in Wedamulla Grama Niladhari division, in Kelaniya Divisional Secretariat, Gampaha District, in Western Province in Sri Lanka with a total of 87 families comprising 73 Buddhist families, 7 Christian families, and 7 Tamil families. There are about 34 Samurdhi beneficiaries in the village and among them, there are 20 individuals receiving the Samurdhi stamps (Preliminary survey 2017). The population of the study is the all families live in Sri Bodhi Gramma village which is 87 (Preliminary survey 2017). The sample is 20 families including 10 Samurdhi beneficiary families and 10 non-beneficiary families purposively selected. Data are collected through semi-structured in-depth interviews and analysed descriptively to examine the effectiveness of the programme from both the beneficiaries' and non-beneficiaries' perspective.

Background of the Study: The Village of Sri Bodhi Gramma

The village is Sri Bodhi Gramma located in Wedamulla Grama Niladhari division in Gampaha District in Mahara Divisional Secretariat in Kelaniya Municipal Council area in Sri Lanka (Mahara Division Secretariat 2017). The community in this village which had been pre-located near the Kelaniya Railway Lines has now been relocated at the Sri Bodhi Gramma Lane (preliminary survey with Mahara Division Secretariat office and Wedamulla Grama Niladhari office, 2017).

Fifteen years ago, railway lines to Sapugaskanda oil refinery had been constructed through Wedamulla and Nungamugoda areas. Due to that, the government had cleared the area by removing the houses near the railway line. As a solution for that, on 27 March 2003, the government has resettled 87 families in the location of the Wedamulla Grama Niladhari area of Gampaha District, Kelaniya Divisional Secretariat, giving two perch for each family, but there are families living in one perch as two families live in one house (Preliminary survey with Mahara Division Secretariat and Wedamulla Grama Niladhari office 2017).

In 2004, this village has been named as "BODHI GRAMMA " by planting a Boo Tree at the entrance of the village by a Monk of the Temple and building a small place to worship (Preliminary survey with Mahara Division Secretariat and Wedamulla Grama Niladhari office 2017).

Community Structure of the Village

According to the preliminary survey with Mahara Division Secretariat office and Wedamulla Grama Niladhari office (2017), the community is in line settlement pattern. There are 76 houses and 87 families living in the community and the population is around 500. Most of the houses are permanently built and only a few houses still remain as temporary houses. There are 34 families of Samurdhi beneficiaries. This community consists of diverse ethnic groups, namely, Sinhala, Hindu, and Christians. Among them, 73 families are Sinhalese, and 7 families are Christians. Among them, most of the Sinhala people are Buddhist and few families are Christians and Tamil Christians. Also, Tamil Hindus mixed with the community. They use Sinhala as the main language and only a few families use the Tamil language. They maintain a mixed culture while celebrating every religious festival without conflicts and live in a peaceful manner. This community is in the liner house pattern. There are six lanes and canal reservation go around the community border. There are well-built houses and also temporary wood and bricks houses. Almost all houses are very small because they have built their houses within one or two perch in three parts, a room, sitting room, and kitchen.

Social Characteristics of the Village

Bodhi Gramma village has a mixed culture and they do not have any cultural or religious discrimination. Most of the community members are less educated and most of them have a negative attitude and dependency mentality. They always try to depend on external parties rather than using their own strength to uplift their lives (Preliminary survey with Mahara Division Secretariat and Wedamulla Grama Niladhari office 2017).

They are not too much adherent to religious institutions. Even religious leaders around the village tried to do religious programmes for their spiritual well-being, they have not been successful. Most of the children go to school up to the General Certificate Examination (Ordinary Level). After that, they try to go for a job because parents have no financial strength to give education to their children. Girls get married after the General Certificate Examination (Ordinary Level) and parents also have less education, they allow children to do as they wish. Some of them are living together without getting married but having children. Their social relationships with community members are very poor. They have very less participation in common events of the community (Preliminary survey with Mahara Division Secretariat office & Wedamulla Grama Niladhari office, 2017).

Economic Characteristics of the Village

Most of the community members occupy as daily wage workers while few of them have monthly income employment. Most of them are labourers or drivers. They have daily income and most of them do not have a practice of savings. Most of the families are a low-income category. Some of the women do self-employment to support their family by earning extra income such as making short eats, sewing clothes, and making small toys. In Bodhi Gramma village, there are 34 families getting the Samurdhi benefits including 14 family stamps and 20 individual stamps as government assistance (Preliminary survey with Mahara Division Secretariat and Wedamulla Grama Niladhari office 2017).

The Political Background of the Village

Bodhi Gramma community members have a lower level of political participation. Some politicians help village activities at the request of the community. Some other people have received personal benefits to uplift their personal lives. (Preliminary survey with Mahara Division Secretariat and Wedamulla Grama Niladhari office 2017).

Education of the Village

In Bodhi Gramma village, no single person has gone to universities or received tertiary education. The community members have a lower level of literacy rate. There are villagers above 50 years old who cannot read or write anything. Now each and every child goes to school but the school attendance is lower. They have a lack of interest to continue higher education. Most of them have left from school after doing General Certificate Examination (Ordinary Level) and some have left before the Ordinary Level and rarely continue to studies until the Advanced Level. Soon after they left their schools for earning money. So they find jobs with a minimum education level with a lower level of wages. This poor education and economic barriers continue as a poverty cycle in the community (Preliminary survey with Mahara Division Secretariat and Wedamulla Grama Niladhari office 2017).

Recourses of the Community

Sri Bodhi Gramma community has considerable physical and human resources. The village has formal service delivery systems including government officers such as Grama Niladhari, Development officer, Midwife, Public Health Inspector (PHI) and the Samurdhi Officer. They have a good connection with the community. These are the internal resources of the community (Preliminary survey with Mahara Division Secretariat and Wedamulla Grama Niladhari office 2017).

As external resources, since it is an urban settlement, mostly these community members have access to services needed for daily life such as private and public banks, secondary schools, hospitals, and other services. Further, a public library, a huge lake which helps to earn money from a large number of lotus flowers and fishery and a company named "AKBAR TEA" which provides many job opportunities for the community (Preliminary survey with Mahara Division Secretariat and Wedamulla Grama Niladhari office 2017).

Literature Review on Poverty

Meaning and Definition of Poverty

Adam Smith (1776 cited by Ranathunga and Gibson 2015) defined poverty as "the inability to purchase necessities required by nature or custom". "Poverty is not just a problem of having access to the basic necessities of life, but also a social handicap" (Ranathunga and Gibson 2015).

According to Amartya Sen (n.d.), "persistence of poverty is the robbing of human potential" and "poverty is not just a lack of money and it is not having the capability to realize one's full potential as a human being". Furthermore, "poverty is the reduction of human contributions to the market, to the commons and, ultimately, to the human condition" (cited by Mincy 2014). This definition has a broader meaning than the income definition of poverty.

Chambers (1983) explained a "deprivation trap", which is developed with five "clusters of disadvantage" which interact with each other to trap people. According to the author, "poverty is one of the five disadvantage clusters and others are powerlessness, physical weakness, isolation, and vulnerability, which are the wider dimensions of poverty". According to the author, these five clusters are interacting with each other.

As a multidimensional phenomenon, poverty is reliably characterized by "a lack of opportunities (i.e., employment, income, health, and education), general deprivation of basic resources (i.e., food, water, and shelter), and powerlessness" (Sen 1999). "There are monetary and nonmonetary approaches to measure poverty levels" (Alailima 2007). "Monetary poverty is twofold and in that there is absolute and relative poverty. In Sri Lanka, absolute poverty is measured using the official poverty line, that is, LKR 3,896 in September 2014" (Department of Census and Statistics, 2014 cited by Bandara 2016).

Approaches to Alleviating of Poverty

According to the United Nations Development Programme (UNDP) (1997 cited by Fernando 2002a), one of the approaches to alleviate poverty is integrated anti-poverty strategies which require convergent action at three fundamental levels as follows:

- Macro-Economic Policies—there should be policies and programmes for increasing both manufactured and agricultural production to cater to local demands of the people.
- Institutional Change—there should be organizations to offer productive assets and resources such as lands, credits, skills to the poor.
- Micro-Level Interventions—the institutional policies should be focused to empower the poor by promoting their livelihoods.

As per the findings of Samad (1989 cited by Fernando 2002a), any poverty reduction policy should include several components, namely direct cash benefits, other human resource development, and productive assets distribution policy.

Another model for poverty alleviation is proposed by Samad and Currey et al. (1989 cited by Fernando 2002a) including four inter-related stages identified as follows:

- Doles—every poverty alleviation programme must have to have a dole component to ensure the physical survival of the beneficiaries.
- Entitlement—the dole can come through an increase in entitlements in which a person has a legal form
- Asset Formation—distribution or redistribution of assets through measures like radical land reforms, tenancy reforms, and fiscal and monetary reforms
- Empowerment/Capacity Creation—empowerment of needy people to establish their independent survival without external support in the form of programme. This comes through human resources development and ends the dependency.

Fernando (2002a) proposed an integrated approach to poverty reduction with reference to the findings of the several studies (UNDP 1997, Samad 1989, Samad and Currey et al. 1989 cited by Fernando 2002a). The approach included the following components:

- Direct cash benefits
- Improving micro-macro linkages
- Implementing the participatory approach
- Getting private sector support
- Social mobilization
- Expanding opportunities
- Capacity building
- Poverty-focused governments
- Decentralization and local governments
- The role of the civil society organization
- Fostering democratic governance
- Protection environment
- Gender empowerment

Samurdhi Programme in Addressing the Issue of Poverty

The Government of Sri Lanka introduced the Janasaviya programme ("strengthening the people") as a poverty alleviation strategy in 1989 (Bandara 2016); due to its degree of politicization, patronage, and mal-targeting (Gamage 2006), in 1995, the government introduced the Samurdhi programme as a major government-sponsored poverty alleviation programme, replacing the "Janasaviya Program" (Bandara 2016). According to the Central Bank of Sri Lanka (2013), "The Samurdhi Program had approximately 1.5 million beneficiary families by the year 2013". The programme is structured with "small grouping" concept; in 2013, there were 400,470 small groups and 2.9 million of members and 27,600 employees (Samurdhi Authority of Sri Lanka, 2011 cited by Bandara 2016). The structure of the Samurdhi programme was redesigned in 2014 and renamed as a Department of "Divineguma" ("uplift of the peoples' lives") in the Ministry of Economic Development (The Government of Sri Lanka, 2012 cited by Bandara 2016). However, today in 2017 the programme again renamed as Samurdhi with some improvements under the authority of Samurdhi.

The Samurdhi Authority has been established by the Act No. 30 of 1995 and functioned with effect from 1 February 1996 under the vision of "building a prosperous Sri Lanka with the least possible poverty" and mission of "contributing towards a stable national development with the least possible poverty through development based on public participation" (Samurdhi Authority 2017). This programme has made a remarkable contribution to uplift the living standard of low-income level families. The Samurdhi Authority added that the Samurdhi programme has been effectively contributed to reducing the local poverty rate from 28.8 per cent to 15 per cent in 1996 as a principal mediator in the matter. Also, "the number of officers of the Samurdhi family who have promised to contribute to this massive national service and who are scattered throughout the island amounts 25,000" (Samurdhi Authority 2017).

Table 3.1 The major components of the Samurdhi programme

Major component	Projects
Welfare	"Food stamp, subsidy for fuel (for lighting kerosene lamps), nutrition package for pregnant and lactating mothers, milk feeding subsidy for children between two and five years".
Microfinance	"Savings, credit, and social insurance schemes".
Infrastructure development	"Infrastructure development in needy areas".
Livelihood development	"Animal husbandry and fisheries, agriculture, industrial, marketing, and services".
Social and spiritual development	"Antinarcotic and antismoking projects, programs for preventing child abuse, women empowerment, cultural and family development".

Source: Samurdhi Authority of Sri Lanka (2011 cited by Bandara 2016)

There are six criteria established by the Samurdhi Authority in 2017 to determine the needy people for offering Samurdhi benefits. Those criteria are education level of the family members, the health status of the family members, economic status of the family such as their income is not exceeding 6000 and expenses are not exceeding 5500, type of the house and its ownership and other property ownership of the people (Samurdhi Authority 2017). Further, the major components of the Samurdhi programme are shown in Table 3.1.

The Samurdhi Authority (2017) provides services under different divisions which are as follows:

Agriculture Division

- Promotion of special projects for agro productions
- Promotion of local food crop cultivation
- Coordination of small-scale plantation crop cultivators through line institutions
- "Asweddumization" (cultivation) of paddy field left fallow
- Home gardening development project
- Post-harvest technology and processing projects

Animal Husbandry and Fisheries Programme

- Milk cow programme
- Erection of cattle sheds
- Establishment of biogas units and processing of biogas units
- Goat keeping projects
- Pig keeping projects
- Poultry keeping projects
- Minor fishing year projects
- Making tanks for ornamental fish
- By-products of milk
- Dry fish, Jaadi, and Maldives fish projects
- Milk sales and milk collection projects

Industrial Development Division

- Development of small-scale industrial projects
- Development of mining villages
- Development of model industrial villages

Sales and Service Division

- Promotion of Samurdhi domestic sales outlets
- Promotion of rice sales
- Promotion of mobile trading
- Promotion of saloons
- Promotion of restaurant/catering services
- Promotion of servicing vehicles
- Promotion of beauty therapy
- Promotion of domestic services
- Promotion of child care centres
- Promotion of communication centres

Banking and Financial Division

- Providing loan facilities
- Promotion of savings

Maha Sangam Division

- Organization of training courses for livelihood development
- Compilation of project reports
- Coordination of support services
- Maintenance of data system

Social Development Division

- "Diriya Piyasa" housing programme
- Model villages programme
- Programmes on children and childcare

The structure of the Samurdhi programme implementing from its bottom to the top level is explained in Fig. 3.1.

Empirical Research

According to Bandara (2016), "even with the many efforts toward poverty reduction, the question persists as to why modern governments are unable to satisfactorily reduce poverty levels", indeed this question has been extensively considered in Sri Lanka. However, the government has a strong vision for ongoing development and prosperity of Sri Lanka, therefore, poverty reduction is a very important element of the public policy in the country. The government of Sri Lanka has been implementing the Samurdhi programme as a major poverty alleviation programme since 1995. Bandara (2016) identified that mal-targeting of the Samurdhi programme has been occurred due to politicization, dependency

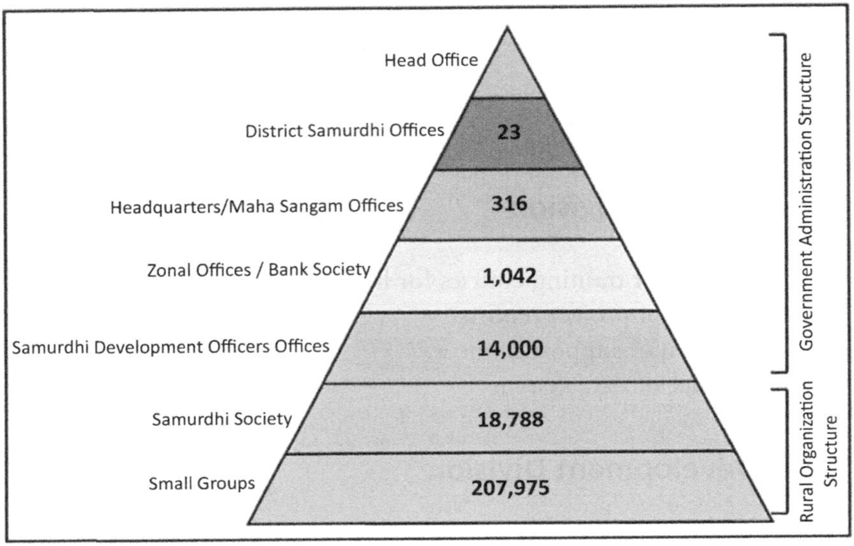

Fig. 3.1 Structure of the Samurdhi programme. (Source: Samurdhi Authority 2017)

mentality of the beneficiaries, and outdated income level cut-offs. It has occurred interruptions to social harmony and effectiveness of the programme. Moreover, the author mentioned empirical evidence. The issue of mal-targeting undermines the value of poverty alleviation programme (AusAid, 2011 cited by Bandara 2016; Bibi and Duclos 2007; Caldes et al. 2006; Bandara 2016).

The Samurdhi programme was designed to uplift people's living standard by providing services under five categories: welfare (provision of food stamps/safety net, milk-feeding subsidy for infants, and nutrition packages for pregnant mothers and lactating mothers), microfinance, infrastructure development, social services, and livelihood development programmes (Bandara 2016). "In terms of food stamps, the government provides monthly cash transfers to beneficiaries on the basis of household income" (Ministry of Finance and Planning, 2012 cited by Bandara 2016) and the programme allocates 80 per cent of the total budget for the welfare component (Gunasinghe 2010).

"In Sri Lanka, the poverty headcount ratio has been sharply declining over the past two" (Department of Census and Statistics, 2014 cited by Bandara 2016). Although the poverty ratio has steadily decreased, the government has continued to support a large portion of the population under the Samurdhi programme (Central Bank of Sri Lanka 2013). It has been observed that even though the Samurdhi programme has many plus points (Bandara 2016), several studies highlighted its inherent weaknesses, namely politicization and corruption (Gamage 2006; Glinskya 2000), mal-targeting (Damayanthi 2014; Salih, 2000 cited by Bandara 2016), and mal-targeting (Gamage 2006; Glinskya 2000). According to Bandara (2016), a large number of needy people are being neglected and left out of the programme, while well-off people have been admitted into the programme as a result of mal-targeting which was formed due to the lack of updated periodical surveys, politicization of the society, following a blanket approach across the entire country, bureaucratic malpractice and a dependency mentality of the people. This mat-targeting practice undermines social harmony and relationships, reduces programme effectiveness, and makes financial losses for Samurdhi banks.

According to the study based on Panadura Divisional Secretariat, Fernando (2009) noted that there is a significant difference between the Samurdhi beneficiary families and the non-beneficiary families in terms of several aspects. When compared to non-beneficiaries, the beneficiaries have increased their loan repayment, increased their income and expenditure, received other benefits, skill development, and improved living conditions. However, the study further revealed that there is no significant difference between both parties on savings levels and the Samurdhi families were not able to improve their investment and market sufficiency and self-development were not succeeded in a satisfactory level. Fernando (2002b) further identified several drawbacks and difficulties in the implementation (Bigman and Srinivasan 2002; Dominique 1998) and planning process of the Samurdhi programme. According to the author, the Samurdhi programme provided a very small amount of cash benefits to a large percentage of families by identifying them as the poor according to the central government. Fernando (2002b) further noted that this cash amount is not adequate for them to escape from lack of consumption and malnutrition but respective authorities did not have a proper mechanism to monitor and evaluate the programme. There are many political appointments within the policy and political interference for identifying target groups, even the policy is based on a bottom-up approach to planning, the implementation process becomes more complex and the success of implementation depends on millions of actors including the beneficiaries themselves (Fernando 2002b).

Analysis and Discussion

Issues and Problems of the Samurdhi Programme

Several issues were identified based on the in-depth discussions conducted with the beneficiaries and non-beneficiaries. Those are presented as follows:

Mal-Targeting

No fair and transparent criterion has been practised for both the selection and exclusion of the beneficiaries from the programme. Majority of the beneficiaries and non-beneficiaries felt that the neediest people such as disabled, old, and non-fixed and very low-income earners do not get the benefits of the Samurdhi, while some people who are well off have received benefits from the programme. Despite the existence of a suitable and stable standard which was introduced by the relevant authority to select beneficiaries, the implementation of those selection criteria had not been practised. Unless those reasons are not properly examined and solved, the benefits of the programme would be misused and break the rights of the needy people.

Lack of Commitment of the Authoritative Officers

It is evident that the Samurdhi officers' commitment was not satisfactory and the level of the service delivery is also very poor. The officers did not take adequate responsibility to serve the people. Normally, the officers have to do periodical visits to identify the real needs of the families, but such interventions were not adequately practised.

Unwanted Political Interference

Political power has been misused as it was revealed that politicians have intervened in selecting the target group, implementing the programme, and distributing benefits. Some of the politicians have used the Samurdhi programme for their personal election campaign to increase their votes and privileges.

Absence of Meeting Ultimate Results of the Samurdhi Programme

Due to the absence of feedback, the results of the programme are not visible. Even though there are some families who have benefited from the programme for more than 10–15 years, they were not able to uplift the living standards to the expected level. Hence, they are still poor as those families could not meet their basic needs as they do not have a proper income source, no proper housing, and other sanitary facilities.

Unsuitable Sub-programmes Under the Samurdhi Programme

The Samurdhi beneficiaries do not like to involve in some activities which are implemented under the main programme. The beneficiaries criticize that those programmes are annoying them (i.e. selling tickets [flags], unfair raffle drawings, beneficiary grouping system, and compulsory savings). These kinds of programmes demotivate people and move people away from the Samurdhi programme.

Lack of Awareness About the Samurdhi Programme

Both beneficiaries and non-beneficiaries do not have adequate knowledge about the Samurdhi programme. They do not know about the purpose of the programme, services and, benefits delivered, the selection criteria of the program, and so on. This ignorance leads to a loss of their rights to have some benefits and their loyalty to the programme.

Effectiveness of the Samurdhi Programme

Evaluation of the effectiveness is done based on the three main components grounded to the Samurdhi programme; "1) income transfer component, which provides consumption support to the poor and funds social insurance support to help protect the poor during extreme (health and loss of life) situations; 2) community development, through investments in economic and social infrastructure, agriculture, nutrition, and small enterprises and; 3) encouraging savings and provide financial markets that serve the poor" (Department of External Resources 2017) with a comparison between the beneficiaries' and non-beneficiaries' perspective (Table 3.2).

It seems that the Samurdhi programme could be considered as well-developed poverty alleviation strategy, but at the implementation stage, several issues and problems were indicated. Thus, the intended objectives of the programme have not been achieved.

There are many activities and services that have been introduced and implemented to reduce the level of poverty under the Samurdhi programme; it is uncertain that all the services are delivered to the beneficiaries. When analysing the beneficiaries' perspective, only very limited services were received. These services are transferring monthly income, compulsory savings, and housing loans at a low-interest rate. There may be some other benefits also unknowingly consuming by them, but they are aware of only three services. The effectiveness of the Samurdhi programme based on its components discussed in detail as follows:

Table 3.2 Analysis of the effectiveness of the Samurdhi programme

Component	Beneficiaries	Non-beneficiaries
(1) Income transfer component and social insurance support to help and protect the poor during extreme situations.	Not satisfied with the income transfer, receiving per month as consumption support and no idea about social insurance support.	Criticize the selection criteria of the Samurdhi and distribution of its benefits. No idea about its services.
(2) Community development, through investments in economic, social infrastructure, agriculture, nutrition, and small enterprises.	No adequate infrastructure development and community development.	Infrastructure development is not at a satisfactory level.
(3) Encouraging savings and providing financial markets that serve the poor.	Not satisfy with compulsory savings but satisfy with housing loans at a low-interest rate.	No idea about savings and financial market opportunities providing from the programme.
(4). Improvement of living standards.	Living standards were able to improve to some extent, but still, they seek for more benefits from the programme.	The neediest families were disappointed as they haven't get the benefits. However, the levels of living standards of both poor non-beneficiaries and some beneficiaries are not much different.

Source: Field Data

Income Transfer Component

The Samurdhi programme transfers a monthly dole as consumption support for the beneficiaries. The amount is different based on their income level. Amount of income transfer is not sufficient to the beneficiaries' even to fulfil their basic needs due to the cost of living in the country. Majority of the beneficiaries receive Rs. 3500 dole per month. But they receive only Rs. 3000 as cash at hand due to the compulsory saving of Rs.

500. Beneficiaries were also not satisfied with the amount they received as it was not adequate.

There are many needy families who were not qualified for the Samurdhi programme. Non-beneficiaries urged that even they are qualified to receive the benefits, but they do not get the Samurdhi. However, some families were selected for the Samurdhi who have better living standards when compared to some non-beneficiary families. Many criticisms were levelled at the programme with reference to the criteria of selection of the target group. All the non-beneficiary families continuously appealed for the benefits of the Samurdhi programme. Some of the non-beneficiaries views are that the beneficiaries were selected based on the Samurdhi officers' political party biases.

Community Development, Through Investments in Economic and Social Infrastructure, Agriculture, Nutrition, and Small Enterprises

Both beneficiaries and non-beneficiaries were not satisfied with the infrastructural development of the village as they sought considerable improvement of the infrastructure in their area. It was observed that physical infrastructure requirements such as suitable houses, toilet facilities, water facilities for sanitary are lacking in the village. But these issues could be addressed through the Samurdhi programme. Several families have interest in starting their own business, but no adequate assistance is from the government or the programme. They complained that some benefits of the programme have been allocated to the families based on the officer's political party attachment.

Encouraging Savings and Providing Financial Markets That Serve the Poor

Beneficiaries were not happy with the compulsory savings programme as they were not aware of what happened to their savings. Therefore, beneficiaries have demotivated on their savings but they were happy with the housing loans. But some of the non-beneficiaries were disappointed with the loan system and sought financial assistance and loans from the programme.

Impact of the Samurdhi Programme on Increasing the Living Standard of People Is Negligible

It was observed that the neediest families have not received the Samurdhi benefits but the families who have better living standards received the benefits from the programme. When compared to the living standards of some beneficiaries with non-beneficiaries, appropriate interventions are required in terms of selecting the target group, distributing services and benefits.

Thus, it has been observed that living standards of the Samurdhi beneficiaries have not been uplifted as it is intended and still they are poor and in dependency mentality and expect the benefits. The living standards of both poor non-beneficiaries and some beneficiaries are not much different.

Discussion

Several studies similarly found the ineffectiveness of the Samurdhi programme. Fernando (2009) revealed that the Samurdhi programme has not achieved its intended objectives at a satisfactory level and there is no significant difference between both the beneficiaries and the non-beneficiaries of Samurdhi. Bandara (2016) found that the success of the Samadhi programme was declined due to its drawbacks. There are several drawbacks and difficulties identified at the implementation and planning process of the Samurdhi programme (Fernando 2002b). Even though the Samurdhi policy has been implemented for nearly 19 years, the issue of poverty is still a major problem in Sri Lanka (Fernando and Anuradha 2015). Mal-targeting was identified as a major issue of the Samurdhi programme and other poverty alleviation policies by different scholars, namely Fernando (2002b), Bandara (2016), AusAid (2011) cited by Bandara 2016; Bibi and Duclos (2007); Caldes et al. (2006), Central Bank of Sri Lanka (2013), Damayanthi (2014); Salih (2000) cited by Bandara (2016), Gamage (2006); Glinskya (2000), and Fernando and Anuradha (2015). Further, Fernando (2002b) and Fernando and Anuradha (2015) identified the lack of authoritative officers' commitment as a drawback of the Samurdhi programme. Moreover, unwanted political interference as a drawback of the programme was identified by Bandara (2016), Gamage (2006), Glinskya (2000), Fernando (2002b), and Fernando and Anuradha (2014). As well the absence of meeting ultimate results of the Samurdhi programme was identified as a drawback by Bandara (2016), Fernando and Anuradha (2015), Fernando (2002a), Fernando (2002b), and Fernando (2009). Further, inappropriate sub-programmes in the Samurdhi programme were identified as a drawback by Fernando (2009), and lack of awareness about Samurdhi programme was identified as a drawback by Fernando and Anuradha (2015).

Strategies to Improve the Effectiveness of the Samurdhi Programme

Even though the Samurdhi is conceptually well developed and could be considered as a better strategy of poverty reduction, when it comes to practice, there are several gaps that limit the success of the programme. Social mobilization through microfinance activities, providing market facilities, capacity building and improving their consumption level is vital in any programme of poverty reduction, since the Samurdhi programme has embodied with all these ingredients but in practice, some of the activities of the programme have not been succeeded as intended (Fernando 2009).

In order to fill that gaps and improve the effectiveness, the following suggestions are made based on the respondents' as well as the authors' perspective.

Establish Proper Targeting Mechanism

Establish a fair and transparent prioritizing mechanism of the target group based on their neediness and the vulnerability. The neediness and vulnerability of people should be surveyed, maintained, and updated records by Samurdhi officers along with Grama Niladhari officers with transparency. That should be implemented continuously and administered independently by superiors to control from external pressures, Income transfer should be given to the most vulnerable parties such as disabled, widows, and poorest children.

Provide Awareness on the Samurdhi Programme

As people do not have enough awareness about the Samurdhi programme and its benefits and services, they missed opportunities for having benefits and achieving the intended results of the programme while misusing the facilities. Therefore, proper awareness should be given to both beneficiaries and non-beneficiaries to understand the actual purpose of the programme. It also may protect people's rights and increase their loyalty and commitment to the programme.

Improve Authoritative Parties Commitment as Efficient Change Agents

Authoritative parties should be properly trained spiritually and physically to serve people under a code of ethics from their heart as it is their responsibility or service more than a job. It may connect the service with exact needy people leading to achieve intended results of the programme.

Restructuring the Programme

The Samurdhi programme should be totally restructured and designed based on experiments and experts' ideas for better targeting to alleviate poverty from its origin. Then it should be divided into sub-targets as objectives to achieve within a limited time frame as it is easy to monitor the progress and to reach the ultimate result step by step fulfilling one objective and another as a hierarchy. That implementation process and its success should be monitored periodically to examine whether the expected outcomes are achieved by the programme and for further modifications and corrections for mistakes.

Give Priority to Assist Self-Employment

If a country really needs to alleviate poverty sustainably, they should be given more priority for investment rather than consumption. For that reason, it should be improved entrepreneurial skills and self-employment of people through every aspect that they need support such as training and education, skill development, instructions and supervision, monetary assistance, market facilities, and required equipment based on their capabilities and capacities. That will be the most valuable investment of the Samurdhi programme as a policy for poverty alleviation.

Provide Specified Services Based on Beneficiaries Needs

In actual sense, poverty has a relative phenomenon. Therefore, the nature of the poverty, level of the poverty, types of poverty, factors affecting poverty, and the impact of poverty are varying one situation to another. For that reason, to alleviate poverty from its origin, solutions for poverty should be designed to address the exact factors of poverty. Consequently, as a poverty alleviation policy at whole designing benefits and services of the Samurdhi should be based on the features of the situation of poverty and background of the beneficiaries (rural, urban, sub-urban poverty etc.), climate of the area, inflation rate, societal/economic background, and most crucial issues that vulnerabilities face. Since the benefits will be highly specified and differed from one situation to another, then it will lead to more success.

Limitation

The study has faced several limitations when collecting primary data through in-depth discussions with the respondents. Some beneficiaries tried to show their vulnerability to have more benefits and try to keep their Samurdhi ownership and benefits. Non-beneficiaries always try to show their need and vulnerability to receive the benefits.

This scenario shows the dependency mind-set of both the beneficiaries and the non-beneficiaries always seeking external assistance. They do not have the confidence to survive without the government support.

Conclusions

This study was based on a specific vulnerable village to explore the success of the Samurdhi programme using a sample including both the Samurdhi beneficiaries and non-beneficiaries. It explored according to the respondents' and the authors' perspective. The programme is conceptually well developed, but at the implementation, the programme has not been a success as there are many weaknesses in the programme. The programme neglected many needy people which is one of the main limitations. However, both beneficiaries and non-beneficiaries do not have enough knowledge and awareness regarding the Samurdhi programme and its benefits. Nevertheless, the Samurdhi programme in this village has not been effective as it is not at the expected level to satisfy both beneficiaries and non-beneficiaries. The identified issues of the programme are the mal-targeting, lack of authoritative officers' commitment, lack of awareness about the programme, unwanted political interference, absence of meeting ultimate results of the programme and introducing improper sub-programmes. The proposed recommendations in addressing these issues are the establishing proper targeting mechanism, providing awareness on Samurdhi programme, improving the commitment of officers, restructuring the programme, giving priority to assist self-employment facilities and providing specified and unique services based on beneficiaries specific issues and needs. These findings will lead to improving the effectiveness of the Samurdhi programme.

References

Alailima, P. (2007). The Conventional Approaches: An Overview of Poverty in Sri Lanka. In A. A. Carder & F. Remnant (Eds.), *Fresh Perspectives: Exploring Alternative Dimensions of Poverty in Sri Lanka*. Colombo: Centre for Poverty Analysis.

Asian Development Bank. (2017). Retrieved June 6, 2017, from the Asian Development Bank Website www.adb.org.

Bibi, S., & Duclos, J. J. (2007). Equity and Policy Effectiveness with Imperfect Targeting. *Journal of Development Economics, 83*, 109–140.

Bigman, D., & Srinivasan, P. V. (2002). Geographical Targeting of Poverty Alleviation Program: Methodology and Applications in Rural India. *Journal of Policy Modelling, 24*, 237–255.

Bandara, N. D. M. (2016). Causes and Consequences of Poverty Targeting Failures: The Case of the Samurdhi Program in Sri Lanka. *Asian Politics & Policy,* 8(2), 281–303. Policy Studies Organization. Published by Wiley Periodicals, Inc.

Caldes, N., Coady, D., & Muluccio, J. A. (2006). The Cost of Poverty Alleviation Transfer Programs: A Comparative Analysis of Three Programs in Latin America. *World Development, 34*(5), 818–837.

Central Bank of Sri Lanka. (2013). *Annual Report-2013*. Colombo: Central Bank of Sri Lanka.

Chambers, R. (1983). *Rural Development: Putting the Last First*. Routledge, Taylor & Francis Group, London and New York. ISBN 978–0–582-64443-4.

Damayanthi, M. K. N. (2014). Good Governance and Poverty Alleviation Programmes in Sri Lanka: Special Reference on Samurdhi Programme. *African-Asian Journal of Rural Development, 47*(1), 43–64. Retrieved June 4, 2017, from https://www.researchgate.net/publication/268152677.

Department of External Resources. (2017). Social protection. Retrieved June 8, 2017, from the Department of External Resources Website www.erd.gov.lk.

Department of Samurdhi Development. (2017). Retrieved June 3, 2017, from the Department of Samurdhi Development Website www.samurdhi.gov.lk.

Dominique, V. D. W. (1998). Targeting Revisited. *The World Bank Observer, 13*(2), 231–248.

Fernando, R. L. S. (2002a, January–December). An Integrated Approach to Poverty Reduction in Developing Countries. In R. S. Gunatunge & H. M. A. Herath (Eds.), *Sri Lanka Journal of Rural Development,* 1(1). Rural Development Training and Research Institute of the Ministry of Samurdhi.
Fernando, R. L. S. (2002b). Factors Affecting Policy Failure: An Analysis of the Implementation of Samurdhi Policy in Sri Lanka. Proceedings of the Third Annual Research Session, Faculty of Management Studies and Commerce, University of Sri Jayewardenepura.
Fernando, R. L. S. (2009). An Evaluation of Samurdhi Programme in Sri Lanka: With Reference to the Villages in Panadura Divisional Secretariat. In A. Singh, K. Kapoor & R. Bhattacharyya (Eds.), *Governance and Poverty Reduction: Beyond the Cage of Best Practices* (pp. 274–296). PHI Learning Private Limited, New Delhi.-110001, ISBN-978-81-203-3698-8.
Fernando, R. L. S., & Anuradha, P. I. (2015). A Policy Framework for Poverty Reduction in Sri Lanka. A Governance Approach. *Journal of Global Studies in Economics.* Central University of Kerala, India. Volume 1, No 1. ISSN 2393–9974.
Gamage, D. (2006). Governance of Poverty or Poverty of Governance. *Economic Review, 31* (10-12), January/March 2006, Colombo, People's Bank, pp. 8–24.
Glinskya, E. (2000). *An Empirical Evaluation of the Samurdhi Program. Background Paper for Sri Lankan Poverty Assessment* (Report No. 22–535-CE). Washington, DC: The World Bank.
Gunasinghe, C. (2010). The Significance of Capital Assets in Moving Out of Poverty: A Case of Sri Lanka. *South Asia Economic Journal, 11*(2), 245–285.
Indexmundi. (2017). Country Facts, Population Below the National Poverty Line. Retrieved June 7, 2017, from the Indexmundi website www.indexmundi.com.
Jayaweera, R. (2010). Better Targeting of Transfers: Samurdhi Programme, Colombo: IPS. Retrieved November 11, 2016 from the IPS Website: //www.ips.lk.
Kelegama, S. (2001). *Poverty Situation and Policy in Sri Lanka.* Paper Delivered at the Asia and Pacific Forum on Poverty: Reforming Policies and Institutions for Poverty Reduction, Manila: Asian Development Banks, 5th–9th February 2001.
Mahara Division Secretariat. (2017). Retrieved November 11, 2016, from http://www.mahara.ds.gov.lk/.

Mincy, G. A. (2014). *Human Potential Actualized*. Website: Center for a Stateless Society: A Left Market Anarchist Think Tank & Media Center. Retrieved November 11, 2016, from https://c4ss.org/content/24175.

Phillips, D., Waddington, H., & White, H. (2014). Better Targeting of Farmers as Channel for Poverty Reduction: A Systematic Review of Farmer Field Schools Targeting. *Development Studies Research, 1*(1), 113–136.

Preliminary Survey. (2017). Preliminary Survey with Mahara Division Secretariat Office & Wedamulla Grama Niladhari Office.

Ranathunga, S. B., & Gibson, J. (2015). The Factors Determine Household-Poverty in the Estate Sector in Sri Lanka. *Global Business and Economics Research Journal, Volume, 4*(1), 17–30.

Sen, A. (1999). *Development as Freedom*. Oxford: Oxford University Press. Introduction. Retrieved November 20, 2017, from http://www.c3l.uni-oldenburg.de/cde/OMDE625/Sen/Sen-intro.pdf

Samurdhi Authority. (2017). Retrieved September 27, 2017, from the Website Samurdhi Authority http://www.samurdhi.gov.lk/web/index.php?lang=en.

Yusuf, M. (2010). *Community Targeting for Poverty Reduction: Lessons from Developing Countries*. The Pardee Papers, No. 8, Boston, MA: Boston University.

4

Vulnerability of the *Charland* Dwellers to Climate Change: Various Adaptation Practices in Bangladesh

Mahfuzul Haque

Introduction

Bangladesh is a live delta and the rivers are changing their courses by the passage of time. The country is endowed with the Ganges-Brahmaputra-Jamuna river systems, with more than 300 perennial tributaries and distributaries. It is also among the world's most vulnerable countries in relation to flooding and riverbank erosion. Rivers in Bangladesh carry an annual silt load of 2.4 billion tons, which is one-fifth of the global sediment load (UNDP 1995, 5; Haque 2013a, 20; Rawlani and Sovacool 2011, 848). The three major rivers as they pass through the country to the Bay of Bengal have a catchment area of 1,559,400 square km, which is 7.5 percent of the total catchment area of these rivers, while the rest is outside the international borders in India, China, Nepal and Bhutan (Hutton and Haque 2003, 406). It's a country of low-lying floodplain, and nearly 50 percent of the country has an elevation of less than 10 m above sea level (Rasheed 2008). In another study (Rawlani and Sovacool

M. Haque (✉)
Department of Development Studies, Dhaka University, Dhaka, Bangladesh

© The Author(s) 2020

Md. N. Momen et al. (eds.), *Building Sustainable Communities*,
https://doi.org/10.1007/978-981-15-2393-9_4

2011, 848), two-thirds of its critical infrastructure is less than 5 m above sea level. In such a deltaic region, where the rivers are frequently changing their courses, *charlands* (raised shoals in the river) are created, which are volatile and vulnerable to floods and riverbank erosion.

Char is a tract of land shoals in a river course or in estuary caused mainly due to accretion. It includes all types of bars, both lateral (point bars) and medial (braid bars) (Nishat et al. 2014, 148). This newly emerged land is also locally known as *diara*. In Bangladesh, a *char* could be considered as by-product of the hydromorphological dynamics of a river (Islam 2010). The total area of *charlands* in Bangladesh is more than 1722 square km, which is 1.16 percent of the country's total area (Rasheed 2008). According to geographical position, one type of *char* is in the form of an island, and the other is adjacent to the mainland. According to the duration of existence, one is permanent or old char and the other one is temporary or erosion-prone char. Because of the braided nature of the Brahmaputra-Jamuna river, there are innumerable chars in the north-western region. The study area is centered on the *charlands* of Brahmaputra-Jamuna river basin in the northwest of Bangladesh in the districts of Jamalpur and Gaibandha. It is also among the world's most vulnerable countries to the effects of flooding and riverbank erosion (Hutton and Haque 2003, 406).

Jamuna *chars* are periodically inundated by floods, and especially during the monsoon season (June–September), lives of the charland people become vulnerable as they face environmental instability and seasonal migration. At times, they have to sell their assets for survival. River erosion has a direct relation with the floods, which takes an acute shape during the floods. Besides floods and riverbank erosion, they face drought, cold wave, *monga* and epidemic in the form of diarrhea. Poverty or *monga*-like situation is a common phenomenon during October–November, when there is no work in the field, although the situations are much improving these days with the introduction of early variety of crops, multicropping and other income-generating activities. Diarrhea is deeply related to flood water and usually takes an epidemic form during and after the floods. The dwellers also suffer from drought twice a year, during October–November, which ruins their *Aman* rice crops, and during March–April, when the other crops like sugarcane, jute and seedbeds are affected. During November–January, the cold waves literally cripple them, affecting their wheat fields.

Cut off from the mainland, the distance from the island chars varies from 2 to 10 km, and boat is the sole means of transportation to the mainland. This physical isolation has had deep-rooted effects on the psyche of the char dwellers (Dipu and Uddin 2015, 58). Away from the main growth centers like markets, they suffer from inaccessibility and seclusion. The *charland* dwellers are the most desperate and vulnerable people living on these newly accreted land in the northwestern region of Bangladesh. The Brahmaputra, being one of the largest sand-bed braided rivers in the world, every year erodes thousand hectares of mainland floodplain, rendering tens of thousands of people landless or homeless. In terms of riverbank erosion, it causes the most damages.

The Jamuna river is the downstream continuation of the Brahmaputra in Bangladesh. It is one of the largest sand-bed braided rivers in the world, and every year it erodes thousand hectares of mainland floodplain, rendering tens of thousands of people landless and/or homeless (Sarker et al. 2014, 1; CCC 2009, 19). The people, however, didn't abandon their lands because of flooding or riverbank erosion. On the other hand, these erosion-prone areas are accommodating a growing population because the newly accreted lands are fertile, containing rich soil nutrient. The focus of the study had been to investigate indigenous knowledge and practices, especially coping strategies of the *charland* dwellers with recurrent floods and river erosion. Based on the secondary sources of literature, the study found that indigenous knowledge and practices of the *charland* people in vulnerable erosion-prone areas provided them with a strong sense of confidence and sustenance, as the areas were hit by repeated floods and riverbank erosion. The study with a general introduction on geomorphological profile of Bangladesh gave an outline of the deltaic nature of the country, existence of the huge floodplain, frequent course-changing rivers and their impacts on the landmass, formation of *charlands* and riverbank erosion and so on.

Due to *Nodi Sikosti* (river erosion), they lose all their belongings—cropland, homesteads, trees, orchards, even graveyards of their ancestors. They are displaced from their ancestral land and become victims of the vagaries of nature. *Char* households often are forced to migrate four to five times in a lifetime, and the poorer people move more (Hassan 2000, 143; Roy et al. 2014, 59; and Haque 2019, 61). Population displacement due to riverbank erosion is widespread in the char areas. People of the *charlands* consider river erosion as a natural phenomenon, and it should be addressed by natural means. Such a displacement of people could also be termed as environmentally induced displacement, because of its association with natural disasters like floods, riverbank erosions and climate change (Bose 2015, 2). After *Nodi Poisti* (char formation, when a strip of land rises out of a riverbed), the people who were originally displaced long back are rarely rehabilitated as they face stiff resistance from the powerful people and land grabbers (Haque 2013b, 45). The focus of the chapter is centered around these newly formed *charlands* inhabited by the desperate and vulnerable people and their indigenous knowledge and practices in coping with the disasters.

Indigenous Knowledge and Practices

Indigenous knowledge and practices (IKP) means functional knowledge of the local community belonging to a particular ethnocultural and agro-ecological condition. IKP is developed through sharing of experiences, and it passes from one generation to another, mostly orally, and it operates in all aspects of community life and persists therein as unwritten forms (Mustafa 2000, 27; Haque 2013a, 275). It is often said that unwritten knowledge runs the risk of being lost due to its vulnerability. Against the threat of disappearance, local wisdom continues to persist in rural areas, villages, old towns, markets and other places. IKP usually connotes a holistic system of knowledge, comprising values, concepts, beliefs and perceptions, which is naturally available among local people (Warner 1991; Sillitoe et al. 1998, 5).

Indigenous knowledge of the *charland* peoples suggests structural measures like repair and rethatching of house roofs during high floods only. While during normal inundation, they raise their homesteads and put mud paste mixed with jute fiber and husks to protect the plinth from direct impact of floods. Usually, short bamboo stakes are placed at regular intervals along the plinth, so that running water cannot hit the plinth directly and erosion is halted (Haque 2019, 65). Before advent of the floods, they strengthen their bamboo poles, and with the level of flood water, they raise their six-legged wooden cots over bamboo platform. Large earthen vats are used to protect seeds and food grains from dampness and rising waters. Portable earthen ovens are put on the raised wooden cots for cooking. The neck of the tube wells is raised in order to ensure availability of drinking water. Sanitary latrines are also raised to cope up with rising flood water. In the northeastern wetlands called *Haor* (saucer-shaped water bodies), people plant a special type of grass weeds in and around their homesteads to avert wave actions hitting their houses.

It is interesting to note that over the years, the people of the *charlands* have developed some local practices based on their indigenous knowledge in the face of recurring floods, often visiting them three times a year. They portray a typical man-environment situation, as they adapt themselves with the whims of the rivers. Indigenous knowledge in agricultural cropping is the adjustment with respect to crops before and after flooding. Considering the weather pattern and flooding intensity and frequency, the *charland* people generally select the types of crops for planting. They plant groundnuts and sweet potatoes at higher level of sandy land out of reach of flood water. As they proceed toward the waterfront, they look for those varieties of *Aman* rice, which are adaptable to high floods. Generally, the deep-water *Aman* can survive in floods and can grow 6 inches a day up to a height of 15 feet as the water rises (Hasan et al. 2000, 38).

The islanders also practice intercropping to avert loss caused by crop failure. They plant short-term flood-sensitive *Aus* together with flood-tolerant *Aman* in the same plot of land. Generally, *Aus* variety of rice and jute are flood sensitive. They try to harvest them before the floods hit the area.

As the floods recede, the farmers go for seedlings in small patches of land. In the southwestern parts of Bangladesh, people have developed an indigenous way of vegetable gardening in water-logged areas. Called *Beira* or "floating garden (Hydroponics)", banana trunks are placed on the water, wherein seedbeds are developed with water hyacinth and mud (Rashid and Khan 2013, 346; Hasan et al. 2000, 39). Vegetables like tomatoes, bitter gourds and eggplants are grown in these floating patches.

People of the *charlands* in northwestern Bangladesh are a group of resilient people as they have been facing the vagaries of nature over the generations. Due to periodic river erosion, they lose their lands and homesteads and shift their homes to safer places on the embankments. They understand their climate very well as they risk living in such a hostile atmosphere, much away from the well-connected thoroughfares, growth centers or market places. Based on their indigenous knowledge, they have developed keen power of observation and can predict the nature of floods. If the monsoon rain is abundant and the water flow is relatively clean and has a current, they become alert to the onslaught of high flood. Depending on the severity of flood, they adopt different coping strategies. All community-based adaptation practices of the *charland* people as they face climate change-related disasters are based on their age-old indigenous knowledge passed on to them over the generations.

Community-Based Adaptation

The Intergovernmental Panel on Climate Change (IPCC 2007) defined adaptation as "an adjustment in natural or human systems in response to actual or expected climate stimuli or their effects, which moderates harm or exploits benefit opportunities". There are various adaptation practices based on scientific findings as well as on age-old community knowledge. Community-based adaptation (CBA) strategies aim to reduce vulnerability to climate change. CBA is intended to bring about some positive changes in agricultural systems. CBA strategies are introduction of hazard-resistant varieties of crops, conducting risk assessments, protection of natural and water resources, provide early warning and raising awareness (UNFCCC 2006; Mercer 2010, 249). Instead of undertaking mitigation measures by reducing greenhouse emissions, the emphasis was given on adaptation. A question was raised whether the developed countries were shying away their responsibilities in disguise of adaptation. Progress on mitigation had been slow. On the other hand, climate change is taking its toll on the low-lying countries like Bangladesh. Adaptation is firmly on international policy agenda (Ayers and Forsyth 2009, 23). The countries are expected to develop their adaptation practices on the basis of grassroots consultations.

CBA looks at the adaptation measures to climate change based on local knowledge and practices. These community people are generally poor and vulnerable to climate change-related disasters. They tax on common property resources and live in resource-poor areas vulnerable to natural disasters like floods and cyclones. According to Dodman and Mitlin (2013, 640), community-based adaptation is based on the principle that local community has the required skills, knowledge and expertise acquired over the generations. These skills made them resilient and helped to face climate-related disasters. CBA functions at the local level in communities that are vulnerable to the adverse impacts of climate change (Ayers and Forsyth 2009, 24). CBA promotes adaptation strategies following a participatory process based on grassroots consultations involving the local stakeholders. While promoting such a strategy, local culture and norms are looked into. Such a resilience to climate change-related disasters is based on disaster risk reduction rather than on relief and rehabilitation.

According to Reid et al. (2009), these practices are based on communities' priorities, needs, knowledge and capacities, which seek to empower people in order to cope with the impacts of climate change. The practices are rooted in participatory development programs in order to strengthen livelihoods and reduce vulnerability, as well as disaster risk reduction targeting to build resilience to climate-related disasters. These are multidisciplinary in nature, encompassing livelihoods, infrastructure, disaster risk reduction, economics, food security, ecosystems and sustainable development.

In the flood- and erosion-hit areas, some of the adaptation practices are raising vegetable over the water-logged areas (known as hydroponics), duck and poultry rearing, goat and sheep rearing and ring-based vegetable cultivation (Rashid and Khan 2013, 346). In order to cope with monsoon and floods, they use portable earthen cooking stoves and flood-proof food storage and build storm-resistant houses. Climate change-induced food insecurity, water scarcity and poor water quality result in waterborne diseases and malnutrition (Sen et al. 2009). They take precautionary measures in case of outbreak of cholera, diarrhea, malaria and dengue. Besides some indigenous healing systems, like the use of medicinal plants to heal common sickness like cold and cough, they go for water-purifying tablets against waterborne diseases, as the awareness is very high.

Challenges Ahead

(a) Indigenous knowledge and practices followed by the *charland* communities are developed over the years. Such a knowledge is relatively a new concept, not widely known outside the development community. It is a classic case of learning by doing or "action research". How to disseminate the new concept and change the mindset of the policy makers remain a big challenge.
(b) Ambiguity exists between adaptation practices being followed as per the national adaptation plan of action (NAPA) prepared by many developing countries and community-based adaptation practices. Often, community-based practices get less priority in the absence of demo pilot projects.
(c) CBA is still in the process of evolution, needed more piloting over the years in different agroecological zones of climate-stress countries. Success stories on CBA would depend on adequate funding by the government and the development partners.
(d) Mainstreaming CBA in macro- and microlevel plans, sectoral plans and policies is a big challenge. It could be only possible if national adaptation plan of action incorporates CBA-related projects based on stakeholder consultation.
(e) *Charland* communities are marginalized, live in inaccessible remote areas and receive limited services or support from the government. National development plans and policies rarely address plights of these marginalized peripheral people.

Conclusion

It is true that the *charland* people of the northwestern Bangladesh could lessen damages to their properties in the face of floods and river erosion due primarily to their coping strategies based on their age-old practices. They believe that river erosion is a natural phenomenon. Instead of structural solution, they have developed practices to live in erosion-hit areas, and the *charlands* are never abandoned. On the other hand, these fertile lands continue to attract landless people from other side of the river.

Promotion of community-based adaptation is a painstaking process, requiring continuous research and development in consultation with the vulnerable *charland* communities. The support of the government and the development partners is necessary in order to develop successful models. Moreover, various initiatives undertaken by local and international NGOs and research bodies are to continue in order to mainstream community-based adaptation practices in all plans and policies at the national and global level. Community-based adaptation practices remain mostly undocumented. It is suggested that the knowledge is documented, owned and promoted by the developing countries. Because of isolation and remoteness of the *charland* dwellers, national policies and plans would rarely affect or benefit these people. Mainstreaming policies and plans for the mainland would be rather ineffective for these people living in inaccessible peripheral lands. Location-specific policies addressing the vulnerability of the communities to climate-induced disasters are required to be adopted and their coping strategies documented.

References

Ayers, J., & Forsyth, T. (2009). Community Based Adaptation to Climate Change: Strengthening Resilience through Development. *Environment: Science and Policy for Sustainable Development, 51*(4), 22–31.

Bose, P. S. (2015). Vulnerabilities and Displacements: Adaptation and Mitigation to Climate Change as a New Development Mantra. *Area*, 168. https://doi.org/10.1111/area.12178.

CCC (Climate Change Cell). (2009). *Climate Change Adaptation Research, Climate Change, Gender and Vulnerable Groups in Bangladesh*, CDMP/DFID/UNDP, Department of Environment, Ministry of Environment and Forests, Dhaka.

Dipu, A. H. M. S. I., & Uddin, M. H. (2015). *Are Land Rights Utterly an Illusion for Landless Char Dwellers?* Proceedings of the First National Char Convention, Dhaka.

Dodman, D., & Mitlin, D. (2013). Challenges for Community-Based Adaptation: Discovering the Potential for Transformation. *Journal of International Development, 25*, 640–659.

Haque, M. (2013a). *Environmental Governance: Emerging Challenges for Bangladesh*. Dhaka: AH Development Publishing House.

Haque, M. (2013b). Indigenous Knowledge and Practices in Disaster Management. In M. Rahman & M. R. Ullah (Eds.), *Human Rights and Climate Justice* (pp. 39–50). Dhaka: Empowering Through Law of the Common People.

Haque, M. (2019). Indigenous Knowledge and Practices in Disaster Management: Experience of the Coastal People of Bangladesh. In B. Zutshi, A. Ahmad, & A. B. Srungarapati (Eds.), *Disaster Risk Reduction: Community Resilience and Response* (pp. 59–72). UK: Palgrave Macmillan.

Hasan, M., Haque, S. M., & Saroar, M. (2000). Indigenous Knowledge and Perception of the Charland People in Coping with Natural Disasters in Bangladesh. *Grassroots Voice: A Journal of Resource & Development, 3*(1–2), 34–44.

Hassan, S. (2000). Indigenous Perceptions, Predictions and Survival Strategies Concerning Cyclones in Bangladesh. In N. A. Khan & S. Sen (Eds.), *Of Popular Wisdom: Indigenous Knowledge and Practices in Bangladesh* (pp. 147–150). Dhaka: BARCIK.

Hutton, D., & Haque, C. E. (2003). Patterns of Coping and Adaptation among Erosion-Induced Displacees in Bangladesh: Implications for Hazard Analysis and Mitigation. *Natural Hazards, 29*, 405–421.

Intergovernmental Panel on Climate Change (IPCC). (2007). Geneva.

Islam, S. N. (2010). Char-Lands Erosion, Livelihoods and Cyclic Displacement of People in Ganges-Padma River Basin in Bangladesh. *Asia-Pacific Journal of Rural Development, 20*(1), 151–174.

Mercer, J. (2010). Policy Arena, Disaster Risk Reduction or Climate Change Adaptation: Are We Reinventing the Wheel? *Journal of International Development, 22*, 247–264.

Mustafa, M. M. (2000). Towards an Understanding of Indigenous Knowledge. In P. Sillitoe (Ed.), *Indigenous Knowledge Development in Bangladesh, Present and Future* (pp. 27–30). Dhaka: University Press Ltd..

Nishat, B., Chakraborty, S. K., Hasan, M. E., & Rahman, A. J. M. Z. (2014). *Rivers Beyond Borders, India-Bangladesh Trans-boundary River Atlas.* Dhaka: IUCN.

Rasheed, K. B. S. (2008). *Bangladesh Resource and Environmental Profile*. Dhaka: AH Development Publishing House.

Rashid, A. K. M. M., & Khan, M. R. (2013). Community Based Adaptation: Theory and Practice. In R. Shaw, F. Mallick, & A. Islam (Eds.), *Climate Change Adaptation Actions in Bangladesh*. Japan: Springer.

Rawlani, A. K., & Sovacool, B. K. (2011). Building Responsiveness to Climate Change Through Community Based Adaptation in Bangladesh. *Mitigation and Adaptation Strategies for Global Change, 16*, 845–863. https://doi.org/10.1007/s11027-011-9298-6.

Reid, H., Alam, M., Berger, R., Cannon, T., & Milligan, A. (2009). *Community-based Adaptation to Climate Change*. Participatory Learning and Action No. 60, London: IIED.

Roy, B. K., Ullah, M. B., & Rahman, M. H. (2014). Climate Change Impact in *Charlands* in Central Area of Bangladesh: Assessing Vulnerability and Adaptation by the Farming Communities. *Journal of Environmental Sciences & Natural Resources, 7*(2), 59–63.

Sarker, M. H., Thorne, C. R., Aktar, M. N., & Ferdous, M. R. (2014). Morpho-dynamics of the Brahmaputra–Jamuna River. *Bangladesh, Geomorphology*. https://doi.org/10.1016/j.geomorph.2013.07.025.

Sen, S., Khatoon, F. Z., Akhter, T., Akter, S., Rahman, M. A., Rahman, M. S., & Lamin, S. (2009). *Climate Change: Impacts and Adaptation Strategies of the Indigenous Communities in Bangladesh*. Dhaka: BARCIK.

Sillitoe, P., Dixon, P., & Barr, J. (1998). IK Research on Floodplains of Bangladesh: The Search for a Methodology. *Grassroots Voice, 1*(1), 5–15.

UNDP (United Nations Development Programme). (1995). *Report on Human Development in Bangladesh, Environment*. Dhaka: UNDP.

UNFCCC (United Nations Framework Convention on Climate Change). (2006). *Technologies for Adaptation to Climate Change*. Bonn: UNFCCC.

Warner, K. (1991). *Shifting Cultivators: Local Technical Knowledge and Natural Resource Management in the Humid Topics*. FAO.

Part II

The State Civil Society Organization and Social Welfare

Rajendra Baikady, Md. Nurul Momen, and Cheng Sheng-Li

Introduction

The concept of Civil Society is one of the most intensely debated and contested issues in the contemporary world. Further, relationship between the state and civil society has been conceptually contested by scholars. While discussing the relationship between civil society and the state, Marxist and liberalist are the two schools of thought, which have two opposite perspectives. Marxists argue that transformation of state and society from a feudal to a capitalist world results in the origin of civil society, whereas Marxists argue that civil society is a bourgeoisie space and an anti-proletarian articulation. Hegel explained the creation of civil society as an "achievement of the modern world" (quoted in Post and Rostenblum 2002, p. 10). Whereas other scholars argued, civil society is the public space between state and its citizen. As a result, most of the discussion in relation to civil society can be seen as contrast to the state and market (Wuthnow 1991; Brown and Korten 1991). Even though there is no agreed meaning for the concept of civil society, in general, civil society organizations (CSOs) work outside the state and creation of civil society is made up of several individuals coming together. According to Edward (2005), CSOs help in developing the values of tolerance and cooperation among the people and therefore, they can be referred to as "associational life of society". Further, he refers to CSOs as the 'Public sphere' that cares about the common good (Edward 2005). CSOs are the 'public sphere', which brings people with shared interest and willingness to accommodate others' views together to work for the common interest (Edward 2005). CSOs work outside the state and aim for achieving the common goals that are beneficial for the larger society and public (Edward 2005). These organizations accommodate people without the barrier of ideology, political affiliation, race, gender, and cultural association. In the view of Heyden (1996), a CSO is a web of associations both public and private and enjoys the right to organize, mobilize and influence political decision. He also states that these rights given to the CSOs are free of state's control and not influenced by the state actors.

Civil society organizations and associations are a group: the neighbourhood, churches, non-governmental development organizations, cooperatives, soccer clubs, choral societies and many other associations. These CSOs in diverse societies will have different values, beliefs, norms and responsibilities. According to Liberal philosophy, the primary function of a civil society is safeguarding the rights and interests of the citizens against the state's interventions (Chatterjee 2002, p. 171). Civil society by contrast strives for achieving the common good by mobilizing the resources among different social groups (Wuthnow 1991). Despite the growing importance of the concept and its role in the modern welfare state, the term Civil Society still remains as partially understood or most of the time rarely understood and frequently contested. Even though these organizations work within the society and help common people obtain their rights and entitlements, the number of these organizations, their work areas, and their source of revenue and policy framework governing their work is largely not available in many countries across the world. In the more recent past, CSOs are clearly segregated from the two dominant actors, that is, the market and the state, and are seen outside the state and market. A clear understanding and conceptualization of CSOs and their contribution is an essential need in light of globalization, liberalization and privatization of means of production across the globe both in developing and developed economies.

Globalization has transformed the social, economic and political environment across the globe, irrespective of the developmental status of countries. While globalization provides opportunities for trade liberalization, foreign direct investment, capital flow, information exchange and technology transfer, the process of globalization and its results, however, are not equal all over the globe. Some countries are not able to cope with the globalization process, especially the underdeveloped and developing countries facing enormous challenges in relation to sustainable development. In addition, over the last two decades, the globalization process has led to adjustment issues among the less developed countries, as they find it difficult cope with the new requirement of a global society. Despite speeding up the process of production and enhancing productivity in many ways, the process of globalization has pushed over a billion people to the edge of poverty. A sub-section of population in every society lives below the poverty line with spending capacity of less than a dollar a day. Children in the poorest countries continue to suffer from malnutrition and majority of them die in the first year of their birth. Access to clean drinking water, sanitation, basic health services and education are out of reach for more than half the people in developing countries.

In contemporary society, the relationship between state and civil society is quite complex, and understanding the nexus between state and civil society is challenging. Many scholars have argued by distinguishing the state and civil society very clearly, however it doesn't mean that civil society can replace the state or civil society is not an alternative to the state (Pietrzyk-Reeves 2008). Furthermore, it is also important that the existence of the state is a prerequisite for any civil society to function. Scholars have also noted the importance of democracy in creating a favourable environment for the growth of civil society in any society (Scholte 2000; Wolf 2001).

Role of Civil Society in Welfare

Research studies over the past 20 years have proved that civil society organizations are helping local communities to build social capital (Onyx and Bullen 2000). Social welfare provisions ensuring education, health, housing and social justice to the deprived and oppressed population were the traditional focus of the third sector. Giving voice to the unaddressed issues and thereby facilitating social inclusion is also an important role played by the civil society organizations (cf. Kramer 1981; Rose-Ackerman and James 1986; Kendall 2003). Civil society organizations play an important role in the process of governance at both national and international levels. Furthermore, the research findings show that the role of the third sector in social welfare varies across countries and depends on the type of state they function. A typology developed by Salamon and his colleagues suggested that civil society plays a larger role as service provider in the countries where the number of third-sector functioning is very large (Salamon and Anheier 1998; Salamon et al. 2000a), whereas civil society in relatively smaller countries tends to fulfil the political, social, and recreational interest of the people (Salamon and Anheier 1998, p. 229).

Discussions pertaining to civil society organizations' role in social welfare have been increasing in the recent past. While CSOs are getting more and more recognition across the globe for their interventions, there is also a growing realization about the plurality of the civil societies (Heinrich and Fioramonti 2007; Edwards 2009). According to Edward (2005), CSOs are based on the cooperative spirit where people come together for a common goal. Further, he states that CSOs made up of the public sector, care for the common good and social welfare (Edward 2005). Whereas Nzomo (2003) includes the informal organizations in the rural areas in the category of civil society organizations and states that CSOs do not include political groups but they include blood associations. Further, it is clearly stated that CSOs' important interest is the 'public sphere' and 'public good' (Edward's 2005). From the observations of comparative civil society research at Johns Hopkins Comparative Non-profit Sector Project (CNP), Salamon and Sokolowski (2004) note that the roles and responsibilities of civil society organizations are different in welfare states,. however, in general, they all contribute to the welfare of society. Scholars have repeatedly noted service-expressive and advocacy as the main societal functions of CSOs (Kramer 1981; Kendall 2003; Salamon et al. 2000a; Frumkin 2002). CSOs are expected to fulfil the service functions, such as delivery of goods and services and expressive functions, such as promoting culture, spirituality, professional or policy values, interests and beliefs among people. Whereas advocacy function is every activity that intends to change polices for ensuring the collective good (Jenkins 1987, p. 297).

Another group of scholars (Kramer 1981; Kendall 2003; Salamon et al. 2000a) noted innovation as one of the important functions for civil society organizations. Further, Salamon et al. (2000a, p. 6) noted that CSOs help society to deal with social problems by innovations and formulating new approaches to problems. Kendall (2003) and Salamon et al. (2000a) further credited community building as a function for CSOs. Community building, a concept developed by Putnam (1993), is an integrative role for the CSOs. Integrative function of CSOs expects them to encourage social interaction, generate sense of community based on mutual trust, common norms and reciprocity. Salamon et al. (2000b) notes the function of CSO in a liberal non-profit regime as substitute to governmental social protection, and in corporatist regimes CSOs cooperate with the state in the provision of social welfare. Further, CSOs play a major role in the democratization process. Countries such as South Korea, the Philippines and Pakistan have gained immense support from the CSOs in their democratization efforts (Cheema 2011; Rafique et al. 2016). CSOs are also playing important role in sustaining and deepening democracy in Pakistan through enhancing the citizen's participation. In the recent past, involvement of CSOs in the developing countries have appeared to play important role in making peoples participation more effective in the process of governance and development.

Scholars have also regarded civil society as an arena of people's participation (Heinrich and Fioramonti 2007; Fioramonti and Fiori 2010). Further, some other scholars refer to CSOs as the locus of communicative action (Habermas 1984; Calhoun 2001). Locke (1955) and Ferguson (1995) argued CSOs are the expression of the modern proprietary class. Tocqueville (2000) and Putnam (2000) treated CSOs as the locus of self-organizations and a breeding ground for democratic practices. While supporting the view of Hegel and Marx, Hunt (1990) and Wood (1990) noted CSOs as the organization between the state and the family and use to perform the task of cultural transfer thought the society. While revising the Marxist perspective, Gramsci explains: CSOs is Kingdom constructed mainly on the basis of consent (Gramsci 1971). Further, CSOs play a very important role in the social, political and the economic development of any society (Edwards and Hulme 1996; Riddell and Robinson 1995; Brown and Korten 1991; Clark 1991).

Civil Society Organizations in many societies also work for discharging the economic, social, cultural and other responsibilities. Even though it is not a part of the state, it performs many responsibilities and roles of the state, such as ensuring rights, providing general welfare provisions and overseeing development of the society. CSOs also play the role of watchdogs and help society to deal with social problems. In the recent days, civil society across the globe is becoming more popular and alert. A country like India with liberal democratic governance encourages civil society, as it helps in educating, socializing and promoting people's participation in the governance. Further, CSOs in Pakistan also play a major role in development, as they are supported by the democratic political structure. Further, CSOs in South African countries and Tunisia, Egypt and Yemen have been gaining more strength. In sum, civil society organizations play a commendable role in the development of any society across the globe. As documented by many scholars, civil society helps in overthrowing the authoritarian government and bringing in the democratic and progressive government system in many countries. Among the South Asian countries, Pakistan had strong civil society support in establishing democratic government system and holding democratic and free elections for constituting a democratic government.

Part II of the book focuses on relevant discussion and shows the connection between the state and civil society and the issues faced by civil society in contemporary society. The section also explores the regulatory framework of civil society organizations, the role of civil society in democratic consolidation process, social audit and its role in ensuring social justice, globalization and its impact on the changes in civil society sector, media and the NGOs and strategic effective governance, democratic practice and good governance in the different countries in South Asia. Chapter 5 makes an attempt (a) to look into the existing status of the governance; (b) to highlight the media and NGOs' strength that are helping the governance to make it more effective; and (c) to suggest ways and means on how to overcome barriers. Against the backdrop of the existing situation, the chapter expects to provide arguments with respect to the approaches of the media and NGOs towards the governance as well as to the people's overall mental state and sentiment. Based on the discussion of the ground realities, the special focus is to look into the coordinated approach of the media and NGOs for improving governance in terms of paving the way between public and private sectors and among various agencies of the government to meet challenges.

Chapter 6 finds out how civil society can contribute to the democratic consolidation process by using a theory of democratic consolidation. Other objectives of the chapter is to discuss the three separate but interrelated points, that is, the conceptual frameworks of civil society and the democratic consolidation process and their interconnection, the approach by which civil society plays a role either positively or negatively in safeguarding or threatening democratic consolidation in Bangladesh, that is, the weak as well as active performance of civil society in relation to democratic consolidation and finally what the role civil society should play for safeguarding Bangladesh's democratic consolidation in future.

Chapter 7 examines the role of civil society organization in the prosperity of the South Asian region. The article argues that despite these limitations, the civil society of South Asia has continued not only to advocate for peace and amity in the region, but also to recommend measures to strengthen non-political integration in the region. Chapter 8 explores the democratic movements and conceptualizes and relates good governance in the context of Nepal. It also tries to analyse the role of the past democratic movements for democracy and show its relation with good governance in Nepal. As regards methodology, it has made the best use of secondary sources. Chapter 9 delineates the dynamics of the civil society organization and state relation. The chapter specially focuses on the relationship between state and CSOs in the current period. It shows how state is withdrawing from various social sector schemes and facilitating the emergence of CSOs. Accordingly, increasing government fund to CSOs has also compromised their autonomy and voluntary character, which has changed the landscape of state–CSO relations. It has changed the way CSOs used to interact with the state traditionally. Chapter 10 is a case study on independent Community Radio Project by a civil society organization called Pulathisi Federation in Sri Lanka. By using the case study method, the author(s) try to understand how lack of legal recognition has affected the collective effort of a civil society from entertaining its right to communication. Findings of this case study is expected to guide similar initiatives around the world, to mobilize civil society to gain legal recognition first, before they kick off other preparatory activities in initiating community media projects. Data has already been collected and the researcher is in the process of writing a relevant chapter.

References

Brown, D., & Korten, D. (1991). Working More Effectively with Nongovernmental Organizations. In S. Paul & A. Israel (Eds.), *Nongovernmental Organizations and the World Bank: Cooperation for Development* (pp. 44–92). Washington, DC: World Bank.

Calhoun, C. (2001). Civil Society/Public Sphere: History of the Concept. In N. J. Smelser & P. B. Baltes (Eds.), *International Encyclopaedia of the Social and Behavioural Sciences*. Amsterdam: Elsevier Science Ltd.

Cheema, G. S. (2011). *Engaging Civil Society to Promote Democratic Local Governance: Emerging Trends and Policy Implications in Asia*. Swedish International Centre for Local Democracy ICLD, Working Paper No. 7, Sweden.

Chhatterjee, P. (2002). On Civil and Political Society in Post-Colonial Democracies. In S. Kaviraj & S. Khilnani (Eds.), *Civil Society: History and Possibilities* (pp. 165–178). New Delhi: Foundation Books.

Clark, J. (1991). *Democratising Development: The Role of Volunteer Organizations*. London: Earthscan.

Edward, M. (2005). Civil society. The Encyclopaedia of Informal Education. Retrieved from www.infed.org/associational/civil_society.htm.

Edwards, M. (2009). *Civil Society* (2nd and Expanded ed.). Cambridge: Polity Press.

Edwards, M., & Hulme, D. (1996). Too Close for Comfort: NGOs, the State and Donors. *World Development, 24*(6), 961–973.

Ferguson, A. (1995). *An Essay on the History of Civil Society (First Published 1767)*. Cambridge: Cambridge University Press.

Fioramonti, L., & Fiori, A. (2010). Civil Society After Democracy: The Evolution of Civic Activism in South Africa and Korea. *Journal of Civil Society, 6*(1), 23–38.

Frumkin, P. (2002). *On Being Nonprofit. A Conceptual and Policy Primer.* Harvard University Press.

Gramsci, A. (1971). *Selections from the Prison Notebooks.* New York: International Publishers.

Habermas, J. (1984). *Theory of Communicative Action, Volume 2.* Boston: Beacon Press.

Heinrich, V. F., & Fioramonti, L. (Eds.). (2007). *CIVICUS Global Survey of the State of Civil Society: Comparative Perspectives.* Bloomfield: Kumarian Press.

Heyden, G. (1996). The Challenge of Analysis and Building Civil Society. *Africa Insight, 26*(2). Institute of Open Learning.

Hunt, G. (1990). The Development of the Concept of Civil Society in Marx. In B. Jessop & C. Malcolm-Brown (Eds.), *Karl Marx's Social and Political Thought: Critical Assessments.* London and New York: Routledge.

Jenkins, J. C. (1987). Nonprofit Organizations and Policy Advocacy. In W. W. Powell (Ed.), *The Nonprofit Sector: A Research Handbook.* New Haven: Yale University Press.

Kendall, J. (2003). *The Voluntary Sector. Comparative Perspectives in the UK.* London and New York: Routledge.

Kramer, R. M. (1981). *Voluntary Agencies in the Welfare State.* Los Angeles and London: Berkely.

Locke, J. (1955). *Two Treatises on Civil Government.* London: J.M. Dent & Sons.

Nzomo, M. (2003). Civil Society in Kenyan Political Transition: 1999–2002. In W. Oyugi et al. (Eds.), *The Politics of Transition in Kenya from KANU to NARC.* Nairobi: A Publication of Heinrich Boll Foundation.

Onyx, J., & Bullen, P. (2000). Measuring Social Capital in Five Communities. *Journal of Applied Behavioral Science, 36*(1), 23–42. https://doi.org/10.1177/0021886300361002.

Pietrzyk-Reeves, D. (2008). Weak Civic Engagement? Post-Communist Participation and Democratic Consolidation. *Polish Sociological Review, 2008*, 73–89.

Post, R. C., & Rosenblum, N. L. (2002). Introduction. In N. L. Rosenblum & R. C. Post (Eds.), *Civil Society and Government* (pp. 1–25). Princeton: Princeton University Press.

Putnam, R. D. (2000). *Bowling Alone: The Collapse and Revival of American Community*. New York: Simon and Schuster.

Putnam, R. D., Leonardi, R., & Nanetti, R. Y. (1993). *Making Democracy Work: Civic Traditions in Modern Italy*. Princeton: Princeton University Press.

Riddell, R. C., & Robinson, M. (1995). *NGOs and Rural Poverty Alleviation*. Oxford: Clarendon Press.

Rose-Ackerman, S., & James, E. (1986). *The Nonprofit Enterprise in Market Economics*. Chur, London, Paris, and New York: Harwood Academic Publishers GmbH.

Salamon, L., & Anheier, H. (1998). Social Origins of Civil Society. *Voluntas, 9*, 213–248.

Salamon, L. M., & Sokolowski, S. W. (2004). *Global Civil Society. Dimensions of the Nonprofit Sector*. Bloomfield: Kumarian Press.

Salamon, L. M., Hems, L. C., & Chinnock, K. (2000a). *The Nonprofit Sector: For What and for Whom?* Working Papers of the Johns Hopkins Comparative Nonprofit Sector Project, no. 37. Baltimore: The Johns Hopkins University Center for Civil Society Studies.

Salamon, L. M., Sokolowski, S. W., & Anheier, H. K. (2000b). *Social Origins of Civil Society: An Overview*. Working Papers of the Johns Hopkins Comparative Nonprofit Sector Project. Baltimore: The Johns Hopkins Center for Civil Society Studies.

Scholte, J. A. (2000). *Globalization: A Critical Introduction*. New York: Palgrave.

de Tocqueville, A. (2000). *Democracy in America*. Chicago: Chicago University Press.

Wolf, M. (2001). *Will the Nation-State Survive Globalization? Foreign Affairs, 80*(1), 178–190. https://doi.org/10.2307/20050051.

Wood, E. M. (1990). The Uses and Abuses of Civil Society. *Socialist Register 1990: The Retreat of the Intellectuals, 26*, 60–84.
Wuthnow, R. (1991). *Between Stares and Markets: The Voluntary Sector in Comparative Perspective*. Princeton, NJ: Princeton University Press.
Rafique, Z., Khoo, S. L., & Idrees, M. W. (2016). Civic Engagement Among the Youth: Empirical Evidence from Kashmir, Pakistan. *Humanomics, 32*(3), 376–388. Retrieved from https://doi.org/10.1108/H-12-2015-0077.

5

Media and NGOs: Strategic Effective Governance

Akbaruddin Ahmad, Rashidul Islam, and Touhid Alam

Introduction

The term 'governance' has been derived from the Greek word 'Kubernan' which means manifold or multiple concepts that comprises of a lot of political, economic, and social issues related to government and administration. Such a changing issue is constantly giving multifarious interpretation in various phases concerning socio-political and socio-economic

A. Ahmad (✉)
Policy Research Centre.bd (PRC.bd), Dhaka, Bangladesh

DarulIhsan University, Dhaka, Bangladesh

R. Islam
Policy Research Centre.bd, Dhaka, Bangladesh

Institute of Cost and Management Accountants of Bangladesh, Dhaka, Bangladesh

T. Alam
Policy Research Centre.bd, Dhaka, Bangladesh

© The Author(s) 2020
Md. N. Momen et al. (eds.), *Building Sustainable Communities*, https://doi.org/10.1007/978-981-15-2393-9_5

status of a country/society. From the time immemorial, especially from the time of Plato and Aristotle, governance has been defined to be the task of running different organs of the government and administration. Later on, it has been defined to include various aspects from different angles of the society and prevailing circumstances full of realities in the particular society. However, there is still no definite concept about the very terminology—governance. It may vary from time to time, place to place, and society to society, both nationally and internationally. We may therefore conclude that governance stands to mean the way by which power is executed in managing the day-to-day affairs of a state in social, political, and economic arenas for the sake of development by ensuring transparency and accountability.

A look at the second half of the 1980s and onward will present us with a peculiar scenario as regards the fashionable term in global politics and economy that helps ensure quality and goodness of governance. Recently, it has been emerging from the forefront of the global agenda for development of the society and a country where the need of good governance is mandatory factor without which no better output can be expected in the arena of socio-economic development of a country and society (Anam 2002).

This is because of the hard realities of life that in spite of huge resources, manpower, and institutional facilities the main objective of economic growth and human development have not yet been made possible in some countries of the world to solve poverty, confront civil frictions, and community development as well. The reasons are obvious and not beyond human capacity to undertake the challenges that we have to face every time. That is why several international organizations like the World Bank, IMF, UNDP, and a large number of NGOs have been functioning emphasizing the need for good governance as one of the preconditions to ensure services and delivery of funds for the purpose (UNDP 2005).

As one of the developing countries of the world, Bangladesh has been pointed out for its failure to extend public sectors, to meet up the growing demand of the common people. Because of the absence of appropriate leadership, lack of transparency and accountability in public administration, and vigorous political influence, proper environment for private sector development could not be achieved.

Governance does not mean a single unit or idea rather it means comprehensively the interaction between the problems created and solution to be made in a befitting manner. For that purpose, the following points may be taken into consideration so that solution can be given in the light of the existing problems.

- Exercise of power
- Legalization of power
- People's rights and authority
- Mobilization of recourses for development of the country and its people
- Ensuring security of life and property

The concept of governance still remains hazy and the term lacks a precise definition. The concept of governance covers various actions by the state and government. The plain meaning of governance means the act, manner, or function of governing. Thus, the act, manner, or the function of governing would have components such as the form of political authority exercised in Bangladesh, which has been changed several times from the inception of Bangladesh. Political authority has been exercised by the parliament that was formed based on a democratic Constitution. The Constitution declares that, "it shall be a fundamental aim of the state to realize through the democratic process in a socialist society, free from exploitation - a society in which the rule of law, fundamental human rights and freedom, equality and justice, political, economic and social will be secured for all the people in the territory of Bangladesh." (World Bank 2006a).

Problems of Good Governance in Bangladesh

Currently, good governance is a major challenge faced in Bangladesh. In many sectors of our national life, its underperformance is evident from views of our international donors and from the pronouncement of the policy makers relating to the realm of good governance. Absence of good governance and efficient political management are now the major constraints on the way to hardening the foundation of democracy and improvement of the quality of governance in Bangladesh.

Now, let us move forward with an attempt to have a look at the problems of good governance in our country with a focus on some major political institutions.

Election Commission

Holding regular elections at both national and local levels is one of the pre-conditions of democracy and good governance. Good governance by the representatives of the people can never be effective unless the people reflecting their sentiments elect their representatives. If the election is not fair, the authorities of the people's representatives shall always be questioned and this will surely weaken the government. Fair and free election plays a very vital role in the formation of the state policies. The election also makes the government accountable and responsible to the people to meet up their needs and expectations. Therefore, to continue the practice of good governance, election must be free, fair, participatory, and acceptable to the people. However, it is a matter of regret that this real practice is not seen in Bangladesh. The election commission in Bangladesh, in the true sense, is not independent of the executive and legislative organs of the government. Directly or indirectly, the election commission is badly influenced by the party in power (If the Parliament is not dissolved) and by the party that was in power in the recent past (If the Parliament is dissolved), to manipulate the election results in its favor.

Therefore, all the elections held since the independence of Bangladesh in 1971 have been characterized by the non-management of electoral policies involved in using intimidation, threat, vote-rigging, and official malpractices. **But it is a matter of good hope that during the regime of the last caretaker government headed by Dr. Fakhruddin Ahmed, some notable changes, like separation of the election commission from the grip of the prime minister's office, took place to make EC more effective and independent and as a result, the holding of the ninth parliamentary election in 2008 was mostly free, fair, and transparent.**

National Parliament

The legislature is no doubt, one of the most significant institutions in a democratic country for promoting good governance in a country. In fact, parliament deals with governance goals of greater accountability, transparency, and participation in the elections by the political parties in a very fair and uninterrupted way. In every country where parliamentary democracy prevails, sovereignty of the people and ensuring all their democratic rights rest in the hands of the legislature, it thus acts as the custodian of the people's sovereignty that makes the legislature the most important organ of the modern democratic state. The very first role that the legislature has to play is ensuring the fundamental principles of governance and embodiment of the institutions entrusted to them. The legislature can ensure the good governance in making constitution and enactment of various laws to fulfill the hopes and aspirations of the people living in the state.

All the concerned dignitaries including the speaker of the parliament have been politicized and used by the ruling party against the major opposition parties of the country. It has become a wrong practice that the ruling party has felt no necessity of consulting with the major opposition parties regarding the appointment of officials to major constitutional positions. It is a fact that parliament and parliamentary committees have been dominated and monopolized by the ruling government.

Therefore, the oppositions are left in the street to face and challenge the government. The unwillingness of the ruling party's law makers to attend the parliamentary sessions, their indifference to abide by the norms and rules of the parliamentary form of democracy, frequent boycott by the main opposition party and reluctant tendency of the government to discuss the major issues in the house have not let the eighth national parliament play an effective role. Comparing to those of previous two parliaments (fifth and seventh national parliament), the eighth national parliament had taken a step backward instead of consolidating the parliamentary democracy in Bangladesh (Ministry of Information 2006).

Independence of Judiciary

The judiciary system involved in making legal decisions is an indispensable part of good governance. Hence, good governance demands necessarily the functioning of judiciary independently and impartially. It is also the responsibility of the judiciary to make the government accountable and responsive to the demand of the common people, which are definitely the essential elements of good governance. The independent and impartial judiciary of a country upholds equal rights of both men and women, which is one of the preconditions to ensure democracy and good governance (Haque 2005).

It is quite evident, that the judiciary can play an independent and impartial role in cleansing the administration of the corrupt politicians and officials by bringing them into the arena of judicial trial for the sake of social justice. Due to the absence of independence of judiciary in Bangladesh, the lower courts have become subservient to the whims of the executives and legislative members.

In the true sense, the independence of judiciary is yet to be ensured in the light of the Masder Hossain Case (1999), although the ruling party had declared when it assumed power in 2009 that good governance should be ensured to harden the foundation of democracy in our journey ahead (Ministry of Information 2006). It should be remembered that the actual dream of our war of liberation was to up hold the democracy and economic equality of the people.

Institutionalization of Political Parties

The major barriers for consolidation of democratic governance in Bangladesh is derived from the distorted growth though there still exist a large number of parties of which very few could really develop clear and cohesive socio-economic programs, stable membership, and democratic practices. Students, laborers, public sector employees, and teachers of government schools and colleges have become politicized to join the ruling party in power.

Bureaucracy and Public Administration

All over the world, bureaucracy is considered to be one of the most powerful instruments of governance. Through bureaucracy, public policies are given shape and put into practice. Therefore, governance inevitably requires efficient bureaucracy to run the whole show of the government-public administration and good governance as they have an inseparable relationship in the concept of a modern democratic state. Of course, there are some universally recognized methods, which have been totally absent in the administration since its inception. **To speak the truth that both civil and military personnel actively participated in the war of liberation.** That is why, the government of Bangladesh inherited political administration. This has resulted in the recruitment of politically influenced, inefficient people and by utilizing those people; good governance cannot be expected in most cases. Their political allegiance is notable in comparison to their professionalism. Now very fast and decisive action is needed at the highest level of the government to overcome the bottlenecks (Ministry of Information 2006).

Local Government

Strong and autonomous local government system is one of the most vital institutions of democratic governance. Local government can be one of the most effective means by which the residents of a local community can accomplish together what neither the national government nor the individual residents can do or accomplish effectively.

The concept of administration in the modern state is not possible without delegation of powers to the local governments with the increasing number of population and massive expansion of governmental activities. Some matters of policy and administration relating to national and international affairs are controlled by central administration and the rest of the functions lie in the hands of local government authorities.

The local government through Article 60 has been empowered to levy taxes for the specific requirements of the local community welfare. The elected members of the local body at the Upazilas (sub district) level unfortunately are not independent and are under the control of the central government. The colonial attitude is still prevalent.

Bangladesh, being a republican state is committed to establish local governments at various levels for the cause of good governance. In accordance with article 59 of the Bangladesh constitution, the local government of every administrative unit shall be entrusted to bodies composed of the provisions of law and the maintenance of public order. They will perform functions relating to administration and work of all public offices and the preparation and implementation of plans and policies as regards public services and economic development of the country (Manik and Zaman 2011; Aminuzzaman 2010).

Media's Role in Governance

Media is the most significant element for conveying information in a society. The media has extended its arena in an unbelievable way and has made its impact on the society through radio signals in villages, newspaper, prime time news shows to know the updates throughout the day or a quick piece of news on the internet through smart phones. Therefore, the part played by media has become quite difficult and critical for the smooth functioning of democracy, since it extends cooperation in forming the public opinion on major issues. This is why media is treated as the fourth estate since it is not a non-state component that aims at protecting the rights of the citizens. As we understand that all specific media have their own selected set of followers so their united effort ensures better governance. History says that print media was basically responsible for conveying information regarding the freedom struggle of the great leaders of the world. It survives to exist popular despite fierce competition and until now, the privileged medium to reach out to masses and classes as well.

It may be mentioned here that the Radio is a very simple portable media that can influence people greatly in rural areas where the residents mostly do not know how to read and write. As a large number of people are still ignorant, this media criticizes government and at the same time it can communicate the urgent information like weather forecasts and the related matters very rapidly. Even television plays a very vital role to draw the attention of the people on important issues that leave a great impact. The latest invention is the internet that plays a very significant role to spread any news within a very short time to millions of people all over the world. That is why the government is now focusing on the internet so that it may reach out to a greater section of the people, especially the youth (Haq 2003; Haque 2005).

Media in Bangladesh

As good governance embodies accountability transparency, responsiveness, equity, justice, efficacy, and effectiveness, media has a great role to play in ensuring criteria, policies in practical life so that the people as a whole benefit. Good governance essentially ensures the management of public resources.

For a developing country like Bangladesh that still reels with poverty, superstition, a range of religious extremism, the involvement of media becomes even more important. Another reason why media is of paramount importance for the country is the immense disparity that is prevalent.

The media in Bangladesh inherits a tradition of independence. In general, it can talk about the issues and its right to do so is guaranteed in our constitution. However, media at the local and national levels are still facing pressure, intimidation, and threat from all levels of administration, commercial, and even criminal bodies. So they have to impose self-censorship to save the own skins.

The media market in Bangladesh has mushroomed over the past ten years, with a large range of highly profitable newspapers and satellite TV channels springing up.

Media, as the fourth estate, aims at protecting citizens; hence, the role of media has become extremely critical for the smooth functioning of effective governance as it helps in forming public opinion on vital topics in a society. All forms of media have a specific set of followers and therefore, they together ensure effective governance. Indeed, all developing countries face myriad issues on a day-to-day, short- and long-term basis, and Bangladesh is no exception. The media is to hold the flag with an open mind for its consistent and transparent role (Iftekharuzzaman 2011).

The process of democratizing the existence of a free and independent media within each nation is essential. Media is thought to be the mirror of the society; for them, it is possible to comment on the affairs of a country's governance, their commitment to democracy, economic and social growth of the country. Surely, media acts as a bridge between the state and the people. The strong and impartial media puts pressure on government to be more transparent and accountable. Nevertheless, it is largely a matter of regret that the media world has been divided and influenced by the government and many of them are observed to be serving their purposes which is badly affecting the ethics of journalism.

It is established all over the world that freedom of media is one of the most significant elements of democracy and governance which is given due importance by various national and international agencies that especially includes the World Bank–governance report vide "Voice and Accountability Component." If the state is in the practice of deregulating its control over the media world, people receive great exposure to varieties of information that help transition to transparent democracy and the media-men play the role of a watch dog. This helps the government to run the administration in a very balanced way (Chowdhury 2005; World Bank 2006b).

The Recent Development of Media in Bangladesh

Despite its growth in number, whether print, electronic, or broadcast, media can enjoy freedom to play their role in the current situation of Bangladesh. This is a very vital issue raised by the critics and experts.

Since the restoration of democracy in 1991, in spite of less restricted environment and more media activity, even the democratically elected governments have upheld the policy of suppression of media when they are faced with critical voices from the said quarters.

Very justly and rightly, "The Freedom House Index" comments that the environment for media in Bangladesh is "Partly free." Even, The World Press Freedom Index 2014 of Reporters without Borders ranked Bangladesh as 146 out of 180 countries. Not only that, the Amnesty International stated that Bangladesh journalists who write about social injustice, human rights, corruption, judicial irregularities, violations, extrajudicial killings, rape in the custody, and gender-based violence are harassed, ill-treated, and detained by police and security agencies on politically motivated charges and tortured inhumanly. Under the banner of media-freedom, the ruling force intends to choke the voice of the media personnel though through the new National Broadcast Policy-2014, the government has claimed to bring all the media outlets into one integrated structure with a view to building up an independent pluralistic accountable and responsive broadcasting system in Bangladesh (Swiss Development Cooperation (SDC) 2013).

Problems of Media in Bangladesh

A responsive, free, and independent media is essentially required to be present to ensure democratic governance in a country. As a precondition to succeed democracy, the role of impartial, free, responsive, and independent media is very much significant so that accountability to the people is established.

Media has a multifarious role to play as a public watch dog on formulating economic policies, monitoring election process, exposing human right abuses, keeping constant eyes on political corruption, and the empowerment of women. Therefore, in any democratic state, the role of media is of first and foremost importance: "An unfettered, professional and virile press is essential in democratic governance" to ensure good governance, transparency, and openness within the government, transaction free and fair press and media is mandatory that acts as a complement to run the administration in a democratic way. There is a saying in our society that a popular government without popular information is nothing but a farce or a tragedy or perhaps both and since our independence in 1971, radio and TV have been used as the mouth piece of the government. That is why, it can be remarked that there appears to be a fairly large audience for non-Bangladeshi radio sources including All India Radio, BBC, and VOA. Bangladesh Nationalist Party (BNP) and Awami League had vowed commitment to the people that they would give autonomy to the state-controlled radio and TV. Nothing has yet been done to this effect (Haq 2003).

Parameters of NGOs in Bangladesh

As one of the developing countries, Bangladesh has experienced a lot and witnessed the downfall of military dictatorship, failure in conventional good governance, speedy globalization, spread of information technology, expansion of inequality, persistence of mass poverty, formation of democratic government, rise of aspiration for alternative development, and a desire for participation in decision making in all levels. It is worth mentioning here that we have also noticed the expansion of democratic space in the midst of disillusionment with the role of state in facilitating coveted progress for the majority of the people. Multiple stakeholders are coming forward to take part in the process of decision-making, protecting essential services to the needy and disadvantaged class of people.

Despite some positive gains in the field of globalization, it is observed with concern that many anti-democratic tendencies have arisen out of concentration of capital at both the national and international levels. From this perspective, some commentators view that the role of civil society that includes NGOs can be very much crucial in democratizing the system and contents of good governance with positive advantages by making new platforms for participation.

A strong and vibrant civil society is necessary for ensuring good governance in the sector of development of Bangladesh. Thus, they may facilitate two-way traffic between the state and the common people; as we have been observing in our day-to-day affairs, we must bear in mind that the civil society is having both social value and set of institutions and thereby they may develop good governance in a developing country like Bangladesh.

NGOs and Civil Society

We have noticed several realities while discussing the evolution of civil society in the context of Bangladesh. Most significantly, in the countries of the developing world, the evolution of civil society lags behind in state building and economic growth the way it should be. A gap is noticed between the old and new though the lives of common people are being transformed fast. An artificial approach or attempt is being taken to patch up the gap and we are not sure how effective this would be, for the "Non-Government Organization" (NGOs) model borrowed from the industrially developed countries of the west is an attempt or initiative to gear up the growth of the civil society. However, this approach may not work in reality. Now, the donors are willing to underwrite the direct provision of services to the poor in Bangladesh. Since many of the NGOs are running after earning money by making investments in various projects/schemes, their ethics that are the "non-profit" organizations and the donors are providing them fund to uplift the standard of the life style of the common people. Even, they do not hesitate to realize money from the common people by applying inhuman and cruel methods. But this is not a conclusive one, we are still optimistic about the role of NGOs in bringing up the standard of life of the distressed common people through alleviation of poverty from the society with that aim in view, a large number of registered NGOs have been functioning all over Bangladesh, we believe and are optimistic about their future.

Government-NGOs Relationship

We must say that the government's attitude towards the NGOs manifested in its five-year plans and other government records has been very positive. The government acknowledges that the NGOs have been playing a very innovative role in uplifting the standard of life of the distressed common people (Haque 2002).

Conclusion

Since the independence of Bangladesh in 1971, it was widely believed that once the task of reconstruction was over, the domestic economy would provide most of the resources for development. However, this concept is now proved a mistake because drainage of the resources of Bangladesh during the British and Pakistani regime had left us with a deficit of food-grains availability. Inadequate internal savings and a major portion of the population living below poverty line led to a state of chronic eternal dependence.

But in the last decade, Bangladesh has made a big turn, under the leadership of Honorable Premiere Sheikh Hasina, to move forward in achieving the status of a middle income country and gradually the government mechanism is also getting more mature in that line of order. To gain proper governance or good governance, media and NGOs are contributing a lot.

The role of media has risen over the time and it would gain more importance in the times to come as many across the world still yearn for better governance or at best governance itself and not autocratic rule. While media does highlight some of the concerns from time to time, it does not delve deeper into the real issues. The checks and balances which media ensures by reporting issues in an objective manner can go a long way in ensuring that governance by government would be fair and fruitful.

There is no denial of the fact that in spite of some management and administrative lapses, NGOs have been playing a very significant role to alleviate poverty in Bangladesh by extending assistance and cooperation for the sake of good governance and we are optimistic about it.

References

Aminuzzaman, S. M. (2010). Lessons Learned and Challenges of Improving Service Delivery of Union Parishad. In *Local Governance Support Project–Learning and Innovation Component (LGSP–LIC)* (pp. 1–112). LGRD–GOB.

Anam, M. (2002). The Media and Development in Bangladesh. In *The Right to Tell: The Role of Mass Media in Economic Development* (World Bank Institute Development Series). World Bank Institute: Washington, DC.

Chowdhury, A. (2005). *Critical Overview of the Growth of the Media in Bangladesh*. Paper Presented at the Launch of 'Governance and Social Trends Project, Media in Bangladesh: Sectoral Evolution and Actor Profiles', Power and Participation Research Centre (PPRC), Dhaka, 22 December.

Haq, S. M. (2003, December 5). Media Is Critical to Good Governance in Bangladesh. *Daily Star*, Op-Ed, Dhaka.

Haque, M. S. (2002). The Changing Balance of Power Between the Government and NGOs in Bangladesh. *International Political Science Review, 23*(4), 411–435.

Haque, S. (2005, June 11). Media, Governance and People's Rights. *New Age*, Post Editorial, Dhaka.

Iftekharuzzaman. (2011). *Implementing Right to Information in Bangladesh: Opportunities and Challenges*. Dhaka: Transparency International.

Manik, M., & Zaman, A. (2011). Union Information & Service Centre (UISC): ICT Enabled One-Stop Service Outlet in Bangladesh. Retrieved April 26, 2014, from http://community.telecentre.org/profiles/blogs/union-information-amp-service-centre-uisc-ict-enabled-one-stop?xg_source=activity.

Ministry of Information. (2006). *Newspaper Registration and Circulation Figures*. Dhaka: Audit Bureau of Circulation.

Swiss Development Cooperation (SDC). (2013). *Supporting Local Governments and Citizens Practice Pro-Poor Good Local Governance: A Handbook Based on the Experience of the SDC Local Governance Programme Sharique in Bangladesh*. SDC: Dhaka. Retrieved April 20, 2014, from http://www.swiss-cooperation.admin.ch/bangladesh/en/Home/Local_Governance/SHARIQUE.

UNDP. (2005). *Human Development Report: International Cooperation at a Crossroads*. New York: UNDP.

World Bank. (2006a). *Bangladesh Country Assistance Strategy (CAS) 2006–2009*. Washington, DC: World Bank.

World Bank. (2006b). *World Development Report: Equity and Development*. Washington, DC: The World Bank and Oxford University Press.

6

Role of Civil Society in Democratic Consolidation Process in Bangladesh

Mostafijur Rahman

Introduction

In Bangladesh, there are many CSOs, and these are involved in poverty alleviation and in the achievement of MDGs by delivering their services to the poor of the country and by working to improve the country's infrastructure and to solve common problems. In the present context of Bangladesh, CSOs are being recognized as effective actors to the donor agencies and at the same time, to both the central and the local government of the country as an implementing partner. But on the contrary, civil society may be a frustration in terms of private and public interests. For example, as election is one of the main components of democratic consolidation in the modern arena, a question may be raised to what extent the elections of Bangladesh are free, fair, and competitive and what role civil society in Bangladesh plays during the national or local elections to achieve a successful democratic consolidation by acting as a pressure

M. Rahman (✉)
Department of Law, Prime University, Dhaka, Bangladesh

© The Author(s) 2020
Md. N. Momen et al. (eds.), *Building Sustainable Communities*,
https://doi.org/10.1007/978-981-15-2393-9_6

group for free, fair, and competitive election. This is why the role of CSOs of Bangladesh in creating democratic consolidation can be measured on the scale of minimal normative standard using some procedural indicators. Bangladesh has become a dynamic developing country in the contemporary globe, and her economy is very resilient with GDP growth rate at an average of 7% over the last years and the poverty rate declining to 31.5%, although Bangladesh is now the seventh largest country in the world with 160 million people. Bangladesh has accepted a parliamentary form of government for its democratic progress after the independence in 1971 and has entered into the process of democratic transition in 1990. During the 1990s, Bangladesh slowly freed itself from the clutches of military rule and the people began to struggle for a new journey in search of democracy. Despite regular parliamentary elections and transfer of powers among political parties from 1990 and onwards, the country continues to face obstacles on the road to democratic consolidation. Democracy is a continuous political process which can be sustained and developed by giving due attention through its responsible citizens. From this point of view, it is being said nowadays in academic texts, studies, seminars, and politics or in every field of the country that civil society can play a role in democratic consolidation. But concern is that, although "civil society in Bangladesh is recognized for its vibrant performance in social development, it is often criticized for its inability to influence good governance and democracy" (Tasnim 2017). Therefore, this chapter assesses the role of the CSOs in the democratic consolidation process in Bangladesh.

Objectives of the Study

The main objective of the study is to examine how civil societies can contribute to the consolidation of democracy and why they cannot play their desired roles in the consolidation of democracy in Bangladesh. Other objectives of the chapter are to discuss some separate but interrelated points and these are the conceptual frameworks of civil society and democratic consolidation and their interconnection, the approach by which civil society plays a positive or negative role in safeguarding democratic consolidation in Bangladesh.

Research Methodology

The study is descriptive and suggestive in nature and mainly based on secondary data and those data have been collected from concerned books, journals, periodicals, websites, daily newspapers, published research work, and other materials related to the research topic.

Definition of Key Concept

Democracy

The concept of democracy has been rooted in Greece which consists of two words: *demos* (the people) and *kratos* (power). In this sense, the term democracy indicates 'power of the people.' US president Abraham Lincoln (1809–1865) defined democracy as: "Government of the people, by the people, for the people"—which demands a high degree of consciousness of the people for its success. According to Gilchrist (1962), "Democracy is the government of the people as distinct from the government of an individual or of a class by giving them a voice in legislation, it educates and enables the individual citizen; it gives each a sense of responsibility which gives a new meaning to his personality." Development and democracy are inter-dependent. In this sense, democracy may be of different forms such as: direct, liberal, industrial, proletarian, representative, social, bourgeois and so on. Direct democracy is established when the people themselves directly express their will on public affairs. Though direct democracy was established in ancient Greek, at present it is impracticable; the reason is, large and complex societies have emerged and the area of the state is very extensive. On the other hand, according to John Stuart Mill "Indirect or representative democracy is one in which the whole people or some numerous portion of them exercise the governing power through deputies periodically elected by themselves." David E. Apter (1981, p. 141) points out some basic principles of democracy and these are power, justice, law, representation, equality, liberty, rights, authority, order, and power. He opines that, justice provides power to law; law provides representation to the equality of access to all citizens. This series of power is compulsory for the better democratic process in any country as democracy gives liberty, liberty gives right to the citizen, and then the state provides order to show its commanding power. If this chain of power is weak in any country, the democratic process also becomes weak in that country. So, it is clear that, democracy has several connotations and there are several features of it.

Democratic Consolidation

Robert Dahl (Dahl 1971, p. 3) in his seminal book *Polyarcyh* sets out the eight criteria as the ideal type democracy and these are: "(1) right to vote, (2) right to be elected, (3) right of political leaders to compete for support, (4) free and fair elections, (5) freedom of expression, (6) alternative sources of information, (7) freedom of association, and (8) public policy institutions depending on votes and expressions of preference" (cited in Gustafsson n.d.). Taking Dahl's theory into consideration, Linz and Stepan (1996) provide the five major arenas of a modern consolidated democracy—civil society, political society, rule of law, state apparatus, and economic society. Democratic consolidation is a matter of ensuring rule of law, human rights, fundamental freedoms, and promotion of political participations within governmental process, free and fair election, and so on. These are the major areas of democratic consolidation. Since democracy is a continuous political process, it can only be sustained and developed by giving due attention to the above-mentioned issues and these issues cannot be ensured relying only on political parties or governmental institutions but also can be ensured by the active performance of CSOs.

Civil Society

The term 'civil society' originated from the Ancient Greece and was used in a Western sense for the first time. It then passed on to many other countries. Coonrod (2009) says that, "the concept of *societas civilas* is thought to have been first applied by Cicero in Rome to mean a 'good society' ensuring peace and order among the people. At the time, no distinction was made between the state and society; rather, it was believed that the state represented the civil form of society and that 'civility' was the requirement of good citizenship." The term 'civil society', translated into Bengali as *shushil shamaj*, which literally means 'gentle society' (Lewis 2016). Considering these concepts, civil society can be defined as the collection of institutions or groups that are free, voluntary, self-reliant, and self-generating as well. Some examples of CSOs are trade unions; charity organizations; religious organizations; community-based organizations; civic movements or advocacy groups; issue-oriented, community and social groups; professional associations; self-help groups; social and political movements; advocacy groups; non-governmental organizations (NGO); independent mass media; think tanks; private voluntary organizations (PVOs); educational interest groups (universities); and social and religious groups. Some of the CSOs are actively engaged in business and some play a pro-active role towards the government.

Understanding the Theoretical Framework of Democratic Consolidation

The concept of democratic consolidation and civil society has stimulated the scholars to debate around various interesting questions. For this very reason, in recent years, a number of models have been proposed. In this chapter, some theoretical models have been presented to understand the theoretical framework of democratic consolidation and civil society.

Apter's Model of Democracy

Apter (1981, p. 170) provides the precondition of democracy and these are population and their participation in democratic government. He also suggests that democracy results from the form of legislation, because legislation provides the societal basis for stability of order and control. From this point of view, Fig. 6.1 describes the Apter's model of democratic government:

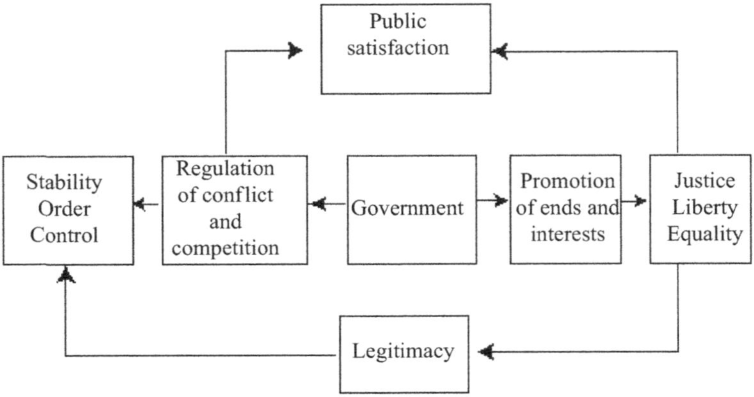

Fig. 6.1 The model of democratic politics

Schedler's Model of Democratic Consolidation

Schedler presents his model based on the different notions of democratic consolidation and these are "avoiding democratic breakdown and avoiding democratic erosion" as negative notion, "completing democracy and deepening democracy" as positive notion and finally "organizing democracy" as neutral notion (Schedler 1997, p. 26). He indicates the first two notions as the original connotation of democracy, the next two notions as the movement towards a high-level of democracy, and the last notion as structural changes of democracy (ibid.).

Merkel's Level of Consolidation

Another concept was formulated by Wolfgang Merkel (1996). This concept includes four different levels of consolidation—one at the macro-level, two at the intermediate-levels and one at the micro-level. As per this theory, democracy is consolidated only if these four levels are consolidated.

Understanding the Theoretical Framework of Civil Society

The functions of CSOs are divided into two broad aspects—vibrant and vigilant. The vibrant civil society ensures that citizens have the means to influence public policies freely. On the other hand, vigilant CSOs have an active oversight role in monitoring and assessing the activities of government. So, CSOs must be both vibrant and vigilant to be able to contribute to democratic consolidation. According to Sievers. B (2009) civil society as a singular social construct has evolved through the seven conceptual streams which reinforce each other in the operation of civil society. The following model places civil society in constant interaction with the other actors in the political system and democratic consolidation (Fig. 6.2).

Apart from the above model, it is clear that democratic consolidation stands on several pillars and these can be developed by the real performance of CSOs. These factors are minimally the necessary indicators that play a vital role for upholding democratic consolidation. According to

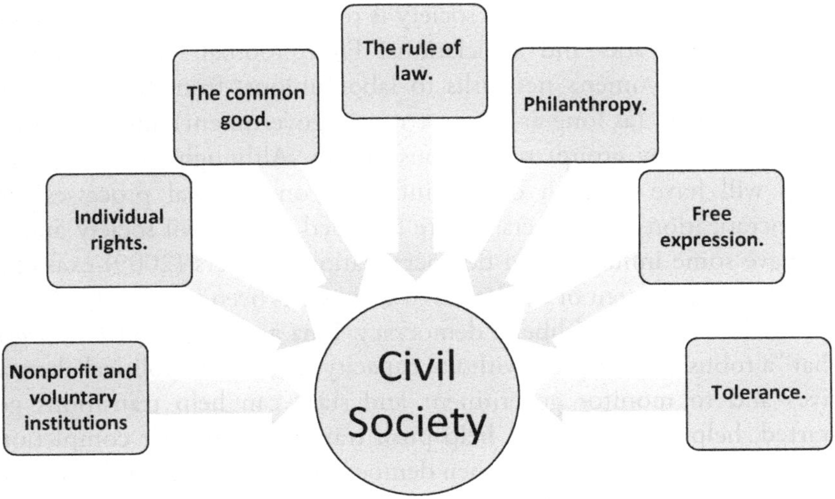

Fig. 6.2 Bruce Sievers' approach of civil society

Ahmed "the conditions for the success of democracy, from that standpoint, are related more too social norms than to political activities and institutional support." Though Bangladesh was founded on the basis of democratic aspirations and democracy (Riaz 2016), despite this, democracy has been hampered since independence in 1971.

Relationship Between Civil Society and Democratic Consolidation

Civil society is necessary to ensure democratic consolidation and the proper functioning of government. Not only that, civil society is an important agent for promoting good governance like transparency, effectiveness, openness, responsiveness, and accountability (Ghaus-Pasha 2005). Where the civil society is strong, the democracy is strong. However, to check the relationship between civil society and democratic consolidation, this can be testified through comparison between democracy in democratic countries (European Countries) and democracy in less democratic countries (Asia, Africa, and Latin America), where a significant upsurge of organized private, non-profit activity are witnessed (ibid.). Enckevort (2010) said, "Civil society is made up by very different organizations, networks and associations. From football clubs to debate groups, from women's networks to labor unions, from book clubs to political parties (as long as they are not in government) and from environmental lobby groups to religious groups. Although some organizations will have a much clearer influence on political processes and democratization than others, all are included in the civil society and all do have some influences on democratization." Sievers (2009) examines that the development of modern civil society has been inextricably linked to the development of liberal democracy. Linz and Stepan (1996) argue that "a robust civil society, with the capacity to generate political alternatives and to monitor government and state can help transitions get started, help resist reversals, help push transitions to their completion, help consolidate, and help deepen democracy. At all stages of the democratization process, a lively and independent civil society is invaluable."

Lionel Bopeg (Bopage 2016) examines that, "civil society has played an extremely active and valued role in fostering civic and political participation with mixed results." For example, he mentions, "Germany's pro-democratic civil society of the 1920s and 1930s failed to safeguard democracy."

Evaluating the Role of CSOs to the Socio-economic Development of Bangladesh

The role of CSOs to Bangladeshi society and the economy are very significant. Many CSOs are involved in poverty alleviation, in improving Bangladesh's infrastructure and common problems. For example, in 1943 when a famine took more than 3 million lives, Ranada Prashad Shaha (R.P. Shaha) built the Kumudini Hospital in 1947 with a view to offer free quality health, free education and employment opportunities, and treatment for rural populations. At the same time, village-based small large cooperative societies were replaced between 1947 and 1970. For instance, the State Bank was established in 1955 that provided agricultural loans and the Bangladesh Academy for Rural Development (BARD) was established in the following year that provided rural development programs supported by major bilateral and multilateral development partners around the globe. In 1970, a typhoon killed more than 500,000 people; in 1971, floods and droughts destroyed on average 1.7 million tons of food crops; this was followed by the independence war that resulted in millions of people's deaths and refugees; in 1974, famine killed another 3 million people; and throughout 1980s, 1.46 million tons of food crops were damaged; however, in response to these phenomena, the civil society began to task with a view to cooperate with the backward section of the country. As a result, after the independence of Bangladesh, the Bangladesh Rural Advancement Committee (BRAC) was founded in 1972 to resettle returning refugees from India. Later, the Association of Development Agencies in Bangladesh (ADAB) to aid in agriculture, the Grameen Bank (GB) to help in economic and social development among the country's poorest, and the Association for Social Advancement (ASA) in 1987 were established. Bangladesh Environmental Lawyers Association (BELA) and Bangladesh Legal Aid and Services Trust (BLAST) (that represent poor people and work to enlarge public interest litigation), Bangladesh Shishu Adhikar (Child Rights) Forum, Bangladesh Women's Health Coalition (BWHC), Transparency International Bangladesh (TIB), Dhaka Ahsania Mission (DAM), Proshika, Thengamara Mohila Sabuj Sangha (TMSS) started their work toward social and economic development of vulnerable groups with a view to the delivery of services in the fields of health, education, environmental conservation, gender equity, trafficking of women and children, good governance, and aid conditionality. Shushashoner Jonno Nagorik, or "Citizens for Good Governance"—known as SHUJAN has become a significant actor in shaping the activities of civil society in Bangladesh. In the health sector, establishing of GK *(Gono Shahshtha Hendra)* as a people's health movement was particularly significant. Transparency International Bangladesh (TIB) highlights corruption in various sectors of society, the inherent strengths, and weaknesses in the functioning of the EC and Parliament. Thus, the civic tradition was reinforced to undertake relief activities in Bangladesh after her independence in 1971 from Pakistan when she got the opportunity to develop socio-political institutions and then in 1990, as the main channel of international aid and delivery of basic services, particularly education, health, and micro-credit when the parliamentary form of government has established. Beyond service delivery, NGOs have also played a role in influencing policy in some key areas (Lewis 2016). For examples, according to Lewis, 'TIB' played a role in helping to pass and implement the Right to Information Act, 2009, 'Action Aid Bangladesh' in challenging the top-down approaches to climate change adaptation, 'Bangladesh Center for Workers Solidarity' in supporting labor rights for garment workers, 'Young Power in Social Action' for laborers in the ship-breaking industry, and 'Flood Action Plan' (FAP), in solution to water control management. He further states that, due to the active role of some NGOs, a new land reform law was passed that required new *khas* land to be redistributed to landless households. Consequently, access to *khas* land by the landless has been ensured. Moreover, many important international NGOs such "Action Aid Bangladesh, Care, Caritas Bangladesh, Concern Bangladesh, Concern Universal, Damien Foundation, Handicap International, Helen Keller International, Leprosy Mission International, Pathfinder International, Plan International Bangladesh, Practical Action, Save the Children USA, Save the Children UK, Save the Children, Sight Savers Royal Commonwealth Society for the Blind, Swedish Free Mission, Terre Des Hommes" (ADB 2008) perform their functions in Bangladesh. Thus, the role of NGOs in Bangladesh society and its economy can be marked as being significant. This must be evaluated positively for the citizens of Bangladesh and its democracy.

Measuring the Role of CSOs in Democratic Consolidation in Bangladesh

There are many factors that can be considered as main aspects of ensuring democratic consolidation, and in these factors, CSOs have a specific role to establish democratic consolidation. According to Linz and Stepan (1996), for democratic consolidation political parties, legislatures, elections, electoral rules, political leadership, interparty democracy, rule of law are, minimally, the necessary yardstick for judging democratic consolidation. Thus, democracy has several connotations and there are several features of it. It is really difficult to find a country in which all features of democracy exist. In case of middle-income countries like Bangladesh, the situation is worse rather than working with all features. This is why, it is important to mention here to identify one or more important features to be representative of democracy in the context of Bangladesh in which CSOs can play roles in different ways to consolidation. So, in this chapter, the focus is on some crucial issues which can be sustained and developed by the active performance of civil society with the governmental institutions and political parties towards consolidated democracy and then it has been identified whether CSOs are able to play their expected role in democratic consolidation in Bangladesh.

Political Participation and Electoral Procedure

In a modern democratic government system, the first and foremost feature of a better democracy is the holding of contestable elections held by the regimes "that are free; that is, contested by meaningful opposition, and fair; that is, not rigged or biased from the process of voting through counting and the declaration of results" (Alamgir 2007). A mature democracy should have an electoral process that ensures free and fair elections. According to John Coonrod (2009), "civil society can play important roles in this regard by providing a *supply* of ideas for electoral and institutional reforms and, at the same time, by advocating or acting as a pressure group for the adoption of such reforms." He further added that, civil society "can create a *demand* for reform by mobilizing public opinion," "collect information about candidates' backgrounds and *supply* the relevant information," "play a *watchdog* role to ensure that everyone concerned adhere to the prevailing laws and thus perform their appropriate responsibilities" (ibid.). In Bangladesh, national elections are held every five years and Bangladesh has been trying to develop democratic consolidation, but the question is whether democratic norms and values are upheld in the election years or whether the general populations can participate in governmental decision making. According to Emajuudin Ahmed, "Elections are held, though not regularly in all cases, and quite often these are rigged. The political leaders, instead of competing for people's votes, try either to purchase them with their black money or cajole them to vote for them through questionable means. In some cases, they use their muscle power, often with hired goons, to force the recalcitrant voters either to abstain from voting or to vote for them, for getting them elected. They take election as the veritable gateway to political power and win they must, by means fair or foul." For example, during the Khaleda regime, a by-election in Magura in 1996 was rigged by the ruling party, (Zakaria 2013) and similarly, the present Hasina government won the 2014 election handily where a majority of the parliament's 300 seats were uncontested (Landry 2016). The elections on 5 January 2014 were boycotted by the major opposition and took place in a climate of unrest; they were not recognized by the international community as having been free, fair, and transparent (Coonrod 2009). Without free and fair elections, the government cannot be held accountable. On the contrary, just holding elections alone do not necessarily ensure an accountable democracy and rule of law. This has been truly proved during many government regimes. Although some of them come into power by free and fair elections, after coming to power they become dictators and appeared to have manipulated many institutions with the judiciary of state. For example, the international community has always put pressure on Bangladesh's politicians to reign in human rights violations committed by government agents. In this milieu, "it is necessary that even when democracy is established, civil society action is needed for keeping its quality intact..." (Bopage 2016). Most probably, for this very reason, free and fair elections require an informed citizenry whose participation in politics or in an election procedure at the national or international level can enhance democratic qualities. It is well known to all that, a participatory civil society is necessary to achieve democratic consolidation. A participatory civil society has three main functions such as: "promoting voluntarism, building social capital, and creating an enabling environment for policy input" (Hyden et al. 2003). In case of voluntarism, the performance of the civil society can be found that CSOs are involved in some types of "collective action coming about as a product of individuals combining their private interests in pursuit of a common good" (Olson 1965 cited in Hyden et al. 2003). Moreover, the role of CSOs can be found also in civic education, advocacy, service in exercising legal and political rights and raising social and political awareness at the grass-roots level. Raising political awareness at the grassroots projects during the 1996 and 2001 elections in Bangladesh gave the opportunity to influence a large section of the voters' decision regarding whom to vote for.

Protection of Fundamental Human Rights and Rule of Law

Rule of law is essential in the advancement and preservation of citizens' rights of a modern democratic country. The rule of law means from the highest in the country to the lowest, all must submit to law and law alone. On the other hand, human rights are those basic rights which are enjoyed by every human being from birth. Fundamental human rights or rule of law can be yardsticks for judging the reality of democracy on the grounds that these are the major criterion of democratic consolidation. Since "modern democratic government is a party government. The party winning majority in the election forms the government. But coming into power, the government may turn itself into a dictatorial one violating the basic rights of people and oppressing the opposition. The aim of having a declaration of fundamental rights in the constitution is to prevent such a possible danger" (Halim 1998, p. 87). However, there are various proxy measures for how rights are protected under a democratic government but the minimal indicator must concern the first and foremost fundamental rights of a man—'*right to life*'. Considering this observation, Mahmudul Islam (Islam 2012) said that "no right can compare with the right to life without which all other rights are meaningless and the rule of law can play its most significant role in this respect." This is particularly a powerful indicator because of three factors, as researcher Jalal Alamgir (2007) stated, "firstly it concerns the right to life. Secondly, it points to violation by those (i.e., an elected govt.) entrusted to protect it. Thirdly, it also provides a measure for the violation of due legal procedure." T*he rule of law and democracy are interlinked.* The rule of law is a critical factor for the advancement of democracy for the protection of human rights of all people and accountability. The advancement, protection, and preservation of the rule of law and human rights are always left in the hands of the organs of the state. Despite the state organs, the CSOs also perform the delicate task of ensuring these. The reason is, CSOs are regarded to be 'watch-dogs' and struggle to ensure the adherence to human rights and rule of law. The violation of human rights and rule of law can be blamed by CSOs, as much as possible, on the grounds that the state organs have failed to protect these rights of the people. CSOs are to focus on particular issues which destabilize the society and to aid the government and its institutions to find out the ways to overcome the situation. Civil Society can create public awareness conducting debates, seminars, and dialogue between people from various backgrounds and government agents for the implementation of the rule of law and human rights and can pressure the government to remove the defects constantly that remain on the path of these. The CSOs can emphasize human rights education and awareness program among the people about the impact of human rights violation. Furthermore, CSOs can utter their voice and work for safeguarding the rights and interests of the general people and creating consciousness among the people about the high consciousness of democracy and the bad impact of human rights violation on democracy. This is why, it is important to mention here to identify respect and protection of human rights and rule of law to be representative of democracy in the context of Bangladesh in which CSOs can play roles in different ways in safeguarding these crucial issues to consolidate democracy. The following scenario appears that, rule of law for the common men in the country exist only in the pages of the constitution (Islam 2012):

The figure shows that average government agents killed almost 41 people extra judicially per year from 1993–1996, 18 from 1996–2001, 207 from 2001–2006, 177 from 2006–2008, and 134 from 2008–2014. In this regard, CSOs of Bangladesh can work towards the achievement of a peaceful society promoting social cohesion, providing services to all and taking active positions on several matters like human rights or rule of law in which all the people whatever their age, race, sex, religion, and so on will have potential human rights.

Challenges of Civil Society

Democracy by CSOs in Bangladesh has not consolidated in real sense. According to some experts, the major reasons are most of the CSOs are interested in service delivery than in playing a part in advocacy, some of them are mainly from western countries, institutionalized, biased, growing process, not national, party supported (conditional), acting as pressure groups, and coming from an elite group. Moreover, "many of the mainstream service delivery NGOs were open to the criticism that rather than contributing to sustainable improvements, they merely let the government off the hook" (Lewis 2016). According to John Coonrod (2009), the main functions of civil society are "to inform the people of their rights, to help them to achieve those rights, and to fight for clean politics and good governance." Bangladesh has no such type of civil society. More organizations take donor support; there are only a few non-partisan groups in Bangladesh that played a significant role in the recent democratic transition. Civil society groups face some challenges in the democratic consolidation process in Bangladesh as follows:

Increasing Number of NGOs

Researcher Zakaria (2013) observes that, since the 1990s, with change in donor policy to assist NGOs, the number of CSOs in Bangladesh increased exponentially. According to him, such a huge number of CSOs coupled with thriving NGO sector gives Bangladeshi civil society a vibrant look. Shakil opined that, there is a huge number of CSOs in Bangladesh, but they have their own weaknesses in democratizing state and society.

Coming from Elite Groups

Bangladesh civil society consists of both elite and non-elite civil society groups, and most of the leaders of civil society are a group of scholars and they are from the elite section who worked as think tanks in citizens groups. For example, as Zakaria (2013) observes, "the national apex association like Federation of Bangladesh Chambers of Commerce and Industries (FBCCI) of the businesspersons receive important attention from the government and political parties."

Foreign Donation

Donations and foreign funds are the key sources of survival for some NGOs, especially, the development NGOs. But "due to their excessive reliance on foreign *funds,* the development of NGOs may focus more on fulfilling the requirements of their major donors than on grassroots democratization, development and empowerment of vulnerable groups" (Quadir 1999, p. 4) on the one hand and on the other, in recent times, foreign aid in Bangladesh has been decreased considerably. Numerous foreign donation recipient-NGOs fall outside the donation privileges and manage by their own funding. If we compare the total landscape of CSOs, the coverage of foreign donations is very limited. In Bangladesh, some "large-size NGOs rely less on foreign donations and are oriented more to market activities. This trend is more linked to a global phenomenon where the GDP of developed countries has shrunk in comparison to that of developing countries" (Ahmed n.d).

Anti-democratic Elements in Civil Society

Harry Blair (2000) observed that, there is also the question of anti-democratic elements in civil society. Various Islamic groups in Bangladesh are part of civil society, just as are human rights and women's advocacy groups. But a democratic state is not required to tolerate organizations trying to subvert it in the name of free speech, nor are international donors wishing to support civil society obliged to assist each and every CSO, including those that want to destroy the state itself.

Less Powerful

Most of the civil societies in Bangladesh are less powerful and their positions are lower than politicians or donors. The reason as identified by Zakaria (2013), emerging as dominant actors, many leaders of civil society groups (such as leaders of trade union) are being influenced by the political elites in Bangladesh. Moreover, considering White's observation, he further mentions that, "the relationship between larger and smaller NGOs is more like particularistic patron-client ties" (ibid.).

Governance Problems in NGO Sector

TIB (2007) finds out some problems on governance in NGOs in its study and these are existence of the relatives of the executive head in the governing body, lack of regular meeting of the governing body, unilateral decision making process by the executive heads ignoring opinions of staffs, lack of financial transparency, lack of accountability to the beneficiaries on financial matters, lack of necessary information about financial details, and so on.

Corruption

There are also corruption allegations against the NGOs in Bangladesh. Tasnim (2007, p. 108) reveals that "the percentage of CSOs contacting the ruling party and opposition party is 64% and 51% respectively and there is an existence of a vertical give and take relationship between the political parties and different CSOs" (cited in Zakaria 2013, p. 25). TIB (2007) also shows corruption allegations in the recruitment and promotion of employees and consultants in NGO sectors, as recruitment and promotion are often influenced by the politicians, donor, and government officials on the basis of nepotism or unilateral choice of the executive head or personal relationship.

Lack of Participation in Advocacy

Tasnim (2007) points out that, the CSO's participation in advocacy, lobby, and policy making with the government is lacking, although they are involved in social service and welfare. Holloway (1993) observed that, a few CSOs lobby or advocate for social change, try to modify government policies, see their role as changing behavior or changing government's policy separately from specific project activities in Bangladesh.

Politicization

A civil society can be termed as politicized when it is politically colored or when it acts as a politically motivated group or organization directed by party leaders or government executives leaving its main aims and interest on which it was formed. In Bangladesh, according to Ahmed (n.d), the depoliticized nature of NGOs in their working relations weakens their position in influencing the government in decision making processes. He reveals that, "labor unions, NGOs, chamber federations and religious civil society are highly affected by politicization. Their autonomy is compromised by this politicized nature while on the other hand many leaders of CSOs participate in party politics" (ibid.). Besides, Zakaria (2013) reveals that, in Bangladesh, "confrontational politics played by the two major political parties through the instruments of clientelism, patronage, nepotism corruption, violence have co-opted, politicized, weakened and polarized the civil society groups and undermined their ability to participate in political process which contribute to governance and democracy"(Zakaria 2013). Such polarization has been used for political expedience that affected the growth of civil society too (ibid.).

Interference on CSO

Tasnim (2007, 2017) points out that civil society is often found to be influenced, used, penetrated, polarized, co-opted, politicized, controlled, divided, and generated by political parties. According to Harry Blair (2000), most of the Bangladesh's higher-profile NGOs choose to concentrate themselves on political agendas rather than on their developmental works. Quadir (2003) argues that "civil society, instead of forming bonds and bridging social capital among different groups, helps sharpen existing political divisions" (cited in Tasnim 2017). Owing to this easy politicization, "the civil society is yet to emerge as a vigilant force to uphold good governance and consolidate democracy" (cited in Zakaria 2013, p. 25).

Other Challenges Revealed by TIB

Transparency International Bangladesh (TIB) has in its study titled "Problems in Good Governance in the NGO Sector: The Way Out" revealed that a large number of NGOs in Bangladesh are associated with mid-level corruption, mis-governance, and absence of regulatory control. The TIB study revealed some challenges in the NGO sector in Bangladesh, and these are "(a) lack of financial sustainability; (b) shortage of efficient employees and high employee attrition; (c) inadequate infrastructure; (d) undue interference and control by the government; (e) lengthy fund release process; (f) low level of inter-sectoral cooperation; (g) inadequate training and low level of true professionalism among employees often aggravated by lack of job security; (h) lack of information and relevant research; (i) religious conservatism and militancy, and threat of terrorism; (j) political pressure and political instability; k) Unfavorable tax regime; and (l) natural calamities" (TIB 2007).

Facts and Findings: In-depth Analysis

The study has been conducted to understand the concept of the democratic consolidation in general and particularly to focus on the role of civil society in that process in Bangladesh. The study shows the inter-link between civil society, development, and democracy in the context of Bangladesh. Bangladesh has acquired a world reputation for its vibrant and robust civil society (Watkins 2015). Today, Bangladesh's civil society spans the world, sharing indigenous expertise and experience, and working together towards a better society for all. As Bangladesh enters middle-income status, it is fitting to recall the best of the country's traditions (ibid.). If we evaluate the role of civil society in Bangladesh, we must see that Bangladesh civil society is playing a pivotal role in the development process by eradicating poverty and economic progress. This development is the precondition of democracy in a country and civil society as key actor of development is playing for consolidated democracy. The study clears that, since its independence from Pakistan in 1971, Bangladesh has developed numerous civil society organizations which are considered as the key actors in development process, and by these efforts, Bangladesh has been engaged in its path to democratic consolidation. This study also revealed some crucial issues—political participation, election procedure, protection of human rights, and rule of law that can be considered as the main aspects of ensuring democratic consolidation and in which CSOs have a specific role to perform with the governmental institutions and political parties. The study also identified some major challenges for which the civil society cannot play in democratic contribution in real sense.

Conclusion and Way Forward: In Search of True Role of Civil Society in Bangladesh

This chapter sketched a picture of democratic consolidation in Bangladesh through the theoretical framework of democracy and civil society and revealed the significance of civil society's involvement in the democratic consolidation process in Bangladesh. Democracy has several connotations and there are several features of it. Among these, there are some more important features to be representative of democracy in the context of Bangladesh in which CSOs can play roles in different ways, to consolidation. Besides, this chapter depicted that the civil society is essential in terms of democratic contribution because it can deal with many of the underlying drivers of consolidation by promoting economic development, alleviating poverty, fighting against corruption, advocating policy change, nurturing democratic values, bringing unity among social cleavages, contributing to good governance, and thus, democratic consolidation process. If CSOs face any challenge to the path of consolidation, they cannot work freely. So, the basic roles of CSOs in democratic consolidation process in the context of Bangladesh are very important as this chapter revealed that the main functions of CSOs are to create public awareness, to raise public concern, to control the state power, to promote political participation, to protect human rights, and to respect rule of law and so on. But the concern is CSOs of Bangladesh face some internal and external challenges in promoting democracy. The general finding of the article claims that the nature of Bangladeshi democracy is confrontational and at the same time, the vigilant nature of civil society is less participatory, politicized, and polarized. For both of these grave conditions, the main influencing factors are political parties and political leaders. The link with the political parties has become the main source of power for the civil society groups. Since almost all time, the political parties penetrate and control the civil society and since most of the civil society organizations are western-based and politicized, they can hardly act independently. And, as a result, the common desires of people have remained unfulfilled. Under the circumstances, democracy could be consolidated fully only when CSOs could do work shoulder-to-shoulder with all the segments of the government overcoming all the challenges in its path towards democracy. In other words, where political parties and CSOs are strong, diverse citizen interests are represented, consensus is built, and economies thrive. In order to improve the role and possible contribution of CSOs in the consolidation of democracy in Bangladesh, the following recommendations could be considered:

1. The major problems that hinder civil society in creating and sustaining stable democracy should be addressed and diminished soon accordingly. The partnership between the state and the media should be increased to address the problems. In Bangladesh, CSOs are still regarded as an unrecognized sector. So ways should be found to mobilize them so that they can do as real actors of democracy. Reforms should be adopted in the education system also for encouraging NGO activities. The CSOs should be allowed to play their actual role in development and democracy.
2. In the Bangladesh context, the CSOs are politicized in line with party politics. Our politicians should pay due respect to civil society organizations. Both the government and opposition should work for the betterment of civil society instead of using them politically. Our civil society should be made positive in playing their role in the society. The economic gap should be reduced, economic prosperity and social justice should be ensured for improving the role of civil society.
3. Socio-economic progress along with political stability is necessary to improve the activities of civil society. The country is facing increasing political and social instability and corruption which are threats for congenial political atmosphere to work independently. Corruption is the main problem in Bangladesh for implementation of any development program. Corruption also has engrained in our civil society. This situation should be redeemed by political reform. Farsighted, tolerant, talented, and dedicated leaders should come forward in politics to reform the overall bad situation in the country
4. Reform within the NGO sector and civil society groups in Bangladesh is much needed. As TIB revealed, a large number of NGOs in Bangladesh are associated with mid-level corruption, mis-governance, and absence of regulatory control work under many constraints and

challenges. So, these constraints should be diminished by improving donor coherence, internal accountability and so on.

5. Most of the CSOs depend on foreign funds instead creating domestic funds. The CSOs need to learn more about fund raising techniques and donor creation in domestic affairs. Government funds could be made also for NGOs and CSOs as government has given funds to some of the government organizations. Unconditional foreign funds could be created more by setting up more organizations like Palli Karma-Sahayak Foundation (PKSF)and others.
6. Another problem is that the CSOs are not specified what activities they are allowed to do and what activities are considered by them to be political. The government should specify their activities and define what activities are allowed to organizations of civil society. The role of civil society should not be limited in local and national activities; it should be promoted to the global activities also.

References

ADB. (2008). *Overview of NGO and Civil Society: Bangladesh,* Civil Society Briefs. Retrieved from https://www.adb.org/publications/overview-ngos-and-civil-society-bangladesh.

Ahmed, S. (n.d.). Civil Society of Bangladesh: Depoliticized in Working Agenda but Politicized in Power Relation, Graduate School of Humanities and Social Sciences, University of Tsukuba. Retrieved October 16, 2017, from https://journal.hass.tsukuba.ac.jp/interfaculty/article/view/16/48.

Alamgir, J. (2007). Democratic Performance in Bangladesh 1991–2006: A Political Measurement. *Journal of Bangladesh Studies, 9*(2), 1–12.

Apter, D. E. (1981). *Introduction to Political Analysis*. New Delhi: Prentice Hall.

Blair, H. (2000). Civil Society, Democratic Development and International Donors in Bangladesh. Retrieved from https://cpb-us-w2.wpmucdn.com/campuspress.yale.edu/dist/b/405/files/2017/08/Bdesh07-HBlair-BdeshCivSoc-RJahan2000-208lmud.pdf.

Bopage, L. (2016). Democracy & the Critical Role of Civil Society: The Lankan Experience. Retrieved from https://www.colombotelegraph.com/index.php/democracy-the-critical-role-of-civil-society-the-lankan-experience.

Coonrod, J. (2009, October 9). Role of Civil Society in Bangladesh's Democratic Transition. *Global Advocacy*. Retrieved from https://advocacy.thp.org/2009/10/09/role-of-civil-society.

Dahl, R. (1971). *Polyarchy: Participation and Opposition*. New Haven: Yale University Press.

Enckevort, E. (2010, February 25). *The Role of Civil Society in Democratising Authoritarian Regimes The Case of Burma (Myanmar)*. Radboud University Nijmegen, p. 15.

Ghaus-Pasha, A. (2005). *Role of Civil Society Organizations in Governance*. 6th Global Forum on Reinventing Government Towards Participatory and Transparent Governance 24–27 May 2005, Seoul, Republic of Korea.

Gilchrist, R. N. (1962). *The Principles of Political Science* (p. 253). Calcatta: Binani Printers Pvt. Ltd.

Gustafsson, M. (n.d.). *Theories of Democratic Consolidation: A Mexico-Germany Comparison*. University of Edinburgh.

Halim, M. A. (1998). *Constitution, Constitutional Law and Politics: Bangladesh Perspective* (M. Yusuf Ali Khan, Ed., p. 352). Dhaka: Rico Printers.

Holloway, R. (1993). *The Nature of Civil Society in Bangladesh, Civil Society-the Non-profit Private Sector: Trying to Categorize in Bangladesh*. Private Rural Initiative Program (PRIP) and UNICEF, Bangladesh, June, 1993.

Hyden, G., Court, J., & Mease, K. (2003). *Civil Society and Governance in 16 Developing Countries*. World Governance Survey Discussion Paper 4 July 2003. Retrieved from https://www.odi.org/sites/odi.org.uk/files/odi-assets/publications-opinion-files/4098.pdf.

Islam, M. (2012). *Constitutional law of Bangladesh* (3rd ed., p. 84). Dhaka: Mullick Brothers.

Landry, D. G. (2016, July 2). The Downfall of Democracy in Bangladesh. *The Diplomat*. Retrieved November 18, 2017, from https://thediplomat.com/2016/07/the-downfall-of-democracy-in-bangladesh/.

Lewis, D. (2016, January). *Non-governmental Organizations and Civil Society in Bangladesh*. The London School of Economics and Political Science. Retrieved from https://www.researchgate.net/publication/299570256_Non-governmental_organizations_and_civil_society_in_Bangladesh/citation/download.

Linz, J. J., & Stepan, A. (1996). *Problems of Democratic Transition and Consolidation: Southern Europe, South America, and Post-Communist Europe* (pp. 7–15). Baltimore: Johns Hopkins University Press.

Merkel, W. (1996). Theorien der Transformation: Die demokratische Konsolidierung postautoritärer Gesellschaften. In C. Offe & K. von Beyme

(Eds.), *Politische Theorien in der Ära der Transformation* (pp. 30–58). Opladen: Westdeutscher Verlag.

Olson, M. (1965). *The Logic of Collective Action*. Cambridge, MA: Harvard University Press.

Quadir, F. (1999). *Democracy, Development and Civil Society in Bangladesh: The Quest for a New Praxis For Sustainability*. Halifax, Nova Scotia: DahouSie University.

Quadir, F. (2003). How Civil is Civil Society? Authoritarian State, Partisan Civil Society, and the Struggle for Democratic Development in Bangladesh. *Canadian Journal of Development Studies, 24*(3), 425–438. https://doi.org/10.1080/02255189.2003.9668930.

Riaz, A. (2016, December 6). What Does Democracy Mean to Bangladeshis? *Daily Star*. Retrieved November 18, 2017, from http://www.thedailystar.net/op-ed/what-does-democracy-mean-bangladeshis-1325527.

Schedler, A. (1997). *Concepts of Democratic Consolidation*. Institute for Advanced Studies, Vienna. This Paper Was Prepared for Delivery at the 1997 Meeting of the Latin American Studies Association (LASA), Continental Plaza Hotel, Guadalajara, Mexico, 17–19 April 1997.

Sievers, B. (2009). What Is Civil Society? *GIA Reader*, Vol. 20, No. 1 (Spring). Retrieved from http://www.giarts.org/article/what-civil-society.

Tasnim, F. (2007). Civil Society in Bangladesh: Rich Grass-Roots Actions but Poor Participation. *Tsukuba University Journal of Law and Political Science, 43*(1), 160–192.

Tasnim, F. (2017). Politicized Civil Society in Bangladesh: Case Study Analyses. *Cosmopolitan Civil Societies: An Interdisciplinary Journal, 9*(1), 98–123. Retrieved November 22, 2017, from http://epress.lib.uts.edu.au/journals/index.php/mcs/article/view/5247/5838. https://doi.org/10.5130/ccs.v9i1.5247.

TIB. (2007, October). *Problems of Governance in the NGO Sector: The Way Out*. Executive Summary. Retrieved from https://www.ti-bangladesh.org/research/ExecSum-NGO-English.pdf.

Watkins, R. (2015). Civil Society as the Key to Development. Retrieved November 28, 2017, from http://www.thedailystar.net/op-ed/politics/civil-society-the-key-development-143335.

Zakaria, S. M. (2013). *Democratic Consolidation in Bangladesh: A Reality Check*. IGS Working Paper Series No. 16/2013, the Institute of Governance Studies, BRAC University, July 2013.

7

Civil Society and Political Change: Globalization and Transforming Relations in South Asia

Anju Lis Kurian and C. Vinodan

Introduction

Over the past decades, civil society has evolved in propelling socio-economic ontogenesis and political mobilization, especially among South Asian countries. This is considered as a result of the much proclaimed neo-liberal interventions which have much-reduced government obligation in political and economic matters. Moreover, these reforms in democracy witnessed in the last decades with intensified emphasis on human rights have fortified the political institutions of people's movements plus organizations with multifaceted ramifications globally (Court et al. 2006). The non-governmental organizations, public drives along with other civil establishments have exerted their connections internationally in the current era of globalization which forms foundations for emerging transnational civil networks (Florini 2000; Mundy and Murphy 2001). In short, civil society is considered as a magnifier to empathize

A. L. Kurian (✉) • C. Vinodan
School of International Relations and Politics, Mahatma Gandhi University, Kottayam, Kerala, India

© The Author(s) 2020

Md. N. Momen et al. (eds.), *Building Sustainable Communities*,
https://doi.org/10.1007/978-981-15-2393-9_7

politics and is positioned as a pivotal component to explicate democratic political transformations and paradigm shifts in development, especially in developing nations (Veltmeyer 2008).

Civil Society

A *Civil society* is represented by small gatherings of public or amalgamation of these groups, for the betterment and benefit of the public, but functioning outside of the governmental and for-profit sectors. Interestingly, people enter into relationships of civil society by appeal rather than compulsion or force which is the cornerstone of differences between civil society and state. Civil society is neither organized for power, which is the outstanding feature of the state, not for profit, which is the pivotal attribute of a business organization/establishment (Kumar 1993; Chandhoke 1995).

Generally, civil society has three main gears:

1. representing a multitude of independent political, economic, intellectual, and religious institutions that are identifiable from the family, clan, locality, or state
2. multifaceted relationships with formal and informal rules and procedures and practices for safeguarding the division between civil society and state, but keeping an efficient nexus among both
3. A plethora of fine-tuned civil manners.

In a democratic perspective, civil society is essential for its fulfillment, as it assists as a parcel in the involvement of the attributes of political society, like involvement in political affiliations, election, and associated engagements which are essential for the health and sustenance of democratic governance.

Though the terminology civil society has Western lineage, the term has entered the South Asian political arena which is well illustrated by leaders in government and political society on various occasions. Moreover, there exist broad differences in excogitation of civil society and mentality and response of governance structures and political leadership to civil society movements across South Asia. The South Asian civil society is a robust and enthusiastic group of people from every walk of life who have contributed to the formation of an effective and efficient flexible alliance or movements which have accented on the importance of promoting people-to-people contacts at all levels beyond borders.

The unprecedented rapidity and extent of civil society's growth in the past few decades in the international scene have resulted in the establishment and conglomeration of "venue" or "forum" which coordinates the global civil society voices to enhance potency, accountability, inclusion, and credibility in international governance. The significance of civil society as a political empowerment structure gained global protuberance since 1980s, coincident with the fundamental global geopolitical and economic changes and numerous transitions all over the world which paves the way to call 'Civil Society' as the 'Third Wave of democracy' by Samuel Huntington (1991).

Civil Society and Globalization

The discourse on globalization and its vibrations on civil society groups are much debatable in the current era of policy issues in the new world order. The diverse nature and dynamics of civil society's relations with the state, as well as collaboration with other non-government/non-profit actors, is in itself suggestive of the depth of strategies and options in alliance with globalization. The role of civil society groups in developing and formulating various policies at the national and international arena is commendable (Tschentscher 2016; CIVICUS 2016). The renowned Nobel Peace Prize winner in 1984, Archbishop Desmond Tutu once said, "*The only way to eat an elephant is piece by piece*". It should be noted that the elephant can only be eaten by a highly organized army of ants, which devours the limbs first and gradually works its way to the head. This makes crystal clear that organized civil society can set off with the strengthening of social movements. O'Brien et al. (2000) well illustrated the realm of civil society which quest for extensive social change and transformations.

Civil Society and Political Change

The past two decades have witnessed rapid economic growth and waves of democratization which encouraged people to develop higher expectations not only for their living standards, but also for government operation and answerability. The dimension and role of civil society in this arena are also highly promulgating with accountable victories. No doubt, organizations dwelling in this realm act as a driving force for impacting establishments and administration in political society with unimaginable implications. It should be noted that the process of political change is often not linear, and the structure and function of civil society radicals in the procedure are attached firmly to alter as circumstances change (Table 7.1).

Table 7.1 Routes of political change via civil society engagement

System maintenance and reform	System reform and maintenance roles may include: • Legitimating system by naturalizing norms associated with it and maintaining favorable balance of power • Strengthening participation in and support base for system • Overseeing and encouraging conformity with rules of the game • Promoting effective functioning of system • Furthering reform and development of system including correction of perceived defects
System change	Promote system change through: • Resisting and denying domestic and international legitimacy for incumbent government and political system • Constructing alternative visions, norms, institutions • Providing refuge and support for political counter-elite • Mobilizing domestic and international support for alternative elite and system • Form social movements and mobilize public at crucial moments to oust incumbent government

Alagappa (2004)

Civil Society in South Asia

South Asia is a conglomeration of a myriad of economies and hierarchies with different regimes and interconnection to civil society organizations. In the majority of the post-colonial world, nexus between social campaigns and political parties is too deep which is looped through their genesis in nationalist movements. It should be noted that, the major political parties in the South Asian region are products of patriot movements from time to time. Mammoth mobilization during the independence contends continued in the backwash of freeing as governments sought more involvement which will boost successful participation and administration. It should be noted that women were often active for the first time in politics in these social campaigns. A panoramic view of civil society movements in Afghanistan is well illustrated (Durand 2015) and it has been noticed that civil society movements are eminent interlocutors, working to defend and promote citizen's rights and interests. In Bangladesh, civil society has a profound role and is an indispensable actor to retort radicalization and also to manage different fundamental gears of radicalization (BEI 2014). Moreover, the relationship with the government and civil society is vital in abstracting the issues of radicalization too in Bangladesh.

The civil society in Bhutan constituting different community affiliations and systems forms an inherent part of the tralatitious Bhutanese society which provides the people with chances to participate in driving decisions related to various activities (ADB 2013). Interestingly, the majority of the traditional and new civil society associations/movements act as a nexus in Bhutan to foster citizen empowerment. Civil society in India gains its strength from the Gandhian tradition of volunteerism, but today, it conveys itself in many different facets of activism with a plethora of interventions impinging on the administrative and political arena like the right to information, human rights, social advocacy, environmental conservation including the elimination of corruption from public institutions with very permeating support from all quarters (Sharma 2013). UNDP (2011) depicted a bird's eye view of the Maldivian Civil Society in a comprehensive way with valid conclusions and recommendations for a brighter future. Though Nepal has a rich history of civil society, the existing is extremely politicized. However, there is an absence of transparent and effective tools to portray the civil society movements (Bhatta 2012). It is quite difficult to have an unassailable assortment of such movements, as they are cross-linked. But, these movements and establishments are bound to face authenticity when these groups lose the humanity/politeness credentials which are more striking in Nepal's case.

Even though the civil society movement in South Asia is considered vivacious, dynamic, and very vocal, it is strangled in Pakistan, especially during the military regime. Pakistan is still at an incipient juncture and needs a certain level of political maturity to offer functioning space for civil society organizations to become an agent for change (Bhattacharya 2016). In Sri Lanka, the civil society custom is dynamic and elaborately threaded with the fabric of the nation. The function of civil society in resettling displaced communities after Tsunami and civil war restored livelihoods, coordinated interventions, co-operated with decision-makers, and cropped towards peaceful declarations are commendable with multifarious outcomes (Yusoff and Sarjoon 2016). In short, the civil society sphere across the South Asian region is well articulated in protecting human rights, transforming relations, political change, promoting economic development, tackling corruption, building peace, and democratic governance.

Democracy and Civil Society in South Asia

The history of democracy deals with the evolution of pathways by which citizens wreak a role in articulating government policies. The effectiveness of democratic tools, therefore, can be evaluated by the degree to which they reduce the gap between citizens and the decisions that affect them. Democracy is consolidated when a reversal to authoritarianism is impossible. Democracy is consolidated when it is made stable, vibrant, efficient, and accountable civil society engagement (Centre for the Study of Developing Societies 2007). The process is complex and the factors leading to democratic consolidation include structural and cultural dimensions too. The South Asian democratic realm exemplifies political go-through matters more in structuring peoples' orientations to democracy than inherited identities such as religion and ethnicity. The role of civil society movements/groups in inducing the people is well established. The role of civil society organizations in avowing the rights and power of the people in several South Asian democracies is commendable. On the other hand, in several new democracies, many such civil society groups are loosely organized and unable to articulate their interests due to failure in pressure building and common policy portfolio (Girling 2002). Moreover, civil society organizations also can grow and nurture future political leaders who are giants in organizing and motivating people, publicize programs, reconcile conflicts, and build alliances that elicit capacity to deal efficiently with political challenges in the region.

The conglomeration of civil society movements in the South Asian region is achieved through myriad overtures, arraying from executing the role of a critical observer to espousing strategies as a norm socializer and creator of alternative human rights dianoetic positions (Rother 2015). However, institutionalized and structural power in the South Asian region remains rather weak now even though endeavors like the ASEAN Civil Society Conference and ASEAN Youth Forum are attempts in this line. It should be noted that transnational, translocal, regional, and global linkages tone up civil society's role and portfolio in the democratic process across South Asia.

Civil Society and Transforming Relations in South Asia

The rapidly changing socio-economic and political medley across South Asia has affected civil society too. The plethora and paradox of religious beliefs existing in the region flanking with idiosyncratic mindsets, tend to differentiate the countries that make up South Asia which is critical while considering civil society and transforming relations in South Asia. The growing political prominence of civil society in the South Asian realm stimulated thoughts about the concept. The analysis of civil society caters to a better discernment of the interface between society and government and that it holds not the only key to the political legitimation, but also transforming relations in South Asia. The 'We Can' initiative in the region is the best example of transforming power in relationships in South Asia in response to gender-based violence against women in the region (Mehta and Gopalakrishnan 2007).

It is difficult to conclude about civil society and transforming relations in South Asia as civil society in the region is constantly being re-illustrated and resuscitated. As emerging superpowers like China, India, and Brazil are enrolling and commencing to remake global discourses, economics and transforming relations often from the perception of their own awareness base and heritage will have marked variations, as civil society theory has emerged from North America and Europe. The concept of civil society, while observed and underpinned, as its own typical conception rather than as the aftermath of state or market failure is more acceptable, especially in the South Asian perspective.

Conclusion

The role of civil society in articulating the gravity of human relations is well established irrespective of the geographic realm. Civil society is an indispensable and inevitably repugned idea that is time tested. The challenge for political leadership in South Asian countries is to unify democratic gains via constitutional and institutional systems and to ensure that the licit interests of historically marginalized segments of the population are assured. Only then the social, economic, and political environment that give rise to conflict in the region can be overcome. All countries across South Asia are interpreted on how civil society nurtures or hampers political switches, particularly democratic development, for the theoretic and empiric knowledge of civil society and comparative politics egressed and fed considerably across the region. Civil society engagements in public spheres are vital for development, opinion creation, and public movement which opens new avenues for collective action for the betterment of individual and state without which effective democratic progression would be impossible.

References

ADB. (2013). *Civil Society Briefs Bhutan*. Asian Development Bank Publication Stock No. ARM135480-2 May 2013.

Alagappa, M. (2004). Civil Society and Political Change: An Analytical Framework. In M. Alagappa (Ed.), *Civil Society & Political Change in Asia* (pp. 25–61). Stanford, California: Stanford University Press.

BEI. (2014). *The Role of Civil Society in Countering Radicalization in Bangladesh*. Bangladesh Enterprise Institute (BEI). September 2014. Bangladesh.

Bhatta, C. D. (2012). Unveiling Nepal's Civil Society. *Journal of Civil Society, 8*(2), 185–199.

Bhattacharya, S. (2016). Civil Society in Pakistan: Functioning and Challenges. *South Asian Journal of Multidisciplinary Studies, 3*(3), 24–45.

Centre for the Study of Developing Societies. (2007). *The State of Democracy in South Asia*. New Delhi: Oxford University Press.

Chandhoke, N. (1995). *State and Civil Society: Explorations in Political Theory*. New Delhi: Sage.

CIVICUS. (2016). *State of Civil Society Report 2016*. http://www.civicus.org/images/documents/SOCS2016/summaries/SoCS-full-review.pdf.

Court, J., Mendizabal, E., Osborne, D., & Young, J. (2006). *Policy Engagement: How Civil Society Can be More Effective*. Overseas Development Institute. ISBN 0850038146.

Durand, M. (2015). *Panorama of Civil Society Organizations in Afghanistan from the Perspective of Coordination*. Agency Coordinating Body for Afghan Relief and Development. Funded by Tawanmandi.

Florini, M. (2000). *The Third Force: The Rise of Transnational Civil Society*. Japan Centre for International Exchange, Tokyo and Carnegie Endowment for International Peace, Washington, DC.

Girling, J. (2002). Conclusion: Economics, Politics, and Civil Society. In D. McCargo (Ed.), *Reforming Thai Politics* (pp. 261–266). Copenhagen: Nordic Institute of Asian Studies.

Huntington, S. P. (1991). *The Third Wave: Democratization in the Late Twentieth Century*. Norman: University of Oklahoma Press.

Kumar, K. (1993). Civil Society: An Enquiry into the Usefulness of an Historical Term. *British Journal of Sociology., 44*(3), 375–395.

Mehta, M., & Gopalakrishnan, C. (2007). We Can': Transforming Power in Relationships in South Asia. *Journal of Gender & Development., 15*(1), 41–49.

Mundy, K., & Murphy, L. (2001). Transnational Advocacy, Global Civil Society? Emerging Evidence from the Field of Education. *Comparative Education Review, 45*(1), 85–126.
O'Brien, R., Goetz, A. M., Scholte, J. A., & Williams, M. (2000). *Contestingglobal Governance Multilateral Economic Institutions and Global Social Movements.* Cambridge University Press.
Rother, S. (2015). Civil Society and Democracy in South and Southeast Asia - An Introduction. *ASIEN, 136*(July), S.05–S.13.
Sharma, C. P. (2013). *Civil Society in India: Making Good Governance a Reality.* CCS Working Paper No. 303. Research Winternship Programme 2013. Centre for Civil Society.
Tschentscher, T. (2016). *Promoting Sustainable Development Through More Effective Civil Society Participation in Environmental Governance: A Selection of Country Case Studies from the EU-NGOs Project.* United Nations Development Programme.
UNDP. (2011). *Comprehensive Study of the Maldivian Civil Society.* United Nations Development Programme in the Maldives UN Building, Buruzu Magu, Male', Maldives.
Veltmeyer, H. (2008). Civil Society and Local Development. *Interações (Campo Grande), 9*(2), 229–243.
Yusoff, M. A., & Sarjoon, A. (2016). Civil Society Activism for Democracy and Peace-Building in Sri Lanka. *The Social Sciences., 11*(5), 661–671.

8

Democratic Practice and Good Governance in Nepal

Girdhari Dahal

Introduction

Nepal is a newly proclaimed federal democratic republic nation of the twenty-first century in the world. The Constitutional Assembly (CA) made the new constitution of Nepal in 2015. The present constitution of Nepal is the people's constitution. It is a dream of people since 1951 to draft their own constitution through the CA. Also, the mandate of mass movement of 2006 was to make a new constitution in Nepal through the CA.

The history of democratic movements in Nepal goes back to the 1940s. The people and the civil society launched a revolution against the autocratic oligarchic regime of the Ranas and established democracy in Nepal in 1951. This was a historic achievement in the entire political history of Nepal. After the establishment of democracy, people got their political rights. But, before this right of the people was fully institutionalized, again the Monarch snatched the rights of people and started partyless

---

G. Dahal (✉)
Department of Political Science, Tribhuvan University, Pokhara, Nepal

© The Author(s) 2020

Md. N. Momen et al. (eds.), *Building Sustainable Communities*,
https://doi.org/10.1007/978-981-15-2393-9_8

Panchayat System in 1961. This was a betrayal for the people of Nepal. The people were deprived of their own political rights that were achieved through the revolution. The partyless *Panchayat System* continued for 30 years. During this period, all the political parties were banned and the fundamental rights of the people were also seized. This tyranny from the monarchy neither could institutionalize democracy nor could it bring any significant development in the nation. Human rights and political rights of the people were highly suppressed. Political parties had been continuously struggling for the restoration of democracy since 1961. However, the political parties had not succeeded in their mission. So, the civil society and the people of Nepal in the leadership of the political parties launched another mass movement for the restoration of democracy in the 1980s. Finally, democracy was restored with constitutional monarch in 1990. People were granted sovereign power of the nation constitutionally through the Constitution of the Kingdom of Nepal 1990.

But the monarch again turned to be autocratic in 2002 and tried to rule over the sovereign people. Democracy could not be institutionalized. Development could not be fostered smoothly in the nation. In the meantime, the peoples' revolt was initiated by the then rebel force called Maoists for safeguarding the rights of people. At the same time, the political parties in the mainstream also initiated a campaign against the autocratic steps of the monarch. Finally, the political parties and the Communist Party of Nepal (Maoist) signed a 12-point comprehensive agreement and jointly conducted another mass movement for the abolishment of Monarchy and the re-establishment of democracy in Nepal (Comprehensive Peace Accord 2006). The mass movement successfully overthrew the monarchy and established a democratic republic in 2006. After the successful completion of the mass movement, the Comprehensive Peace Accord was signed between the government and revolutionary Maoists, the Interim Constitution was formed, and election was held for the CA (Dahal 2015). The first CA was dissolved without drafting constitution and again the election of second CA was held which could successfully write a constitution for Nepal, institutionalizing a Federal Democratic Republic in the country (Constitution of Nepal 2015). The Constitution of Nepal was promulgated by the CA in 2015. This was a remarkable achievement in the entire political history of Nepal.

With different revolutions and democratic movements, Nepal has always attempted to practice good governance in Nepal. Several significant milestones have been achieved in terms of good governance after each movement. However, there is still a long way to go before the concept of good governance is fully implemented. There are different barriers in the path of democratic practice and good governance. The Nepali democracy failed to cross the barriers of kinship, nepotism, and corruption. There is a vast difference in the promise and performance of political leaders both at macro and micro levels. There has been a continuous scarcity of responsible and stable government at national level. Politics has been drenched in greed and vested interest of politicians and their cadres leading the emergence of neo-patrimonialism or a leader-centered political practice. All these malpractices in political arena have severely influenced the good governance in Nepal (Neil 2007).

Objective

The main objective of this chapter is to explore the practice of democratic movements and the concept of good governance in Nepal. It has also attempted to analyze the contribution of democratic movements to promote good governance and consolidate democracy in Nepal.

Methodology

This chapter makes a brief description and analysis of the democratic movements and the practice of good governance in Nepal. So, it is a descriptive, historical, and analytical chapter. It makes the best use of data from secondary sources on democratic movements and good governance in Nepal that are available from both print and online sources.

History of Democratic Movements in Nepal

Different dynasties ruled over Nepal from the ancient period to the modern age. The Gopal dynasty, Mahispalas, Kirats, Lichhavies, and the Mallas ruled over Nepal until the eighteenth century. The democratic movements in Nepal started during the family rule of the Ranas. The credit of democratic movements in Nepal goes to *Prajaparishad* Party in 1937 and *Nagarik Adhikar Samittee* (Sukraraj Shastri). Madhav Raj Joshi, the father of Sukraraj Shastri, was influenced by *Arya Samaj* of Dayananda Saraswati, while Shukraraj Shastri, one of the four martyrs of Nepal, was influenced by Vedic philosophy (Sharma 1995). Shukraraj Shastri had also translated the *Brahmasutra Bhasya* into Nepali. Shukraraj Shastri and another young leader Gangalal Shrestha were arrested and imprisoned by the Ranas for giving a speech on *Gita* philosophy at Indrachowk, Kathmandu. Shukraraj Shastri, Dharmabhakta Mathema, Dashrath Chand, and Gangalal Shrestha were sentenced to death in 1941 in the accusation of the *Parcha Kanda* (Pamphlet Scandal). These death sentences to these heroic leaders of Nepal boosted the protest against the autocratic, tyrannical rule of Rana in the nation.

With the birth of *Parjaparishad*, Nepal witnessed different political parties in Nepal and in India. The *Rastriya Congress Party* was established in 1947 and *Prajatantrik Congress* was established in 1949 by the non-residential Nepalis in India. Puspalal established Nepal Communist Party in 22 April, 1949 (Puspalal Smriti Pratisthan 2013). Then, the National Congress and the Democratic Congress were united to form Nepali Congress in 1950.

Laborers in Biratnagar Jute Mill revolted against the Ranas in 1947 by organizing a strike campaign. So did the students at *Tindhara Sanskrik* hostel Kathmandu for political rights. This movement was named as

Jayatu Sanskritam. All these sporadic movements against Ranas were united after the *Sahid Kanda* of 1941. Finally, the democratic revolution in 1951 put an end to 104-year-old autocratic, family rule of the Ranas. The people were free to practice their civil and political rights after the establishment of democracy in 1951.

History of the Constitution of Nepal

More than half a dozen constitutions were made in the history of Nepal. First written constitution was announced by Rana Prime Minister Mohan Shamsher Rana in 1947. But it was not implemented. It was proclaimed only to continue Rana rule but it was not for the peoples' rights with democracy. It did not address the people's political rights as well as human rights. The political parties of Nepal did not accept it; rather they carried on their movements against the Rana oligarchs.

Democracy was established in Nepal after the revolution of 1951. After that Late King Tribhuvan announced Interim Constitution in 1951 and also proposed for a new constitution by constituent assembly. It was the second written constitution in the history of Nepal. The third Constitution of the Kingdom of Nepal 1959 was announced by the late King Mahendra. The fourth constitution of Nepal in 1964 was announced by the late King Mahendra with a partyless Panchayat system. It lasted for 30 years and came to an end along with the overthrow of the system. The mass movement of 1990 restored democracy in Nepal and a new constitution (Constitution of the Kingdom of Nepal 1990) was promulgated with the provision of constitutional monarchy and multi-party democracy. But unfortunately, constitutional monarchy tried to turn into autocratic rule. The latest people's movement started in 2006 finally ended the 240-year-long Shah Dynasty in Nepal. After this movement, two CA elections have been held and the second CA could draft a new constitution of Nepal 2015. It is a constitution created by the people. It has followed the norms and values of constitutionalism and good governance in Nepal. The people of Nepal have become sovereign after the people's movement of 2006.

In the constitutional history of Nepal, the Constitution of Nepal—2015 is made by the peoples' representatives in the CA. They were elected by the people. They had the mandate of the people bestowed by the mass movement of 2006. All the other constitutions were drafted by kings or constitution recommendation committees.

Democracy in Nepal

Democracy is a system of government in which everyone in the country can vote to elect its representatives to the government. A democratic country has a government elected by the people of the country (Longman Dictionary 1998: 360). This means that it is a system in which everyone is equal and has the right to vote, make decisions, and so on (ibid.). The central idea of this saying is that democracy is the rule of law and order with three pillars of democracy viz. liberty, fraternity, and equality. In Nepal, democracy is referred to as *Loktantra*. 'Lok' means people or soul and 'Tantra' means rule or body. So, democracy can be compared with soul and body (Adhikari 2014). The welfare of an individual as body and soul is also a democracy. As said in *Veda*, it is सर्‌वभुतहतिरता (*sarwabhutahite rata*), that is, eat together, live together, and so on. A revolution or democratic movement has been a fundamental change in political power or organizational structure that takes place in a relatively short period of time when the population rises up in revolt against the current authorities (Krause 2016). According to Aristotle, it is a complete change from one constitution to another, modification of an existing constitution. Revolutions have occurred through human history and vary widely in terms of methods, duration, and motivating ideology. Their results include major changes in culture, economy, and socio-political institutions. Aristotle explained in great detail the theory of revolution. It is his study of nearly 158 constitutions that helped him understand the implications of revolutions on a political system. In his work, *Politics*, he discussed at length all about revolutions. Based on his study, Aristotle gave a scientific analysis and expert treatment to the subject of revolutions. He gave a very broad meaning to the term 'revolution' which meant two things to him. Firstly, it implies any major or minor change in the constitution such as a change in monarchy or oligarchy and so on. Secondly, it implies a change in the ruling power even though it did not lead to a change in the government or the constitution. He further stated that a revolution could be either direct or indirect, thereby affecting a particular institution. (Mondal 2016)

The definition of revolution is the movement of one object around a center or a forceful overthrow of a government by the people or any sudden change. An example of revolution is movement of the earth around the sun (Merriam-Webster, dictionary).

Through a declaration, King Tribhuvan in 1951 stated the establishment of democracy and proclaimed to draft a new constitution through the constituent assembly. But the declaration for drafting the constitution through a constitutional assembly was never put into practice. The immediate successor of King Tribhuvan, King Mahendra also did not implement the declaration of constitution drafting through the CA; he instead formulated the constitution drafting committee and promulgated the Constitution for the Kingdom of Nepal 1959. As per the provisions set in the constitution, the first general election was held in 1959. The political parties participated in the election. People elected their representatives and the government was also formed. However, King Mahendra dissolved the parliament, abandoned the political parties, and took all the powers into his hands in 1961 by using the article 55 of the same constitution. The Nepali Congress Party and the Nepal Communist Party members who were against the bad practices of the king were arrested and imprisoned. The Partyless Panchayat System was imposed in the nation. So, the political parties again started revolting against the Partyless Panchayat System (1960–1990) in different ways. Taking lessons from the previous political movements, another milestone was created in the Nepali political history in 1980 that combined movement of the democrats and communist parties along with their sister organizations and compelled the Panchayat government to hold a referendum. The referendum was held in 1980 and concluded with 55 percent of the total votes in favor of amended Panchayat system and 45 percent in favor of the multiparty democracy (Shaha 1993). The amended Panchyat system also could not address the democratic thrusts of Nepali people.

The communist front was organized under Sahana Pradhan in 1990. Ganeshman Shingh who led the democratic groups was made the supreme commander of the democratic movement of Nepal. In the collective efforts of the political parties, a mass movement was launched in the country which abolished partyless Panchyat system and established the multi-party democracy with constitutional monarchy in 1990.

The Democratic Movement, 2006, in Nepal is similar to the French Revolution because both revolutions overthrew the monarchy, and both were for the establishment of a republic nation and sought for political liberty and equality. The Seven Party Alliances (SPA) and Maoists collaborated to end the autocratic monarchy and to establish democracy. In addition, there were some bourgeoisies both in the French and in the Nepal revolution who stayed loyal to the king. The French called them the loyalists, while Nepalese identified these traitors as the royalists. The revolution was aggressive because arms and ammunitions were used. All in all, both the French and Nepal revolutions took place because the people wanted to make their country a republic nation. Nepal's 2006 revolution was the result of people's desire for freedom and democratic rights. The king seized all political power in his hands in order to control the armed forces that was creating political unrest in the country as it is mentioned in the following lines:

> The Seven Party Alliances (SPA) and the Maoists hunted for emancipation and to make Nepal a republic, giving rights to all citizens. Everything seemed to be falling into place as the royalty was pushed out from their position. Then in just a moment, Nepal was once again crowded with strikers and armed opponents as the Maoists stormed out of the government. Democracy in Nepal is on the edge of trouble. An agreement must be made now and a solid union should be created to make Nepal a secure nation. (Juliana 2008)

The 1990 constitution of the Kingdom of Nepal was drafted following the mass movement of 1990. For the first time in the history of Nepal, the sovereignty of the nation was vested on people by the article two of the constitution of the Kingdom of Nepal (The Constitution Kingdom of Nepal, 1990). After this, the general election was held in 1992, mid-term election in 1995, and general election in 2000. People had expected a drastic change in the social, economic, and cultural aspects of the nation following the mass movement of Nepal; however, the hope of Nepali people was just in vain. So *Rastriya Janamorcha* led by Baburam Bhattarai put forward 40 points demand before the government, the 40 points included the agendas concerning the lives of people and issues of nationality. The then government did not pay any attention toward the 40 points demands. So, the Maoist Party started armed insurgency in 1996. During the insurgency, Royal massacre took place in 2002 that left none alive from King Birendra's family. Then, the younger brother Gyanendra became the new king of Nepal. King Gyanendra snatched the executive powers from the elected body into his hands by overthrowing the government led by Prime Minister Sher Bahadur Deuba. Following the incident, many governments were formed and dissolved using the article 127 of the Constitution of the Kingdom of Nepal 1990. This nature can be termed as *Duryodhan prabitti,* that is, self-centeredness.

But the king could not achieve his objectives with any of these governments, so he again formed another government of Prime Minister Sher Bahadur Deuba and dissolved that too in 2005 and started his direct rule. People were not satisfied with this act of the king and the political parties strongly disagreed with the move. The parliamentary parties were continuously struggling for the restoration of the parliament whereas the revolutionary Maoists were conducting armed insurgency which they called the "people's war".

Twelve-Point Understanding and the People's Movement of 2006

The main target of both the parliamentarians and the revolutionary Maoists was to end the autocratic rule of the monarch. Realizing the importance of the united campaigns, the seven parliamentary parties and the revolutionary Maoists signed the 12-point understanding to collectively conduct protest against the monarchy. The initiating point and the meeting point for the understanding were restoration of parliament and election of constituent assembly.

The mass movement of April, 2006 was organized jointly by the seven political parties and revolutionary Maoists with the same spirit and understanding for restoration of parliament and election of constituent assembly. Finally, the united force was successful and the king was forced to lie down on his knees and the parliament was restored. The reinstated parliament heavily cut down the powers of the king and also drafted a new constitution for the interim period—the Interim Constitution in 2007. The revolutionary Maoist Party also joined the legislature, that is, parliament and government. The first ever election of CA in the history of Nepal was held in 2008. Then, the first meeting of the constitutional assembly in 2009 formally declared Nepal as a federal republic without any bloodshed. The people's representatives elected their president and the prime minister. The first CA did many significant activities for institutionalization of peace process and changes in the nation. However, the first CA was finally dissolved without promulgating new constitution. So, the election of CA was held again in 2013. This could finally draft a new constitution in 2015. This constitution of Nepal is in the course of implementation. The democratic movement of Nepal is still in the process of consolidating its political system.

Constitution Assembly Drafted Constitution

The constitution of Nepal is the result of various mass movements and revolutions in the political and democratic history of Nepal. This constitution is also a dream of Nepali people to write their own constitution through the CA which sustained through a 65-year-long struggle for democracy. This constitution has vested the sole sovereign power in the people of Nepal. There were two CAs case in the history of Nepal. The first CA (2008) was dissolved without drafting any constitution. The second CA (2013) could finally draft new constitution for Nepal. It is an exceptional case in the world. This constitution was totally made by people's representatives. It was passed by the chairman of CA and declared by the first president of Nepal as the representatives of the people of Nepal. The constitution of Nepal was passed by more than 90 percent CA members, so this particular constitution is also called people's constitution. The constitution of Nepal has the preamble with 35 divisions, 308 articles, and 9 schedules.

The CA was composed of 601 members. These members are comprised of representatives selected inclusively from several areas. During the time of the declaration of the constitution of Nepal, the total CA members were 598 in number and among them 537 members signed the new constitution. The chairman of the constitution assembly verified it and finally the constitution was declared by the head of the state. So, it is called people's constitution of Nepal.

Democracy and Good Governance

According to the eastern philosophy, democratic norms and values are आत्मानाःप्रतिकुलानि परेसामन समाचरेत् *or Sarve Bhawntu Sukhina sarve santu niramya*. Democracy means the condition in which you should bear in mind the fact that things you like are liked by others and the things you don't like is not liked by others too. All the people should be healthy, happy and prosperous. According to the western philosophy, democracy is liberty, equality, and fraternity. Liberty means freedom of speech, publication, and so on. Equality means equal rights to participate in rule and to choose representatives of the government as well as of the parliament. Federal democratic republic of Nepal has been institutionalized through the promulgation of the constitution of Nepal in 2015. The elections of the Speaker of the Parliament, the President, the Prime Minister, and the Chief Justice have been held under the new constitution of 2015. The new constitution theoretically supports the concept of good governance in Nepal. Democracy and good governance have a close relation. Democracy follows the rule of law and the rule of the law is the most essential element of good governance. Also, legitimacy of the government is another element of good governance and the most important factor of democracy.

Good Governance

In addition to the devotion for democracy, people have also been equally concerned with good governance in the nation. Good governance includes people's rule, rule of law, participatory, equitable and inclusive democracy, effective, responsive, transparent, accountable and efficient institutions (Dahal 2011). Legitimate government, transparent rule, accountability to the people, free and fair election, independent judiciary, fundamental rights, and duties of the people are the essential elements of good governance as well as democracy. This is described in the following lines:

> Governance can be used in several contexts such as corporate governance, international governance, national governance and local governance. Since governance is the process of decision-making and the process by which decisions are implemented, an analysis of governance focuses on the formal and informal actors involved in decision-making and implementing the decisions made and the formal and informal structures that have been set in place to arrive at and implement the decision. (UNESCAP)

Good governance has eight major characteristics. It is participatory, consensus oriented, accountable, transparent, responsive, effective and efficient, equitable and inclusive and follows the rule of law. It assures that corruption is minimized, the views of minorities are taken into account, and the voices of the most vulnerable in society are heard in decision-making. It is also responsive to the present and future needs of society.

According to the World Bank, governance refers to (a) the form of the political regime; (b) the process by which authority is exercised in the management of a country's economic and social resources for development; and (c) the capacity of governments to design, formulate, and implement policies and discharge functions. In the similar way, this is mentioned here in this way:

> The political agenda of good governance has been developed and maintained primarily by national and international technocrats, administrations

> and 'knowledge' institutions. Before turning to the substance of the policy issues on the agenda of good governance, a remark on the form of these problems is needed. An absolutely crucial part of good governance is its interpretation of policy problems as wicked problems. (Bang and Esmark 2013)

Democracy as a modern political regime widespread in the twentieth century has in its core political representation. There is no democracy without parliament, whose main task is to represent the people as a single political entity. Governance that utilizes the capabilities of information technology maximally to achieve its objectives of good governance is now called the electronic governance or simply, E-Governance and sometimes, E-Democracy (Karki 2001). The constitution of Nepal makes it amply clear that the state shall protect "life, liberty and property of people", render social justice, and maintain internal cohesion (Constitution of Nepal). Good governance is a process of executing a coherent governing plan for the nation based on the interests and priorities of the people (Dahal et al. 2002). Good governance requires the establishment of a rule-governed system to guarantee the equitable, just, and rational distribution of the advantages among all holders of interests in common (Dahal et al. 2002). The term "governance" is closely associated with control, rule, or administration of state through implementation of laws. In recent years, it entered the vocabulary of development in a rather episodic way (Dahal 1999). Good governance is also associated with encouragement of participatory democracy, local leadership, civil society, and decentralization of power to the grass roots.

The term "good governance" is coined by United Nations Development Program (UNDP) and World Bank. UNDP defines "governance" as the exercise of political, economic, and administrative authority in management of a country's affairs at all levels. They have defined it in new ways such as rule of law, legitimacy of the government, accountability to the people, transparent economic sector, corruption-free activities of government and non-governmental sector. Good governance is corruption-free government based on rule of law and efficient in public sector management (Dahal 1999). The main root of good governance is *Ram Rajya* meaning ideal state (Dahal 2011). Vedic norms and values of Nepal were guided by this philosophy. Lord Ram was a model king in Eastern society. He was very famous and popular among the people. He did not even hesitate to send his wife to the jungle to address the public voice (Ramayan). Thus, Nepal is a country which is influenced by the Vedic culture *Ram Rajya* as well as new term "good governance". Governance comprises the mechanisms, processes, and institutions through which all the stakeholders, namely, the state, private sector, civil society, and citizens articulate their interests. Both of these concepts have the same theme, that is, to be responsible to the people. New constitution of Nepal has tried to strengthen *Ram Rajya* as well as good governance in Nepal. The conceptual frame work in figures explains in detail the qualities of good governance (Fig. 8.1).

Democracy and good governance are closely related to each other. Democracy can be strong only if there is good governance. Governance that utilizes the capabilities of information technology maximally to achieve its objectives of good governance is now called e-governance and sometimes it is even called e-democracy. Thus, good governance can be realized only in a democracy. Good governance and democracy are very much related to one another. The people and the civil society of Nepal have continuously fought for the establishment of democracy and good governance in the country. Finally, the country has become a federal

Fig. 8.1 Conceptual frame work of good governance. (Source: https://www.gdrc.org/u-gov/escap-governance.htm)

democratic republic state. People have been the sole sovereign power of the nation (Constitution of Nepal-Article 2). The people and the civil society have continuously devoted themselves for democracy and good governance in the country.

Democracy and good governance are two sides of a coin. They go side-by-side. Good governance, on the one hand, strengthens democracy; on the other hand, it is the pre-requisite to good governance. The success of each democratic movement is promoting good governance in Nepal. There has been improvement in the rule of law, accountability, transparency, and responsibility toward people. New laws, provisions, policies, and practices have been practiced to ensure good governance in Nepal. Theoretically, Nepal has made several provisions in its constitution and laws for good governance, but practically, people are not enjoying as much good governance as it has been written in legal documents; apart from this, the promising point is that Nepal is heading towards strengthening the quality of good governance for its people.

Development of Good Governance

The development of good governance in Nepal can be categorized into three different phases. The first phase of evolution of good governance covers the time period from 1951–1990, the second phase from 1990–2006, and the third phase from 2006 onwards.

The first major democratic movement in the history of modern Nepal is the revolution of 1951. After this revolution, good governance started to evolve in Nepal. In the period from 1951–1990, public service commission was established; administrative reform commission was formed; civil service act and rule was drafted and implemented; and the Supreme Court was established.

The second major democratic movement in the history of modern Nepal is the mass movement of 1990. After this mass movement, some other series of events happened for good governance in the period from 1990–2006. During this period, higher public administrative reform commission under the chairmanship of Prime Minister Girija Prasad Koirala was formed, and the local self-governance act 2055 was drafted.

The third phase of development of good governance in Nepal took place after the democratic mass movement of 2006. During this period, good governance act and rule was drafted and a new constitution was drafted.

In this way, good governance gradually evolved in Nepal and it is still in the process of evolution. As a result of which, there is a slight improvement in the quality and quantity of services delivered to the people of Nepal. However, still there are many barriers on the path to this.

Challenges to Good Governance in Nepal

Good governance has not been fully exercised in Nepal. There are many hurdles to its successful implementation. Several factors are directly and indirectly responsible for the poor performance of good governance. Corruption, no local level election for a long time, lack of timely political consensus among major political parties, political instability, weak civil society, problems in bureaucracy, unregulated private sector, underdevelopment and unemployment, and lack of awareness are among the major challenges to good governance in Nepal. Corruption is prevalent everywhere in Nepal. Political and bureaucratic corruption is even more severe. Officials from the grassroots level to the top level are involved in corruption. Corruption is the most complex challenge to good governance. There has been no local level election for more than 15 years. The last local level election was held in 1998. After a long time, the local government election was held in 2017. Political instability has been a characteristic feature of Nepali politics. There is no political consensus among the political parties on key political issues. Even if they come to a common understanding, it would take relatively longer time for them to reach that point.

Civil society plays crucial role in promoting good governance. But civil society is not strong and organized properly in Nepal. Even the civil society is not impartial. They are highly influenced by major political parties. So they are more like sister organizations of the political parties. Bureaucratic system is the main implementing body of good governance, but in Nepal it has several weaknesses and loop holes. It is traditional.

Bureaucrats do not show professionalism in their work. There is the practice of red-tapism, nepotism, favoritism in the bureaucracy.

In the modern era, private sectors are also the major service providers to the people of a nation. But in Nepal, the private sector is not thoroughly regulated by the government. They do not obey any rule or provision that is intended to regulate them. People are suffering from their malpractices and bad functioning. Underdevelopment and high level of unemployment are also negatively affecting good governance in Nepal. Also, low level of civic consciousness among the people and ignorance is creating disturbance in smooth practice of good governance in Nepal.

Conclusion

Three major democratic movements and other movements were carried out in the political history of Nepal. As a result of these democratic movements, different national and local level elections were held at different times which helped to strengthen democracy. Good governance also gradually developed after each democratic movement. Good governance evolved in Nepal in three consecutive phases: from 1951–1990, 1990–2006, and 2006–present. Now, Nepal is a federal democratic republic and good governance is becoming an increasingly prioritized task of the government. In this way, democratic practice and good governance are being institutionalized steadily. However, achieving the goals of good governance is not an easy talk due to the prevailing, patrimonial corruption, political volatility, greed of the politicians and bureaucrats, and the grievances of the common people of Nepal.

Acknowledgments I would like to express my sincere thanks to Prof. Taranath Baral, Department of Political Science, Tribhuvan University, Prithvi Narayan Campus, Pokhara, for making a thorough reading of this chapter and for the valuable advice. Similarly, I would like to thank Dr. Prakash Upadhyay, Department of Anthropology, Tribhuvan University, Prithivi Narayan Campus for his valuable suggestions to bring out this chapter in this form and Dr. Min Pun, Department of English, Tribhuvan University, Prithvi Narayan Campus, Pokhara, for the language edit.

References

Adhikari, Narayan Psd. (2014, February 6). Personal Communication. Pokhara.

Bang, H., & Esmark, A. (2013). A systems theory of Good Governance. ICPP, Grenoble, 45.

Comprehensive Peace Accord. (2006). The Comprehensive Peace Accord Concluded between the Government of Nepal and Communist Party of Nepal (Maoist). Retrieved from http://www.lawcommission.gov.nppdf.

Constitution of Nepal. (2015). The Government of Nepal. Kathmandu: Kanun Kitab Bebastha Samittee.

Dahal, R. (1999). Prospects of Good Governance. In G. B. Thapa (Ed.), *Decentralization and Good Governance in Nepal.* Germany: Political Science Association of Nepal Friedrich-Ebert-Stiftung (FES).

Dahal, G. (2011). Good Governance: Theoretical Perspective. *Pragyik Prabhaha, Pokhara, Bi-annual Research Journal, 2*(Year 1st), 1–7.

Dahal, G. (2015). Constitutional Assembly of Nepal Milestone for Peace, Development and Political Stability. *KMC Journal of Interdisciplinary Studies, 4*(1), 70–77.

Dahal, D. R., Uprety, H., & Subba, P. (2002). Defining Good Governance. In *Good Governance & Decentralization in Nepal.* GDS Center for Governance & Development Studies in cooperation with Friedrich-Ebert-Stiftuing (FES).

http://soscbaha.org/downloads/mcrl2014/mcrl2014.pdf.

http://www.dictionary.com/browse/revolution.pdf.

http://www.gdrc.org/u-gov/escap-governance.htm.

http://www.ipa.government.bg/sites/default/files/pregled-dobro_upravlenie.pdf.

http://www.yourdictionary.com/revolution%20pdf.

https://modernrevolutionsc.podbean.com/e/2006-democracy-movement-in-nepal-by-kristel-and-juliana.

https://modernrevolutionsc.podbean.com/e/2006-democracy-movement-in-nepal-by-kristel-and-julianapdf.

https://plato.stanford.edu/entries/aristotle-politics.pdf.

https://www.merriam-webster.com/dictionary/revolution.

Juliana, K. (2008). *2006 Democracy Movement in Nepal.*

Karki, B. B. (2001). A Decade of Governance: Constitutional practice, *Issues of Governance in Nepal.* In A. P. Shrestha & S. R. Dahal (Eds.), *Nepal Foundation for Advanced Studies (NEFAS) Coalition for Action on South Asian Cooperation (CASAC).* Germany: Friedrich-Ebert-Stiftung (FES).

Krause, C. (2016). Retrieved from https://www.facebook.com/mikeklinglerfans/posts/10153915257159289.

Longman Dictionary of Contemporary English. (1998). India: Longman Group Ltd.

Mondal, P. (2016). Aristotle's Theory of Revolution: Causes and Methods to Prevent Revolution. Retrieved from http://www.yourarticlelibrary.com/politics/aristotles-theory-of-revolution-causes-and-methods-to-prevent-revolution/40126/.

Neil, T. O. (2007). *Neopatrimonialism and Public Sector Performance and Reform.* Retrieved from https://www.odi.org/sites/odi.org.uk/files/odi-assets/publications-opinion-files/4393.pdf.

Puspalal Smriti Pratisthan. (2013). Document of Communist Party. Kathamandu.

Shaha, R. (1993). Politics in Nepal 1980–91. New Delhi: Manohar Publishers and Distributors.

Sharma, R. H. (1995). Great man shree Sukrraj Sastri. Martyr Shukra Memory Book. Kathamandu: Ganendrabhadur Kayasth.

The Interim Constitution of Nepal. (2007). Nepal Government, Kanun Kitab Babastha Samittee. Kathmandu

The Constitution of Nepal. (2015). Nepal Government, Kanun Kitab Babastha Samittee. Kathmandu.

United Nations Economic and Social Commission for Asia and the Pacific. Retrieved from http://www.unescap.org/sites/default/files/good-governance.pdf.

9

Taxonomy of State-CSOs Relations in India

Binod Kumar

Introduction

There are varieties of organizations working actively outside the scope of state and market. These organizations include non-profit organizations (NPOs), non-governmental organization (NGOs), community based organizations (CBOs) and government organized non-governmental organizations (GONGOs). They offer program and services in partnership with central, state and local governments. These organizations provide wide range of services outside the domain of state and market.

The non-governmental organizations (NGOs) occupy a large space of the gamut of civil society organizations which are characterized by registration and recognition with the state. Hence, NGOs are legally constituted entities to resolve problems of society. The term 'NGO' initially

This work has been part of my MPhil thesis at the Centre for the Study of Law and Governance, Jawaharlal Nehru University, New Delhi.

B. Kumar (✉)
Department of Social Work, Dr. Bhim Rao Ambedkar College, University of Delhi, New Delhi, India

© The Author(s) 2020
Md. N. Momen et al. (eds.), *Building Sustainable Communities*,
https://doi.org/10.1007/978-981-15-2393-9_9

used by the United Nations refers organizations that are not part of the government and market. NGOs are registered with state under the Societies Registration Act, 1860. These forms of organizations are primarily involved in development activities and service delivery in collaboration with state. Within CSOs, there are other forms of organizations which may not be registered with the state and functions independently of state. Likewise, voluntary organizations are another form of organizations within the umbrella of civil society which contains some additional characteristics, as they are non-obligatory, unpaid, carried out for the benefit of others in society and they are organized.[1]

Booming civil society sector in India can be attributed to the failure of the Nehruvian model of development and an upsurge of anti-people policies in the 1970s (Kumar 2010). However, the importance of civil society organizations in India acquired salience only in the 1990s, when India departed from the welfare centric to the entrepreneurial model. Hence, the emergence of a large number of civil society organizations in India was not a product of civic culture; instead, it was a political decision which paved the way for the emergence of this sector. During the 1990s, the state started withdrawing from various social sector programs, and the gap created in due course has been filled by NGOs. They have responded to the state in two ways: where the majority of organizations collaborated with the government to deliver services on behalf of government, whereas some organizations are incorporated as supporting organization to their profit making body. For example, Azim Premji Foundation has been started and funded by Wipro, a profit making body.

The interaction between state and CSOs underlines the crucial role of NGOs at the micro and meso level of governance. The micro and meso level analysis of the work of CSOs allows us to know the work and function of the agency. The macro-level perspective does not clear as to who is doing what at the grassroots level. However, CSOs are seen as the passive recipients of aid at the macro level and in turn, state is seen as a victim of political pressure to decentralize and share power with CSOs (Eade and Ligteringen 2001).

---

[1] http://www.grupcies.com/boletin/images/stories/PDFBoletin/ArticuloI_Edic_59.pdf.

Of late, the increased dependency of state on CSOs has led to the absence of interdependency. The reverse approach to examine the interaction between state and CSOs recognizes: CSOs increasing dependence on state. The same is also applicable in case of interaction between intergovernmental organizations and NGOs (Edward and Hulme 1996). The changing interaction between state and CSOs requires micro, meso and macro level analysis to understand it in entirety. By the passage of time, the interaction between state and CSOs has metamorphosed across the globe. Where government is getting smaller, CSOs are becoming bigger and stronger. The global imperative to have smaller and regulatory state has transformed the character of CSOs. Now, civil society is characterized by 'private' in effort but 'public' in nature, consequently filling the vacuum created by the withdrawal of state. The NGOs in current forms are performing the role of public agencies and paid by public funds (Hasan, Onyx and Lyons 2008).

Edward and Hulme's (1996) analysis of state-CSOs interaction seems very convincing and interesting. The analysis is reminiscent of Jurgen Habermas's colonization theory. According to this theory, the state has succeeded in intruding, controlling and instrumentalizing the public sphere. Edward and Hulme (1996) do not refer to this theory, but their approach to view CSOs as instruments and puppets of the state provokes associations with Habermas's colonization theory. Edward and Hulme (1996) reflect little on process of interdependence between state and CSOs and even less on state's dependence on CSOs. Despite that it would not be plausible to interpret dependence of state on CSOs as victory of civil society. Meaning thereby, there is little to witness inverse of colonization theory that suggests how CSOs have intruded, controlled and instrumentalized the political state (Steiner-Khamsi 2008).

Young (2000) has suggested a triangular model of CSOs-government relations (Fig. 9.1) and argues that to a varying degree all three types of relations are being interplayed at the same time; however, some assume more importance during some periods than do others (Edward and Hulme 1996). Initially, the relationship between CSOs and state were adversarial; however, it has changed significantly in the wake of the adoption of new economic policy and structural adjustment program. On the

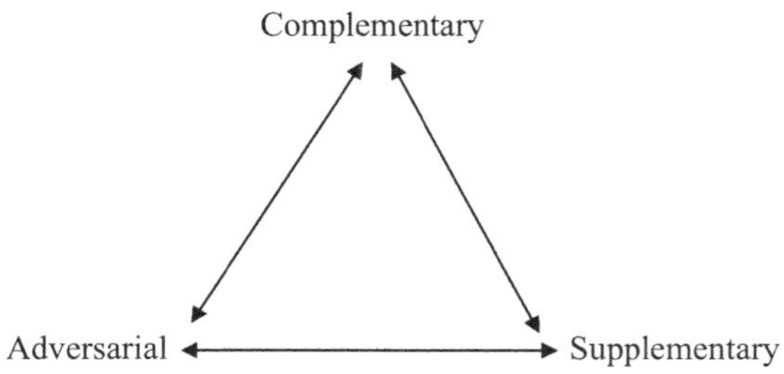

Fig. 9.1 Modes of interaction between state and CSOs (Source: Anheier 2005: 285)

basis of the above analysis, state-CSOs interactions could be categorized under three broad rubrics:

Supplementary: CSOs are providing voluntary services which are mostly not covered by the state, and there has been extension of several services by the voluntary sector in the response of government cutback.

Complementary: This covers contracts and partnership between government and CSOs in the response of new public management and outsourcing (transaction costs and greater efficiency are prime agendas).

Adversarial: Civil society organizations (CSOs) are involved in advocacy for the rights of marginalized people left unserved or underserved by the state. CSOs are in adversarial role because they lobby for higher budget allocation. More pertinent example to this relationship between CSO and state has emerged in the form of dam construction and environment-ecological protection where state argues in favor of displacement and dam construction for various purposes. In contrast, civil society organizations stand against the same.

State-CSOs Interface

There has been an increasing reliance on a partnership of CSOs with state and market. The increased convergence of the work has resulted in blurring boundaries among state, market and civil society (sometimes referred to as the third sector). In the changed scenario, there is a real reason to think about governance of the third sector. There have been pressing needs to regulate the third sector not only to establish a framework for them, but also to create a level playing field for the development of the third sector itself. NGOs are one of the foremost components of the third sector which needs attention in the current context.

In the backdrop of intrusion of state to control the third sector, there is state's imperative to establish a framework of governance to ensure accountability. Accountability has been integrated part of the governance framework; hence, when we talk about governance, it is imperative to talk about accountability. The NGO's upward accountability to external donors misplaces the local priority and tries to transplant the donor-driven development agenda. At the domestic level, NGOs upward accountability lies with the bureaucrats and administrators rather than to the system to ensure flow of funds. In this way, the inherent foundational attributes of the NGOs like community orientation, efficiency and political activism are in the phase of crumbling. Hence, NGOs have not only increased in number and size across the country, but they have also become more institutionalized, dependent and ideology-driven. Consequently, NGOs have been co-opted and corrupted by their funders, and it is leading to misplaced or ghost community priorities.

CSOs transformation process from state independent to state dependent entities has made them more vulnerable. They are functioning at the mercy of state, meaning thereby if they have good relationships with government officials or have political clout, only then can they get funds from the state. To get rid of external influences and dependence, it is imperative to understand the state-CSOs dynamic relations rather than analyzing the conditions under which they have been able to avoid the trap of external dependency (Dongre and Gopalan 2008).

The changing landscape and nature of welfare state have necessitated more and more number of CSOs joining hands with central or state government to implement the specific project. The partnership is based on the premise of ensuring the state's traditional functions like health and education. Nonetheless, NGOs have de facto been metamorphosed into Quasi-Non-Governmental Organizations (Choudhury and Ahmed 2002).

CSOs in Liberalization and Free Trade Regime

With increasing schism between rich and poor in the society, there has been an emergence of various non-state actors. Non-governmental organizations (NGOs) are one among them, which aim to raise the voices of poor and advocates in their interest. Liberalization and free trade have rendered the people jobless at the bottom of the pyramid. In this backdrop, civil society organizations found legitimacy and acceptance in society. In a liberal democracy like India, people see them as a ray of hope and alternative to government; however, government visualizes them as potential partners in development. This partnership has repercussions for governance not only at the national level but at the international level too. In *Government, Politics and the State* (Pierre and Peters 2000), the model for governance at three levels has been provided: (1) *moving up*: emerging role of international organizations which refers to the role of international agencies in governance, (2) *moving down*: regions, localities and community which refers to decentralization of state authority to local and regional institutions, and finally, (3) *moving out*: NPOs, corporatization and privatization which refers to the delegation of state functions to non-state institutions. Most of the advance democracies have helped to set up a large number of NGOs in public service delivery if they have not privatized the functions altogether. The idea of creating satellite institutions for enhancing public service delivery has gained massive popularity and is currently used at all levels of government (Pierre and Peters 2000). Meaning thereby, the government can use for-profit or non-profit organizations to fulfill the gap in service delivery. In some cases, these organizations have existed before and now the government is using them to reach out to the community. Moreover, in other instances, government has fostered the creation of these organizations and then became significant funders (Pierre and Peters 2000). If we compare it with the Indian scenario, various NGOs have been working for poverty alleviation without any collaboration with government and now the government is partnering with them and making use of their expertise. For example, to achieve the goal of financial inclusion of marginalized people, the government has facilitated the creation of micro-credit institutions and self-help groups (SHG) with regulation and governance through the Reserve Bank of India (RBI).

Genealogical Excavation of State-CSOs Relations in India

In a country like India, charity has been an integrated part of the religious belief system since time immemorial. In other words, the voluntary sector in India owes its origin to religious obligations, philanthropy and traditional charity. Traditionally, the welfare of poor and marginalized has been co-shared between state and religion in India. By the passage of time, there has been emergence of other social institutions for the welfare of poor; however, it was not systematic, organized and institutionalized. The evolution of organized philanthropy and voluntary organizations are newer phenomena. Notwithstanding, voluntary work in India has been heavily influenced by contemporary religious, social and cultural values.

The paradigm shift on the evolution of voluntary sector can be witnessed in the form of (1) traditional and (2) modern. Traditional voluntary work revolves around religion and charity attached to religion, whereas the modern charity got its recognition only in the eighteenth century with the enactment of the English Literary and Scientific Institutions Act, 1854 of UK. It is *Society Registration Act, 1860* in India which provided a legal basis to the voluntary work on institutionalized pattern. The enactment of the said act was done after the mutiny of 1857. Meaning thereby, it was enacted partly to provide the legal basis to the voluntary work and partly to regulate the flow of funds and keep a tab on organizations and collectives working for the independence of the country (ARC 2008). Despite that numerous organizations were established in pursuant to their objectives.

The voluntary sector in the modern era is sub-divided into pre-independence and post-independence, whereas post-independence phase can easily be categorized into Nehruvian Phase (1947–1964), Indira Gandhi Phase (1964–1984), Rajeev Gandhi Phase (1984–1989), Post liberalization Phase (1991 and after) and Post-2014 phase.

Traditional Era

Traditional voluntary works in India stemmed from religious strictures cutting across all religions; in as much as the concept of salvation was very closely associated with charity and philanthropic giving. *Daan* (giving in charity), *Dakshina* (giving to guru in return for knowledge), *Bhiksha* (giving to monk) in Buddhism and *Zakat* in Islam, all connote the very conception of religious charity in some way.

During this phase, voluntarism was limited to cultural and religious activities; however, in limited sense, it was also extended to education, medicine and assistance in the time of distress like famine and disaster which continues to dominate today also. Education was imparted through *mathas*, *pathshalas* and *ashram* attached to Hindu temples. Likewise, Jainism and Buddhism promoted their version of volunteerism in the form of education and medicine. Moreover, a new kind of philanthropy emerged with the advent of Mughals in India which was limited mainly to food, education, hospital and shelter.

Modern Era

Pre-independence period

The advent of East India Company (EIC) in India witnessed a departure from the traditional to the modern form of volunteerism characterizing the arrival of Christian Missionaries largely engaged in the fields of education, hospital care, the welfare of downtrodden and the marginalized (Nair 2007). In the early stage, as state and religion interacted closely with each other, voluntary organizations manifested with religion were primarily engaged in welfare of poor and downtrodden and hence existed in harmony with state. However, there was no established principle to regulate the relationship between state and voluntary organizations and the same was largely influenced by the policies of the individual ruler.

From the early nineteenth century to the end of the colonial period was marked by intervention of church and church-associated organizations moreover process of socio-religious reform in different parts was initiated by educated Indians in the same period only. Subsequently, the Charter of 1813 removed all restrictions from Christian Missionaries functioning in India paving way to the establishment of various church and church-associated organizations. Although, their motive was to spread Christianity, however, they chose to do the same through the establishment of schools and hospitals in remote rural areas. Simultaneously, they also tried to organize the rural communities in cooperative credit groups and inculcated self-reliance in them.

During the same period, Indian social reformers like Raja Rammohan Roy and Ishwar Chandra Vidyasagar were influenced and inspired by the work of missionaries. Various forms of voluntary societies and collectives, namely the Brahmo Samaj, Arya Samaj, Ahmediya Movement, Aligarh Movement and Ramakrishna Mission, and so on were witnessed during this period. They imbibed scientificity, modernity and progressive thought in their approach. On the one hand, these voluntary collectives focused on the agenda of social reform at the grassroots level and advanced programs of mass mobilization for self-rule and self-reliance on the other. These were the foremost indigenous and organized CSOs of the country. Despite being specific to a region or community, they were common in ultimate manifestations. Some of them continue to exist and flourish even today.

Rapidly emerging social and educational organizations necessitated the enactment of *Societies Registration Act of 1860* to regulate and oversee the activities of the organizations. The act, which continues to govern the majority of CSOs in the country even today, provides the legal basis to these organizations. This was the first attempt to provide legal status, organize and oversee the voluntary works in India. The codification of law governing voluntary organizations limited the scope of work for the first time as a defined boundary did not exist earlier. It fixed the horizon under which voluntary organizations are supposed to work. Nevertheless, a firm foundation for the secular and current form of voluntary action was not laid until the 'Servants of India Society', an NGO, was established in 1905 in Pune by Gopal Krishna Gokhale (ADB, 2009).

The advent of Mahatma Gandhi in the freedom struggle paved the way for a new voluntary regime which witnessed the emergence of demand for *Swaraj* or Self-Rule. It invigorated Indian masses with socio-political awareness diffused with nationalist identity and turned the masses into a politically aware collective. Gandhian volunteerism thus facilitated a paradigmatic shift from socio-religious based reform to awareness about political process and the spirit of nation-building. In other words, Gandhian volunteerism was a hallmark of mass participation and political sensitization which led to various successful initiatives like Non-Cooperation Movement of 1921, Civil Disobedience Movement of 1930 and Quit India Movement of 1942 during the freedom struggle. Notwithstanding, Tandon (2002) observes, the legacy of volunteerism of political process and nation-building was not carried forward primarily because those who earlier worked in the voluntary sector subsequently became part of the government. Hence, post-independence India saw stagnancy in the vibrancy of volunteerism, but it was revived during emergency and post-emergency period (Tandon 2002).

Post-independence period

The post-independence period saw passive cooperation between state and civil society organizations which turned hostile after the imposition of the infamous emergency of 1975. The post-independence period witnessed the evolution of a large number of voluntary organizations based on the ideology of Mahatma Gandhi and state in the formative years promoted these organizations for development work. However, Tandon (2002) cited the reason of emergence of a large number of CSOs in post-independence era to the incapacity of the state to incorporate all leaders within the government who emerged out of the freedom struggle. In this phase, state supported the Gandhian voluntary initiatives with generous disbursement of the funds inasmuch as that in many cases government took control of these organizations through their governing bodies. However, in this process, state supported voluntary organizations lost their autonomy and consequently became the parastatal agencies, what is popularly referred as government organised non-governmental organizations (GONGOs). This was the first phase when state started co-opting the work of voluntary organizations, which ultimately compromised their autonomy.

The early support to the voluntary sector in post-independence India came up in the form of establishment of Central Social Welfare Board (CSWB) in 1953. Imperative of the new government in India was to fast track the economic growth, and in this process, social issues like health, education and sanitation were left behind. Consequently, civil society organizations (CSOs) were encouraged to fill the gaps. Until this period, NGOs were in passive support of the state and the state facilitated the much-needed platform. The government initiatives like the National Community Development Program (NCDP) and the National Extension Services (NES) revitalized the scope of the voluntary sector in free India.

Meanwhile, three tiers of Panchayati Raj Institutions (PRIs) were introduced in 1958 to promote decentralization in governance. To strengthen the voluntary sector in India, the Association for Voluntary Agencies for Rural Development (AVARD) was founded in 1958 as a consortium of voluntary agencies. However, the imposition of emergency drew the breaking point in the history of the volunteer sector in India, and there was a sudden shift in CSOs-state relationship in the post-emergency period when it turned into a confrontation from collaboration.

The period of 1970–1980 saw a turbulent relationship between state and CSOs because of the socialist movement started by Jaiprakash Narayan and the imposition of emergency by the then Prime Minister Indira Gandhi. In this period, the voluntary organizations were alleged to have nexus with foreign powers which wanted to destabilize the national government of India which consequently reduced the scope of work for the CSOs. Nevertheless, in-depth analysis of the reason for the strained relationship between the state and CSOs has been multifarious. State failure to meet the aspirations of the people and the inability to reduce poverty and inequality in independent India created disillusionment among intellectuals in particular and common people in general. This resulted in the exploration of alternate ways to state policy consequently leading to establishment of large number of non-profit organizations. The state system came under massive attack from various organizations from various parts of country in the form of student movement and socialist movement. Non-governmental organizations (NGOs) without any affiliations with political parties were seen as the answer to the failure of the democratic process, and these organizations mainly targeted the issues of the marginalized and the reduction of poverty. However, these organizations retained their Gandhian spirit of work and identity. Shaped by the need of hour, there was emergence of two different types of NGOs: one which directly targeted the reduction of poverty through the adoption of appropriate technology and collaborated with the government in the reduction of poverty, and the other which adopted the right-based approach with a focus on awareness and empowerment and assumed a confrontational position with the state.

Dissenting voices of the civil society against the policy of Congress government were quieted by the imposition of emergency in 1975–1977. The growing rift between civil society and government had apparent implications on resource base, workspace and autonomy of civil society organizations. This was the initial period when the government seriously started to think about regulating the work of these organizations and enacted the Foreign Contribution Regulation Act (FCRA), 1976. While the act was enacted ostensibly to regulate the foreign funds of NGOs whose activities were likely to be interpreted as pernicious to the sovereignty and integrity of the country, however indirectly, it tried to regulate the funding of political parties and CSOs having political affiliations (Nair 2007).

The non-Congress government in the post-emergency period gave a new lease of life for voluntary organizations. Janata Party Government at center introduced tax concessions for commercial companies for their voluntary initiatives. It finally led to the establishment of various successful innovative ventures and infused professionalism in non-profit organizations. Moreover, it was further strengthened by policy regarding promotion of voluntary sector in the Sixth Five Year Plan (1980–1985) document.

The easeful relationship between the government and CSOs was truncated by the subsequent Congress government in 1980. The government withdrew the tax concession provided by the Janata Party government. Moreover, the government also started to regulate the work of CSOs as the Act of 1976 enabled the government to keep a tab on their work and account. The increased tussle between the government and civil society organizations resulted in increased regulation and surveillance by the government. The government ensured that no CSO is involved in political mobilization on the one hand and political parties are not involved in mass mobilization in the garb of non-governmental work on the other. The active role played by CSOs in safeguarding the rights of people during the emergency turned the contemporary government hostile to them. Subsequently, the government instituted an inquiry commission known as the Kudal Commission (1980) against a large number of Gandhian organizations which caused irreparable damage to their reputation and work. In addition to that, the central government also imposed heavy regulation and control on CSOs through the institution of FCRA, and the financial act of 1983 curtailed exemption to corporate agencies against donations to NGOs. It removed all tax exemptions on income generation activities by NGOs. The government also established a national fund for rural development to channelize the corporate funds and prevent NGOs from directly getting funds from corporate houses. It also proposed for the establishment of national and state councils for voluntary agencies with code of ethics for NGOs; however, it never materialized after that.

Notwithstanding, the curtailment by the government, Sixth Plan and subsequent five-year plans took a call for voluntary sector participation in the process of development, which resulted in increment of funds to voluntary organizations by different government departments. As government fund to CSOs started ballooning, state started directing and controlling NGOs initially through PADI (People's Action for Development in India) and CART (Council for Advancement of Rural Technology) which were subsequently merged and constituted as CAPART (Council for Advancement of People's Action and Rural Technology). As the 1980s saw the state deflecting toward pro-market and stepping toward structural reforms, CSOs gained ground as 'third sector' complementary to state and market (Kohli 1990).

In the backdrop of politico-economic changes of the 1990s in the form of the Congress losing ground to regional parties and introduction of structural adjustment, there was a metamorphic change in the forms and nature of CSOs. Implementation of policies of transparency, accountability together with decentralization by the government had such an impact on NGOs that there was a paradigm shift in economic and social development approaches adopted by them. Due to the adoption of the structural adjustment programme, there was massive pressure in the government to perform better, become more visible and increase social capital, which consequently resulted in greater spaces of work for the third sector. After 1991, there has been an overall increase in collaboration between state, market and civil society organizations. We can witness the trend which suggests more and more NGOs are working in service delivery which traditionally was an exclusive domain of the sovereign state. The growing partnership between state and CSOs indicates the failure of the state in improving the reach and quality of essential services (Nair 2007).

However, Kohli (1990) observed that Ms. Indira Gandhi had become severely disillusioned about the state's potential for social change (Jenkins 2011). Therefore, she stressed the connection between contemporary voluntary organizations and a political class which caused her political downfall on the one hand, and she promoted the same directly or indirectly through liberalization on the other. She started the process of liberalization which got concrete shape under her son Rajiv Gandhi and finally under Prime Minister Narasimha Rao in 1991, which in turn created a space for CSOs under a broader framework of third sector in complementarity with state and market. The improving relationship between the state and the voluntary sector saw a new light in the 1990s, when the state apart from collaboration with CSOs started to open a forum for dialogues. Through a series of conferences in 1992, 1994, 2000 and 2004, Planning Commission of India initiated the state-voluntary sector interface. In the year 2000, the commission was entrusted to facilitate the dialogue between voluntary sector and state in order to identify the areas for collaboration. In addition, during the Eighth Five-Year Plan (FYP) government facilitated the creation of nationwide network of voluntary organizations which continued till the Tenth Five-Year Plan. In the same period, CAPART was decentralized into eight regional offices to facilitate the work of the voluntary sector in poor and remote areas of the country.

In addition to that, the decade of 1990s saw another significant development that is decentralized governance in the form of 73rd and 74th Constitutional Amendments which enlarged the working space for CSOs. Consequently, it legitimized their work at the local level in collaboration with local governance structures and Panchayati Raj Institutions (PRIs). In post-1990s, civil society organizations have grown enormously and have taken a different shape. In a changed scenario, these organizations are closely linked to development and engaged in a host of activities like community organizations, advocacy, human rights and development interventions. The paradigm shift in the nature and work of voluntary organizations has changed them into the agency of state.

The availability of a large amount of external as well as internal funds to the NGOs has created mixed results. In some cases, committed and visionary organizations took leverage of the resources and were able to achieve their objectives. However, numerous organizations emerged to siphon off the public funds. Diversification of voluntary sector continued in later half of 1990s because of liberalization, globalization, structural reform and decentralization together with ostensible state withdrawal from the service delivery. In the same period, some NGOs followed the public service contractor model. In contrast, others worked as intermediaries, and grassroot mobilization took the back seat as the strategy and agenda of these organizations.

In the post-1990s phase, the rapid growth necessitated a sense of social responsibility in 1990s, which was termed as 'new economy philanthropy' by Sidel (2000). The new kind of philanthropy is urban and has emerged due to successful economic innovation which in turn tries to create innovation to solve the social problems. In terms of sharing relationships with the state, the 'new economic philanthropy' is focused and working in tune with the state apparatus to solve the social problems (Sidel 2000). The focus of their intervention is limited to non-controversial issues like health and education in consonance with the requirements of the state. These organizations composed of the large economic base with well-resourced social institutions; hence, they have a huge capacity to usher the social change in comparison to the traditional philanthropic organizations. Where most of the conventional initiatives have been unorganized with unclear vision, the 'new economic philanthropy' is focused, organized and managed by professionals. The new philanthropic model has changed the landscape of "third sector"; moreover, they have more chances to succeed and lead the local and national initiatives due to their resource base and political clout.

Emerging Layers of Interaction Between State and CSOs

The kind of interaction which exists between state and CSOs directly depends upon the dominant strategy (of confrontation and cooperation) prevailing between them. Nevertheless, CSOs cannot be insulated to the outer world, and they have to respond to the socio-political and economic environment of the state to maintain their relevance. The interaction between the state and CSOs is a product of the space created by the 'state'. The dominant ideology of the government sets the agenda, whereas CSOs are expected to respond to that. Sometimes, state works as an enabler and provides a conducive environment for institutionalization of initiatives of CSOs and provides the legitimacy. In a federal system like ours, it is more complicated for CSOs to survive and operate when the ruling party at provincial level is different from federal level. The state expects the development organizations to respond to the changing priority of society. In post-2014, when the government shifted its focus from anti-poverty programs to skill building and Swacch Bharat, NGOs directly changed their priorities. They shifted their focus to the Clean India Mission and skill-building (Kumar 2010). Subsequently, when the state faced the global pressure from multilateral donor agencies and adopted structural adjustment programmes, NGOs in India scaled up their activities and changed their approaches. The outcome was cooption of CSOs representative in various bodies of the government. Mimicking and veering toward the governmental approach of development has impacted the functions of NGOs in such a manner that even NGOs are trying to scale up and assess their work on various quantitative methods as the government does. In a growth-rate oriented development regime where the government is trying to evaluate every impact on quantitative parameters, NGOs are also employing the same methods and apparatus. Current trends have prompted bureaucratization in NGOs, consequently incurring high transaction cost.

There are various weaknesses of the voluntary sector in India; notwithstanding, it has its strength also. The government has recognized their strength through various documents and has shown interest in collaboration with voluntary sector not only as a passive implementer of government policies and programmes but also as an active partner in policy formulation. The recognition is witnessed through the establishment of Voluntary Action Cell within Planning Commission, the apex policy-making body of India. The Planning Commission has been appointed as a nodal agency to interact with voluntary organizations regarding preparation of the Five-Year Plan (FYP) documents. Report of the '*Public Private Participation (PPP) Sub Group on Social Sector*' constituted by the Planning Commission has also affirmed the need to collaborate with the voluntary sector in the implementation of various government schemes. The report scrutinizes the poor performance of the public utilities and social services in general. It concludes that Public-Private-Partnership (PPP) brings in greater professionalism to voluntary organizations through introducing business practice on the one hand and ensuring better quality services on the other. It reiterates that PPP in the social sector is a promise of a better quality of services through clear focus. It contemplates that introduction of PPP would reverse the chronic under investment in social sector through mobilizing public and private capital; however, experience in this regard shows that it did not open the gate to private sector participation (GOI 2004).

The current debate on the role of CSOs indicates toward replacing the state as a representative of democracy. In the period of rapid economic growth and withdrawal of state, CSOs are filling the gap created by the withdrawal of state. In the changed scenario, CSOs are performing many tasks which were earlier the exclusive domain of the state. There has been a mechanism to hold the state accountable; however, there is a lack of accountability mechanism for NGOs. To leverage the role of NGOs in the process of development, there is a requirement of a balanced partnership to further the need and aspiration of the society (Kamat 2003). The changing role of CSOs from social change agents to the franchise of the state has changed the character of the sector. The change in functional approach of CSOs from activism to managerialism would have far-reaching consequences because technical staff take the work as apolitical and tend to disconnect from socio-political and economic processes. The increasing state funds for NGOs have necessitated them to professionalize their staff and work in the way state asks. They are mimicking the state apparatus, which has a pernicious effect on the functions of NGOs. In other words, state funding has a significant role in depoliticizing the work of NGOs and consciously state is facilitating the same.

Kamat (2003) in her study, '*Development Hegemony: NGOs and the State in India*', has shown that there are linkages between professionalization and de-politicization of NGOs. She also reiterates metamorphic transformation in the nature of work NGOs are dealing and the kind of organizational set up they are adopting. In her research in Western India, she indicates how NGOs moved away from confronting state through mobilization of poor for their empowerment, and instead, they took the skill-training approach to mitigate poverty and inequality, which suits the ideology of state (Kamat 2003).

Conclusion

In forging the tripartite partnership among state, market and civil society organizations (CSOs), NGOs have a very critical role. It provides a social basis for democracy by highlighting the critical issues of governance. CSOs' interaction with state and market determine its relationship with them. The civil society organizations collaborate with the government for policy formulation and depose its expertise to the government on a particular issue and sometimes acts as a whistleblower to ensure accountability and transparency in the government. The new development process is characterized by cooperation and constructive partnership among state, market and civil society. The nature of these three institutions makes them distinct in their approach and work; however, there are some areas where they need to engage and negotiate with each other for the betterment of society. However, cleavages have been orchestrated in such a manner that common ground seems abnormal, and conflicts seem very common and natural.

References

ADB. (2009). *Overview of Civil Society Organizations: India*. ADB.

Anheier, S. (2005). *Non-profit Organizations: Theory, Management and Policy*. Oxon: Routledge.

ARC, 2. (2008). *Social Capital: A Shared Destiny*. New Delhi: GOI.

Choudhury, E., & Ahmed, S. (2002). The Shifting Meaning of Governance: Public Accountability of Third Sector Organizations in an Emergent Global Regime. *International Journal of Public Administration, 25*(iv), 561–589.

Dongre, Y., & Gopalan, S. (2008). Third Sector Governance in India. In S. Hasan, & J. Onyx, *Comparative Third Sector Governance in Asia* (pp. 227–251). New York: Springer.

Eade, D., & Ligteringen, E. (2001). NGOs and the future: taking stock, shaping debates, changing practice. In D. Eade, & E. Ligteringen, *Debating Development NGOs and the Future* (pp. 11–18). Oxfam International.

Edward, M., & Hulme, D. (1996). Too Close for Comfort? The Impact of Official Aid on Non Governmental Organizations. *World Development*, 961–973.

GOI. (2004). *Report of the PPP Sub-Group on Social Sector.* Government Report, New Delhi.

Hasan, S., Lyons, M., & Onyx, J. (2008). Third Sector Organisation Governance: Introducing the themes and the chapters. In S. Hasan, & J. Onyx, *Comparative Third Sector Governance in Asia: Structure, Process and Political Economy* (pp. 1–18). Springer.

Jenkins, R. (2011). NGOs and Indian Politics. In N. G. Jayal & P. B. Mehta (Eds.), *The Oxford Companion to Politics in India* (pp. 409–426). New Delhi: Oxford University Press.

Kamat, S. (2003). NGOs and the New Democracy: The False Saviors of International Development. *Development and Modernization, 25*(1).

Kohli, A. (1990). *Democracy and Discontent: India's Growing Crisis of Governability*. Princeton: Princeton University Press.

Kumar, A. (2010). Question of Poor. In R. Taylor (Ed.), *Third Sector Research*. New York: Springer.

Nair, P. (2007). *Whose Public Action? Analysing Inter-Sectoral Collaboration for Service Delivery*. London: International Development Department.
Pierre, J., & Peters, B. G. (2000). *Governance, Politics and the State*. New York: St. Martin's Press.
Sidel, M. (2000). *The New Economy Philanthropy in the High Technology Community of Bangalore and Hyderabad, India: Partnership with the State and Ambiguous Search for Social Innovation*. generousgiving.org.
Steiner-Khamsi, G. (2008). What's left for NGOs to do? In I. Silova, & G. Steiner-Khamsi, *How Ngos React* (pp. 247–284). USA: Kumarian Press.
Tandon, R. (2002). *Volutary Action, Civil Society and the State*. New Delhi: Mosaic Books.
Young, D. (2000). Alternative Models of Government-Nonprofit Relations: Theoretical and International Perspectives. Nonprofit and Voluntary Sector Quarterly, 149–172.

10

The Civil Struggle to Access Airwaves: Study on SARU Community Radio Initiative in Sri Lanka

M. C. Rasmin and W. A. D. P. Wanigasundera

Introduction

"SARU" was an independent Community Radio (CR) project, started in 2005, in Sri Lanka, by a civil society organization called Pulathisi Federation operating out of the Polonnaruwa District. This has been identified as the very first civil society-led CR initiative in Sri Lanka. The World Bank (WB) supported this initiative, through its Livelihood Development Project, implemented by Gemidiriya Foundation. The civil society members, predominantly farmers, from at least 27 villages worked for nearly 10 years, in preparing this project with the support of the Federation. During this period, the community was mobilized, a team of 32 youngsters were capacitated, program guidelines were developed, sustainability plans were established, and a test transmission was completed

M. C. Rasmin (✉)
Assam Don Bosco University, Guwahati, Assam, India

W. A. D. P. Wanigasundera
University of Peradeniya, Peradeniya, Sri Lanka

© The Author(s) 2020
Md. N. Momen et al. (eds.), *Building Sustainable Communities*,
https://doi.org/10.1007/978-981-15-2393-9_10

with the highest level of participation of community members. After such a lengthy struggle, the SARU team realized that the existing legal system with regards to radio broadcasting does not accommodate the rights of a civil society group such as the Federation, to own a radio license and frequency. Alternatively, the Federation made an agreement with the state broadcaster, the Sri Lanka Broadcasting Corporation (SLBC) on September 09, 2010 to use one of its spare frequencies. However, after signing an agreement, SLBC did not act according to the agreement and the project could not reach its goal, despite a decade of civil struggle. This study emphasizes the importance of having a legal recognition for the people's rights to own a communication medium like CR and the need for an action-based, consistent, and participatory advocacy plan to make policy changes to accommodate CR that is led by civil society.

What Is Community Radio?

The term "Community Radio" is used widely and understood differently across the world (Wilkins et al. 2014) In Latin America, it is known as "grassroots" or "civic" radio. In Africa, it is known as "rural" or "bush" radio. In some parts of Europe, it is referred to as "free" radio (Jankowski 2002). In Australia, "public" radio and in the United Sates it is known as "community" radio (Offer 2002). This indicates that CR is often conceptualized and perceived differently by people in various parts of the world. However, it has historically remained as a medium of the people who are often excluded by the mainstream media, or otherwise do not have a voice. CR is a medium of the people, by the people, for the people (Mtimde 2000; Tabing 2002; Jallov 2012). The non-profit nature, community ownership, participation, and self-management are the key characteristics of CR (Barlow 1988; Fraser and Estrada 2001; Offer 2002). The CR has potential to reach marginalized, hold governments accountable, advocate for their rights, and drive social change. It is an autonomous entity—owned, managed, controlled by community (Bosch 2014; Pavarala and Malik 2007).

Community Radio in Sri Lanka

It is widely believed that the CR in Sri Lanka started with the Accelerated Mahaweli Development Scheme—a massive irrigation development project, implemented by the Government of Sri Lanka (GoSL), in the late 1970s. A participatory radio project called Mahaweli Community Radio (MCR) started to facilitate the development of the people who were newly re-settled along the Mahaweli Project in 1979, with the support of Danish International Development Agency (DANIDA) and UNESCO (David 1992, 2008; Jayarathne et al. 2005; Rasmin 2011). However, scholars believe that it had never been owned, managed, controlled by the people in Sri Lanka and the entire MCR should be considered as a state driven, community-based broadcasting system not truly CR as such (Jayarathne et al. 2005; Gunawardene 2007). It should be noted that all the MCR stations are now shut-down and none of the stations were owned by civil society groups that functioned in Sri Lanka. Jayarathne (2012) argues that CR has never been started as result of a need felt by the people in Sri Lanka. Under such a complex history, the SARU community Radio project can be considered as the first civil-society-led-CR project, started by the people, in the Polonnaruwa District of Sri Lanka.

SARU Community Radio

SARU was initiated to use CR as medium to provide a voice for the members from a specific community, pre-dominantly farmers in Polonnaruwa District, to facilitate their social economic growth, to promote local knowledge and culture, and to remain as an independent people's voice (Swarnalatha and Batuvitage 2012). During the WB pilot study on 'Community Radio in Sri Lanka', a research team visited Polonnaruwa twice and discovered an overwhelming interest by the people in Polonnaruwa to start a CR (Jayarathne et al. 2005). Eventually, the WB appointed local experts to support the initiation of a radio called SARU. Swarnalatha and Batuvitage (2012) noted that the WB consultants used participatory approaches to engage the community at large preparing for SARU CR. In the initial stage, the team concentrated in capacitating the community. They worked on forming and planning committees, formulating policies and plans, identifying program types, training broadcasters and building community consensus getting the community. WB and the Federation formed an assembly of 135 people from 27 villages; 11-member SARU management for SARU and 32 young people were selected as trainee broadcasters. The Federation was able to mobilize 30,000 signatures of the people in this region expressing their need for a CR station.

Subsequently, in the second phase, a test transmission was carried out and the Federation applied for Community Radio Licenses. Eventually, in 2010, the Management Team of the Federation and *Gemidiriya* Foundation submitted application to the Ministry of Media and Telecommunication Regulatory Commission of Sri Lanka (TRCSL), seeking CR license and a dedicated spectrum. As indicated in the introduction, after nearly 10 years of struggle, the SARU team came to realize that the existing legal system does not allow organizations such as the Federation to own a Radio station. By now, it has been more than 15 years, and people in the Polonnaruwa District, where the SARU CR was initiated, are waiting for a favorable response from the Government of Sri Lanka (GoSL).

Objective of the Study

This study is an attempt to understand the need of the policy recognition and constant, action oriented, goal-driven, civil intervention to secure their rights to communicate through CR. While this is a specific case to Sri Lanka, the learning of SARU could definitely be applied in a larger context around the world as well.

Methodology

In order to achieve the objective, the explanatory qualitative case study methodology is used in this study. Such methodology has been widely used to understand a specific phenomenon (Joia 2002). While this methodology allows the use of multiple data sources (Baxter and Jack 2008), it provides an opportunity to explore a larger context and a scenario (Yin 2003). Jankowski (1991) supported qualitative method despite a variety of other methods, including participant observation, descriptive case studies, theoretical discussion, and contribution to policy debate which are the approaches that could contextualize to generate arguments.

For this study, five Focus Group Discussions (FGD) were conducted, of which four FGDs were held in Polonnaruwa district, with the members of the Federation, trained young broadcasters, people from the village where the test transmission was carried out, and people from a neutral village. The Fifth FGD was conducted in Colombo with broadcasting policy experts. In addition, 23 in-depth interviews were conducted with the people who actively worked for various committees in the SARU. The Federation committee provided full access to the project documents during the field survey and analysis. A policy consultation with media freedom advocates and community media policy experts was organized to understand the broader media policy environment. Following results table is presented based on the response collected through in-depth interviews that provides a quantifiable perspective. However, data collected from other sources—FGD and consultations—are carefully used to inform the discussion, results, and the conclusion.

Results

The close analysis of data collected discovered that several factors that affected the SARU in its plight in different degrees. The Table 10.1 shows the factors and their impact on the sustainability of SARU CR project.

Table 10.1 Issues affecting SARU community radio

Issue	No effect (Weightage = 0)	Some effect (Weightage = 2)	High effect (Weightage = 3)	Achieved marks	Total marks	%
1. Complex project structure	3	17	6	26	46	56.5
2. Poor policy interventions in the planning stage	3	5	30	38	46	82.6
3. Lack of legal recognition and rigid media policy	0	1	44	45	46	97.8
4. Poor public engagement in activism and in policy lobbying	2	11	20	33	46	71.7
5. Political unwillingness	0	10	26	36	46	78.3
6. Poor public knowledge on media policy	1	2	40	43	46	93.5

Effect of the Lack of an Inclusive Policy

The close examination of data discovered, despite higher-level inclusivity and people's participation in the planning process of SARU, it could not go-on-air permanently, because of lack of legal recognition. Table 10.1 proves that the lack of an adequate policy framework to accommodate CR and a rigid broadcast policy are the two key factors affecting the attempt of the Federation and people in Polonnaruwa in obtaining a legitimate license and a dedicated frequency. It was discovered that there were only two legal instruments—SLBC Act, No 37 of 1966 (as amended) and TRCSL Act No 27 of 1996 (as amended) that existed pertaining to broadcasting in Sri Lanka. It was found that these two instruments control various aspects of independent broadcasting in Sri Lanka while they contain no provisions directly or indirectly on community broadcasting that is entirely owned by the community or civil society groups.

Exclusion of People in the Policy Dialogue

As discussed in the previous section ("SARU Community Radio" section), SARU planning process was very comprehensive which included starting from creating a community need to conducting the test transmission. However, there was no evidence found to substantiate the inclusion of policy dialogue in the planning stage. Table 10.1 confirms that the exclusion of policy discussion in the initial stage affected the ability of the Federation to secure legal recognition for SARU. Both document review, FGD, and individual interviews confirmed that neither public were educated about the policy requirements, rigid nature of Sri Lankan broadcast media policy, and challenges nor did they engage in lobbying for license and frequency. It was revealed that in most of the meetings with state authorities very few people representing SARU participated—there was no inclusive and active participation from the civil society.

Passive Activism

The activism led by the Federation with its constituencies in convincing the Government of Sri Lanka and the authorities was found to be very passive—not action oriented; consistent; and goal driven. In reviewing the Federation documents, around 10 formal letters were found to have been submitted (from 04.01.2006 to 26.03.2010) to various parties—politicians; government authorities—Ministry of Media and Telecommunication Regulatory Commission of Sri Lanka (TRCS); other Ministries, and Presidents. However, none of these formal communications failed to make any impact while there was no substantial evidence available on the involvement of people in organizing policy forums, discussions, campaigns, media activism, and protests, and so on in support of their letters.

Poor Knowledge of Relevant Policy

The knowledge level of people who were involved in planning SARU on media policy mechanism was found to be very low. There have not been adequate attempts by the Federation, WB, and other concerned parties to educate people about the larger policy context in which SARU began its journey. However, it is noteworthy that the SARU team and the Federation made immense efforts to build the capacity of the young broadcasters and the community to run the radio. The section below discusses the results presented in Table 10.1, in a broader context of the implementation of the SARU project. Based on these results and discussions, the recommendation will be presented later in this chapter.

Discussion

Inclusion of Policy Discussion in the Planning

As briefly indicated in the previous section, the entire planning exercise was heavily focused on mobilizing community, developing program formats and guidelines, developing sustainability plan, capacitating the village level broadcasters, and preparing for test transmission and so on.

Community Building

In the planning process, building a community consensus was given priority. Close analysis of the responses confirmed that the majority of the people in Polonnaruwa strongly believed that they needed a CR for two strong reasons. Firstly, CR as an alternative to the Mainstream Radio (MR). Secondly, as a tool to amplify their voices and culture. The responses observed during the FGDs and in-depth interviews are presented in Table 10.2.

The analysis of the responses discovered that the people think a medium like SARU CR can facilitate their own development. A respondent from the village of Nissankamallapura mentioned, "We have our

Table 10.2 People's perspectives on the need of SARU Radio

As an alternative to mainstream radio	As a medium to amplify the people's voice
• MR excludes marginalized; • Heavily focus on problem not so on the solutions; • Does not extend sense of 'ownership' to the people; • Does not have consistency in investigating people's issues; • Higher elite and competitive nature and less focus on people's knowledge; • Give too much attention on politics with less relevance in the life of marginalized.	• CR can provide vital information to the people on their own development; • It can promote local talents; culture; heritage; • Address issues related to health, environment, climate, agriculture, education; • It can amplify the voice and help in keeping governance system accountable to the people.

own knowledge; our voice; our stories; our concerns; our talents and our rights and our development". The majority of the respondents revealed that they needed a radio that belongs to them.

Program Planning

While building community consensus, the Federation and the WB team engaged the community in developing inclusive programs, ethics, and guidelines. The responses revealed that the SARU radio programs were planned to inform the people on their own development; promote people's rights and co-existence and governance. Table 10.3 presents the program strategy developed by the SARU.

Sustainability Plan

The members of SARU and the Federation came up with a comprehensive sustainability plan for CR that included aspects of both financial and social sustainability. From the responses, it became apparent that this plan was developed through a wider consultation with people from various parts of Polonnaruwa. Respondents noted that they were ready to

Table 10.3 Program strategy of SARU

Provide information	Promote community rights	Promote democracy
• CR program should enhance community efforts in their own development; • Provide cutting edge knowledge, new information; • Provide information about new marketing opportunities; • Provide and facilitate access to development planning and activities	• Program should provide space for people's stories; • Should remain independent; • Promote freedom of speech and expression; and • Provide space for people to raise their voice	• Programs should promote unique, individual and collective identities; • Promote local culture and heritage; • Inclusion and diversity • Foster freedom of expression; and • Strengthen democratic process, good governance and justice

share the financial burden that was a great strength observed of SARU. The majority of the respondents confirmed that they were ready to make donation from their harvests. Under the social sustainability plan, people looked at possibilities of bringing community support, volunteerism, knowledge, guidance in strengthening and sustaining SARU broadcasting. The key observations of the SARU sustainability plan are presented in the Table 10.4.

Training Communicators

The training programs offered for the young people have made significant impacts on the life of the young communicators. The respondents confirmed that it enhanced the interpersonal skills; ability to articulate complex problems; ability to identify community issues; ability to produce creative radio programs and the production capacity of young boys and girls. Young boys and girls have learned not only broadcasting skills but also gained deeper knowledge about various concepts such as community, participation, gender, inclusion, freedom of expression, and so on. The adults found visible change in the knowledge and behavior of the young boys and girls who were trained. In learning production skills, youngsters enhanced their ability to scripting, community interviewing,

Table 10.4 Sustainability plan of SARU community radio

Financial sustainability plan	Social sustainability plan
Collecting membership (Rs.10/= per month)	Regular trainings and recruitment of interns
An external mechanism to collect monthly fee	Connect SARU with schools' media clubs
Donations from donors and businessmen	Using mobile technologies for programming
Organize film-shows and drama events	Provide adequate time for communities
Establish a street theatre group to collect funds	Encourage local talents of different kinds
Run an audio visual; production service	Develop system to sustain participation
Collect donations soon after local harvesting	Generate constant community contents
Airing obituary notices; public announcements	Seek community volunteerism
Produce sponsored programmes	Connect community organizations

field production, making one-minute programs, and so on. Most of the young people who worked for SARU are presently married. Most of them have children. It was noted that several young people have been using their communication skills gained during the SARU trainings, in their present employment.

Test Transmission

SARU CR hosted a 10 days of initial test transmission from 14th May 2009, using a temporary frequency—FM 96.1., provided by SLBC with the temporary approval of TRCSL. The test-broadcasting took place at *Nishshankamallapura,* which is one of the villages where the *Gemidiriya* project was being implemented. The data revealed that the test transmission discovered collective strength and hidden talents of the youth, children and adults. People recognized various issues affecting women, children, youth, and broadly, farmers through the programs. More women, the elderly persons, even the ones who were jobless stayed around the live studio and supported logistics. Men helped in building the studio, setting up antenna and other aspects of logistics. People provided meals for the organizers; hosted the guest resource persons and made milk rice to celebrate the test transmission. During the night time, even after the end of the transmission, people did not go home, some remained at the studio and some stayed around singing songs.

Absence of Policy Discussion

Having discussed the entire planning process and its impact, it should be noted, that policy preparation was not part of the planning; people were not sensitized and or educated during the preparation on potential policy challenges; and people were not engaged in lobbying for dedicated CR policy or separate license. The section below will further discuss this. It was learned that the one year of WB training does not include any session on media policy. Neither WB nor its consultants explained to the people the legal scenario with regard to community broadcasting in Sri Lanka. It became very obvious that the planning of SARU happened in a policy vacuum. The planning did not capitalize on the collective strength of the people to make an effective appeal to the Government, seeking legitimate CR license.

Struggling for Legitimacy

Although the SARU project was initiated in 2005, the Federation took nearly four years to submit an application to the Ministry of Media, seeking CR license. During this period, the Federation wrote several appeals requesting license and frequency to various authorities of which ten were found. It was revealed that the intention of the Federation was to articulate the need of the CR and to explain its potential. This study discovered two key findings with regard to lobbying for license by the Federation. Firstly, the whole process of making formal communication with authorities was observed to be highly passive with no adequate interaction, strong demands, and call for action. These passive communications did not transform into a community movement, as in India, Nepal, and Bangladesh. There has also been an underlying perception that CR policy can just be granted by the Ministry of Media and TRCSL, which was not true.

"Policy development process has to be very inclusive, constructive, and open. The NEFEJ and ACORAB in Nepal, made one-to-one contact with several government entities, organized public awareness forums, mobilized ground level support base, and stayed engaged with key stakeholders and policy makers". (*personal communication with a CR policy expert from Nepal*—July 22, 2017)	"Government did not want to free airwaves. Judiciary made a good ground for us. We sensitized people; capacitated them; explained them the policy challenges; involved them in meetings; included them in policy consultations. The Dalit women who were part of DDS, were fully aware of our necessity of legal recognition for CR". (*personal communication with a CR expert from India*—July 25, 2017)	"We cannot just create any animosity with the government. Convincing and educating government is part of the advocacy. We worked closely with people, organizations, academia, NGOs, individuals, political activists and several others. Any policy development cannot be achieved in isolation". (*personal communication with a CR advocate from Dhaka*—August 1, 2017)

NEFEJ: Nepal Forum of Environmental Journalists
ACORAB: Association of Community Radio Broadcasters Nepal
DDS: Deccan Development Society

Secondly, in the absence of adequate provisions in the existing broadcasting law, authorities could not proceed with the application from the Federation. Radio broadcasting in Sri Lanka has historically been controlled largely by the Government of Sri Lanka (Weerasinghe 2018). The private broadcasting in Sri Lanka was recognized in the 1980s (Jayarathne and Kellapotha 2015). There was no policy or legal instrument to differentiate state or public service broadcasting and commercial broadcasting. Apparently, there was no three-tier broadcasting system too (Jayarathne et al. 2005). As indicated in the "Results" section, both the key legal instruments—SLBC Act, no 37 of 1966 (as amended)—TRCSL Act No 27 of 1996 (as amended) control various aspects of independent broadcasting and the room for the establishment of a decentralized, independent community broadcasting system which is owned by non-profit people's entities is narrowed or not in existence.

It was learned that the licensing and spectrum allocation process is not transparently happening in Sri Lanka. Often, Radio licenses were given without any standard criteria. Even though the frequencies are limited public property, there is no accountability involved in the process of allocating frequencies and they are distributed within limited groups. As per the responses and available data, no license has been given to independent community group. Ownership plurality was very much compromised due to political influence (NSMR 2016; IREX 2017; Jayawardena and Gunatillake 2015; Weerasinghe 2018). Under such rigid media policy system, it was observed that the Federation, with its weak advocacy, could not get the legal recognition—license and frequency.

Memorandum of Understanding with State Broadcaster

At some point, the Federation realized that their efforts to convince the Ministry of Media and TRCSL could not continue. A collective decision was made by the Federation to sign an MoU with the state broadcaster—SLBC. Accordingly, an MoU was signed on 09 September 2010. This was to be effective for two years. As per the MoU, SLBC agreed to provide one of its spare frequencies FM 100.7 to SARU. Federation agreed to pay the SLBC to purchase transmitter and relevant radio equipment on its behalf. This surprising arrangement, however, is completely contradictory to WB feasibility study produced by Jayarathne et al. (2005). One of the major findings of the feasibility study was that SLBC is not a supportive and a feasible entity for the independent and community broadcasting. What was even more problematic was that, MoU made it compulsory for SARY to seek prior approval from SLBC, before airing any programs (memorandum of understanding with state broadcaster). The most problematic provision in the agreement was that—if the SLBC wished, it could request the scripts of SARU programs for prior scrutiny. As per the MoU, it was obvious that SLBC was empowered to increase the monthly fee, control editorial independence, terminate the service at any time with or without stating reason and block the transmission, and so on. Majority of the respondents noted that this was the best possible solution at that time—to survive with SLBC satisfying its wills, until getting a dedicated license. Though the MoU was signed between the two parties, SLBC did not act on its promise. Not even for a single day, the

SARU could go on air, as per the MoU. It was evidenced that the money paid by the Federation, to SLBC, to purchase a transmitter for SARU was not returned for some years in several installments. The Federation has to almost prepare a legal case against the SLBC.

People's Realization on Policy

MoU became yet another unsuccessful attempt by the Federation. With MoU being unsuccessful, the majority of the people attached to SARU realized that the existing legal system did not support the idea of an independent and community led-broadcasting.

During the FGD with the Federation, the people who were in the forefront of the SARU project firmly noted that lack of a policy provision was the major cause for the failure of SARU. They also believed that a policy level intervention was needed to create a viable environment for CR in Sri Lanka, and working with SLBC temporary frequency was not a long-lasting solution. A Federation member noted that, "If there is an equitable law, we should have owned a radio. License is for big people; frequencies are for big people; tower is for big people; all the government assistance is only for big people; we have everything but no radio". Another added, "We do not know where to go now. There is no legal system, so nobody would want to provide a solution for our problem. If there is a policy, the secretaries of the media ministry would have reacted accordingly". It is noteworthy that people in Polonnaruwa believed the government was afraid of people's freedom of speech and therefore not responsive to their demand for a radio license.

Way Forward

Having a voice and right to communication is a fundamental entitlement of people in any community (Buckley 2008). Globally and in South Asia, such civil-entitlement has been legally recognized in different ways. In Nepal, the Broadcasting Policy has given space for community groups and civil society organizations to own, manage and run CR stations

(Dahal and Aram 2011). In India, after a decade of activism, locally based, registered community organizations can own community Radio from 2006 with the revised CR policy guidelines (Pavarala and Malik 2007). In Bangladesh, the CR as third sector is legally recognized through 2008 CR installation guideline (Reza 2012). However, in Sri Lanka, it has become impossible for civil society to push a policy reform on CR. It is therefore, vital for the people in Polonnruwa and any other civil society groups, who are envisioning launching a CR Project like SARU to strategize a rights-based advocacy plan to gain access to airways as the very first step.

A right-based advocacy plan can be designed keeping few existing legal provisions in context. Firstly, Sri Lankan constitution could be a strong basis to establish an argument demanding CR for civil society. Constitution firmly guarantees the freedom of expression (FOE). According to Article 14(1) (a), every citizen is entitled to freedom of speech and expression, including publication. Despite this provision is somewhat compromised with several other laws—national security law, emergency regulation, Sri Lanka press council law, official secrets act, act related to contempt of court, parliamentary privileges, in Sri Lanka (IREX 2017; CPA 2011), the FOE can still be a legitimate ground make an advocacy plan for Sri Lanka.

Secondly, the Sri Lankan Supreme Court has given two prenominal judgments in the past to uphold the independence broadcasting sector. In the year 2008, a Ministerial Gazette, dated 10th October was produced by the then government. According to the gazette, companies with multi-million-rupee investments must apply for new licenses by 10th November. This was ruled out by the Supreme Court and highly criticized by the local experts. This is a strong precedence where the judiciary wanted to uphold the independence of the broadcasting.

Similarly, a controversial Broadcasting Authority Bill was tabled in the parliament in April 1997. As per, the proposed bill, a new broadcasting authority was to be established. The Bill stipulated that the proposed Authority was to be directly appointed by the Media Minister. As per the bill, the Media Minister was to be empowered to issue guidelines in respect of the operation of stations through state-issued licenses. Moreover, it afforded the Minister the power to dictate policy and program content in a manner that would have 'rendered the electronic media completely vulnerable to the whims and fancies of politicians and any partisan interests they might represent. Jayawardena and Gunatillake (2015) observed that the Supreme Court's ruling against the entire Bill was a seminal judgment and a reflection of the independence of the judiciary during that period. This judgment can be considered as a trigger to push independent broadcasting and space for CR in Sri Lanka.

Finally, SARU Federation and other similar civil groups need to come up with a holistic, goal driven, constant, action-oriented, policy-focused lobbying with the support of a mass community movement for policy level change to seek legal recognition for people to access airwaves.

References

Barlow, W. (1988). Community Radio in the U.S.: The Struggle for a Democratic Medium. *Media, Culture and Society, 10*(1), 81–105.

Baxter, P., & Jack, S. (2008). Qualitative Case Study Methodology: Study Design and Implementation for Novice Researchers. *Qualitative Report, 13*(4), 544–559. Retrieved from http://nsuworks.nova.edu/tqr/vol13/iss4/2.

Bosch, T. E. (2014). Community Radio. In K. G. Wilkins, T. Tufte, & R. Obregon (Eds.), *The Handbook of Development Communication and Social Change*. [Online]. Chicester: John Wiley & Sons.

Buckley, S. (2008). *Community Broadcasting: Good Practice in Policy, Law and Regulation*. Paper prepared for UNESCO for World Press Freedom Day.

CPA. (2011). *Freedom of Expression on the Internet in Sri Lanka*. Colombo: CPA.

Dahal, S., & Aram, A. (2011). Crafting a Community Radio 'Friendly' Broadcast Policy in Nepal, Observatorio (OBS*) Journal. V.5. Anna University, Chennai, India.

David, M. J. R. (1992). Mahaweli Community Radio. In B. Girard (Ed.), *A Passion for Radio: Radio Waves and Community*. Montreal: Black Rose Books.

David, M. J. R. (2008). Sri Lanka. In C. A. Arnaldo & L. N. Tabing (Eds.), *Community Radio in Asia: Putting People in Charge* (pp. 13–30). Tambuli Community Radio, Manila: CRDL Enterprise. https://doi.org/10.1080/10402659908426238.

Fraser, C., & Estrada, S. R. (2001). *Community Radio Handbook*. Paris: UNESCO.

Gunawardene, N. (2007, June 4). Lions and Community Radio: Part of Sri Lanka's Mythical Lore. Retrieved from http://movingimages.wordpress.com/2007/06/04/lions-and-community-radio-part-of-sri-lankas-mythical-lore.html.

IREX. (2017). *Media Sustainability Index*. Washington, DC: International Research Exchange Board.

Jallov, B. (2012). *Empowerment Radio - Voices Building a Community* (1st ed.). Gudhjem: Empower House.

Jankowski, N. (1991). Qualitative Research and Community Media. In K. B. Jensen et al. (Eds.), *A Handbook of Qualitative Methods for Mass Communication Research* (pp. 167–174). London: Routledge.

Jankowski, N. W. (2002). The Conceptual Contours of Community Media. In N. W. Jankowski & O. Prehn (Eds.), *Community Media in the Information Age: Perspectives and Prospects* (pp. 1–18). Cresskill, NJ: Hampton Press.

Jayarathne, T. (2012). Sri Lankan's Struggle to Establish Community Radio. In K. Seneviratne (Ed.), *People's Voice. People's Empowerment* (pp. 309–319). Singapore: AMIC.

Jayarathne, T., & Kellapotha, S. (2015). The Political Economy of the Electronic Media. In W. Crawley, D. Page, & K. Jayawardena (Eds.), *Sri Lanka, in Embattled Media: Democracy, Governance and Reform in Sri Lanka*. SAGE Publication.

Jayarathne, T., Jayawardena, K. P., & Gunaratne, A. (2005). *Community Radio in Sri Lanka*. Colombo: Law and Society Trust.

Jayawardena, K., & Gunatillake, G. (2015). One Step Forward and Many Steps Back; Media Reform Examined. In W. Crawley, D. Page, & K. Jayawardena (Eds.), *Embattled Media: Democracy, Governance and Reform in Sri Lanka*. SAGE Publication.

Joia, L. (2002). Analyzing a Web-Based e-Commerce Learning Community: A Case Study in Brazil. *Internet Research, 12*(4), 305–317. https://doi.org/10.1108/10662240210438434.

Mtimde, L. (2000). Radio Broadcasting in South Africa: An Overview. *International Journal of Cultural Studies, Vol., 3*(2), 173–179.

NSMR. (2016). *Rebuilding Public Trust, National Secretariat for Media Reform*. Colombo: NSMR.

Offer, J. O. (2002). *Community Radio and Its Influence in the Society: The Case of Enugu State, Nigeria*. Frankfurt am Main: IKO-Verlag fur Interkulturelle Kommunikation.

Pavarala, V., & Malik, K. (2007). *Other Voice: The Struggle for Community Radio in India*. London: SAGE.

Rasmin, M. C. (2011). *Community Radio - A Medium for Human Development*. Colombo: Chemamadu Publication.

Reza, S. M. (2012). From Elite Perceptions to Marginal Voices: Community Radio in Bangladesh. In J. Gordon (Ed.), *Community Radio in the Twenty First Century*. Bern: Peter Lang AG, International Academic Publishers.

Swarnalatha, S., & Batuvitage, G. (2012). Sri Lankan Experiment Community Driven. In K. Seneviratne (Ed.), *People's Voice. People's Empowerment* (pp. 309–319). Singapore: AMIC.

Tabing, L. (2002). *How to Do Community Radio, A Primer for Community Radio Operators* (p. 9). Delhi: UNESCO.

Weerasinghe, P. (2018). *Democratizing Airways: A Multi-Stakeholder Problem Analysis on Broadcast Regulation in Sri Lanka*. NSMR: Colombo.

Wilkins, K. G., Tufte, T., & Obregon, R. (2014). *The Handbook of Development Communication and Social Change*. John Wiley & Sons, Inc.

Yin, R. K. (2003). *Applications of Case Study Research: Design and Methods* (2nd ed.). Sage Publications.

Part III

Civil Society Initiatives for Women Welfare

Rajendra Baikady, Md Nurul Momen, and Cheng Sheng-Li

Introduction

The South Asian region is characterized by high population, dependence on agriculture and poor income. The region is also recognized as deprived region in the world (HDSA 1997). South Asia is also known as the "patriarch belt" (Caldwell 1982) and the kinship-ordered social structure system followed in the region makes women subordinate to the men (Mathema 1998). Women in the region also have low status, and are denied access to property and land (Bardhan 1986). In addition, the work done by women is not recognized and also largely unpaid. Strong patriarchal ideology still prevalent in the region controls the socio-cultural practices in many of these countries. These socio-cultural practices restrict the mobility of women and prevent them from accessing opportunities for development and enhancing their abilities. Women in most South Asian countries face discrimination in many spheres of their lives. Even though it is not possible to quantify the gender discrimination faced by women in all areas of their lives, some areas such as labour market, political participation, education participation, health and nutrition, socio-cultural participation are the most affected and influenced by the strong patriarchal ideology.

The terms equality and equal opportunity refer to non-discriminatory access to sources of livelihood, health, and education, social, economic and political participation. Whereas gender discrimination restricts women's access to power and authority, the institutions of class, religion and ethnic hierarchies and socio-cultural traditions, customs and norms further keeps women away from public life (Kabeer 1994; Carr et al. 1996). Therefore, empowerment of women is a process that ensures transformation of women's lives by changing the structure of these institutions and ensuring equality and accessibility.

Women in South Asia: A Situational Analysis

Despite challenges, the status of women in the South Asian region in the recent decades seems to be better off in a number of well-being indicators across the life cycle. There are improvements in terms of life expectancy, education and working conditions and many women have also entered politics at least at the local government levels in many countries. The region is also showing a greater positivity towards the issues related to women and an understanding of the need for women empowerment, which in turn helps in bringing equality of opportunity. Women's economic contribution in the household chores is rarely paid and acknowledged in the region. The nature of work performed by women and the flawed definition of economic activity makes women's contribution statistically invisible. Women in the region play a predominant role in the care sector. The reproductive responsibilities and care work, at least most of it, is undertaken primarily by women, which is unpaid. All tasks that are associated with bearing, rearing and nurturing children are not considered as productive work and not economically accounted for in many South Asian countries. While even though these activities performed by women in the region are largely considered as unproductive domestic chores, which leads to undervaluation and non-recognition of their economic contribution, they play a very crucial role in maintaining family stability and nurturing the environment of any society. It is because of the non-recognition of women's contribution to domestic economy and non-monetary benefits that women continue to face low status in the South Asian region.

Women's contribution in the South Asian economy is largely not acknowledged. Women in Sri Lanka contribute extensively in the agricultural productivity and food production, which is the main economic sector for rural population in the country. In addition, women in the country are also extensively involved in the plantation sector (Weerahewa and Ariyawardana 2003). In general, rural women in the South Asian region contribute abundantly in activities such as crop farming, animal husbandry and a host of off-farm activities. Looking after livestock, including collecting fodder and water, farming eggs and milk, ensuring the health and safety of animals, and taking care of animal comfort which is associated with these activities, consumes extensive time of women in the region.

Despite the region's economic growth and development, the situation of women has not improved. The region has witnessed a downward trend in the child sex ratio at birth and excess female mortality rate. There are many studies conducted in relation to sex ratio at birth, gender differences in nutrition intake, and gender differences in mortality rate (UN 1998; Bhat 2002; Krishnaji and James 2002; Waldron 1998) in an effort to empower women. The studies have also examined the linkages between fertility rates, sex preferences, contraceptive use and abortions in many South Asian countries, including Nepal and Bangladesh; however, the imbalance is not significant (Arnold 2001; Karki 1992; Leone et al. 2003).

The overall lower social status of women in the South Asian region results in deteriorated health as well as overall well-being and quality of life for women. Women's health requirements are always given low priority because of their subordination to men, low recognition and unequal pay for their work, low participation in the formal economy, lack of labour participation, lack of access to economic resources and decision making, lower level of participation in education and low social participation, all keeping women in low health status and vulnerable. As a result of these subjugated rights, women themselves do not realize and articulate their health care needs in the region. Low self-esteem and lack of adequate public health services constrain women from accessing health care resources (Baru 2003; Gopalan and Shiva 2000). On the other hand, South Asia is reported to be one of the poorest regions in the world and accommodates more than 40 per cent of the world's poor (HDSA 1997). Given this level of poverty, availing private health care services by women is likely to be limited. Furthermore, some research found that excessive importance is given to specific reproductive health, whereas other general health needs are utterly neglected (Qadeer 1998; Datta 2003). On the other hand, even though women tend to experience more mental illness and there is high prevalence of sexual violence in the region, studies focusing on these areas are limited (Goudet et al. 2018). Maternal nutrition programmes have very poor performance in the low- and middle-income countries (Goudet et al. 2018), which is evidence that there is less preference for women's health. Women in the region face severe anaemia and calcium deficiency which makes them vulnerable and increases the risk of postpartum haemorrhage and hypertensive disorders (Rahman et al. 2016; World Health Organization [WHO] 2013). Postpartum haemorrhage and hypertensive disorders are the most common reasons for global maternal deaths worldwide (Say et al. 2014). South Asia accounts for over one-third of the world's anaemic women and no country in the region is showing progress towards reducing anaemia in women of reproductive age by 2025 (Development Initiatives 2017).

The need for comprehensive maternal nutrition initiative in the region has been well recorded in the research. It has been argued that comprehensive maternal nutrition interventions during pregnancy will help boost pregnant women's health and thereby contribute in healthy fetal growth and development (Bhutta et al. 2013; Dewey 2016). Low education, lack of family support and inadequate coverage of nutritional programmes are the barriers for women in the low- and middle-income countries for availing and adapting to recommended nutrition behaviour (Kavle and Landry 2017; Victora et al. 2012). This in turn also affects maternal health in the region. One out of 10 women in the South Asian region is reported to have low stature and few has very low body mass index (BMI) (Goudet et al. 2018). The issues related to low maternal stature and BMI need severe attention because studies have been proving these to be the contributing reasons for child stunting in South Asia (Kim et al. 2017). It is also important to note that the region should make considerable efforts in improving maternal and child nutrition to achieve the targeted nutritional level for mother and child as agreed in the global nutrition targets of the Sustainable Development Goals and the World Health Assembly. These developments in return are said to enhance the developmental potential of children and the economic development of the South Asian nations (Aguayo and Menon 2016; Vir 2016; Development Initiatives 2017).

Female labour force participation in the region is quite low and influenced by many factors. Education plays a considerable role in female labour force participation (Tansel 2001) and even human capital theories recognize the importance of education in the employment outcomes. Educational attainment of an individual plays a major role in the decision to participate in the labour market (Tansel 2001). The literature on human capital predicts that high labour force participation is influenced by higher level of education, which results in higher productivity (Ejaz 2007; Psacharopoulos and Tzannatos 1989; Tansel 2001). A number of studies have shown higher returns to education for women (Duraisamy 2000; Psacharopoulos 1994). Further, the importance of women's education and its influence on human development and women empowerment is well accepted by the South Asian countries. Sri Lanka and Maldives have the best performance with respect to women's education in the region (Jayaweera and Sanmugam 2002). However, even in Sri Lanka and Maldives, women's educational participation at the higher educational, professional and non-traditional course levels is curtailed by gender bias (Jayaweera and Sanmugam 2002).

Cultural and societal norms across the globe and especially in the South Asian region are expected to have a significant influence on women's decision to participate in the labour market. These customs and traditional practices also influence their choice of work and mobility. The region, caste and the religion further make these processes complicated. Research has widely recognized that women have been restricted to the role of care givers and constantly constrained by taking up paid employment (Das and Desai 2003; Desai and Jain 1994; Goksel 2012; Jaeger 2010; Panda 1999). Women's rights and their engagement at the workplace are restricted by cultural factors and religion. Further, religion still plays a major role in determining gender-specific social norms in many countries, including the South Asian region. Restricted participation by women in the workforce is seen more among the South Asian countries, where women are further constrained by traditional gender roles and familial relations (Das 2006), restricted mainly to caregiving work (Das 2006).

A study focusing on the situation of women in urban India found that women labour force participation is negatively affected by higher social status and the "Sanskritization" process (Klasen and Pieters 2012). Religious and cultural conservatism is restricting women's workforce participation in Turkey and Chile (O'Neil and Bilgin 2013; Pastore and Tenaglia 2013). However, the studies have also found that attractive job opportunities with higher wages encourage better educated women to participate in the labour market and the stigma attached to labour market participation among these women is found to be less (Klasen and Pieters 2012). This relationship in South Asia has been documented by several earlier studies (Das 2006; Olsen and Mehta 2006) that have found a negative relationship between the two (Das and Desai 2003; Dasgupta and Goldar 2005). However, there are some other studies that found this relationship to be positive (Bhalla and Kaur 2011; Faridi et al. 2009; Hafeez and Ahmad 2002). Ejaz (2007); Klasen and Lamanna (2009) noted the importance of declining fertility rate and women's participation in the labour market, which in turn leads to positive impact on economic development. It is important to note that unchanged patriarchal mindset, cultural and traditional structure and norms of the South Asian societies are the major constraints in achieving gender equality and overall women empowerment in the region (Agarwal 2001). However, women empowerment is the route by which women enhance their decision-making power, that is, a distribution and transformation of women's position from the powerless to the powerful. This process of empowerment also ensures the economic, political and social participation of women in any society. Research studies across the region and even the world shows that sexual and gender-based violence are the criminal activities that primarily target women, and these criminal activities and the crime committed against women show the marginalization of women in any society (Balve 2015). Subordination of women is largely prevalent in the South Asian region and studies have found that women in society in relation to men have poor positions in all socio-economic, cultural and political affairs (Gupta 2003). Even in India, a democratic political system where the Constitution ensures certain privileges and rights for women in the country, women still face discriminatory practices (Sharma 2002).

Part III of the book includes chapters focusing on programmes related to social welfare issues, such as women empowerment, women's rights, child welfare, and globalization and its effects on civil society in South Asia. Chapter 11 is an attempt to familiarize the different waves of women's movement spearheaded by civil society agencies, and the role of non-governmental agencies, activists raising topical issues for women's life and dignity will be deliberated upon in this section. At the end, the chapter provides some policy implications to strengthen the role of civil society for an enhanced role as a champion and protector of women's human rights in contemporary India. Chapter 12 elaborates a case study of SKDRDP's model of women empowerment and its impact on women and involvement of local women as village level workers in the process of planning and implementation of development programs in India. Chapter 13 presents a case study that illustrates the role of non-governmental organizations in solving the social problems in Pakistan. In order to do so, the authors analyze the services, functions and projects of All Pakistan Women's Association (APWA). Chapter 14 illustrates the civil society's contribution with respect to women empowerment in Bangladesh. The main objective of this chapter is to provide a bird's eye view on the contribution of civil society in the case of women empowerment in Bangladesh. The findings are expressed through the results of qualitative research with a purposive sampling technique. Chapter 15 explores the impact of Third Sector Organizations (TSOs) on women empowerment through promotion of entrepreneurship. The interpretations in this chapter are based on the analysis of data captured through a well-structured questionnaire that was administered to the beneficiaries to understand the extent of enablement that the interventions from these organizations provided them before and after the interventions.

References

Sharma. (2002). *Women in Indian Religions*. Oxford University Press.

Agarwal, S. P. (2001). *Women's Education in India (1995–98) Present Status, Perspective, Plan, Statistical Indicators with Global View* (Vol. III). New Delhi: Concept Publications Co.

Aguayo, V. M., & Menon, P. (2016). Stop Stunting: Improving Child Feeding, Women's Nutrition and Household Sanitation in South Asia. *Maternal & Child Nutrition, 12*(S1), 3–11. https://doi.org/10.1111/mcn.12283.

Arnold, F. (2001). Son Preference in South Asia. In J. F. Sathar & Z. A. Phillips (Eds.), *Fertility Transition in South Asia* (pp. 281–299). Oxford: Oxford University Press.

Balve, S. (2015). Women Entrepreneurship Towards Women Empowerment in India: Plan Initiatives. *Southern Economist, 54*(3), 11–16.

Bardhan, K. (1986). Women: Work, Welfare, and Status: Forces of Tradition and Change in India. *South Asia Bulletin, 6*(1), 3–16.

Baru, R. V. (2003). Privatisation of Health Services – A South Asian Perspective. *Economic and Political Weekly, 38*(42), 4433–4437.

Bhalla, S., & Kaur, R. (2011). *Labour Force Participation of Women in India: Some Facts, Some Queries*. Working Paper No. 40. London: Asia Research Centre, London School of Economics and Political Science.

Bhat, P. N. M. (2002). On the Trail of 'Missing' Indian Females – I: Search for Clues and II: Illusion and Reality. *Economic and Political Weekly, 37*(51–52), 5105–5118, 5244–5263.

Bhutta, Z. A., Das, J. K., Rizvi, A., Gaffey, M. F., Walker, N., Horton, S., & Black, R. E. (2013). Evidence-Based Interventions for Improvement of Maternal and Child Nutrition: What Can Be Done and at What Cost? *Lancet, 382*(9890), 452–477. https://doi.org/10.1016/S0140-6736(13)60996-4.

Caldwell, J. (1982). *Theory of Fertility Decline*. London: Academic Press.

Carr, M., Chen, M., & Jhabvala, R. (Eds.). (1996). *Speaking Out – Women's Economic Empowerment in South Asia*. New Delhi: Vistaar Publications.

Das, M. B. (2006). *Do Traditional Axes of Exclusion Affect Labour Market Outcomes in India?* Social Development Papers, South Asia Series, No. 97. Washington, DC: World Bank.

Das, M. B., & Desai, S. (2003). *Why Are Educated Women Less Likely to Be Employed in India? Testing Competing Hypotheses.* Social Protection Discussion Paper Series, No. 313. Washington, DC: World Bank.

Dasgupta, P., & Goldar, B. (2005). *Female Labour Supply in Rural India: An Econometric Analysis.* Working Paper. New Delhi: Institute of Economic Growth.

Datta, A. (2003). Articulation of an Integrated Women's Health Policy Using the Life Cycle Approach. *Indian Journal of Gender Studies, 10*(1), 25–43.

Desai, S., & Jain, D. (1994). Maternal Employment and Changes in Family Dynamics: The Social Context of Women's Work in Rural South India. *Population and Development Review, 20*(1), 115–136.

Development Initiatives. (2017). *Global Nutrition Report 2017: Nourishing the SDGs.* Bristol, UK: Development Initiatives.

Dewey, K. G. (2016). Reducing Stunting by Improving Maternal, Infant and Young Child Nutrition in Regions Such as South Asia: Evidence, Challenges and Opportunities. *Maternal & Child Nutrition, 12*, 27–38. https://doi.org/10.1111/mcn.12282.

Duraisamy, P. (2000). *Changes in Returns to Education in India, 1983–94: By Gender, Age-Cohort and Location.* Centre Discussion Paper, No. 815. Connecticut: Economic Growth Centre, Yale University.

Ejaz, M. (2007). Determinants of Female Labor Force Participation in Pakistan: An Empirical Analysis of PSLM (2004–05) Micro Data. *The Lahore Journal of Economics, 2*(Special Edition), 204–235.

Faridi, M. Z., Malik, S., & Basit, A. B. (2009). Impact of Education on Female Labour Force Participation in Pakistan: Empirical Evidence from Primary Data Analysis. *Pakistan Journal of Social Sciences, 29*(1), 127–140.

Goksel, I. (2012). *The Reasons of Decreasing Trend of Female Labour Force Participation in Turkey: The Role of Conservatism.* Working Paper No. 5. Balcova Izmir: Izmir University of Economics.

Gopalan, S., & Shiva, M. (2000). *National Profile on Women, Health and Development.* Voluntary Health Association of India Compiled for WHO.

Gupta, N. L. (2003). *Women's Education Through Ages.* New Delhi: Concept Publications Co.

Hafeez, A., & Ahmad, E. (2002). Factors Determining the Labour Force Participation Decision of Educated Married Women in a District of Punjab. *Pakistan Economic and Social Review, 40*(1), 75–88.

HDSA. (1997–2002). *Human Development in South Asia.* MahbubulHaq Human Development Centre, Karachi: Oxford University Press.

Jaeger, U. (2010). *Working or Stay-at-Home Mum? The Influence of Family Benefits and Religiosity.* Working Paper No. 84. Munich: Information Institute, Leibniz Institute for Economic Research, University of Munich.

Jayaweera, S., & Sanmugam, T. (2002). *Graduate Employment in Sri Lanka in the 1990s.* Colombo: Centre for Women's Research.

Kabeer, N. (1994). *Reversed Realities – Gender Hierarchies in Development Thought.* New Delhi: Kali for Women.

Karki, Y. B. (1992). Sex Ratio in Nepal. *Economic Journal of Nepal, 15*(1), 30–37.

Kavle, J. A., & Landry, M. (2017). Addressing Barriers to Maternal Nutrition in Low- and Middle-Income Countries: A Review of the Evidence and Programme Implications. *Maternal & Child Nutrition, 14*(1), e12508. https://doi.org/10.1111/mcn.12508.

Kim, R., Mejia-Guevara, I., Corsi, D. J., Aguayo, V. M., & Subramanian, S. V. (2017). Relative Importance of 13 Correlates of Child Stunting in South Asia: Insights from Nationally Representative Data from Afghanistan, Bangladesh, India, Nepal, and Pakistan. *Social Science & Medicine, 187*, 144–154. https://doi.org/10.1016/j.socscimed.2017.06.017.

Klasen, S., & Lamanna, F. (2009). The Impact of Gender Inequality in Education and Employment on Economic Growth: New Evidence for a Panel of Countries. *Feminist Economics, 15*(3), 91–132.

Klasen, S., & Pieters, J. (2012). *Push Or Pull? Drivers of Female Labor Force Participation During India's Economic Boom.* IZA Discussion Paper Series, Working Paper No. 6395. Bonn: Institute for the Study of Labor.

Krishnaji, N., & James, K. S. (2002). Gender Differentials in Adult Mortality with Notes and Rural-Urban Contrasts. *Economic and Political Weekly, 37*(46), 4633–4637.

Leone, T., Mathews, Z., & GianpieroDallaZuanna. (2003). Impact and Determinants of Sex Preference in Nepal. *International Family Planning Perspectives, 29*(2), 69–75.

Mathema, M. (1998). Women in South Asia: Pakistan, Bangladesh, and Nepal, in Stromquist (ed). Matlab, Bangladesh. *International Family Planning Perspectives, 27*(3), 137–143.

O'Neil, M. L., & Bilgin, M. H. (2013). Religion as a Factor Influencing Turkish Women's Decisions to Work. *Journal of International Women's Studies, 14*(3), 163–173.

Olsen, W., & Mehta, S. (2006). A Pluralist Account of Labour Participation in India. Working Paper (Global Poverty Research Group). Retrieved October 14, 2014, from http://economics.ouls.ox.ac.uk/14041/1/gprg-wps-042.pdf.

Panda, P. K. (1999). *Poverty and Young Women's Employment: Linkages in Kerala.* Working Paper. Thiruvananthapuram, Centre for Development Studies.

Pastore, F., & Tenaglia, S. (2013). *Oraet Non Labora? A Test of the Impact of Religion on Female Labor Supply.* IZA Discussion Paper Series, Working Paper No. 7356. Bonn: Institute for the Study of Labor.

Psacharopoulos, G. (1994). Returns to Investment in Education: A Global Update. *World Development, 22*(9), 1325–1343.

Psacharopoulos, G., & Tzannatos, Z. (1989). Female Labour Force Participation: An International Perspective. *World Bank Research Observer, 4*(2), 187–201.

Qadeer, I. (1998). Reproductive Health – A Public Health Perspective. *Economic and Political Weekly, 33*(41), 2675–2684.

Rahman, M. M., Abe, S. K., Rahman, M. S., Kanda, M., Narita, S., Bilano, V., & Shibuya, K. (2016). Maternal Anemia and Risk of Adverse Birth and Health Outcomes in Low- and Middle-Income Countries: Systematic Review and Meta-Analysis. *American Journal of Clinical Nutrition, 103*(2), 495–504. https://doi.org/10.3945/ajcn.115.107896.

Say, L., Chou, D., Gemmil, A., Tuncalp, P., Moller, A. B., Daniels, J., & Alkema, L. (2014). Global Causes of Maternal Death: A WHO Systematic Analysis. *Lancet, 2*(6), e323–e33z. https://doi.org/10.1016/S2214-109X(14)70227-X.

Goudet, S., Murira, Z., Torlesse, H., Hatchard, J., & Busch-Hallen, J. (2018). Effectiveness of Programme Approaches to Improve the Coverage of Maternal Nutrition Interventions in South Asia. *Maternal and Child Nutrition, 14*(S4), e12699. https://doi.org/10.1111/mcn.12699.

Tansel, A. (2001). *Economic Development and Female Labor Force Participation in Turkey: Time-Series Evidence and cross-Province Estimates*. ERC Working Papers in Economics, Economic Research Centre. Ankara: Middle East Technical University.

UN. (1998). *Too Young To Die: Genes Or Gender?* Department of Economic and Social Affairs, Population Division, New York: United Nations.

Victora, C. G., Barros, F. C., Assuncao, M. C., Restrepo-Mendez, M. C., Matijasevich, A., & Martorell, R. (2012). Scaling Up Maternal Nutrition Programs to Improve Birth Outcomes: A Review of Implementation Issues. *Food and Nutrition Bulletin, 33*(2 Suppl), S6–S26. https://doi.org/10.1177/15648265120332S102.

Vir, S. C. (2016). Improving Women's Nutrition Imperative for Rapid Reduction of Childhood Stunting in South Asia: Coupling of Nutrition Specific Interventions with Nutrition Sensitive Measures Essential. *Maternal & Child Nutrition, 12*(S1), 72–90. https://doi.org/10.1111/mcn.12255.

Waldron, I. (1998). Sex Differences in Infant and Early Childhood Mortality: Major Causes of Death and Possible Biological Causes. In UN (Ed.), *Too Young to Die: Genes or Gender* (pp. 64–83). New York: United Nations.

Weerahewa, J., & Ariyawardana, A. (2003). Impact of the WTO on Women Workers in Sri Lanka. In V. Jha (Ed.), *Trade, Globalisation and Gender – Evidence from South Asia*. New Delhi: United Nations Development Fund for Women in Collaboration with UNCTAD.

World Health Organization (WHO). (2013). *Guideline: Calcium Supplementation in Pregnant Women*. Geneva: World Health Organization.

11

Civil Society as the Guardian of Women's Human Rights in India

Bharti Chhibber

Introduction

Vibrancy of democracy in a state is known by the strength of civil society in the country. The chapter examines the role of civil society in women's struggle for equality in the post-colonial India. Women have been suppressed for long in a male-dominated social order that has created stereotypes putting women in secondary position. Inhuman social evils like child marriage, domestic violence, dowry deaths, female foeticide and infanticide, sexual harassment at workplace and sexual assault violate women's human rights to live a dignified life.

At the international level, many initiatives like UN Convention on Elimination of All Forms of Discrimination Against Women (CEDAW) adopted in 1979, Fourth World Conference on Women held at Beijing in 1995 and Millennium Development Goals 2000 recognized the issue of gender equality calling for remedial action at the country level. In post-colonial India, Article 15 of the Constitution states that there can be

B. Chhibber (✉)
Department of Political Science, University of Delhi, New Delhi, Delhi, India

© The Author(s) 2020
Md. N. Momen et al. (eds.), *Building Sustainable Communities*,
https://doi.org/10.1007/978-981-15-2393-9_11

no discrimination on the grounds of religion, race, caste or sex. Likewise, the Constitution has also banned practice of forced labour and human trafficking. Further, Article 39 of Part IV of the Constitution comprising Directive Principles of State Policy reflects that states shall secure equal right to livelihood for women and equal pay for equal work. However, in practice women are still relegated to second-class citizens in many cases within the patriarchal society.

Women were part of the Indian independence movement both as volunteers and leaders. After independence women's grassroots movements with increased emphasis on economic and social rights of women came to forefront. This was followed by campaign against specific evils like dowry deaths, rape and sex-selective abortions ultimately forcing the governments to act. Relatively recent example is the civil society outrage in the Nirbhaya case in 2012. We have a robust legal system to ensure protection of women's rights including Dowry Prohibition Act and (anti) Sati Act and Protection of Women from Domestic Violence Act. However, their proper implementation is the call of the hour. Further, laws act only as a code of conduct to be followed, and they require public initiatives to bring about a change in the dominant social structure. It is imperative that legal and societal reforms go hand in hand.

The chapter is in two sections. After a brief theoretical understanding of the concepts of civil society and conceptualizing women's rights under gender studies, the first section examines the critical issue of women's human rights in India. The second section analyses women's struggle for equality and the demand for human rights in India. Civil society played a significant role in demanding and promoting the cause of women's rights through socio-legal reforms in post-colonial India. The different waves of women's movement spearheaded by civil society agencies, role of non-governmental agencies and activists raising topical issues of women's life and dignity will be deliberated upon in this section. Finally, the chapter provides some policy implications to strengthen the role of civil society for it to play an enhanced role as a champion and protector of women's human rights in contemporary India.

Notion of Civil Society

As a term 'civil society' denotes a sphere of autonomous groups and organizations functioning independently of the government (Heywood 2014). It is a sphere where non-governmental organizations and groups interact. These associations are voluntary and activism and social movements are the critical elements of civil society with a definite focus on rights of the people.

Charles Taylor highlighted two major thoughts in the understanding of 'civil society'. Although both aim at containing an absolutist state, for John Locke, society exists prior to the government. The second tradition flows from the eighteenth-century French thinkers, Montesquieu, and his disciple Alexis de Tocqueville. They believed in a constitution, rule of law and independent groups to restrain state despotism (Taylor 1990). Later theorists like Jürgen Habermas saw civil society as a mechanism of developing common discourse and keep a check on the state. On a different line, Marx and Gramsci linked state with bourgeois in civil society protecting the interests of the propertied class (Elliot 2003).

Historically, 'civil society' gained momentum with the Solidarity movement in Poland. It was popularized by the liberals during the eighteenth-century Europe, but lost sheen later in the nineteenth century with the state-centric approach. The 'civil society' again became prominent in the last quarter of the twentieth century with the disintegration of the communist states and political activist movements for freedom of expression against repressive states (Elliott 2003). Charles Taylor reflects that civil society again became a talking point in the late 1970s and early 1980s. However, civil society is no longer synonymous with the political society, but as in the Hegelian terms, civil society exists against the state and is partially independent of it (Taylor 1990). In the context of relatively later entrants in the modern nation-state system, Khilnani emphasizes that new states have come under tremendous pressure as they are expected to guarantee their own security together with legitimizing their existence within the modern democratic system working for the welfare of their citizens (Khilnani 2001).

In contemporary times civil society is increasingly identified with social movements that is closely associated with the issue of human rights. Human rights are intrinsic to the very existence of human beings. As Chris Brown argues, the key role of the government is to ensure that human rights are not violated. Acceptance of the political system is based on its performance and functioning. Consequently, political life can be seen as a direct or indirect agreement between the people and the government (Brown 2008). Civil society provide platform for information and deliberation over critical issues affecting the masses and making it heard at the governmental level (Putnam 2000).

Conceptualizing Women Rights Within Gender Studies

Civil society initiatives have been closely associated with the gender issues. Gender refers to the socially and culturally formulated distinction between femininity and masculinity. These differences instead of being biological are the result of interplay of many societal, technological and material factors. Gender is an outcome of internalization of perspective and mindset of the community regarding divergent role-plays and obligations for women and men. The traditional view is based on public-private dichotomy where women are confined to household tasks and men are expected to go out and earn. This socialization process can be seen from childhood itself. You must have also noticed that girl child is usually given toys in the form of kitchen utensils to play with, whereas boys are given toy cars and toy guns to prepare them for their future roles. This conforms to masculine characteristics for men like power and autonomy and feminine characteristics like weakness, dependence and emotions associated with women. Since these characteristics are social constructions, not biological ones, they indicate gender inequality representing power relationships.

Western theorists have pointed out that women's work in household goes unpaid, their role as producers is ignored signifying power relations and their subordination. Sylvia Walby discusses three approaches to gender subjugation—socialization theory, neo-Freudian psychoanalytic theory and discourse analysis. Socialization approach emphasizes that masculine and feminine identities are due to the socialization process from early childhood. Here masculinity is characterized by aggressiveness and assertiveness, while femininity is associated with passive and gentle behaviour. The media feeds into these perceptions through images of women as being physically attractive and glamourous, while men are shown as authoritarian and dominant. Psychoanalytic approach stresses on psychological thinking forwarded from one generation to another creating gendered identities in the process. Discourse approach points out to the ideological basis of patriarchy responsible for women's oppression. Language, religion, mythology, cinema and literature are seen as part of patriarchal discourse. This approach is criticized for negating the role of economic and political institutions in constructing gendered identities (Walby 1990).

Liberal feminists argue that by eliminating legal hurdles, the issue of women's subjugation can be taken care of. But post-liberal approaches believe that patriarchy is structural, entrenched in societies, and legal changes will not be suffice to remove gender inequalities. Marxist and socialist feminists approach explains women's subordination in economic terms which recognizes only paid work in the public realm negating women's household contribution which is not valued. Women perform both reproductive and labour work in households which amounts to double burden. Post-colonialists stress on colonial relations of control and subservience under European imperialism in the eighteenth and nineteenth centuries.

Post-modernist examines relationship between knowledge and power highlighting that those who create knowledge get power in the process. Post-modernist feminists show that men are considered as people who know everything and knowledge is actually based on men's lives in the public realm. They challenge the universal assumptions highlighting the need for the recognition of diversity and differences. Also, within women as a category there can be diverse identities and experiences which need to be taken into account. Hence, in the nineteenth century feminine discourse was centred on polarization with women confined to separate spheres of private world and men belonged to the public world. Later on, debate included arguments on unity and diversity among women.

Gender Discourse in India and Human Rights

Cultural anthropological studies in India have criticized ethnocentrism and class bias of Western gender narratives trying to project a universalistic theory. In fact, coloured women in the West have themselves criticized Western notions for ignoring discrimination on the basis of race and class. Chandra Mohanty denounces Western feminists' description of women of developing world as poor, illiterate, always discriminated against and lacking agency. She argues that women's subservience does not require universal understanding of women's demands, rather must be dealt with in specific cultural situation (Mohanty 1988).

In India gender discourse has highlighted women's struggle in varied but interrelated realms of patriarchal society, power relations, interaction of caste, class and gender relations and impact of globalization. Owing to diverse cultures, languages and regional specificities of India, gender narratives have divergences, debates and disagreements. Bandana Purkayastha, Mangala Subramaniam, Manisha Desai and Sunita Bose (2003) in their article on 'The Study of Gender in India: A Partial Review' have highlighted three overlapping dimensions in gender approaches in India. Firstly, there is an underlying current of conflict between writers who have prominently focussed on gender and those who have a more convergent perspective. Secondly, the epistemological origin of work on gender is not necessarily constructed on feminism. Thirdly, the function of the nation-state and international structures in fostering gender inequities is substantially stressed (Purkayastha et al. 2003). Writers like Butalia (1993) and Ray and Basu (1999) foreground gender, whereas others like Kelkar (1992), Gandhi and Shah (1992) and Kumar (1993) have reflected on the intersectionality of class, religion, caste, gender and regional dynamics. Gail Omvedt in her work, *We Will Smash This Prison*, reflects on 'personal is political' to highlight discrimination against women. She questioned the patriarchal values of power and control over women and confinement of women to the private domain to subjugate her by the male authority (1980). Gender discourse in India has also touched other spheres. Vandana Shiva has worked on environment in the context of women in the form of ecofeminism (1993). Bagchi (1996) and Chatterjee (1992) emphasized on the marginalization of women further due to mechanisms of neocolonialism, neoliberalism and globalization.

Kamla Bhasin underscores that family, religion, media and law perpetuate patriarchal system. They legitimize and naturalize patriarchy. With a male head in the family, boys and girls learn 'appropriate' behaviour as demanded by society with boys learning to control and girls to behave obediently and be subservient. Media's portrayal of women as inferior further legitimizes patriarchy. With the technological finesse and artistic projection, media messages and images become compelling (Bhasin 1986).

International Concerns for Women's Rights

At the international level, issue of women's rights has been raised and discussed in various organizations including the United Nations (UN). On the recommendation of the Committee on the Status of Women, the UN Economic and Social Council in 1990 adopted a resolution pointing out that violation of women's rights in the public and private spheres is related to unequal position in the society (United Nations 1995b: 131–132). The 1979 Convention on the Elimination of All Forms of Discrimination Against Women (CEDAW) does not specifically talk about violence against women. But the 1992 statement stated that violence against women curtails their human rights and states are required to mitigate violence both by administration and in the private realm (United Nations 1995b: 131–132).

The Vienna Declaration of the 1993 UN Conference on Human Rights called for the appointment of a special rapporteur on violence against women and drafting of a declaration eliminating violence against women. In 1994 the General Assembly of the United Nations adopted the Declaration on the Elimination of Violence Against Women. Article 1 defines violence against women as physical, sexual and psychological harm or threats of harm in public or family. The Declaration mentions that gender violence is due to unequal power relations between men and women which is socially formulated and is defended on historical grounds (Coomaraswamy and Kois 1999: 183).

The Fourth World Conference on Women in Beijing held in 1995 in its policy document, the Platform for Action, explained violence against women in terms of gender cruelty culminating in physical, sexual or psychological distress to women, coercive action and denial of freedom (United Nations 1995a). It further states that brutality against women deprives them of basic opportunity to enjoy their human rights. The failure of state to ensure women's rights is a matter of serious concern (United Nations 1995a).

These international conferences may not have binding power on states; however, they do build public opinion and channelize people's attention to the crucial issues affecting humanity as a whole.

> Human rights declarations and conventions contribute to cultural change. They articulate new norms legitimated by the fact that they were produced through a process of international deliberation that resulted in consensus. Communications systems such as the Internet disseminate these documents and bring global attention to behavior that is defined as a human rights violation. Although these networks of communication and the consequent mobilization of public opinion cannot be called law, they can serve as powerful motivators for local actors concerned about their public image on the international stage. Mobilizing transnational public opinion through reports, information exchange, and the media and the arts contributes in important ways to strengthening the quasi-legal system of human rights. (Merry 2009: 83)

Similarly, declarations have the moral power of the world behind them (Coomaraswamy and Kois 1999: 182).

Women Movements in India

In India, we can trace the first wave of contemporary women's movement to the pre-independence period. Women were part of the freedom movement both as volunteers and leaders. Social and religious reforms and women's education were contributory factors in this development.

After independence, Indian constitution guaranteed equality to women. Even in the Fundamental Duties, Article 51A specifies that every citizen should come forward to eliminate behaviour disparaging to the honour of women. However, even after seven decades of independence and growth in India, women still suffer from subjugation. At times, due to various personal laws, right to equality of women is not enforceable. Owing to religious and cultural connotations, these become difficult to reform. Agnihotri reflects that the women's movement in India is a critical effort to reclaim political and public space as an equal citizen of the country. Violence, however, is perpetrated through the state system and its machinery as well as the society and family. It draws strength from ideologies, age-old customs, traditions, religious and caste identities propagating status quo (Agnihotri and Majumdar 1995).

Women are underrepresented in the political sphere in governance, in parliament and bureaucracy as well as in other administrative units like judicial services, armed forces and in police services. It is indeed ironical because it is often argued that women political leaders would contribute to a more peaceful global order (Fukuyama 1998). Denial of economic and political opportunity exacerbates subjugation of women. Further, violence against women in unequal relations highlight continuance of ideologies that endorses violence during conflict. 'Aggressive behaviour is considered part of daily life and dominant conceptions of masculinity valorise tough, aggressive behaviour' (Vickers 2002).

Likewise, many practices banned by the Constitution are still continuing including *begar* (forced labour) and women and child trafficking. In Part IV, Directive Principles of State Policy, Article 39, talks about equal wages for women on par with men, but this is openly flouted. The Maternity Benefit Act 1961 guarantees maternity leave with wages to women but women working in the unorganized sector are deprived of this right. According to an estimate by the National Commission on Self-Employed Women, 94% of women workers are part of this informal sector. Therefore, a large number of women are deprived of social security even in cases where they are legally entitled to it (Anker 1998). This is in spite of the fact that if women join the mainstream workforce, country's GDP growth would touch our true potential. The Maternity Benefit (Amendment) Act, 2017, provides for 26 weeks of fully paid leave for the first two children. It applies to any establishment having ten or more employees. The Act introduces leaves for mothers having a baby through adoption or surrogacy. It further states that establishments with more than 50 employees must set up or provide a crèche facility for children (GOI 2017).

In addition, owing to the development projects, environmental disasters and in uprooting due to conflictual situations, it is the women who face the maximum brunt of the situation. In the rural sector, strong class-caste relationship in the ownership of land is also disadvantageous for women. Besides, women's reproductive work has always been undervalued, it is not considered a work at all. Additionally, women's productive work like food production for the family again goes unpaid. It is a sad situation that modern means of communication including media with their wide reach are used to project women as consumers and reproductive beings rather than producers.

In the 1946–1951 Tebhaga movement for land reforms, women participated with conviction. *Nari Bahini*, women volunteers, reported to have guarded villages at night. In the early 1970s, women organized local movements to deal with immediate issues affecting their daily lives. For example, in Maharashtra, tribal women organized *Shramik Sangathana* against alcohol-related violence by drunken husbands and illegal alcohol production and consumption. To highlight their right to live without fear of violence, women destroyed illegal liquor shops in villages. In then Bombay, anti-price rise movement is another example. In 1972, Self-Employed Women's Association (SEWA) was formed from women's wing of the Textile Labour Association in Ahmedabad. It catered to the women in the informal sector who were not protected by the labour laws. It organized women into cooperatives based on production and distribution so that they do not have to deal with middlemen and also have greater control over their production.

The Chipko movement in 1972 highlighted women's close relationship with environment, right to common resources and important role in sustainable development. The movement showed that any action leading to adverse impact on their sources of livelihood water, fuel or fodder would be strongly dealt with. Women protested tree-felling by private contractors by hugging the trees. Thus if women are victims of environmental degradation, they also are agents of environmental protection.

Section II

Women's Rights and Socio-Legal Reforms

Towards the second half of the 1970s emerged autonomous phase of women's movements, that is, movements led by women themselves dealing with gender-specific issues and not subordinated to other issues or organizations. Women involved were from liberal, educated upper and middle strata of society. Till now more or less women organizations were either women wing of political parties or their offshoots. Committee on the Status of Women in India, appointed in 1971 by the Ministry of Education and Social Welfare, published a report in 1974 entitled 'Towards Equality' which underscored the fact that the position of Indian women has declined since 1911. Gender disparities have increased in various sectors like job, health, literacy and political representation. So, gender issues became significant part of the development debate. The report further highlighted the question of the Uniform Civil Code in the context of gender equity and justice. The nonexistence of the Uniform Civil Code even after so many years of independence is an inconsistency which cannot be explained in consonance with the ideals of secularism, scientific thinking and modern rationalism. The existence of various personal laws validating prejudice against women is a violation of fundamental rights of the people (Committee on the Status of Women 1974).

Issues of dowry deaths and rape in police custody were vehemently taken up by women's movement in the 1970s (Kumar 1993: 128–129). Mass demonstrations took place as part of major anti-rape campaign from 1980 to 1983. Rape is seen as an act of power through which patriarchy maintains its control over women. State and its machinery can also be perpetuators of violence. Mathura rape case in police custody and subsequent judicial verdict acquitting the accused drew widespread protests. A campaign calling for retrial as well as to raise public awareness was launched along with the need to change laws relating to rape and violence against women. These women's movements succeeded in bringing about changes in rape laws. The government introduced a bill defining custodial rape, in camera trials and now onus of proof rested on the accused.

Further, bride-burning cases until the early 1980s were termed as suicide that too are not due to dowry harassment. Widespread mass movement under the aegis of *Dahej Virodhi Chetna Manch* forced the government to act. In 1983 the Criminal Law (Second Amendment) Act was passed, making abuse of wife a non-bailable offence and redefined cruelty to include mental and physical intimidation. Section 113 of the Evidence Act was amended. This shifted the burden of proof from the complainant. Further, death of a woman within first seven years of marriage would be probed. The National Commission for Women recommended more stringent dowry laws.

Campaigns were also launched against sex-selective abortions and sati. In 1987, Roop Kanwar sati incident brought the issue to the forefront. Glorification of sati and state inaction was highly condemned. In 1985, the Forum Against Sex Determination and Sex Pre-Selection was formed in Bombay. Further Maharashtra banned the use of amniocentesis for sex-selective abortions. Between 1986 and 2001, five million girls were eliminated due to unethical foetal sex determination (George 2007). In 1984, parliament enacted Prenatal Diagnostic Techniques Act to prevent its misuse. However, the law was not implemented. In 2001 the Supreme Court directed the government to implement the Act. Finally, the law was amended to make it more stringent. It is imperative that instead of ad hoc measures, both the government and the civil society should adopt a more proactive approach.

This period also saw maturing of Indian women's movement. Along with campaigns, protests, creating awareness through street plays, they decided to work for providing aid and support to women in need in individual cases. Women centres were created to facilitate healthcare and legal aid. Moreover, they collaborated with state in setting up women's cell in police stations, educational and awareness programmes and recommendation committees. The issue of women's empowerment came to the forefront with state-sponsored development schemes. A full chapter on Women and Development was incorporated in the Sixth Five Year Plan

(1980–1985). It was realized that without economic independence, education and family planning services, constitutional guarantee of equality cannot be achieved. All future anti-poverty programmes were to include women as targets.

Around the mid-1980s mass-based grassroots women's movement emerged related to the issue of development. *Stree Mukti Sangharsh* (SMS) and *Shetkari Mahila Aghadi* (SMA) in Maharashtra, Coastal Fisher Women's Association in Kerala and Jharkhand *Nari Mukti Samiti* are some of the movements. The Coastal Fisher Women's Association has been protesting overfishing by factory ships of multinational corporations. They argue that liberalization and export orientation of Indian economy has not only led to depletion of the seas, but as fish become unavailable, it leads to the question of food security of the poor. Protests culminated in banning of mechanized fishing for a year. The third wave has provided 'alternatives for poor rural women based on critique of contemporary industrial, consumer-oriented development. These concrete alternatives are giving women control over resources, especially community resources such as land, water and waste land' (Desai 2001).

On the legal front, the Supreme Court issued guidelines in a writ petition on the Vishaka vs State of Rajasthan case for the sake of gender equity. Violation of dignity of woman is an infringement of her fundamental rights under Articles 14, 15 and 21 of the Constitution. Freedom to join any profession is also determined by congenial and secured working space. In pursuance to this, the National Commission for Women formulated a code of conduct at workplace to be respected by all managements. Likewise, Protection of Women from Domestic Violence Act also aims at gender parity and gender justice. Passed in 2006, the new law made domestic violence a crime. According to this law, domestic violence includes sexual, verbal, economic, emotional and physical abuse. It took six years of activism and mobilization by the Indian women's movement to frame this law and get it passed. Also, the Hindu Succession Act for a long time denied right to equality in inheritance rights and the right to the ancestral home and property to women. This has been amended to give daughters of the house share in ancestral property.

However, in spite of these measures and amendments, violence against women continues. Rather they have become more ghastly such as the 16

December 2012 Delhi gang rape, or many other such cases which may not have got similar media attention. After such incidents

> inquiries are held, committees are set up, scapegoats are found. Yet rarely do we see a realistic discussion on the factors contributing to the culture of rape as is evident from the steady increase in the number of crimes against women being reported. What are the factors fuelling the growth and spread of this culture? Why has so-called modernity not even touched the patriarchal and misogynist attitudes that encourage successive generations of Indian men to believe that they are entitled to do as they wish with women. (Lessons from a Hanging, EPW 2014)

It is critical that we try to find answers to these questions which lie in our own society.

After the civil society outrage over the gang rape and death of a student in New Delhi in 2012, the central government established a commission to suggest law reforms to deal with the issue of gender violence. Following the commission's report, parliament adopted amendments expanding definitions of rape and sexual assault with the scope of death penalty in rape cases, and criminalizing acid attacks. Yet we still need stringent laws in cases of 'honour killings' and witness protection programmes. Likewise, following the people's protest in the aftermath of rape and murder of a young girl in Kathua in Jammu and Kashmir, the Union government brought about a Criminal Law (Amendment) Ordinance 2018 which was promulgated by the President of India. The ordinance introduced death penalty for rape of a girl below the age of 12 years (Bhandari 2018).

Some scholars have argued that law reforms can bring about limited change, and in spite of new laws, women continue to be subjugated. However, we should remember that laws by themselves cannot change patriarchal structure and economic system, but it can certainly give legal and economic power to women to some extent. The free legal aid process should be improved so that it can be accessed by all the women in need.

Issue of Political Rights

If women are not included in the policy formation, society and the country as a whole lose on vital human resources in terms of their capabilities, expertise, aptitude and outlook. 'Yet despite all the compelling arguments in favour of increasing the participation of women in politics, women are still underrepresented among the political elite' (Conway et al. 2005).

No doubt, women are still underrepresented in politics but they are certainly making inroads in political arena. This will further fructify as women leaders in political groups will not only be able to make the functioning of these organizations hopefully more transparent, culminating in more people-centric policies, but will also have a dynamic influence on political behaviour of the women, society and larger arena of the state. Women's political participation can further contribute to empowerment of women in many ways. It will open ranks for other women to participate. Issues affecting women primarily will be taken seriously. This may further be instrumental in bringing about change in public attitude towards women who have to acknowledge women as equal partner in development.

Usually, it is documented that a significant gender gap exists in various fields like health, labour market, education and political empowerment. But it is heartening to know that a study actually finds decline in gender bias in voter turnout in India. As Kapoor and Ravi rightly pointed out, equal voting rights are equivalent to human rights in a democratic system. By exercising this right, they choose their representatives which exhibits their choice as individuals. As more and more women are exercising their constitutional right as citizens of the country, it can be seen as an expression of their empowerment which itself is an extraordinary achievement (Kapoor and Ravi 2014).

On the issue of political representation, opposition of some parties to the 33% reservation of women in parliament has shown the patriarchal ideas of these male leaders. As Sen argues, women commanding political control are important as feminist call for not only sharing political power but transforming the essence of power and basis of governance making it truly representative. Political authority is paramount for women empowerment. Hence, in order to be cardinal actors in the delivery of public services and policy formation, women have to stake claim to state power (World Bank 2000).

Conclusion

Equal rights for women in practice are crucial for a progressive and truly democratic environment and civil society has a critical role in it. If women were to contribute and make a difference to the country's future, we have to give them equal opportunities to do so. We do have critical laws and constitutional guarantees to safeguard women's rights, but we require their actual administration. Further, laws are only a set of code of conduct. For a positive change in social structure, we have to bring about a change in the mindset of the people. Law reforms and society reforms should go hand in hand. Legislations may not end societal disparity, but they can surely contribute to women's empowerment. These will also go a long way in promoting societal interests at large. In fact an engaging and dynamic civil society is a reflection of a resonating democracy highlighting the true participation of the people. An enterprising civil society will further the cause of democracy by limiting the state, providing domain for demands articulation and maintaining a crucial balance between the state and the society.

References

Agnihotri, I., & Majumdar, V. (1995). Changing Terms of Political Discourse: Women's Movement in India, 1970s–1990s. *Economic and Political Weekly, 30*(29), 1869–1878.

Anker, R. (1998). *Gender and Jobs: Sex Segregation of Occupations in the World.* Geneva: International Labour Organisation.

Bagchi, J. (1996). Ethnicity and the Empowerment of Women: The Colonial Legacy. In K. Jayawardena & M. Alwis (Eds.), *Embodied Violence: Communalising Women's Sexuality in South Asia*. Kali for Women: New Delhi.

Bhandari, V. (2018, May 7). Kathua and After: Why the Death Penalty for Child Rape Will Not Deter Offenders. *The Wire*. Retrieved June 7, 2018, from https://thewire.in/law/kathua-and-after-why-the-death-penalty-for-child-rape-will-not-deter-offenders.

Bhasin, K. (1986). *What Is Patriarchy?* New Delhi: Kali for Women.

Brown, C. (2008). Human Rights. In J. Baylis, S. Smith, & P. Owens (Eds.), *The Globalization of World Politics* (pp. 506–521). New York: Oxford University Press.

Butalia, U. (1993). Community, State, and Gender: On Women's Agency During Partition. *Economic and Political Weekly, 28*(17), 12–24.

Chatterjee, P. (1992). The Nation and Its Women. In R. Guha (Ed.), *A Subaltern Studies Reader: 1986–1995*. Minneapolis: University of Minnesota Press.

Conway, M. M., Steuernagel Gertrude, A., & Ahern, D. W. (2005). *Women and Political Participation: Cultural Change in Political Arena*. Washington: CQ Press.

Coomaraswamy, R., & Kois, L. M. (1999). Violence Against Women. In K. D. Askin & D. M. Koenig (Eds.), *Women and International Human Rights Law* (Vol. 1, pp. 177–217). New York: Transnational, Ardsley.

Desai, M. (2001). India: Women's Movements from Nationalism To Sustainable Development. In L. Walter (Ed.), *Women's Rights: A Global View* (pp. 99–11). London: Greenwood Press.

Elliot, C. M. (2003). Civil Society and Democracy: A Comparative Review Essay. In C. M. Elliot (Ed.), *Civil Society and Democracy: A Reader* (pp. 1–39). New York: Oxford University Press.

Fukuyama, F. (1998). Women and the Evolution of World Politics. *Foreign Affairs, 77*(5), 24–40.

Gandhi, N., & Shah, N. (1992). *The Issues at Stake: Theory and Practice in the Contemporary Women's Movement in India*. New Delhi: Kali for Women.

George, S. (2007, March 7). Hidden Genocide. *The Times of India*. New Delhi.

Government of India, Committee on the Status of Women. (1974). *Towards Equality*. New Delhi.

Government of India. (2017). Ministry of Labour and Employment, Maternity Benefits Amendment Act 2017.

Heywood, A. (2014). *Global Politics*. Basingstoke, Hampshire: Palgrave Macmillan.

Kapoor, M., & Ravi, S. (2014). Women Voters in Indian Democracy: A Silent Revolution. *Economic and Political Weekly, 49*(12), 63–67.

Kelkar, G. (1992). *Violence Against Women*. New Delhi: Manohar.

Khilnani, S. (2001). The Development of Civil Society. In S. Kaviraj & S. Khilnani (Eds.), *Civil Society: History and Possibilities*. New Delhi: Foundation Books.

Kumar, R. (1993). *The History of Doing: An Illustrated Account of Movements for Women's Rights and Feminism in India 1800–1990*. London: Verso.

Lessons from a Hanging. (2014). *Economic and Political Weekly, 49*(25), 7.

Merry, S. E. (2009). *Gender Violence: A Cultural Perspective*. Malden, MA: Wiley Blackwell.
Mohanty, C. (1988). Under Western Eyes: Feminist Scholarship and Colonial Discourses. *Feminist Review, 30*, 61–88.
Omvedt, G. (1980). *We Will Smash This Prison: Indian Women in Struggle*. London: Zed Books.
Purkayastha, B., Subramaniam, M., Desai, M., & Bose, S. (2003). The Study of Gender in India: A Partial Review. *Gender and Society, 17*(4), 503–524.
Putnam, R. (2000). *Bowling Alone*. New York: Simon and Schuster.
Ray, B., & Basu, A. (Eds.). (1999). *From Independence Towards Freedom: Indian Women Since 1947*. New Delhi: Oxford University.
Shiva, V. (1993). The Chipko Women's Concept of Freedom. In M. Meis & V. Shiva (Eds.), *Ecofeminism*. London: Zed Books.
Taylor, C. (1990). Modes of Civil Society. *Public Culture, 3*(1), 95–118.
United Nations. (1995a). *Beijing Declaration and Platform for Action: Platform 3*. The IV World Conference on Women, 1995-Beijing, China: Official Documents.
United Nations. (1995b). From Nairobi to Beijing: Second Review and Appraisal of the Nairobi Forward-Looking Strategies for the Advancement of Women, New York.
Vickers, J. (2002). Thinking About Violence. In V. Dhruvarajan & J. Vickers (Eds.), *Gender Race and Nation: A Global Perspective* (pp. 222–246). Toronto: University of Toronto Press.
Walby, S. (1990). *Theorizing Patriarchy*. Oxford: Basil Blackwell.
World Bank. (2000). *Sen, Samita, 'Towards a Feminist Politics? The Indian Women's Movement in Historical Perspective'*. Working Paper No. 9, New Delhi.

12

Role of Civil Society Organisations in Women Empowerment: A Study of SKDRDP, India

Naveen Naik Karje and Anjali Kulkarni

Introduction

In light of globalisation, the focus of developing countries is on the advancement of science and technology and improving communication. Development, however, is not measured only on these criteria. Human development is equally important. Women's empowerment is crucial for development as they make up almost half of the population. However, in a male-dominated, patriarchal Indian society, women lack the freedom of choice and are treated as subordinates in every sphere of life. Women are deprived of access to resources like money, credit facilities, as well as opportunities for education, training, and developing their capacities. In India, women live with multiple fears and face numerous hurdles. They carry the burden of neglect and discrimination, of household work, of looking after their family members, and of work outside the house (Kilby

N. Naik Karje (✉) • A. Kulkarni
School of Rural Development, Tata Institute of Social Sciences, Tuljapur, India
e-mail: anjali.kulkarni@tiss.edu

© The Author(s) 2020

Md. N. Momen et al. (eds.), *Building Sustainable Communities*,
https://doi.org/10.1007/978-981-15-2393-9_12

2011). Their contribution to household activities is undervalued, and remains unpaid and unrecognised. Gender-based discrimination starts since birth and continues throughout life with inadequate attention, nourishment, medical care, and education. In many villages, towns, and cities, girls who have reached reproductive age are sometimes sold in the name of marriage, child labour, and prostitution. Sex ratio, the ratio of males to females in a population, is a measure of gender discrimination, which portrays a systematic societal bias towards men and boys in all aspects of life. Given that the life expectancy for women is greater than men, sex ratio should ideally measure 1000 women per 1000 men (Kilby 2011), but as per the 2011 census, the sex ratio in India is as low as 940 females per 1000 males. Mortality rates for women are higher as compared to men owing to the discrimination and violence experienced by women throughout their life. Nobel Prize winner Amartya Sen, in *Missing Women*, has cited sex-selective abortion as one of the reasons for the disparity in sex ratio. Various data have shown a continuous rise in crimes against women. Through the 73rd amendment, there was a push for women's political empowerment with 33% reservation in local self-governance institutions.

In the Indian context, "the civil society comprises individual and collective initiative for the common public good" (Tandon 2002). This definition covers both informal and individual initiatives as well as those of more formally structured organisations. Civil society organisations have responded to the government's socio-economic development agenda over the years. Thus, when the government changed its priorities from capital-oriented growth to anti-poverty programmes, civil society organisations made a distinct shift from welfare and service delivery interventions to a direct attack on poverty (Tandon 2002). Thus, civil society organisations or the third sector play an important role in the welfare and development of people. Civil society organisations are keen observers of people's needs. Not only do they raise the concerns of people, but they also monitor the state's responses to people's needs. A role of civil society organisations that is the most unrecognised is their demonstration of innovative approaches towards development-related issues. Based on some successful innovations, the state can formulate larger policies.

Daniel Siegel and Jenny Yancey (1992) pointed out that civil society organisations have the potential to create a deeply rooted network of organisations and institutions that mediate between the citizens and the State. According to them, these organisations perform several functions. They provide platforms for the expression of varied and complex needs of society and further actively try to address these needs. They motivate individuals to act as citizens instead of depending on state power or benefits. Alternatives to state agencies providing services are also created with more independence and flexible approach.

Civil society organisations have recognised that women's issues need to be addressed and they should have a key role in the development process. The approach that women should be participants in their development process and not just passive recipients of development programmes has emerged through the work done for women's empowerment. The third sector has shown a successful demonstration of women's participation in development processes.

In a country like India where patriarchal domination exists, women's empowerment and their participation in development become a challenge. It is fundamental to create women's access to resources and provide hand-holding when they take up entrepreneurial activities. Here, civil society organizations play an important role in poverty alleviation and empowerment, particularly through the supply of microcredit. The provision of microcredit involves initiatives to be taken on part of the state and non-state organisations in making small amounts of credit available to poor clients (Burra et al. 2005). This amount may be used for production or consumption purposes, in other words, credit sought for basic requirements such as food or income-generation activities. Civil society organisations offer micro-finance interventions for better gender equations in favour of women which includes improved health and nutrition, improved education, increased self-reliance, the standard of living, and better management of natural resources. According to Krishna (2003), empowerment means increasing the capacity of individuals or groups to create effective development, to make better life choices, and to transform these choices into desired actions and outcomes.

Theoretical Framework

In a patriarchal society, women are dependent on men and do not have the right to live their lives on their terms. Their productive powers are exploited. Their household work is never counted in economic terms and remains invisible and unpaid. When they work outside, they are underpaid. These societies subjugate women in all spheres of life. Kabeer stated that the social, political, and economic structures of human society are important phenomena in examining gender-based oppression and the domination by the powerful in our society (Turner and Maschi 2014). In this context, the feminist economist Martha MacDonald has argued that "the economy is not limited to market; it is also an analysis and fundamental categorising gender principle, challenging and differentiate between formal and informal economy, paid and unpaid work, market and non-market activities, productive and reproductive labour." Feminist and empowerment theories are important for understanding and assessing the impact of interventions for individual and socio-political empowerment (Turner and Maschi 2014). Feminists believe that the major reasons for women's poor socio-economic status are inequality and discrimination against women. The status of women's empowerment is measured based on political, economic, and social power relations. Therefore, women should have equal access to all forms of power.

LEO (Leveraging Economic Opportunities Project) proposed a theory which focuses on gender equality and economic empowerment of women. LEO also suggested some indicators for women empowerment such as women's choice of mobility, access to rights, freedom to act and fulfil their potential (LEO 2014). Empowerment of women in rural areas needs to begin at the household level, since a rigid village structure constructed by societal norms, prejudices, cultural values, patriarchy, and social institutions created barriers for women empowerment. However, to bring empowerment, some other elements are equally necessary for an intervention. This framework further states that civil society organisations and local governments can intervene through community organisation work, address barriers and problems, and enable women's access to resources, opportunities, as well as provide services to women for empowerment. These organisations need to identify and understand women's problems in different subsystems—household, community, and workplace. Through various programmes, organisations create access to economic opportunities, but a need for planning, monitoring, and evaluation of empowerment programmes and processes still exists.

SKDRDP has demonstrated a model of women's economic and social empowerment which relates to women's economic empowerment theory stated above. The process followed by SKDRDP for women empowerment also fits into the intervention framework.

The present chapter elaborates a case study of SKDRDP's model of women empowerment, its impact on women, and involvement of local women as village level workers in the process of planning and implementation of development programmes. This chapter is based on a study conducted in Udupi district in Karnataka state of India in the year 2016–17. The study particularly explores women empowerment programmes like Self-Help Group (SHG) and SHG with Janan Vikas Group (JVK) implemented by SKDRDP.

Objectives of the Study

The major objectives were:

1. To understand the socio-economic empowerment of women through the SHG programmes, and
2. To find out the impact of capacity building and JVK activities in terms of women empowerment.

Methodology

The study was conducted in three blocks namely Udupi, Kundapur, and Karkal in Udupi district in Karnataka. For the present study, both quantitative and qualitative research methods were used. Multistage stratified sampling method was employed. Various criteria like number of SHGs in

Table 12.1 Details of sampling for the study

District	Block	Village[a]	Number of group[b]	Women member[c]
Udupi	Udupi	4	8 (4 JVK + 4 SHG)	40 (5 × 8 = 40)
	Karkal	4	8 (4 JVK + 4 SHG)	40 (5 × 8 = 40)
	Kundapur	4	8 (4 JVK + 4 SHG)	40 (5 × 8 = 40)
Total		12	24 (12 JVK + 12 SHG)	120 (3 × 40 = 120)

[a]From three blocks, four villages were selected based on Highest & Lowest presence of SHG group

[b]From four villages, four jvk and four shg group have been selected based on five to eight years old group

[c]From 24 groups, five individual members were selected based on who had income generation activity

villages, duration of SHG groups (functioning for five to eight years), the background of individual members and their participation in income generation activities, and coverage under health insurance were studied. The total number of SHGs was 10,412. There were 277 JVK groups in these three blocks. The data was collected from 120 women through interview schedule and Focus Group Discussion (FGD) conducted in 24 self-help groups. The tools for data collection were focused group discussions and structured interviews. The data was collected from September to December 2016 (Table 12.1).

Approaches of SKDRDP

In 1991, “Shri Kshethra Dharmasathala Rural Development Project (SKDRDP)” was promoted by Dr D Veerendra Heggade in Karnataka. It is a charitable trust popularly known as SKDRDP. The ideas effected by SKDRDP are considered to be pioneering efforts for women’s empowerment. Women are guided to form their self-help saving groups with the provision of microcredit facilities. These groups are also supported for self-employment opportunities, insurances, and so on. There are two groups—SHGs and SHGs with JVK. While SHGs save and lend, SHGs with JVK do this along with training for capacity building for starting business activities.

Every centre covers three to four SHG groups who decide to join the centres. Generally, in every block, 25 Jnana Vikas centres are established. These groups have monthly meetings in which issues related to health, family welfare, hygiene, sanitation, and children's education are discussed. Local women who are village coordinators manage these centres. SKDRDP train the village coordinators at National Institute for Self Help Group Training [NIST] and RUDSET (Rural Development and Self Employment Training Institute). The content of training focuses on family relationships and coordination, health and sanitation, education, self-employment, and government schemes and programmes. Trained village coordinators further train the JVK groups and develop their entrepreneurial skills related to dairy, tailoring, skills in the production of phenyl liquid, soap, plates, artificial jewellery, cloth bag making, and so on.

Socio-Demographic Profile of Respondents

Table 12.2 and 12.3 provides a socio-demographic profile of SHG members in three blocks.

The above table shows that the majority of women were in the age group of 31–45 years (55.8%) and 46–60 years (27.5%) showing that the majority of the women belonged to the working-age group. The caste composition shows that most of the women were from socially and economically deprived communities who are likely to have scarce resources in terms of starting a business. Out of the total women covered, 85% were married.

Table 12.2 Socio-demographic profile of the SHG members

Age in years	15–30	31–45	46–60	61–75	Total
No of women	18	67	33	2	120
Percentage	15	55.8	27.5	1.7	100
Caste	General	OBC	SC	ST	Total
No. of women	19	80	6	15	120
Percentage	15.8	66.7	5	12.5	100
Marital status	Single	Married	Widow	Disperse	Total
No. of women	8	102	9	1	120
Percentage	6.7	85	7.5	0.8	100

Table 12.3 Educational status of women

Level of education	Illiterate	Primary	Secondary	PUC	Graduate & above	Total
No. of women	14	49	42	9	6	120
Percentage	12	41	35	8	5	100

Table 12.4 SHG members' occupation (%)

Occupation	No. of women	Percentage
Business	25	21
Agriculture	5	4
Non-agriculture	40	33
Daily wage	32	27
Service	10	8
Student/education	2	2
Housewife	6	5
Total	120	100

Education is an important factor in women's empowerment as it gives access to knowledge and information. The data shows that 12% of women were illiterate and very few women, that is, 5% had an opportunity for higher education, 41% of women had primary schooling, and 35% had secondary level school education. SKDRDP focuses on ensuring basic literacy and thus those who were illiterate learnt to write their name and signatures after joining SHGs.

Occupational Status of Women

Table 12.4 shows that 4% of women were engaged in agricultural work, 33% of SHG members were engaged in non-agricultural occupations, and 27% of SHG members were daily wage workers. Twenty-one per cent of members were engaged in business activities like tailoring, shops, catering services, phenyl, and soap making. Women who had jobs like primary school guest teacher, *anganwadi* helper, midday meal cook, *Seva Prathinidhi* consisted of 8% of the population. Only 2% were students and5% of the women covered were housewives and were not involved in wage-earning which means that 93% of the women covered had some earnings.

Socio-Economic Empowerment of Women through the SHG Programme

Micro-Finance Programme

The SKDRDP initiated a micro-finance scheme known as *Pragathinidhi* [meaning 'fund for progress'] in 1996, which provides loans to members to promote their livelihood and develop infrastructure. In 2003, the SHG bank linkages were expanded on a larger scale. Therefore, *Pragathinidhi* has been stopped and SKDRDP has started with direct bank linkage of SHGs. Presently, the SKDRDP is implementing a financial inclusion plan like Banking/Business Correspondent (BC) and Business Facilitator (BF) in all districts of Karnataka. Under this program, SKDRDP is promoting access to banking facilities to SHGs.

SHG Member Loan Status and Purpose of Loan

Table 12.5 shows that consumption purpose loans (50.5%) were higher than the productive purpose (18.7%). The remaining 30.8% of the members had not taken loans from SHGs in the last two years.

However, attention needs to be drawn to the fact that the data is about the loans taken in the last two years. It is possible that the women who had taken higher amounts of loans for productive purposes were in the process of repayment and would not get loans unless earlier loans are cleared.

The FGDs revealed that through SHGs, women could take care of the basic needs of their families. They experienced that after joining the SHGs, their families were economically better off when compared to the situation before joining SHGs. Since the basic needs are now fulfilled, as shared by one woman, the focus is on providing education to children—*"now we are giving more importance to children's education because they should not face any problems in their future as we faced."*

High amounts of loans and high recovery rates suggest that women earn an adequate income. Income-generation activities financed through SHG loans have facilitated purchases such as vehicles for family members, investments in hair salons, and so on.

Table 12.5 Status of loan amount and purpose of loan (%)—last two years

Loan amount in Rs		0	1–50,000	50,001–100,000	100,001–150,000	150,001–200,000	250,001–300,000	Total
Loan purpose	No loan	100	0	0	0	0	0	30.8
	Productive	0	23.1	34	33.3	50	0	18.7
	Consumption	0	76.9	66	66.7	50	100	50.5

SKDRDP Programmes and Their Impacts in Terms of Women's Empowerment

In a patriarchal society like India, it is difficult for women to access saving and credit facilities. Besides, there are cultural hindrances if women want to start and manage entrepreneurial activities. However, SKDRDP's programmes related to micro-finance and entrepreneurship promotion showed that women were able to break these barriers. It also brought positive impacts on women and their families' economic conditions and facilitated social empowerment of women.

This chapter gives a detailed account of the economic and social empowerment of women through SKDRDP programmes.

Economic Empowerment

Improvement in Saving Capacity of Women

The assessment of pre- and post-SHG saving capacity of women can be one effective indicator in evaluating the access created to financial services through the SKDRDP programme.

The above Table 12.6 shows that 90.8% of women did not save earlier and very few had occasional savings. As confessed by women themselves, it is due to the SHGs that they have developed a habit to save.

Table 12.6 Pre- and post-savings by women SHG members

Pre-SHG		Post-SHG	
Amount in Rs.	Women (%)	Amount in Rs. (monthly)	Women (%)
0	90.8	40	8.33
25	1.7	80	50
30	0.8	100	4.17
50	4.2	120	12.5
100	0.8	200	25
150	0.8		
300	0.8		
Total	100	Total	100

"*Before joining SHG we used to keep money in tiffin box in kitchen room and we never visited banks due to fear and hesitation,*" a woman responded during FGD. Some women pointed out that earlier there was neither any awareness about the need for saving nor opportunities due to the lack of knowledge about postal and bank services.

Entrepreneurial Activities by SHG Women

As loans were made available to women, they could take up entrepreneurial activities which proved to be a positive indicator of women empowerment considering women find it difficult to take up business activities due to socio-cultural and economic barriers.

SKDRDP had a well-planned programme for capacity building of women members of JVK groups to promote entrepreneurship. Training related to various entrepreneurial activities was given. It was supplemented by exposure or study visits to successful units. They were also taught how to enhance traditional activities through JVK. The above Table 12.7 suggests that 52.5% of women had started their entrepreneurial activities. Investments for these activities were made possible due to

Table 12.7 Entrepreneurial activities undertaken by SHG members

Initiated entrepreneurial activity	No. of women	Percentage
Yes	63	52.5
No	57	47.5
Total	120	100.0
Entrepreneurial activities undertaken	No. of women	Percentage
Animal husbandry	36	30.0
Animal husbandry + Jasmine agriculture	1	0.8
Artificial jewellery making	1	0.8
Catering services (Dosa)	1	0.8
Catering services (Idli)	1	0.8
Cloth bag production	1	0.8
Jasmine agriculture	1	0.8
Phenyl liquid	1	0.8
Phenyl liquid + Artificial jewellery	1	0.8
Shop	3	2.5
Tailoring	16	13.3
Total	63	52.5

SHG loans. Thus, SKDRDP was successful in fostering entrepreneurship among women leading to their overall empowerment. The JVK further trained them to sustain these activities. These inputs for effective management of business activities are required as women traditionally do not get such kinds of skill training.

Increase in Women's Income Levels

Income is one of the traditional indicators of women empowerment. The following Table 12.8 shows women's incomes in pre- and post-SHG stages.

It is very clear from the above data that women's income levels had gone up after joining SHGs, as 50% of women covered in the study had zero income before joining SHGs.

Post joining SHG, only 5% of women had no income. Other income ranges suggest that almost 52.5% were earning a substantial income.

During FGDs, women shared that earlier due to very few earning opportunities their families did not have a sufficient income. Inadequate income impacted not only food consumption patterns but also the education of their children.

Improvement in Agricultural Production

Women's membership in SHGs proved to be a major cause in boosting agricultural production. Many times, the farmers find it difficult to take loans from financial services for agricultural development. Those women

Table 12.8 Women's income in pre and post SHG stage [per month]

Income pre-SHG in Rs.	No. of women	Percentage	Income post SHG in Rs.	No. of women	Percentage
0	60	50.0	0	6	5.0
1–1500	33	27.5	1–3000	43	35.8
1501–2000	17	14.2	3001–6000	51	42.5
2001–3500	7	5.8	6001–9000	12	10.0
3501–4000	1	.8	9001–12,000	7	5.8
4001–5500	2	1.7	12,001–15,000	1	0.8
Total	120	100.0	Total	120	100.0

who had taken loans from SHGs were able to purchase pump sets, tractors, and other agricultural inputs. The use of quality seeds and new technology, new irrigation practices, organic farming, mixed farming, and so on helped to increase production. Some women invested the loan amounts for nut plantation and owing to the demand in the markets, earned healthy profits.

In these areas, paddy and vegetables are grown. The Yantradhara Programme under the SKDRDP provided agricultural equipment at an affordable cost. The SKDRDP introduced the cultivation of jasmine flowers and created a direct link between producers to market by eliminating the middlemen. Such practices enhanced women's income.

Social Empowerment

Social empowerment can be measured at the family and societal level.

Empowerment in Family

Social empowerment in a patriarchal society starts from the family as it is the place where inequality begins and is promoted. Generally, when women get empowered within the family, their socio-political empowerment can be further promoted.

Women empowerment is not a short-term process; it's a long-time process. Therefore, civil society organisations also effectively use the platform created by micro-finance activities in innovative ways.

The women shared that after joining SHGs many changes were noticed within their families and individual lives. Generally, there is resistance to women's programmes in patriarchal societies. However, the women shared that now the family members have realized the benefits of their membership to SHGs and hence there is support from family members. They are even reminded by the family members about SHG meetings.

The women also pointed out that their involvement in family decision-making had increased.

According to the women, this was because they could bring about some changes within their families after joining SHGs. They could undertake the construction of houses, digging of wells, ensuring food for the family, and education of children with the support of SHGs. As 52.5% women were able to generate income and 95% women had an income after joining SHGs, their visibility in terms of contribution to family income was very clear and it helped women get empowered at the family level. It is the greatest achievement in terms of women empowerment in a male-dominated rural society.

Change in the Position and Role of Women in the Family

Before joining SHG, the only goal in a woman's life was to take care of her husband and children and engage in household chores. After joining SHG, women have earned a stand in their family's decision-making processes. Women are involved in making decisions related to children's education, their marriage, purchase, or sale of a property, and other major family decisions. Due to their economic empowerment, they are being treated as equals by the men in their family.

Accessing Information

Due to women's low literacy and subordinated position in the society, their access to information is limited. Gaining knowledge and information is essential for being empowered. The major source of information for the women was the SHG weekly or monthly meetings. In addition to information about self-employment, government schemes and programmes, health and sanitation, water management, managing daily expenses, they also received information about awareness on social issues related to the importance of children's education, child labour, marriage, abortion, and so on. These inputs gave the women confidence and enhanced their self-esteem which are important indicators of empowerment.

Improved Family Life and Health Status

Due to increased awareness, women started giving importance to nutritious food, immediate treatment for ill-health or injury, and so on. Some women had husbands addicted to alcohol and tobacco. The de-addiction program organized by SKDRDP helped give treatment to alcoholic family members and brought about some positive changes in the family life of women. They were able to avail health insurance for treatment which helped to improve the overall health status of everyone in the family.

Developing Skills in Financial Management

In the FGDs, women discussed their learnings in financial management and managing expenses like children's pocket money, water bill, loan repayment, savings amount, food expenses and electricity bill as well as ensuring savings and repayment of the loan amounts.

Social Empowerment at the Societal Level

Economic empowerment alone is not enough. For the overall development of women, social empowerment is equally needed. Women's participation at larger societal level needs encouragement. Following are some of the indicators of social empowerment.

Developing Conflict Resolution Capacities

In the FGDs, women shared that in the time of crisis, they are now able to support their family members or anybody in the SHG who requires help. For example—at the family level, they discuss and deal with problems together; and at the SHG level, if a particular member has difficulty in repayment. Such issues are also sorted out by discussions and by offering some kind of support in the form of contribution by other members.

Participation at the Societal Level

Once women get recognition at the family level and freedom of expression, their participation at larger social processes increases.

Women's Participation in Social Awareness Programme

SKDRDP initially created awareness related to social issues among women in the SHGs. These women become pivotal in further generating awareness on certain social issues (Table 12.9).

In the FGDs, the women shared how they created awareness among people through street plays, dramas, skits, and songs on the above-mentioned issues. One member mentioned the drama named *Kathale Enda Belekinedege* meaning "journey of darkness to lights," which focused on the discrimination women faced within the family and society and the need for women's empowerment. The women actively participated in blood donation campaigns as well.

Recognition at the Societal Level

One of the JVK groups received the best group award from SKDRDP. The national radio station "Aakashvani" interviewed them. In this interview, the group discussed their lives and how SHGs had impacted them,

Table 12.9 Participation of SHG members in social awareness

Social awareness issues	Media used
Abortion	Drama
Effects of alcohol consumption	Devotional song/Folk song
Child labor	Quiz
Child marriage	Street plays
Self-employment	Songs
Blood donation	Skits
Education	Slogans
Hygiene	
Nutrition	

success in self-employment, and so on. They, in turn, became role models for and motivated others to join SHGs and take up self-employment.

Use of Financial Services

Women using facilities available at larger societal levels like banks, and their presence at public events also indicate their level of empowerment. They have started going to banks and can handle bank transactions for their SHGs independently.

Women's Participation in Political Processes

As the women got empowered at the family level and had enhanced confidence of working independently through SHGs, their presence in political processes increased.

The gram sabha is the general assembly of voters of the village under the Panchayat Raj system where major issues related to village development are discussed.

In FGDs, women shared that before joining SHGs, they used to not attend gram sabha and thought that it was a waste of time. They believed that women did not have any role in these meetings. After joining SHGs, this has changed and as shown in the above Table 12.10, 68.3% of women were attending gram sabha. The women raised issues such as water facility, housing, roads, school, hospital, special facilities for deprived community groups, and so on. They did not hesitate or fear to talk to government officers asking about development schemes.

Table 12.10 Women's attendance in Gram Sabha

Attending Gram Sabha	No. of respondent	Percentage
Yes	82	68.3
No	38	31.7
Total	120	100

Table 12.11 Women contested in gram panchayat election

Election contested/year/position	No. of respondent	Percentage
Not contested	118	98.3
Got membership (2005, 2010, 2015)	1	0.8
Got post of Vice President (2010)	1	0.8
Total	120	100

Contesting Gram Panchayat Election

Another indicator of women's political empowerment can be women contesting elections.

Table 12.11 illustrates that very few women took part in direct political processes. One woman had nominated but withdrew at the last moment due to personal reasons. The other two members contested elections in gram panchayat (a village level self-governance body) and occupied membership. One even became the vice president. Additionally, one woman contested elections for three terms in 2005, 2010, and 2015 and won. The woman who became vice president was an active member of "Mahila Morcha." These two women actively worked on social issues and encouraged other women to attend gram sabha meetings and motivated them to raise questions.

Though the number of women is not a lot, however, in a male-dominated political environment in rural India, getting such a space is an achievement for women.

Awareness and Utilization of Government Schemes

Traditionally, since women do not have access to information, they are unable to use many government schemes meant for them (Table 12.12).

SKDRDP provides information about government schemes and programmes through resource persons in monthly meetings. This information helps SHG women to get benefits from government schemes.

Table 12.12 SHG members' awareness and utilization of government schemes

Government schemes known to SHG women	
Employment Guarantee Scheme	Widow Pension Scheme
Self-employment scheme	Health card
Ashraya Yojana	Aadhaar card
Indira Awas Yojana	Water facility scheme
Ganaga Kalyana Yojana	Education scholarship and fellowship
Free Toilet Scheme	Pradhan Manthri Gram Sadak Yojana
Pradhan Mantari Ujawala Yojana	Jan Dhana Yojana

Major Findings

1. Women developed savings and repayment habits through SKDRDP programmes and it has promoted their empowerment through a variety of ways.
2. There was a substantial increase in the income levels of women, thus, suggesting the economic empowerment of women.
3. Nearly 53.5% of women covered under the study started independent income-generation activities.
4. Post joining SHGs, only 5% of women were not involved in any income generation or wage-earning and it clearly showed that through SKDRDP the women gained economic empowerment.
5. The women started using formal financial services like banks to which earlier they had no access.
6. The economic empowerment of women and availability of resources from SHGs like health insurance, support for children's education, and so on improved their status at the family level. They started being treated as equals.
7. Women were more aware of government schemes and could avail the same.

The SHGs promoted women's access to knowledge and information and developed leadership skills. Further, women have actively started working on various social issues that have negative impacts on society.

Thus, SKDRDP was successful in making women noticeable at the larger societal level.

Due to SKDRDP activities linked with SHGs and JVKs, women developed leadership skills which have not only enhanced their self-esteem and confidence but also assured them a space in the political sphere as well.

SKDRDP has demonstrated an integrated model of women empowerment. It shows that for women empowerment, a multi-pronged approach is needed. Mere provisions of loans cannot lead to the economic empowerment of women. They also need access to information and support in their roles at the family level. Through the SKDRDP model, empowerment was possible at the individual level, family level, and at the larger social level as well.

Conclusion

SKDRDP is one of the civil society organisations who recognised that for the overall development of the society, women's development needs to be focused. It noticed that economic independence would bring many changes. Its model also made efforts to facilitate women's access to information and knowledge which helped in enhancing their self-esteem and confidence. SHG membership and its benefit for the family also helped the women to create a space for themselves at the family level. Further, women also started getting a space in the political environment although at a smaller scale.

The efforts of SKDRDP in creating a successful model of women empowerment need to be noted. It shifted the focus from welfare to a participatory approach and established that women can be equal partners in the development process. Its model is more sustainable as it created local resources and developed local human capacities to boost the development programme. Thus, the present case study of SKDRDP model based on empirical data confirms that civil society organisations play an important role in women empowerment and development.

References

Burra, N., et al. (2005). *Micro-credit, Poverty and Empowerment Linking the Tried.* New Delhi: Sage.

Kilby, P. (2011). *NGOs in India the Challenges of Women's Empowerment and Accountability.* Oxon: Routledge.

Krishna, A. (2003). *Social Capital, Community Driven Development and Empowerment: A Short Note on Concepts and Operations.* World Bank Working Paper.

LEO. (2014). *Women's Economic Empowerment: Pushing the Frontiers of Inclusive Market Development.* Washington, DC: USAID.

Siegel, D., & Yancey, J. (1992). *The Rebirth of Civil Society.* New York: Rockefeller Brothers Fund.

Tandon, R. (2002). *Voluntary Action Civil Society and the State.* New Delhi: Mosaic Books.

Turner, S. G., & Maschi, T. M. (2014). Feminist and Empowerment Theory and Social Work Practice. *Journal of Social Work Practice: Psychotherapeutic Approaches in Health, Welfare and the Community, 29,* 1–12.

13

Role and Contribution of Non-Governmental Organizations in Women Empowerment: A Case Study of the All Pakistan Women's Association

Nasreen Aslam Shah

Introduction

In the process of nation building and socio-economic evolution, development plays a vital role as a procedure of growth towards advancement of the country and its people. The major objective of development is to improve the standard of life for all. In this development process, all the stakeholders of society are involved as governments can't perform all duties on its own when it comes to the well-being of the citizens. The representation of the civil society is very important in the national

N. A. Shah (✉)
Department of Social Work & Director, Centre of Excellence for Women's Studies, University of Karachi, Karachi, Pakistan

© The Author(s) 2020
Md. N. Momen et al. (eds.), *Building Sustainable Communities*,
https://doi.org/10.1007/978-981-15-2393-9_13

process of decision making as well as mainstream development. Through non-profit organizations, governments become aware of the opinion of common people. This also creates pressure on governments in taking important decisions, and also helps with regard to providing voluntary services to people, changing thoughts and behaviours through training programmes and workshops, providing loans and services to women for empowerment, and so on.

According to Mr. K.S. Dadzie, United Nations Director General for Development and International Economic Cooperation, "the final aim of development must be the constant increase of wellbeing of the entire population on the basis of its full participation in the process of development and a fair distribution of benefits there. The development cannot be achieved by Government departments alone. Besides the Government agencies, NGOs can play an important role in improving the quality of life of our people" (Goel and Kumar 2005).

The development process can't be carried out by government institutions alone, but also with the help of nongovernmental organizations, as they play an important role in this process. Here, the question arises—what is meant by non-governmental organizations?

Basically, the term means those non-profit associations which work to bring changes in the society and provide social and charitable services even in extremely political, national, and international contexts through political influences. Although, it has become common among NGOs to work with the corporate sector and governments to make affirmative influence on the policy of governments and private sector approaches. NGOs are currently struggling to build inventive programs that can change or redirect the activities of the government and the private sector in a way that may possibly meet larger social or economic goals. That's why NGO performance results in expected improvement of the participatory organizations and has absolutely improved the community attitude and status of numerous organizations (Doh and Teegen 2003). As Dr. T. Adeoya Lambo, Deputy Director—General, WHO, in World Health rightly observed:

> "What is happening around us shakes our complacency, challenges our faith in human progress and imbues us with an intense feeling of shame,

doubt and guilt. In a world where the gigantic scientific and phenomenal technological achievements command our administration and almost fetish acceptance, we are witnessing an intolerable degradation of man. Our pride is belonging to a generation that for the first time since the genesis of men has set foot on another planet cannot, however, disguise the awful truth that it may be easier to travel to the moon than to erase from the surface of the earth the image of inevitable poverty, human exploitation, injustice and the degradation of human welfare." (Goel and Kumar 2005)

History of Non-Governmental Organizations

From ancient times, generous people have given of their objects to needy in the form of private gifts. They appointed almoners to distribute their rewards or people whom they trust to give out their patronage. History has evidence that generous people have made contribution in this arena. Before the dissolution of the Monasteries, most of the offerings for charity were distributed by them. Subsequently the church wardens were often nominated for this purpose. Moreover numerous trusts were also known for this as such. These were the founders of our modern-day non-profit organizations. Many good-hearted people were deeply motivated to help large numbers of poor who were living in miserable conditions, and they made a platform with other likeminded people to establish new associations for the assistance of the deprived people. Since then, this practice has continued, but due to poverty, the miseries of the previous century persist. However, this has created many other types of miseries; thus, es is still rising (Rizwi 2008).

Nature, Scope, and Role of NGOs

Nowadays, the role of an NGO is not only confined to provision of few benefits to the community. It also aims to create motivation in people, mobilize funds, leadership development in the community and encourage people's participation in development programmes for autonomy. An

NGO works as an enabler, and as soon as the society becomes autonomous, its role is moved to another place where the services are required. Although, NGOs works for the assistance of people and provision of public goods as direct providers, many development NGOs are working for productive activities. The major role of NGOs as an activist or catalyst and as provider or implementer is where bureaucracy is unresponsive or incompetent, or when the programmes are not flexible, which leads to inefficiency and the deprivation of the poor. The functions of voluntary development organizations have changed over the years, such as welfare of people, sustainable development, development of community, and movements for rights. Non-profit organizations also work as power brokers, safety valve, resource mobilization, criticizers, unifying force, and provider of special needs. NGOs are supposed to reduce tension and coercion in society and integrate the people (Prasad 2000).

The roots of professional social work and organized social services are initiated from the charitable stance and religious outlook. The charitable action is an instinct in the mind of the people who through their humanitarian impulses grant services for the people who are suffering (Sarker 2015).

The NGO structure varies, depending on the needs and requirements of the society. They can be work on national level as well as worldwide. Even grass root organizations have become active and recognized for their work for the betterment of communities on an international level. Often, this happens through the built up of international linkages and support. Different NGOs are provided an institutional structure by International NGOs (Shetty 2013).

The Commission on Sustainable Development (CSD), in Agenda 21, categorizes the civil society in (a) Women, (b) Children and Youth (c) Indigenous Peoples and Communities, (d) Non-governmental Organizations (e) Workers and Trade Unions (f) The Scientific and Technological Community (g) Business and Industry and Farmers.

These major groups are officially recognized by the United Nations, and to improve the conditions of these groups, mechanisms have been developed specifically for NGOs. NGOs and other civil society representatives are partners in administration, as well as a main thrust behind more noteworthy global collaboration through the dynamic preparation of open help for universal declaration (Pace 2002).

These voluntary organizations are making a great contribution by supporting a huge participation in conducting activities worldwide.

Their details are as follows:

The advocacy of people, their awareness raising and identification, researching and analysing issues, and most importantly, informing people about these problems. Other than that, mobilizing people in utilizing media campaigns and other kinds of activism and develop awareness among business leaders and policy makers. NGOs act as a bridge between government and the masses as these organizations convey the opinion to higher levels. Another important task of organizations is to provide quality education for building capacity, prevalence of information, and giving training to masses and trainers, provision of social and operational services on human grounds for development, and also in assessment and observation as a third party, whether they are asked or not, about performance of public and corporate sector (Nelson 2007).

Commencement of the Non-Government Organizations in Pakistan

In Pakistan, voluntary organizations are growing and are recognized for their role to create awareness. Hence, they are purposely approached by the Government to encourage participation of people to provide social services to the needy people. The long-lasting partnership of public and private sectors has been clearly demonstrated in the umbrella of social welfare activities, explicitly, in community development, child welfare, women's development, health, education, and working with the disabled.

Pakistan was not established as effortlessly as was expected. A number of harsh realities came up at that time, although people had the courage to cope up with these difficulties. However, for many other people, the struggle for reorientation was hard due to numerous social and psychological problems that were created. Therefore, in the early stages, social work and social thinking was developed. Spontaneous efforts were made during the period from 1947 to 1951 to overcome the immediate national crisis. From 1952 onwards, forms of difficulties were changed due to this nature of spontaneous good will to serve the humanity. The work being done was transforming into systematic yet simple, organized, and long-term social services.

A few welfare organizations existed at the time of the creation of Pakistan. Most of which served the impoverished, the blind, the hard of hearing, and the sick. But these organizations practically did nothing to cure the social ills. Therefore, it was an important assignment to develop voluntary agencies and prepare trainees of all the in-service social welfare training programmes throughout the country. The training of professional social workers later became a part of the curriculum in the universities.

The concept of welfare originated at the time of independence. It leads to the creation of many charitable organizations for women's socio-economic wellbeing. During the 1980s, the term "non-governmental organizations" gained popularity in Pakistan in comparison with the term "voluntary agencies"; which implied conception at a specific time to engage in charitable contributions of time and money thus also reflecting values designed to improve human lives (Rehmatullah 2002). Although the number of NGOs has increased, (Khalid 2001) the situation of non-profit sector is not that good in Pakistan. In the decade of 1990s, some promising growth resulting from a number of favourable factors like the return to democracy, the prevalence of deregulation and privatization, globalization process, civil society's international linkages, and the fall of financial strength of governments limited the public delivery of social services (Pasha and Asif Iqbal 2002).

The nongovernmental organizations (NGOs) and women in development both effectively work together for change in society. There are few women NGOs, Behbud Association with branches in Lahore, Karachi, and Rawalpindi—Maternity and Child Welfare Association of Pakistan, and Voluntary Health and Nutrition Association (PVHNA) (Banuri et al. 1997). The major problem is that all women NGOs and activist paid attention to gender issues, and ignored the wider societal background in which such gender inequalities occur. Women from upper class and middle class led female activists and established organizations of women. The agenda formulation and the approaches of NGOs for accomplishment of objectives showed the leadership's class background. The economically advantaged and empowered class were also moderate in vision, having the opportunities and resources to stand for and lead (Rashid 2006).

Women NGOs like the All Pakistan Women's Association (APWA) support the gradual changes in the background of Islamic culture. The two main objectives of APWA were to make women participate in mainstream development of country, provision of education, and the creation of employment opportunities, mostly in the traditional handicrafts sector. APWA's major task is not to confine women to their homes as mothers or housekeepers, but to contribute to their economic growth. The initiative of APWA is to educate the people and to stop them from seeing women as just their spouse, mother, and house keeper. This is a difficult task as the society is used to seeing them as such. Such contrasts in belief can be a potential source of contention between association like APWA and customary researchers whose lessons are still largely accepted (Shah 1989).

As the oldest women organization of Pakistan, APWA has been actively working after the establishment of Pakistan, not only for the welfare of women but also in the formation of laws, women empowerment, and health. Therefore, APWA was selected as a case study for this chapter.

It is essential to mention Begum Raa'na Liaquat Ali Khan who established APWA and was a leading lady of that time. APWA's success was due to the efforts of her hard work and dedication.

Begum Raa'na Liaquat Ali Khan

Begum Raa'na Liaquat Ali Khan, originator of APWA and the better half of Pakistan's First Prime minister Liaquat Ali Khan, was a woman of outstanding accomplishments whose commitment to civil rights and societal wellbeing was acclaimed broadly and universally. Her endeavours to activate women inside the charitable services picked up numerous esteemed honours for her. Her enthusiasm for helpful causes and spearheading of women is the reason why women rights have furnished women with a plan so that they can act collectively, (APWA 1987).

Begum Raa'na Liaquat is known to be the pioneer of the women's development in Pakistan. Her first chance to compose Muslim women introduced itself around the same time as when she framed a volunteer nursing and first aid services in Delhi. Then again, from 1947, as the evacuees poured in from over the edge, in the midst of the most pitiable conditions with cholera, diarrhoea, and small pox being rampant, she called upon women to approach and gather sustenance and therapeutic supplies from government workplaces (Khan 1995). During this period in Pakistan's history, there weren't many medical attendants in Karachi. So Begum Liaquat requested that the armed forces prepare women to give injections and first aid. Women were subsequently prepared in three to a half year courses and in that capacity, the Para-military powers for ladies were formed. During this period, young women were additionally urged by Begum Liaquat to take up nursing as a calling.

They were likewise trained in rifle shooting, writing, and a large group of different obligations so they could be valuable amid national emergencies, like the displaced people's emergency of 1947. As the founder and lifelong president of APWA, Begum Liaquat played a significant role in the development of women in political, instructive, financial, and many other different fields (Mulki 2013). To organize women in a systematic way, Begum Liaquat created the platform of APWA to bring welfare services to women and children without any help from donor agencies.

The All Pakistan Women's Association

APWA has been a particularly influential organization from its foundation. It has branches in 56 areas across Pakistan, both in natural and urban fringe areas. APWA is a patronage association depending upon the contributions for finance. APWA got the UNESCO Adult Literacy Prize in 1974 and later the Peace Messenger Certificate in 1987.

After the independence of Bangladesh, the association in that nation was renamed Bangladesh Mahila Samiti (IAW Communication Unit 2016). APWA was particularly the brain child of Begum Raa'na Liaquat

Ali Khan, who perceived requirements intended for a nationwide relationship to be united and to facilitate women's' practices for their greater benefit. Under the sponsorship of APWA, Mrs. Liaquat set up schools, dispensaries, maternity homes, and family planning centres in both urban and country zones. Her essential ideology was well-being, instruction, and training. By mid-1950s, 32 branches of APWA having an aggregate participation of around 1200 ladies, had around 800 occupied with social work. APWA developed contacts by building 20 modern homes with other women, where approximately 40,000 women went through different phases of preparing yearly. Apart from these efforts, through its 13 basic education centres, 100 social welfare centres, six dispensaries, APWA extended its assistance to another 1500 were women.

Activities

Works of APWA

- The up-to-date and quick participation of women of Pakistani towards equality, development, and peace in the country.
- The development and well-being of the Pakistani women throughout the development of lawful, civil, legitimate, societal, and monetary status and rights.
- Endorsement of societal, learning, and cultural with economic programs.
- The Health and education of the people of Pakistan in the home, the family, and the country.
- Cooperation with government and UN socialized agencies UNICEF and Economic and Social Council (ECOSOC).

Status

APWA has an advisory position with the Pakistani Government; APWA has consulting position Category B with the United Nations.

Structure

General membership throughout Pakistan, governing body, executive body, national secretariat, national president, national vice presidents, secretary general, joint secretary general, treasurer, chairperson (APWA 1987).

Women from all divisions of life, rich and poor, highly educated to totally illiterate, came together at the call of Begum Liaquat and devoted themselves whole heartedly to the task of uplifting the status of Pakistani women.

The historical background of APWA is the historical background of women in Pakistan. APWA was framed to handle the outcast emergencies that developed because of migration amongst India and Pakistan. Begum Raa'na Liaquat Ali Khan accepted the mission of women development voluntarily and willingly without politicizing it. She strived for acknowledgment of a nationalized connection for managing as well as facilitating the women's movement for the societal and monetary strengthening of women and youngsters in Pakistan. Following more than 60 years of its reality, APWA has risen and is one of the most seasoned and regarded non-governmental association in the country. This association is extremely useful for women, particularly in earning own wages thus helping their families financially.

Objectives of the Research

Every research has its aims and objectives to proceed that help to find reasons and consequences. This study has the following objectives:

(a) To find out if APWA still work as a caucus group for government as have been the case in the past
(b) To find out the role of APWA for bring change in women's status in society
(c) To find out the difficulties they face during organizational work

Research Methodology

In Pakistan, many NGOs are playing an active role in solving problems of society and bringing a change in the status of women. Similarly, APWA is the oldest non-profit organization that has done much more work and has brought a big change in women's lives in Pakistan. The aim of this research is to focus and highlight the efforts and functions of such an old non-governmental organization of Pakistan. Furthermore, this subject is selected to enlighten the welfare work done by NGOs and also promote the philosophy to make better changes in society, particularly for women.

Interviews were conducted from APWA Karachi headquarter office. Non-probability sampling method and purposive techniques have been used to get the accurate measures on this research problem. In non-probability, sampling size is not determined, and it often covers a few cases and is used for qualitative methods (Sarantakos 2005). This research is conducted by the case study method. Case study is defined as the collection of information about an individual, a family, a group of persons or organizations. Basically, it is an in-depth study to investigate about particular phenomena. As P.V. Young said, "Case study is a method of exploring and analyzing the life of a social unit, be it that a person, a family, an institution cultural group or even entire community (Ghosh 2015)". The case study is used in many situations as a research method (Yin 2014). Hence, the interview schedule was designed to investigate the non-profit organizational work functions and welfare work of APWA. Interview schedule consists of both close and open-ended questions. This interview schedule found out in-depth work of the APWA towards Pakistan and their women. Qualitative method is used to collect data for such document studies.

During study, the concerned authorities cooperated with humility and tried to provide knowledge as per their approach. The analysis of data is done by simple observational and qualitative analyses and derived conclusions.

In Pakistan, numerous NGOs are assuming the dynamic part in tackling issues of society and to bring about a change in the status of women. APWA, being the most matured non-benefit association, has done significantly more work and has resulted in a major change in women's lives in Pakistan. The research issue expected to feature is about the endeavours and elements of such an old and reputable non-profitable organization of Pakistan. Moreover, this subject is chosen to enlighten the welfare work done by NGOs and further elevate the theory to improve change in the public eye, especially for women. The researcher arranged a meeting in APWA Karachi headquarter office. Henceforth, the meeting plan was intended to research the non-benefit hierarchical work capacities and welfare work of APWA. Meeting plan comprised of both close and open-ended questions. This meeting plan discovered inside and out work of the APWA towards Pakistan and their women. Subjective technique was utilized to gather information from such archives deliberately. The preparing of information was led by talking through contextual investigation. Meetings with the women were held in order to know their thoughts on how to explore issues and doing philanthropic work for improvement of women and their condition. Throughout the study, the concerned authorities extended their full cooperation and provided all available information. The investigation of information was finished by basic observational and subjective examinations and determined conclusions.

Interview

The researcher made contact with the focal person of APWA for the purpose of data collection and interview about the performance, service, and functions of APWA. After her permission, researcher visited APWA Karachi headquarters to study, analyse, and interview the people who worked at this oldest and respected organization of Pakistan. The central headquarter of APWA is situated at Garden West region of Karachi, near Saddar (central region of city consider as the heart of the city). Karachi is the metropolitan city of Pakistan, the industrial hub of the country, and the first national capital of Pakistan. Therefore, it is an area of the old city of Karachi, near the zoological Garden on one side and Saddar from the other side. The headquarters office is located at the place where the foundation of this organization was laid by Begum Raa'na Liaquat Ali Khan.

The building of APWA headquarters at Karachi is owned by APWA and was bought during her tenure as Governor of Sindh Ms. Farzana Rehman is the chairperson of public relations. She introduced me to the chairperson of APWA—Begum Mehar Afroz Habib. Begum Afroz briefly clarified about the association and its points and destinations. She said that the association was developed in 1949 by ladies of national stature with an aim to enable women by imparting training and improving their health.

After that, she referred the researcher to Ms. Farzana Rehman. Ms. Farzana Rehman illuminated about the APWA, its points, capacities, and history. APWA is a seventy one years old association, established by Begum Raa'na Liaquat Ali Khan, wife of the first Prime Minister of Pakistan Nawabzada Liaquat Ali Khan and her likeminded affiliates in February 29, 1949 at a meeting held at her home. She was lifetime leader of APWA. The affiliation was enrolled by the Voluntary Social Welfare Agencies Ordinance 1961. In the early days, six to ten individuals worked in affiliation. Now, more than 15 individuals are working in the Karachi home office and different branches additionally have more than 15 workers. Majority of workers are women in this NGO.

APWA has more than 100 branches all through Pakistan and globally. Every region has such a significant number of branches, for example, in Sindh, Punjab, Baluchistan, and Khyberpukhtunkhwa. Likewise, the national level NGO has global branches in the UK, Canada, and America. When inquired about the nature of social work for women, she said that APWA was created to aid poor and penniless individuals; consequently, they confront a huge number of issues. In such a manner, we additionally confront such a large number of issues for taking care of the issues of women and society, for example, socio-cultural and social issues. As an association involved in women development, it faces additional difficulties owing to predominant patriarchal culture. Resultantly, the association is finding it difficult to sustain and enhance its outreach owing to lack of funds and donations. Another major obstacle is the mentality of individuals. We do social welfare for women, she said. Ladies have worked here for the entire night for advancement of affiliation and settling women issues. During winters, they even spent complete nights while making shawls and knitting sweaters. She said that individuals support them for accomplishment of their objectives.

According to her analysis, following the hardships and too much efforts of association, the structure of society has changed to some extent. APWA has not only pursues legislation for improving the plight of women and children but also provides legal aid services. We have passed the Family Law Ordinance 1961, Child Marriage Act, through which women became eligible for property rights and land ownership, marriage and divorce registration, along with second marriage permission by wife and arbitrary council.

By maintaining legal aid centres, she said, it has modified the society by making people aware, making women empowered and confident, appraisal of feelings of philanthropic work and self-help. And we can observe these changes at a glance pour on thousands of beneficiaries of APWA. She further explained in detail with interest about the services of association and NGOs through analysing the women's issues. She said that though at the time of establishment it was the only NGO of its kind, there was no hindrance in extending its activities across the country. However, working on women issues was always considered to a difficult task in the longer run. Likewise, APWA also faces hindrances. Now, the association does not confront any sort of obstacles except fiscal issues. Ms. Farzana looks quite satisfied about the activities of the APWA.

She also said that by keeping up legal aid centres, it has changed the general public through making individuals mindful, making women enabled and sure, examination of sentiments of humanitarian work, and self-improvement. Furthermore, we can watch these progressions pour on a large number of recipients of APWA. She enthusiastically explained that collaborative efforts of Gos and NGOs had been very fruitful after effective categorization of women issues into functional sub domains. APWA was managed by incredible women in the past and has been in great hands since then but still faces multiple challenges. Yet, now the affiliation does not face any kind of obstructions aside from monetary issues. Ms. Farzana looks much fulfilled from exercises of APWA: "I am so much satisfied with the performance of APWA to bring social change because it is comprehensively paying attention on women's problems and children," she said.

She at that point informed about the capacities and the administration of APWA. She told that for the most part ladies are working here, the gatherings of governing body continue week by week, and after that, a methodology is figured out for capacities and administrations. This technique is about the improvement and social welfare of women and children. The technique additionally examined the lesser labourers for improvement of approach, administration, and usage. Working timings of establishment are from 9 a.m. to 4 p.m. APWA individuals work here voluntarily, just specialized staff does paid work. Upon asking, "How to function in this NGO, and is there a prerequisite of any capability to work?" She clarified that "there is no criteria or qualification barring anyone to work at APWA. The only things required are sensitivity, to human values supplemented by corresponding passion to work for welfare of poor individuals especially for women and children. Therefore, related NGO work here, we go to other cities and towns and gives their administrations to rustic urban individuals who require help and generosity, our affiliation have region focuses in various urban communities of Pakistan."

Ms. Farzana Rehman illustrated the three focal objectives of association: to educate women, to empower women, and to improve their health. APWA, at the beginning, worked only for women but with the passage of time, the aims have changed to the social welfare of women and children. On questioning about ways of funding, she said that APWA does not receive funding from any donor agency. She gave a detailed account of projects of the association. Major projects include health clinics for women and children, establishment of education centres and vocational institutes and efforts to improve plight of imprisoned women. Some of these projects been successfully completed while others are in progress. APWA has contacts with other NGOs and participates in different programmes of other NGOs. All the NGOs are cooperative to each other in Pakistan and all other NGOs consider APWA to be a respected NGO and a community-based organization. They record data of their services in documentation and soft copies. Association publishes annual reports and brochures; they have also published a report named APWA 50 years of service, in which history, projects, and efforts are compiled.

The association gives internship opportunities to students to their future improvement and advancement. They prepare their volunteers and labourers through expert specialists. They prepare and educate individuals about the program coordination, operations like how to instruct, how to do work, and how to do function at an NGO. She assessed that Pakistan has been very fruitful for accomplishing Millennium Development Goals, now women are viewed as enough enabled and certain and NGOs are currently assuming their part for accomplishing these objectives as women's strengthening. Pakistani NGOs are powerfully taking part in the satisfaction of global understanding and targets. But there is sufficient time required to achieve the objectives. Although viewpoints in Pakistan confirm to the worldwide understanding but very little progress has been made so far. Consequently, APWA is likewise taking an interest in accomplishing their objectives and targets. As indicated by Ms. Farzana, APWA works for rustic and urban zones in various ways. Administrative responsibilities of rural projects are assigned to ladies with strong passion to achieve the goal of creating awareness amongst women about their rights. While in urban regions, they work for women empowerment. While responding to question regarding access to government departments / organizations, she said there are no issues as APWA is a non-benefit organization.

The most important thing she mentioned very proudly again and again is that they work voluntarily. They have no linkages from donors nationally or internationally. The association has no affiliation with US-AID, but internationally affiliated with United Nations, International Alliance of Women, International Council of Women, and many others. At the end of the interview, she suggested some approaches for every sort of organization; she suggested that they create groups and work gladly, voluntarily, and without any self-indulgence. She identified health sector, women education, and women empowerment as the three domains that APWA should focus more in the future.

Discussion and Analysis

Pakistan is an Islamic Republic, established for the sake of Islam and its belief systems. Despite this, immense issues exists which have social, conservative, political, and religious context. Moreover, people living in Pakistan generally follow a few taboos which stifle women's rights. To take care of these issues, government and charitable associations have been taking measures. The most enduring classes are of the poor, women, and children. The chosen NGO of the city do work their best for the welfare and strengthening of women. These selected NGOs in Karachi endeavor to protect and enact social rights while responding to incidents of violations and assisting the victims through designated authorities. For the most part, NGOs in Karachi chip away at the premise of social rights while gathering the data of the infringement of these rights, enactment, counteractive action of women rights, mishandle and restoring the casualties by the compassionate help through support, and awareness raising projects.

The All Pakistan Women's Association, abbreviated as APWA, the oldest organization of Pakistan, has been working since 65 years. It has been providing services for a long time, and has made history in Pakistan. It established a network of industrial homes, girl's schools, and colleges. It runs clinics and also outlets for crafts made by women who learn at their industrial homes. APWA plays a very important role to improve the social welfare conditions by utilizing political, economic, social, and technological factors in Pakistan (Patti 2000).

During the interview, Ms. Farzana Rehman indicated that the organization faced many problems during work, particularly the conservative mind sets of people and patriarchal society. "Purdah" in the beginning created many problems for the welfare of women. In Pakistan, gender relations are based on two perceptions (a) women are subordinate to men (b) family honour resides in women's actions. Another important issue is created that Purdah made different spheres of male and females (Blood 1996). Ms. Farzana said that although the situation now has been changed to transform attitudes and values due to the influence of NGOs, media, and education (Asian Network of Women in Communication 1996), it was very problematic to work in these circumstances.

By the creation of legal aid centres, people become aware about the rights of women and women feel more empowered and confident. Violence against women is a serious issue of Pakistan. Gender-based violence is insidious and constant, despite class differences, issues of caste, age, and society. In many different forms especially in family, community, and state levels, gender-based violence is perpetrated. Women with lack of confidence due to fear of violence develop a sense of insecurity.

Gender-based violence is continued at a wide range of levels, that is, at the family, group, and state levels, and in a wide range of structures (Shakil et al. 2013). Women have specific problems and suffer every day from intimate partner violence, which causes injuries, unwanted pregnancies, abortions, and sexually transmitted diseases. Many clinic-based studies showed that 52% women are victim of physical violence and 82% of women suffer from psychological violence (Zakar 2012). In Punjab, females often remain quiet due to family honour. To reduce the intensity of the issue, APWA launched Free Legal Aid Cell with the objective to provide free legal advocacy/awareness programmes (APWA Punjab 2016).

For the urban educated women, the 1950s and 1960s were hopeful decades, as society had all the allocations of being pushed towards an advanced dynamic future. The momentum for APWA's protest which started in 1955 was the second marriage of the Prime Minister Mohammad Ali Bogra that time, since the Family Law Ordinance places (among different conditions) certain controls on polygamy and divorce (Committee on Women's Studies in Asia 1995).

Muslim Family Laws Ordinance (MFLO) 1961 and Family Courts Act 1964 were the result of the continuous struggle of APWA. The endorsement of these laws was a holistic process followed by the Muslim Laws of marriage and divorce. The Family laws introduced essential marriage registration, set confinements on the act of polygamy, and changed the law identifying with dower and support in marriage and divorce. In addition, APWA without any assistance activated the women and the society and campaigned with the then government to order the family laws which gave lawful strengthening to women and children. It was because of the dynamic and tireless campaigning of APWA that three clauses maintaining women's equal status were embedded in the 1973 Constitution.

As APWA has the credit to take an interest fundamentally to pass the Family Law Ordinance 1961, Child Marriage Act, marriage and separation enlistment alongside second marriage authorization by wife and arbitrary council. With APWA, United Front for Women's Rights (UFWR) pressurized the government to bring in a new ordinance/bill for the protection of rights of women including marriage, polygamy, divorce, maintenance, inheritance, and children's custody (Maqsood 2016).

During emergencies, APWA always provides relief, looks after the orphans, has established an APWA college in Lahore, and created some vocational training centres. APWA's approach was interactive, progressive, and towards the betterment of society. According to Ayesha Jalal, APWA has gained extensive support from government and women's rights activists (Haider 2000).

Elite class women founded APWA to solve the problems of poor and middle class women. APWA's founding members actually could not differentiate between the problems of privileged and under privileged class women and had the opinion that all the women had the same problems irrespective of class, religion, or culture. However, this theory was proved wrong afterwards. Though APWA representative said that APWA did not receive any funding from donor agencies, it's also a reality that due to good relations with the government, APWA always received government funding because of its non-threatening status, as we can easily say that the organization's focus is on welfare and development of women including education and income generating schemes to strengthen the women economically.

While APWA struggled for political and legal reforms and achieved some as in 1953, APWA suggested ten reserved seats for women in the National and Provincial Assemblies.

Another criticism APWA faces in spite of a primarily non-political and welfare approach is by the religious sector of the country due to the unveiling. The ladies of APWA listened to many harsh comments as The Majlis-e-Ahrar, a right-wing orthodox party, tagged them as prostitutes. The Jamaat-e-Islami (JI) and Jamiat-ul-Ulema-e-Islam (JUI) also dislike them, even with having opposed the creation of Pakistan. The clash and stress between women's rights activist and right-wing religious lobbies has a long history, but at that time, it did not turn into an open war as afterwards. APWA also did not challenge the military rule; therefore, as APWA explained itself, it did not challenge the military dictatorship which defined itself as a generous, modest, and contemporary organization (Saigol 2016).

APWA also could not separate women's issues from peace and social inequality as they failed to project them as independent issues. At that time, famous women's organizations were occupied by the thinking of only charitable work among women. On the whole, class position of these groups reflected the basic cause of not changing social status of women was that the association merged philanthropy with reforms and with participation of lower and middle class women these reforms applied from the top to change the system in to tolerable but without the understanding of ground realities and inequities (Rouse 1988).

No doubt, All Pakistan Women's Association played a very important role in the development of women in early years of Pakistan. But afterwards APWA lost the spirit and now a days, the organization is not as active as it was before. They continued the welfare approach, but with the passage of time the leading role which it played in the formation of women protection laws remained unsuccessful. APWA opened many schools, clinics, in far flung areas, but the scope is not improving as compared to the past.

It has been observed that in Pakistan, the understanding of women's rights isn't only a question of women's liberation; it is a development that ought to be mainstreamed and urged to dispose of brutality against women and children. The debate on women's rights about regard, respect, and human rights has been transparently recognized yet not rehearsed. APWA should work on these lines to motivate it and others so it can again revitalize its disposition.

Acknowledgement The researcher is thankful to Ms Shazia Kanwal, student of BS 4th Year, Centre of Excellence for Women's Studies, University of Karachi, for her help in conducting the interviews from APWA.

References

All Pakistan Women's Association. (1987). APWA 50 Years of Service, Karachi.

All Pakistan Women's Association (APWA). Retrieved on June 1, 2017, from http://www.pakistanherald.com/profile/apwa-1143.

APWA Punjab. (2016). Retrieved July 25, from http://apwapunjab.com/legal-cell.html.

Asian Network of Women in Communication. (1996). Paper from Regional Workshop on "Women Rights in the Media", Lahore, Pakistan, 9–11 April, Himmat Society.

Banuri, T. J., et al. (1997). *Just Development: Beyond Adjustment with a Human Face* (pp. 174–175). Karachi: Oxford University Press.

Blood, P. R. (1996). *Pakistan a Country Study*. Washington, DC: Library of Congress, Federal Research Division.

Committee on Women's Studies in Asia. (1995). *Changing Lives: Life Stories of Asian Pioneers in Women's Studies*. New York: Feminist Press.

Doh, P. J., & Teegen, H. (2003). *Globalization and NGOs, Transforming Business, Government, and Society*. London: Praeger.

Ghosh, B. N. (2015). *Scientific Method and Social Research*. New Delhi: Sterling Publishers.

Goel, S. L., & Kumar, R. (2005). *Administration and Management of NGOs*. New Delhi: Deep & Deep Publications.

Haider, N. (2000). Islamic Legal Reform: The Case of Pakistan and Family Law. *Yale Journal of Law & Feminism, 12*(2), Article 5. Retrieved June 1, 2017, from http://apwa.way2expert.com/about_apwa/begum-raana-liaquat-ali-khan/retrieved.

IAW Communication Unit. (2016). Retrieved from http://womenalliance.org/bangladesh-mahila-samity.

Khalid, M. (2001). *Social Work Theory and Practice: With Special Reference to Pakistan*. Karachi: Kifayat Academy.

Khan, K. S. (1995). Setting Health Care Priorities in Pakistan. *Journal of Pakistan Medical Association, 45*, 222–227.

Maqsood, N. (2016). The Debate over Polygamy in the Context of Muslim Family Law Ordinance 1961. *Specialty Journal of Politics and Law, 1*(1), 49–59.

Mulki, M. A. (2013). The First Lady. Retrieved August 3, 2017, from https://www.dawn.com/news/785779.

Nelson, J. (2007). *The Operation of Non-Governmental Organizations (NGOs) in a World of Corporate and Other Codes of Conduct. Corporate Social Responsibility Initiative*. Working Paper No. 34. Cambridge: John F. Kennedy School of Government, Harvard University, p. 2.

Pace, W. (2002). Governance and Civil Society. Paper read at UNEP Civil Society Consultation on International Environmental Governance. Retrieved May 2015, from www.env-net.org/wp-content/uploads/2013/09/gemmil.pdf.

Pasha, A. G., & Asif Iqbal, M. (2002). Non Profit Sector in Pakistan: Government Policy and Feature Issue. *The Pakistan Development Review, 41*(4), 879–908.

Patti, R. J. (2000). *The Hand Book of Social Welfare Administration*. New Delhi: Sage Publications.

Prasad, K. (2000). *NGOs and Socio-e-Economic Development Opportunities*. Delhi: Deep & Deep Publications.

Rashid, T. (2006). *Contested Representation: Punjabi Women in Feminist Debate in Pakistan* (p. 202). Karachi: Oxford University Press.

Rehmatullah, S. (2002). *Social Welfare in Pakistan*. Pakistan: Oxford University Press.

Rizwi, A. K. (2008). *The Future of Social Welfare*. New Delhi: Mohit Publications.

Rouse, S. (1988). *Women's Movement in Pakistan: State, Class, Gender*. Retrieved from http://www.wluml.org/node/241.

Saigol, R. (2016). *Feminism and the Women's Movement in Pakistan, Actors, Debates and Strategies*. Islamabad, Pakistan: Friedrich-Ebert-Stiftung.

Sarantakos, S. (2005). *Social Research*. New York: Palgrave Macmillan.

Sarker, C. P. (2015). Voluntarism to Social Work and Social Services in Bangladesh. *Pakistan Journal of Applied Social Sciences, 1*, 33–43.

Shah, N. M. (1989). *Pakistani Women: A Socio-Economic & Demographic Profile* (p. 31). Islamabad: Pakistan Institute of Development Economics.

Shakil, M., et al. (2013). *Violence against Women: Behavioral Shift Because of Education in Violence Against Women in Pakistan*. Saarbrücken, Germany: Lap Lambert Academic Publishing GmbH KG.

Shetty, M. K. (2013). *Breast Gynecological Cancers: An Integrated Approach in Developing Countries* (p. 260). New York: Springer.

Yin, R. K. (2014). *Case Study Research, Design and Methods*. Los Angeles, CA: Sage Publications.

Zakar, R. (2012). *Intimate Partner Violence Against Women and Its Implications for Women's Health in Pakistan*. Hamburg: Disserta Verlog.

14

Civil Society's Contribution to Women Empowerment: Bangladesh Perspective

Tania Afrin Tonny and Jewel Ahmed

Introduction

How can we empower others? Empowerment is a "bottom-up" process and cannot be imposed; it is only possible to provide the conditions in which it might occur. It requires sociocultural support and cannot be achieved individually. In the meantime, women in Bangladesh are still struggling in all phases of their life (Dialogue on Concepts of Women's Empowerment 2008). Women should be empowered so that they can free themselves from all kinds of violence and oppression and can contribute to a sustainable society based on equity and justice. Bangladesh is a small country of South Asia where there are thousands of non-government organisations (NGOs) contributing to the society as development partners with the government and the private sector. In a democratic country like Bangladesh, the constitution has guaranteed the rights of her citizens to join in association with freedom of speech. Articles

T. A. Tonny (✉)
Jatiya Kabi Kazi Nazrul Islam University, Mymensingh, Bangladesh

J. Ahmed
World Child Cancer, Dhaka, Bangladesh

© The Author(s) 2020

Md. N. Momen et al. (eds.), *Building Sustainable Communities*,
https://doi.org/10.1007/978-981-15-2393-9_14

38 and 39 of the Constitution have declared the freedom of association and freedom of thought. There are numerous civil society organisations (hereafter CSOs) in addition to NGOs, in such forms as voluntary organisations, professional bodies, religious institutions, sports, and cultural/recreational clubs. Additionally, there are media (print and digital) that are supposed to be working for the country's overall development and progress. The present study is a little endeavour to know the contribution of civil society organisations in case of women empowerment in Bangladesh. The civil societies are mainly social service-oriented in Bangladesh. Tasnim (2008) noted that civil society of Bangladesh mainly works towards the development and social welfare of the country and it has little contribution towards the country's democracy. It was observed by Tasnim (2007) that civil societies are highly participative in grass-roots social services and women empowerment is notable amongst them. Ahmed (2011) pointed out that civil societies of Bangladesh are depoliticised in working objectives and functions and in providing social services, while they are politicised in advocacy, lobbying, network-building, and influencing of policy. On the Civil Society in Bangladesh: An update of the Application of BD-JIGS data (2007) observed that the actions of civil societies are based on social service whereas advocacy is a new dimension and leads to democracy in Bangladesh. The reviewed literature for this study shows that CSOs are able to engage with communities and contribute to the development of effective and accessible services. Those organisations can take critical information to vulnerable groups who need it the most and help in fostering trust within communities to improve the functioning of the justice system of the country. Malhotra et al. (2002) noted that empowerment can take place if the methods of participatory development have been established; upon that, civil society at both macro and meso levels of society can contribute to empowerment as a political mechanism. The general objective of the chapter is to investigate the contribution of civil society in supporting the empowerment of women in Bangladesh. The specific objectives include (a) identifying the relationship of civil society and women empowerment and (b) identifying the strategy of civil society in contributing towards women empowerment in the country. The study used a qualitative approach (i.e., with subjective information) and explored new research areas that can serve as the basis

Components of Civil Society	Number of Respondents
NGOs	10
Professional Associations	05
Media (Print and electronics)	05
Women Organizations	03
Voluntary Organizations	02
Total	25

Fig. 14.1 Sample size of the study

for crafting new theories. The study used both primary and secondary data; the collection tool for the primary data was in-depth interview (IDI). This approach offers an opportunity to capture descriptive data about people's behaviours, attitudes, and perceptions. Newspapers, journals, books, published reports, and the internet are used as secondary data sources for the present study. The study followed a purposive sampling technique due to the nature of the research design and objectives of the study. The sample size of the study is outlined in the figure (Fig. 14.1) A total of 25 respondents from the different components of civil society were interviewed. The components include NGOs, professional associations, media (print and electronics), women organizations, and voluntary organizations.

Civil Society

In the late seventeenth century, the concept of civil society entered into the discourse of intellect with the writings of Locke and Harrington. The notion was developed as an umbrella concept that was taken up by Smith, Rousseau, Hegel, and later de Tocqueville, which incorporated some institutions from the outside of the state. It accumulated "voluntary regions" which was denoted by de Tocqueville, covered by both the private and public organisations, social interactions, and relationships is based on cooperation and can create trust among people and may form public opinion, institutions and even political parties. In the 1820s and 1830s, the British political economist Hodgkins pointed out that, inhumanity, dominating tendency, and exploitation leads to capitalism. In 1840, Karl Marx criticised this view. Between 1842 and 1845, Marx's writings influenced the concept of civil society and its transformation. Marx believed that cooperative ties would provide the basis for collective bindings and social organisations (Rahman 1999). The concept of civil society was an important phenomenon of capitalism in the mid-nineteenth century. Though civil society and capitalism have been seen as different terms. It was considered as a solitary sphere in which various kinds of community comes into being gradually. The industrialised market economy puts many blocks in the road to civil society. Class divisions in society, differential housing, dual labour market, unemployment, and poverty can be considered as constraint factors to the development of civil society; when a society is free from bureaucratic control and encourages civil engagement supremacy, then social capital can flourish. Additionally, civil society can play an important role with social capital to promote democratisation and participation. The NGOs are a major component of civil society; as such, they can play a positive role in enhancing the social and economic empowerment of poor people (Rahman 1999). Civil society has many other components, as displayed in the figure (Fig. 14.2). The figure shows the several components of civil society includes NGOs, professional bodies, voluntary organizations, Media (print and electronics), women organizations, trade and labor unions, cultural and recreational bodies, sports clubs and youth organizations, social welfare organizations, religious institutions, industrial associations and registered cooperatives.

Generally, civil society is composed of some groups and organisations, works for the general interest of the citizens and operates into the outside of the commercial and governmental sectors. Ibrahim and Hulme (2010) noted that civil society is the combination of individuals, organisations, and institutions situated among the family, state, and market for advancing the common interest wherein people work and cooperate voluntarily (Anheier 2004). According to Lester Salamon (1994), there are some specific causes for the rapid growth of civil society: the limitations of the

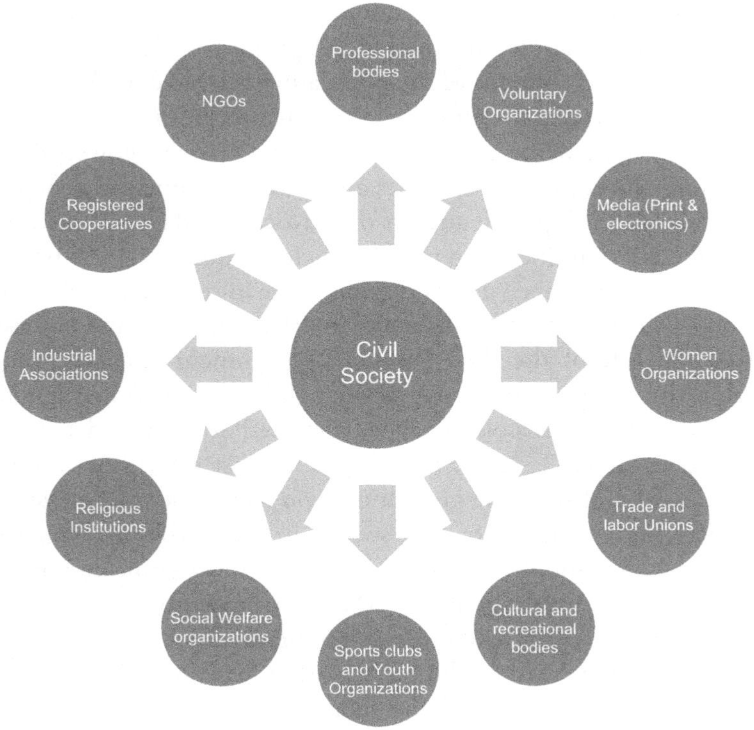

Fig. 14.2 Components of civil society. (Source: Researcher's own compilation)

modern state to deliver services by NGOs, environmental issues, limitations of socialism, advancement of communication, development of education, and growth of the global economy. Tasnim (2008) pointed out the different types of grass-roots programs in which civil society organisations are involved in Bangladesh. Those are rural developments by government-organized non-government organization (GONGO) cooperation, group network development, grievances ventilation and local representation, women empowerment, increasing literacy rate and educational development, human rights and health services, and social/political awareness and participation. Finally, Guerrero (2003, p. 212) stated that "CSOs and NGOs are voluntary, non-profit, value-driven formations committed to bringing about social change." Ahmed (2008) found that foreign-funded NGOs have a relation with their target group population that is similar to clients. A large amount of the population is treated as clients and beneficiaries, as opposed to being treated as "members" by the foreign-funded NGOs. In influencing public policies, these are very powerful, whereas religious institutions like mosques are powerful enough to generate capital for social development and for the advocacy and those institutions are less powerful to influence in public policies. Rahman (1999) pointed out that the Bangladeshi civil society plays a role in strengthening democratisation of the country. For instance, in 1952, students and activists protested the rule of the then Pakistani government and shed blood for establishing their mother tongue "Bangla" as the national language of Pakistan which was considered as the seed for the independence of Bangladesh in 1971. Later, the struggle against the rule of the military government and then the growth of civic engagement of citizens led to broadening the concept of civil society in Bangladesh in which NGOs played the most significant role. Women are now significantly participating in the different programs of NGOs in the history of rural development of Bangladesh, which is a landmark. The concept of women participation is related to the participation of women in several NGOs and to improving their social and economic conditions for a better livelihood. Participation sometimes leads to empowerment.

Empowerment

Empowerment as a process tries to obtain some opportunities for the marginalised group of people, directly or through the help of others who are not marginalised and want to share their access to the different opportunities. Mayoux (2000) stated that empowerment is a process of power within, power to, power with, and power over, where "power within" is an integral change process, "power to" is capacity augmentation, "power with" is men's and women's collective mobilisations and "power over" is questioning the pattern of subordination connected to gender. Generally, to achieve one's goals, to exercise power, and to control own circumstances by any individual, community, or group is called empowerment.

It is a process and a capacity by which, an individual or a group can achieve their interests and help others for maximising the quality of their living. Batliwala (1994) said that the overall welfare of the population depends on their influences over the external actions whether it is much or less. Kabeer (2001) has given a definition of empowerment as follows, in a given context, when peoples are able to make their strategy for their life choice which was denied previously is considered as an expansion of empowerment.

Women Empowerment

Women empowerment and gender equality are the agenda of global development which is one of the remarkable points for the sustainable development goals (SDG) of the United Nations. Achieve gender equality and empower all women and girls is identified as the goal number 5 of the SDG. United Nations (UN) viewed that, women empowerment has five components: self worth sense of women, having rights and the determination of choices, right to access of opportunities and resources, right to control over their own lives both at home and outside, and finally, the ability to influence the society for making its changes and establishing justice nationally and internationally as cited in Islam and Dogra (2011). It has been observed that, the role of NGOs in women empowerment is significant. NGOs are plying a substitute as well as a harmonising role along with the government and market in the women empowerment process of Bangladesh. Women empowerment as an issue of NGOs has emerged recently. It is nothing but a process by which women can be more powerful in their life and they may be able to manage any challenges in their life. Basically, women empowerment means having the power to take any decisions in life and women should have an access to information and resources as well. Ahamad et al. (2015) noted that, women empowerment contributes to the sustainable development by uplifting the socioeconomic and political conditions of women in India, whereas empowerment leads to control over material resources, intellectual properties, and ideology and which also facilitates the life of women at multiple levels, such as, family and community level, market and state level. It plays an important role to convert an idle society into a self-sustainable society where women must be empowered in sociopolitical, economic as well as from legal aspects. The study also argued that, a radical change has been found in the living standard and the life style of women after joining NGOs. Women were found to start their own earnings and they were becoming more independent and also were self-motivated. They were also found to take their own decisions and also it has been seen that they were giving suggestions and opinions to their families. A small number of women living countryside were found to participate actively in the village meeting or *gram sabha* and those were involved in the decisions of their family in India. Awan (2012) stated that, women empowerment process is not only limited to control over the financial matter or reproductive roles; it is a compilation of three interrelated dimensions: literacy, health, and employment. Women empowerment represents women as active agents and not the passive recipients only in the development of a community as a whole. The study noted that the notion of women empowerment leads to the overall changes in the dominating patriarchal structure of a society like India (Awan 2012). It is also equally true for Bangladesh. The study found that, in Pakistan, gender discrimination is a dominating factor which is hampering the overall process of women empowerment and established a linkage between the contribution of civil society and women empowerment in Pakistan. The study also noted that the patriarchal mind-set, out-dated traditions and cultural norms confined civil society organisations like NGOs, whereas NGOs bringing awareness among women and played a significant role in Pakistan. The study found that the hard work of civil society can only be successful in Pakistan if the root of patriarchy is minimised. Much research has showed that, in Bangladesh, the growth of NGOs and civil societies has failed due to the lack of better politics and the dysfunction of its market system like many other developing countries of the world. The failure of the traditional market system has created a vacuum which has been filled by the civil societies in Bangladesh (Ahmed 2011). The study pointed that, in the last 50 years, agricultural, industrial, and accommodating civil societies of Bangladesh have been raised significantly and individuals were becoming involved in civil societies because of economic issues where social and political issues were not so important. The study identified that the mid-1940s was the first phase of growth of civil societies in Bangladesh; the second phase began in the 1970s. The study noted that women are not comprehensively included at all levels and civil societies of Bangladesh; predominantly, however, NGOs are taking affirmative action to reduce the gender gap. The study found that civil societies have perceived their members as clients rather than citizens. Parveen and Leonhauser (2004) identified three important dimensions of empowerment: socioeconomic, familial, and psychological. The socioeconomic aspect includes economic contributions of both firms and non-firms to household wellbeing, as well as access to socioeconomic resources and possession of productive and non-productive assets. The familial dimension includes participation in household decision-making, that is, increasing the role of women in house-level decision making must lead to their wellbeing and for their children. The psychological aspect includes the perception of gender awareness and basic women's rights as well as the capacity to cope with diverse domestic works (Parveen and Leonhauser 2004). The study also found that women empowerment is merely satisfactory at the household level, with traditional beliefs, attitudes, and practices intensely entrenched in females' lives, thus hampering the overall empowerment process.

Civil Society's Contribution to Women Empowerment in Bangladesh

Kabeer (2001) identified the three interrelated components of women empowerment process: resources, agencies, and achievements. Chen (1992) identified four components of women empowerment: resources, perceptions, relationships, and power. Yount (2017) has conceptualised women empowerment, based on Kabeer (1999), as a dynamic process. By this process, women acquire resources that enable them to develop their own voices, articulate preferences, and gain the capacity to make their own decisions in terms of fulfilling their life aspirations. By reviewing the literature and analysing the primary data, the present study has found that civil society of Bangladesh is contributing towards women empowerment in three aspects; firstly, raising awareness among women; secondly, ensuring participation of women in decision making process; and finally, taking action in the true sense as depicted in the figure (Fig. 14.3). The figure (Fig. 14.3) shows the relationship of civil society and women empowerment in Bangladesh and illustrates the contribution of civil society to women empowerment through the three interrelated aspects: awareness, participation and action.

It has been found that the civil society organisations, basically, are contributing by a three-stage process of women empowerment. Firstly, CSOs are engaged for increasing "awareness" regarding women's rights (which includes socioeconomic and political rights of women); secondly, encouraging women to "participate" in a true sense (which includes affiliation and engagement with development activities for their own betterment as well as for the society as a whole); and finally, these are leading towards the "action" (which includes positive changes in sociocultural behaviours and attitudes and the mind-sets of men and women). The overall process includes women making their own choices and decisions, promoting freedom of speech with improving social networks, and reducing the level of marginalisation of women in Bangladesh. The study also found that CSOs are citizen-led organisations in a state system that aims to correct the state actions or mechanisms in favour of the most marginalised citizens and people living in the state. Therefore, it is a corrective force to

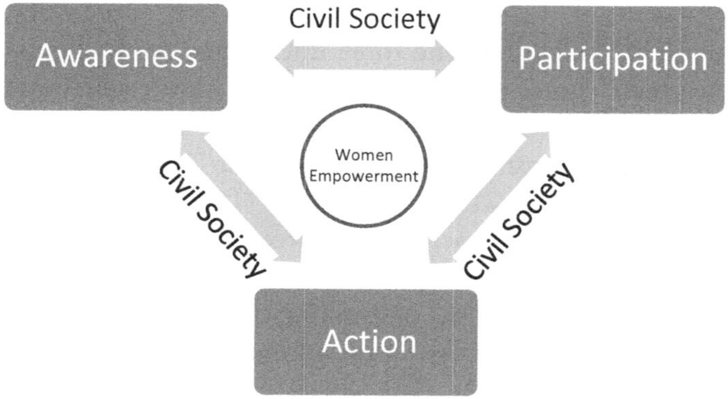

Fig. 14.3 Civil society's contribution to women empowerment process

support the state and to improve its functions. All the citizens not serving as duty bearers in any public institutions or forces like bureaucrats, armies, and judges (and who are not direct members of any political party) is called civil society member. A CSO must be based on some principles that include freedom from any bias, race, sex, ethnicity, religion, class, and so on. These organisations are not against the state but rather endeavour to ensure that the state works equally for all its citizens, irrespective of their identity. It stands for the demand side and can be united for issue-based movements. The key function of CSOs is to put pressure on the state machineries to ensure justice for all. This pressure can take many forms; however, not violating the rules or distorting law or state orders. Giving intellectual advice is one of the key means of contributing to the state. Other functions of CSOs include articulating demands, criticising policies and laws, proposing alternatives, negotiating for people's demands, and so on. It has been found that civil society is contributing considerably towards women empowerment in Bangladesh, especially fighting against gender-based violence and breaking the gender stereotypes and demanding affirmative action for gender equality (e.g., quotas for women in government machinery). Since women are a deprived category in Bangladesh and a civil society organisation has something to say or do for women regarding their deprivation, there must be a connection. The study found that the CSOs of Bangladesh are contributing positively towards women empowerment, whereas making women equal to men is a key focus of CSOs. Basically, the advanced urban group of civil society has articulated the movement for equal rights of women in post-liberation regime in Bangladesh. The roles of CSOs in empowering women have been remarkable to this point. There are many instances in Bangladesh, for example, forming national women advancement policy is a result of continuous movements of the CSOs in Bangladesh. Achievements in the movement thus far in the area of women empowerment, largely came from the different movements of CSOs. They lead in forming women's rights-focused NGOs, developing networks of like-minded organisations, movement for affirmative action like, reserved seats in Parliament and local government institutions and in public services, movements against gender discriminations, building capacities of women in different areas of socioeconomic and political context of Bangladesh. The CSOs in

Bangladesh have a multiple role to play in empowering women in the country. On the one hand, it works as a "think tank" to deliver intellectual support to the policy-makers; mobilise masses of people; and, on the other hand, work to make them aware as well as put pressure on the state machinery. Thus, it plays a three-tiered role, micro-meso-macro contributions to ensure voices and participations of all segments of society especially for women in Bangladesh. The women's movements, over the years, reflect such roles of civil society actions. All the strategies mentioned above have more or less effects on the progress of women's rights in Bangladesh—and every aspect is equally important. The NGO, like Bangladesh Rural Advancement Committee (BRAC), has changed its focus over decades but keeps the same principles in mind: to make women equal to men and make them free from all forms of discriminations existing in the society. It has a vision and mission which guide the organisation what to do and how to make a gender-responsive society and to empower women in Bangladesh. The initiatives range from economic to political empowerment to reflect a sense of "inclusiveness" in changing women's lives. The reviewed literature has widely discussed such roles of the NGOs too. CSOs are enhancing capacities in different areas, including economic, social, and political, research, and advocacy (enhancing capacities and doing effective advocacy) and public movements for political empowerment of women. However, women empowerment refers not to economic empowerment only, but also to a dignified life to be necessarily supported by the cultural landscape. In general, all the past efforts of CSOs made a good sense although "learning by doing" has always been a focus of the civil society organisations. The past endeavour of civil society in Bangladesh is appreciated too. The present study has found that, for the contribution of CSOs, women are getting preferences for the recruitment process of government institutions and in the NGOs. In most cases, at least 30% of staffs are women in Bangladeshi NGOs like Bangladesh Rural Advancement Committee (BRAC), the world largest and the number one NGO of the world. This organisation's female staffs have enough flexibility to do field visits and official work. The Bangladeshi government and NGOs are now providing six months of maternal leave with pay for the women employees. The CSOs are working for gender-sensitive workplaces also. For improving gender sensitivity within organisations (intra) and partner and stakeholders as well (inter). The CSOs can talk, initiate dialogue, or raise their voice regarding women empowerment; they are able to identify barriers to women empowerment. Thus, CSOs are very significant and make a major contribution to women empowerment of Bangladesh.

Conclusion

Women empowerment is an agenda for making the society better based on social justice. Therefore, along with government, CSOs should continue their efforts in favour of a common interest of the society. A comprehensive strategy of CSOs is needed to make changes in gender roles (depending on global and national contexts). These organisations should adopt a holistic approach considering both the economic and cultural complexities in societies of Bangladesh. CSOs should work for increasing the awareness among the mass of people, especially those who are more vulnerable to creating a social movement towards women empowerment and CSOs should work as a pressure group. CSOs should have a strong advocacy tool to cope with governmental bodies such as the Ministry of Women and Children Affairs (MoWCA) and develop proper strategies in this regard. CSOs should engage more men supporting women empowerment as patriarchy is prevailing all over Bangladesh. A proactive role of policymakers and the positive roles of some different actors like think tanks, government agencies, political parties, and professional bodies are important. As well, all the CSOs need to cooperate with each other, and all should be united so that women empowerment can easily be established in Bangladesh. Last but not least, the government and CSOs should keep their concentration on issues creating obstacles to the path of women empowerment. Further, a comprehensive plan should be outlined immediately for ensuring women empowerment practically in Bangladesh.

References

Ahamad, T., Hemlata, & Narayana, A. (2015). *Role of NGOs in Women Empowerment: With Special Reference to Uttar Pradesh.*

Ahmed, S. (2008). *Dichotomous Structure of NGO-Based Inputs Versus Islam Based Inputs to Civil Society: The Case of Bangladesh.*

Ahmed, S. (2011). *The Civil Society of Bangladesh: Depoliticized in Working Agenda But Politicized in Power Relation.*

Anheier, H. K. (2004). *Civil Society: Measurement, Evaluation and Policy.* London: Earthscan.

Awan, S. Z. (2012). *Role of Civil Society in Empowering Pakistani Women.*

Batliwala, S. (1994). *The Meaning of Women's Empowerment: New Concepts from Action.*

Chen, M. (1992). Conceptual Model for Women Empowerment *Dialogue on Concepts of Women's Empowerment.* (2008).

Guerrero, D. M. (2003). *The Growing Role of Civil Society Organizations (CSOs) in the Third World and Government Policies Towards Them.*

Ibrahim, S., & Hulme, D. (2010). *Has Civil Society Helped the Poor? A Review of the Roles and Contributions of Civil Society to Poverty Reduction.*

Islam, M. S., & Dogra, S. (2011). *Women Empowerment in Bangladesh: The Rise of the Other Half.*

Kabeer, N. (1999). Resources, agency, achievements: Reflections on the measurement of women's empowerment. *Discussing women's empowerment: Theory and practice. Swedish International Development Agency, 30*, 17–54.

Kabeer, N. (2001). *Reflections on the Measurement of Women's Empowerment.*

Kabeer, N. (2012). *Women's Economic Empowerment and Inclusive Growth: Labor Markets and Enterprise Development.*

Malhotra, A., Schuler, S. R., & Boender C. (2002). *Measuring Women's Empowerment as a Variable in International Development.*

Mayoux, L. (2000). *Micro-finance and the Empowerment of Women: A Review of Thee Key Issues.*

Parveen, S., & Leonhauser, I. (2004). *Empowerment of Rural Women in Bangladesh: A Household Level Analysis.*

Pathways South Asia. (2008, January 21). *Dialogue on Concepts of Women's Empowerment – Bangladesh.* Retrieved from http://archive.ids.ac.uk/pathwaysofempowerment/www.pathwaysofempowerment.org/archive_resources/dialogue-on-concepts-of-women-s-empowerment-bangladesh.html

Rahman, A. (1999). *NGOs and Civil Society in Bangladesh.*
Salamon, L. M. (1994). *The Rise of the Nonprofit Sector.*
Tasnim, F. (2007). *Civil Society in Bangladesh: More Grass-Roots Actions But Poor Participation.*
Tasnim, F. (2008). *Civil Society in Bangladesh: Vibrant But Not Vigilant.*
Yount, K. (2017). *A Framework for Measuring Women's Empowerment at Multiple Levels.*

15

Women Empowerment: Impact Assessment of Select Interventions by Various Third Sector Organizations in India

K. N. Veena, Shashidhar Channappa, and V. J. Byra Reddy

Introduction and the Context of the Study

"If by strength is meant moral power, then woman is immeasurably man's superior," said Mahatma Gandhi.

Leaders around the globe have promoted the idea of gender equality and women empowerment. Yet, it is true that it is not completely

K. N. Veena (✉)
DayanandaSagar Business Academy, Bangalore, India

S. Channappa
Department of Social work, The Oxford College of Arts, Bangalore, India

V. J. Byra Reddy
College of Management & Economic Studies, University of Petroleum and Energy Studies, Dehradun, India

© The Author(s) 2020
Md. N. Momen et al. (eds.), *Building Sustainable Communities*,
https://doi.org/10.1007/978-981-15-2393-9_15

practiced. Gender inequality knows no social, economic, or national boundaries. Statistics on gender-based abuse show horrifying numbers about the gender-based violence and abuse women face across the world. Worldwide, an estimated one in three[1] sexual abuse in her lifetime. One in five women worldwide is a victim of rape or attempted rape in her lifetime.

Another area of concern is "gender equality." Gender equality index calculated by Social Watch, a network comprising national coalition of civil society organizations, calculates the gender inequality index based on the gap between women and men in education, economy, and political empowerment. According to the 2012 gender equity index, India is placed at the 145th position with index 0.37 with Norway in the lead with 0.89.[2]

World Economic Forum has assessed the current size of the gender gap by measuring the extent to which women in 58 countries have achieved equality with men in five critical areas: economic participation, economic opportunity, political empowerment, educational attainment, and health and well-being.

The Gender Empowerment Measure (GEM), a measure of inequalities between men's and women's opportunities in a country, observes inequalities in three areas: political participation and decision making, economic participation and decision making, and power over economic resources.

Empowerment is meant to bring out the power inside and contingently requires the opportunities to do so; women, more than men, suffer from lack of opportunities to bring out the power in them owing to the predominance of patriarchal relations in society during the development of human civilization. It is heartening to note that in recent decades, the issue of women empowerment has received much attention from the society as such, and consequently, the body of research focusing on this important aspect has only grown. Several of the recent researches argue that women empowerment requires a systemic transformation in not just any institutions, but fundamentally in those supporting patriarchal

[1] http://www.unfpa.org/gender-based-violence#sthash.9LVPUYLP.dpuf.

[2] http://www.socialwatch.org/node/14367.

structures (Kabeer 2001; Bisnath and Elson 1999; Sen and Grown 1987; Batliwala 1994). And also that the concept of empowerment should be studied in the context of prevailing sociocultural, political, and economic conditions (Malhotra et al. 2002). Women still remain at disadvantage in all aspects of life. Millions of women remain in abject poverty, often being subject to labour exploitation, and even gender violence at work.

One of the effective tools toward women empowerment is providing an enabling environment for women to develop entrepreneurial abilities (Kushnir et al. 2010; Ismail et al. 2012) and the approach toward women empowerment need not necessarily be through big organized entrepreneurship; the approach of empowering women at the grassroots both in the urban and in the rural setting could be very effective and in terms of capacity could reach wider geography (Kuppusamy et al. 2010; Ba 2013; Global Entrepreneurship and Development Institute 2013).

The grassroots level Third Sector Organizations (TSOs) have been playing an important role in providing this enablement to women in terms of capacity building activities for the women to take to entrepreneurship. Further, NGO interventions positively contribute to women empowerment. NGO's central goal is empowering the powerless women folk or helping them to realize their hidden potentialities, through the power of thought, power of the word, and power of organization, with a view to helping them to participate in the socioeconomic development of the self and society. However, it is of interest to see how these TSOs have been empowering women through entrepreneurship development at the grassroots level, through small enterprises and organized small and medium scale enterprises.

In this context, it is proposed to understand the impact of various Third Sector Organizations working on women empowerment through entrepreneurship development.

Methodology

To Study the impact of TSO's on women empowerment through promotion of entrepreneurship, a three-tier TSOs have been chosen in such a way that an NGO—BUZZ India—working at the grassroots level and another TSOs like NEN working on encouraging women to take up small and medium scale entrepreneurial ventures and sustaining them continually, and at the third tier, Goldman Sachs CSR division working on skill augmentation of established women entrepreneurs to improve their skills through training.

The objective of the present study is to look at the benefits or otherwise that the women entrepreneurs have reaped from the interventions at the micro, medium, and the organized medium level.

Structure of the Entrepreneurship Development Programs (EDPs) by TSOs

Organization	Focus	Design & delivery
BUZZ India	• Women at the grassroots level with/without formal education • Capacitate women in their entrepreneurial ventures • Make entrepreneurship education mobile and localized, accessible, and meaningful	• Short-term training program • Mobile Academy—Take the B-school to them • Tailor-made financial consultation—Powerful solutions for personal and business growth • Capsule model of training—Counter lack of time and cultural constraints • Homework assignments—Involve the family in the learning cycle • Simple financial consulting and training that helps personal growth as well

(*continued*)

(continued)

Organization	Focus	Design & delivery
National Entrepreneurship Network—NEN **Dream 2 Destination—D2D**	• **Specialized entrepreneur support program** for 100 women entrepreneurs spread throughout Bangalore, Hyderabad, Jaipur, Kolkata, and New Delhi. • The program leveraged a blended approach of **one-to-one counselling, workshops, self-service learning, and targeted mentoring**, aimed at enabling participants to reach independent business milestone goals	• 18 months of program execution, with 10 ecosystem partners, 7 advisors, 6 educators, 20 mentors – 100 participants admitted from 182 applications – seven workshops administered at the start of program to individually profile • Participants and customize program offerings – 59 official mentor-mentee connections completed – five months of progress-monitoring, with five standardized progress • Trackers to provide valid monitoring and evaluation
Goldman Sachs—**10,000 women entrepreneurship program**	• Specialized entrepreneurship program for graduates with business ownership • Providing businesswomen with a business and management education, thus helping them grow the business. • Mentoring facilities for Business which are ready for scaling-up	• three months program with three modules of (seven days each) class room interaction • Between modules, entrepreneurs to implement classroom learning and report progress • Tracking progress to provide post EDP mentoring and evaluation

A well-structured questionnaire, containing behavioural aspects, technical aspects, and soft skills that impacted from the interventions of these TSOs, was administered to the beneficiaries before and after the interventions to understand the extent of enablement provided. The administration of questionnaire was further followed by personal interviews of the beneficiaries to look at the socioeconomic issues that they faced at all levels and TSOs enablement in overcoming the challenges. In some cases, beneficiaries were interviewed to gather the required information.

The data obtained using the questionnaire and the interview was analyzed to arrive at conclusions about the impact of TSO interventions on the empowerment of women.

Results and Discussion: NEN Initiatives

NEN D2D Program

The participants were phased out as per the design of the program

- 100 participants were admitted to D2D
- 66 were deemed eligible for mentorship
- 36 completed the first progress tracker
- 23 completed the third progress tracker
- 13 completed the final progress tracker

Participants experienced the following

- Improved confidence in personal entrepreneurial skill sets
- Improved ability to identify new customer segments
- Enhanced clarity in tracking cash flow and in writing/reading financial statements
- Achievement of revenue consistency through the establishment of sales targets
- Restructuring of a business plan to remove operational inefficiencies

- five participants pitched for investment from Indian Angel Network investors through an online pitch event hosted by NEN
- Four generated investor interest
- Two were nominated to participate in the 2015 TiEcon Kolkata fund-raising pitching conference.

Results and Discussion of BUZZ India and Goldman Sachs

From Table 15.1 it could be seen that there are fewer younger women entrepreneurs under the age of 30 in both the cohorts of BUZZ India as well as the Goldman Sachs programs compared to many who are in the age group of 30 and above. The combined total of 49 respondents shows that there are five women entrepreneurs in the age group of 20–30 and the remaining 44 are in the age group of 30 years and above. It is interesting to note that almost half of the respondents are in the age group of 31–40.

Unlike the table on the age profile of the respondents, the Table 15.2 on education profile brings out an interesting distribution. There are highly qualified women in the cohort of Goldman Sachs program who mostly are small and medium scale entrepreneurs. The cohorts of women in the BUZZ India program are illiterate or school-educated women entrepreneurs. Another interesting feature that comes out in the above table is that among the 31 women entrepreneurs in the BUZZ India program, 13 have no formal education at all while the other 10 have just middle education. Only four of these 31 women have received education up to matriculation or the tenth grade. These women obviously have got

Table 15.1 Age profile of participants in EDPs

Profile of respondents		EDP conducted by		Total
		BUZZ India	Goldman Sachs	
Age (years)	20–30	3	2	5
	31–40	12	8	20
	Above 40	16	8	24
Total		31	18	49

Table 15.2 Education profile of participants in EDPs

Profile of respondents		EDP conducted by		
		BUZZ India	Goldman Sachs	Total
Education	No formal schooling	13	0	13
	Primary School	2	0	2
	Middle School	8	0	8
	High School	4	0	4
	Tenth standard/matriculation	4	0	4
	Under graduation	0	8	8
	Post-graduation	0	3	3
	Professional qualification	0	5	5
	Diploma	0	2	2
Total		31	18	49

into entrepreneurship owing to some dire compulsion rather than some enterprise flossing out of educational qualification and these women mostly are from the lower socioeconomic strata of the society.

The women entrepreneurs subjected to intervention by the Goldman Sachs are all well-educated and most of them, 16 out of 18 have some graduation or a higher qualification; only two of them have a diploma. It could be said from their educational qualification that they are from the higher socioeconomic strata of the society.

Table 15.3 shows the type of business owned by these 49 women entrepreneurs from both the BUZZ India and Goldman Sachs Program. It could be inferred from this table that there are exclusive businesses/enterprises that the women from both these programs have got into. For instance, it is exclusive for the women from the BUZZ India program to indulge in businesses like tailoring, milk vending, food vending, owning petty shops, selling flowers, and laundry. Businesses like health and wellness, hotel, insurance, day-care centres, online training, technology firms are exclusive to the women from the Goldman Sachs Program.

It could be further inferred from this table that their educational qualifications and the skill sets acquired through further training, enabled these women to start their business enterprises.

Table 15.4 shows the motivation behind these women starting their enterprises; it is clear from this table that excluding a major motivation such as the financial need or necessity that is being shared by about 75%

Table 15.3 Type of business owned by participants of EDPs

Profile of respondents		EDP conducted by BUZZ India	EDP conducted by Goldman Sachs	Total
Type of business	Tailoring	4	0	4
	Health and wellness	0	2	2
	Branding and design—virtual	0	2	2
	Consulting and training	0	2	2
	Food vending	9	0	9
	Hotel	0	5	5
	Insurance & investment/Day care centre	0	3	3
	Milk vending	10	0	10
	Online training/Technology based	0	4	4
	Selling flowers/Petty shop/Ironing	8	0	8
Total		31	18	49

Table 15.4 Motivation to start business

Profile of respondents		EDP conducted by BUZZ India	EDP conducted by Goldman Sachs	Total
Motivation	Financial need/necessity	29	8	37
	Joy of working/to create value	0	6	6
	Natural emotion/urge	0	4	4
	Self-respect	1	0	1
	To increase savings	1	0	1
Total		31	18	49

of women of both cohorts, rest of the motivations are exclusive. While the joy of working/creating value and natural emotion/urge is the motivation for starting an enterprise for the women belonging to the Goldman Sachs cohort; self-respect and the need to increase savings of the family are the major reasons for the women belonging to BUZZ India program to start their enterprises.

Table 15.5 brings out the source of investments from which these women belonging to both of the cohorts have started their enterprises. It could be seen from this table that the two sources of investments such as the support from the government agency and the source of funds from

Table 15.5 Source of investment to start business among participants in EDPs

Profile of respondents		EDP conducted by BUZZ India	EDP conducted by Goldman Sachs	Total
Source of investment	Self-funded	8	12	20
	Bank loan	15	5	20
	Support from Govt. agency	2	0	2
	Private money lender	3	0	3
	Others	3	1	4
Total		31	16	49

the private money lenders are exclusive to the women belonging to the BUZZ India program. The other sources of investments such as self-funding, bank loans, and others are common to women from both of these cohorts.

While self-funding is seen to be prevalent among the women entrepreneurs from the Goldman Sachs cohort, bank loans are the most sought-after the source of investment for the women belonging to the BUZZ India cohort.

Table 15.6 shows the improvement that the BUZZ India participants have had on 16 critical parameters. It is heartening to see that these women have had improvements in all the parameters. The magnitude of improvement that these women have seen before and after the training imparted by BUZZ India is also very welcoming, as the magnitude of improvement is 1.19 at the least and 1.77 at the highest. The top three parameters that these women have seen improvements in include awareness about available facilities to run a business, ability to think beyond today—long-term business planning—and ability to network in the same order. The three parameters where these women have seen improvements in are: support from family, accounting personal and business finances separately, and the ability to take financial decisions.

Table 15.7 shows the improvement that the Goldman Sachs participants have had on 16 critical parameters. It is heartening to see that these women have had improvements in all of the parameters. However, compared to the women from the BUZZ India program, all of these women have seen improvements from a far higher level of 2.22 on a four-point scale. The

Table 15.6 Average scores for parameters—women at grassroots level: participants of BUZZ India EDP

		Average score		
Sl No.	Parameter	Before EDP	After EDP	Improvement
1	Self-confidence	1.32	2.90	1.58
2	Business knowledge	1.35	2.81	1.45
3	Ideas to expand business	1.39	2.81	1.42
4	Ability to network	1.26	2.97	1.71
5	Awareness about available facilities to run a business	1.26	3.03	1.77
6	Support from family	1.97	3.16	1.19
7	Ability to take financial decisions	1.55	2.87	1.32
8	Accounting personal and business finances separately	1.58	2.87	1.29
9	Knowledge about borrowing and saving	1.58	3.06	1.48
10	Efficient management of business finances	1.35	2.87	1.52
11	Ability to think beyond today—long-term business planning	1.19	2.94	1.74
12	Long-term planning for family needs	1.32	2.87	1.55
13	Social acceptance	1.42	2.90	1.48
14	Self-pride	1.39	3.00	1.61
15	Ability to perform marketing and sales activities	1.29	2.71	1.42
16	Handling operational and execution issues	1.39	2.84	1.45
	Overall	1.41	2.83	1.50

magnitude of improvement that these women have seen before and after the training imparted by the Goldman Sachs program is also very welcoming as the magnitude of improvement is 0.61 at the least and 1.33 at the highest. Though this magnitude is small, it is to be borne in mind that these women started off at higher scores of 2.22 and above. The top three parameters that these women have seen improvements in include handling business operations, ability to perform marketing and sales activities, and ability to think beyond today, that is, long-term business planning in the same order. The lowest three parameters where these women have seen improvements in are efficient management of business finances, awareness about available facilities to run a business, and ability to network.

Table 15.7 Average scores for parameters—women at the middle level: participants of Goldman Sachs EDP

Sl No.	Parameter	Average score: Before EDP	After EDP	Improvement
1	Self-confidence	2.44	3.44	1.00
2	Business knowledge	2.22	3.33	1.11
3	Ideas to expand business	2.67	3.67	1.00
4	Ability to network	2.67	3.44	0.78
5	Awareness about available facilities to run a business	2.46	3.22	0.78
6	Support from family	2.67	3.67	1.00
7	Ability to take financial decisions	2.44	3.56	1.11
8	Accounting personal and business finances separately	2.56	3.67	1.11
9	Knowledge about borrowing and saving	2.22	3.11	0.89
10	Efficient management of business finances	2.61	3.22	0.61
11	Ability to think beyond today—long-term business planning	2.11	3.44	1.33
12	Long-term planning for family needs	2.44	3.78	1.33
13	Social acceptance	2.44	3.50	1.06
14	Self-pride	2.44	3.78	1.33
15	Ability to perform marketing and sales activities	2.44	3.78	1.33
16	Handling operational and execution issues	2.89	4.00	1.11
	Overall	2.48	3.48	1.06

Conclusions

It was observed that the behavioural aspects of the respondents at the grassroots level showed a marginal change before and after interventions; also, the confidence to continue and expand the business further had grown significantly. Technical aspects, particularly the ability to handle finance and the awareness about borrowing support from the state institutions had markedly increased among the entrepreneurs. The respondents indicated an improvement over a period of time in soft skills, in dealing with the financial agencies and in networking skills.

At the tier two levels, among the small-scale women entrepreneurs, the survey results indicated that the intervention of AWAKE had a profound impact on the technical skills to sustain the ventures. The continued support of the TSO also made these women entrepreneurs overcome challenges around financial difficulties, manpower planning, and inventory planning. However, the behavioural aspects and the soft skills were acquired as a matter of experience, after running these ventures for longer periods. The personal interviews carried out with these women entrepreneurs surprisingly brought out the familial challenges that they have had to go through in carrying out their businesses as against the grassroots level micro-entrepreneurs. They also indicated that the nature of problems that they go through from the 'social context' is far heavier than the problems that the micro women entrepreneurs go through.

At the tier three level, among the organized women entrepreneurs, the survey results indicated that the systematic training organized by Goldman Sachs CSR initiative, helped them get the managerial knowledge and technical skills required to make their organizations more efficient. The behavioural skills and the soft skills part had a very minor role to play in enhancing the entrepreneurial abilities of these women entrepreneurs.

References

Ba, S. (2013). Boao Forum Report on the Development of SMB Financing. Department of Statistics (2010).

Batliwala, S. (1994). The Meaning of Women's Empowerment: New Concepts from Action. In G. Sen, A. Germain, & L. C. Chen (Eds.), *Population Policies Reconsidered: Health, Empowerment and Rights* (pp. 127–138). Cambridge: Harvard University Press.

Bisnath, S., & Elson, D. (1999). Women's Empowerment Revisited. Background Paper, Progress of the World's Women. UNIFEM. Retrieved from http://iknowpolitics.org/sites/default/files/progress_of_the_world_s_women_2000.pdf.

Global Entrepreneurship and Development Institute. (2013). The Gender Global Entrepreneurship and Development Index. Retrieved from http://www.thegedi.org/.

Ismail, H. C., Shamsudin, F. M., & Chowdhury, M. S. (2012). An Exploratory Study of Motivational Factors on Women Entrepreneurship Venturing in Malaysia. *Business and Economic Research, 2*(1), 1–13.

Kabeer, N. (2001). Reflections on the Measurement of Women's Empowerment. In *Discussing Women's Empowerment-Theory and Practice*. Sida Studies No. 3. Stockholm: NovumGrafiska AB.

Kuppusamy, J., Ganesan, J., & Rosada, S. A. (2010). Leadership Styles and Management Techniques: An Analysis of Malaysian Women Entrepreneurs. *Communications of the IBIMA, 2010*, 1–10.

Kushnir, K., Mirmulstein, M. L., & Ramalho, R. (2010). *Micro, Small, and Medium Enterprises Around the World: How Many Are There, and What Affects the Count?, in MSME Country Indicators 2010*. Washington, DC: World Bank & International Finance Corporation.

Malhotra, A., Schuler, S. R., & Boender, C. (2002). Measuring Women's Empowerment as a Variable in International Development, Background Paper Prepared for the World Bank Workshop on Poverty and Gender: New Perspectives. Retrieved from http://siteresources.worldbank.org/INTGENDER/Resources/MalhotraSchulerBoender.pdf.

Sen, G., & Grown, C. (1987). *Development, Crises, and Alternative Visions: Third World Women's Perspectives*. New York: Monthly Review Press.

Part IV

Building Sustainable Communities: Civil Society Response

Md. Nurul Momen, Rajendra Baikady, and Cheng Sheng-Li

Introduction

South Asia remains one of the poor regions in the world. It is also home to the world's largest number of poor people, and the reason the region is called a hunger hot spot. South Asia faces substantial challenges related to hunger, extreme poverty, gender disparity, political conflict, and because of these problems the region lags behind in terms of development (Ghani 2010). With the introduction of 17 sustainable development goals (SDGs) by the United Nations (UN) to deal with the global problems faced by our planet, various nations across the world have started to focus on sustainable development (United Nations General Assembly n.d.). While there are different aspects to sustainability, the social aspect of sustainability concentrates on welfare at the grassroots (Olawumi and Chan 2018). Sustainability is also linked to mobilization of resources from the developed economies to the developing economies. Even though South Asia has experienced a long period of rapid economic growth, there are numerous limitations. Being a less developed region, its economic activities, linked to globalization and privatization, do not contribute much to the development in the region. Further, the region is mostly affected by the failure of the market and by state policies.

Key Issues for Achieving Sustainable Development in South Asian Region: Civil Society Interventions

On 25th September 2015, the United Nations General Assembly adopted 17 SDGs with 169 targets with an aim to achieve sustainability by 2030. The agenda had three core dimensions: economic, social and environmental development. The list of 17 SDGs and the 169 targets included therein resulted from the experiences, and unfinished works of the Millennium Development Goals (MDGs) that was in operation since 2000. Further, SDG-1 to SDG-7 is about fulfilling the basic needs that are necessary for human development. This process started with the implementation of the MDGs in the year 2000. Despite both state and non-state (NGOs) interventions, the South Asian region today represents population with severe deprivation in education and health, poverty and hunger. MDGs aimed for eradication of poverty, and hunger and reduction in deprivation, whereas member states through SDG have committed to 'leave no one behind', and provide a life of dignity for all peoples in all nations. In order to achieve sustainability and improve the standard of living, along with economic and social development, the region needs to perform reasonably well in the areas of eliminating extreme poverty and reducing inequalities, eradicating hunger and achieving food security and ensuring quality of access to education and health for all.

Eliminating Extreme Poverty and Reducing Inequalities

There is no doubt the process of globalization and neoliberalism has contributed significantly to the economic and industrial growth. However, along with the economic growth, income inequality in societies has also been increasing, and the gap between people based on wealth is widening by the day. Income inequality has risen across the globe over the last decades, and according to the literature, the widening gap between income groups is mainly because of globalization, skill-based technical change and the decreasing bargaining power of workers. As noted by the literature, although economic growth across South Asia has resulted in reduced poverty, the total number of poor, that is, people living on less than $1.25 a day, has increased along with the rich, from 549 million to 595 million between 1981 and 2005 (Ravallion et al. 2009). The present data reveals that people below poverty line in the South Asian region account for 42 per cent of the world's poor, earning less than US$ 1.25 per day (IFPRI n.d.). Another 21 per cent of the population is undernourished, and more than 41 per cent children are underweight (IFPRI n.d.).

The UN aims for reducing poverty and accelerate human development in the developing economies. In order to achieve the stated goal, the developing countries and economies in transition need financial resources in the form of investments for industries and service sector. Foreign Direct Investment (FDI) is one such source of bringing resources from developed countries to developing countries, and helping them to achieve human and economic development gains. According to available statistics, FDI flow to developing countries reached US $765 bn in 2016 (UNCTAD 2012; UNCTAD 2016). Furthermore, Asia is the largest recipient of FDI in the world (UNCTAD 2016). In order to enhance the investment in economic activities and to lift people out of poverty in the developing countries, countries need to emphasize on a set of effective policy measures (United Nations n.d.).

Recently, the World Bank statistics and the traditional way of measuring poverty have been much discussed and debated (Moatsos 2017). On the other hand, there is a demand for new and complementary ways of measuring poverty in terms of both monetary and non-monetary assets (Saunders 2018). It is also argued that neoliberalism failed in bringing development to most people (Ostry et al. 2016), and as a result, mass poverty and income inequality still persist in the developing world (San Juan 2017). The South Asian region must also think in terms of providing enhanced social security provisions to its unorganized (i.e. informal) sector workers, which is highly imperative in order to achieve sustainable development in a timely manner in the region. Children and women's issues are also worthy of attention. Chronic under-nutrition among the children in South Asia is rampant, while women face iron-deficiency anaemia and issues related to employment opportunity and safety at the workplace.

Eradicating Hunger and Achieving Food Security

Almost one-quarter of the world's population resides in the South Asian region. According to recent estimation, all nations in the South Asian region account for 1.7 billion people of the world's 7 billion population (UNFPA 2011). According to a report by the food and agriculture organization (FAO), one-sixth of the world's population (1.01 billion) suffers from acute hunger, and the South Asian region is home to half these people (FAO 2009). Further, South Asia continues to be one of the largest hunger hot spots in the world. Malnutrition remains one of the critical health issues in children below five years who are most at risk, caused by multiple interlinked factors (WHO 2013; Bomela 2009). Sub-Saharan Africa and South Asia are the major regions affected by malnutrition in all its forms, and malnutrition in turn leads to several diseases among the poor in these regions (The United Nations Children's Fund 2015; Black et al. 2013; FAO 2010). India, Pakistan and Bangladesh are the most affected by malnutrition in South Asia (Headey et al. 2016). Over the last five decades, many studies and much research have been conducted on food poverty (Crawford and Thorbecke 1980; Sanyal and Babu 2009; UNICEF 2006). In the South Asian region, calories intake in people is less because of poor social and economic conditions. People living below the food poverty line, defined in terms of calories intake by an individual, face many health concerns (Hassan and Babu 1991).

Nutrition is a very crucial requirement in developing the productivity of people which contributes to the economic growth or to development in general. According to World Bank estimation, children suffer from malnutrition during their first 2 years of childhood are expected to have 10–17 per cent lower income than the well-nourished children (World Bank 2009). Studies have also found that food price inflation is contributing to the malnutrition, as it impacts on the purchasing power of the poor. Further, rising food price in the South Asian region has seriously affected both urban and rural consumers and farmers (ADB 2011). According to a report published by World Bank, children under 5 years in Bhutan, Bangladesh, India and Nepal face iron deficiency (55 per cent) and vitamin A deficiency (28–57 per cent), by far the first cause of nutritional deficiency (World Bank 2009). According to a recent study, micronutrient deficiencies are more severe in the region (Harding et al. 2018).

In order to achieve the SDGs that focus on eradication of hunger, South Asian countries need to develop agricultural productivity. Enhanced agricultural productivity ensures the availability of food and thus accelerates the reduction of poverty and hunger in the region. Many small-scale and marginal farmers have low agricultural production, which is not even enough to meet their household consumption. In order to feed their children and family, these farmers need to purchase food from the market, and when the food price goes up it makes them vulnerable to hunger and privation (FAO 2010). Therefore, it is important to develop agriculture in the South Asian region, as it supports the income of more than half the population. Any development in the agricultural sector of the region will lead to poverty eradication and hunger reduction. In the recent years, the United Nations Development program (UNDP 2010) began projecting poverty as a multidimensional problem, and started measuring poverty with the help of the multidimensional poverty index (MPI) (see Alkire and Santos 2010; Anand and Sen 1997; Kanbur and Squire 2001; Micklewright and Stewart 2001; Pogge 2009); however, despite using the multidimensional poverty index, the South Asian region (especially, Bangladesh, India and Pakistan) continues to reduce poverty at a slow pace.

There is substantial evidence available that focuses on different aspects of nutritional deficiency. Some scholars have explored the intra-household biases of nutritional deficiency (Pande 2003; Jayachandran and Pande 2013), whereas other studies have shown the impact of genetic predispositions on nutritional deficiency (Nubé 2009). Studies have also documented low-quality diets and nutritional problems (Deaton and Dreze 2008; Headey et al. 2012), poverty and household food insecurity and its effects on nutritional food consumption (Menon 2012). Some authors have studied the inefficiency of nutritional programmes and strategies in nutritional deficiency (Das Gupta et al. 2005; World Bank 2005), and others have also highlighted the poor sanitation and hygiene practices (Spears 2013; Spears et al. 2013). All these studies importantly focus on South Asia's inability to perform better in terms of provision of nutrition and in the reduction of poverty and hunger.

Ensuring Quality Access to Education and Health for All

South Asia has been successful in meeting the target related to universal primary education enrolment of the MDGs. Universal education is a key strategy to enhance the well-being of the community, and achieving economic and social development in developing societies. Education will lead to enhanced employability, and employment leads to economic empowerment. Education and economic empowerment together lead to better standard of living. Unfortunately, a large number of children of school-going age do not attend school, and there is a huge disparity between the gender and social class. In keeping with the MDGs' (Goal 2) primary aim to achieve universal primary education by 2015, it achieved a considerable success of 97 per cent school enrolment (UNDP 2015). Education being one of the most basic requirements for human development receives very little attention in developing countries. Education is a basic human right, as it not only helps the people to improve the standard of living, and involve in the economic activities, but it also helps people in making decisions, and equips them with indispensable life skills and critical thinking ability.

South Asian countries have the highest out-of-school numbers in the world; hence, the region needs to focus more on girls' education. Despite the achievement of MDGs in universal education, the gap between the genders still persists. A large number of school-aged girls are not attending school in developing countries. Approximately, one-third of girls in South Asian countries do not attend school, and in some regions the ratio is one in four that attend primary school. For the development of the region and achieving SDGs, education of women is a very important aspect for their healthier and happier life. The experiences and evidences from the developing countries show that faster economic growth is possible through educating girls and women (Herz and Sperling 2004). Studies have also found that women with secondary or higher education work in non-manual jobs and perform wage work (Duraisamy 2002). However, the substantial employment prospect depends on how well the women are educated (Malhotra et al. 2003). Furthermore, increase in the level of female education leads to reducing birth rate and thus reduction in poverty and economic productivity (Herz and Sperling 2004).

Increasing and accelerating the level of female education may also help in closing the gender gap, and also promote per capita income growth (Herz and Sperling 2004). Other studies have argued that women with education seek parental care, and also avail of medical support during child birth (Malhotra et al. 2003). Hence, better educated women tend to ensure health for the family and also for children. Evidences also show that each additional year of education for women leads to 7–9 per cent reduction in mortality for children under the age of 5 years (Institute for Health Metrics and Evaluation Education 2016). Another study by Behrman et al. (2008) found that in Pakistan, income increased with the increases in education, and gross domestic product increased with the amount of increases of female education (Hanushek and Woessmann 2008).

Further studies predicated that, high prevalence of reproductive health issues in the South Asian region are mainly because of low status of women (Sheikh and Loney 2018). And it is expected that eliminating gender disparities in school enrolment may promote improvements in female education and reproductive health. Further, education also results in imbalance in the labour market; unskilled and non-educated labour migration is less when compared to educated and skilled labour (Docquier et al. 2014).

Summing up, South Asia needs to develop pro-growth policies and development-oriented programmes. When policies related to development are not pro-growth oriented, they may not effect poverty reduction and human development (Panagariya 2009). South Asia is at the critical stage of historical transformation, and needs well-handled developmental policies for the inclusive growth and prosperity of the region.

Part IV makes an attempt to organize the chapters that address specific issues related to sustainable development in the South Asian region, issues such as social justice, good governance, problems and challenges of the farming community, issues related to civil society functioning in different territories, women empowerment, political participation of women, etc.

Chapter 16 is based on a research study conducted in Buldhana, Jalna and Aurangabad districts of Maharashtra. The chapter is written based on the field data collected through various data collection tools, such as interviews, participatory observation and survey through social audit process. It investigates how the social audits became an effective tool to access and promote justice in contemporary Indian society. Chapter 17 discusses the issues and challenges that have emerged from the Rohingya refugee crisis. The chapter is based on a study that focused on the anticipated social system breakdown of Bangladesh, especially in the hill tracts and the possible solution to overcome the crisis. Chapter 18 explores the impact of organic farming on sustainable livelihood of farmers in the Indian context. The chapter is based on a primary qualitative research mainly through interview that was conducted with the progressive organic farmers. The study reflects on and explores the differences of social livelihood between organic and non-organic farmers. Chapter 19 examines the managerial efficacy of Jagratha Samithi cum Family Empowerment forum. The study concludes that vision-driven social projects should plan for managerial viability to ensure its sustainability post NGO exit. As findings, designing managerial intactness into projects facilitate success. Chapter 20 discusses intervention strategies and approaches for rehabilitation of differently abled people in light of the social work profession. Social workers can play an important role in wealth redistribution to enable as well as empower differently abled people to be more independent and to improve their living standards. Chapter 21 examines the status of poverty and social inequality in Bangladesh.

References

Alkire, S., & Santos, M. (2010). *Acute Multidimensional Poverty: A New Index for Developing Countries*. Human Development Research Paper 11. New York: UNDP.

Anand, S., & Sen, A. (1997). *Concepts of Human Development and Poverty: A Multidimensional Perspective*. Human Development Report 1997. New York: UNDP.

Asian Development Bank (ADB). (2011). *Asia 2050: Revitalizing the Asian Century*. Madaluyong City.

Behrman, J. R., Ross, D., & Sabot, R. (2008). Improving Quality Versus Increasing the Quantity of Schooling: Estimates of Rates of Return from Rural Pakistan. *Journal of Development Economics, 85*(1), 94–104.

Black, R. E., Victora, C. G., & Walker, S. P. (2013). Maternal and Child Under Nutrition and Overweight in Low-income and Middle-income Countries. *Lancet, 382*, 427–451.

Bomela, N. J. (2009). Social, Economic, Health and Environmental Determinants of Child Nutritional Status in Three Central Asian Republics. *Public Health and Nutrition, 12*, 1871–1877.

Crawford, E., & Thorbecke, E. (1980). The Analysis of Food Poverty: An Illustration from Kenya. *Pakistan Development Review, 19*(4), 311–35.

Das, G. M., Lokshin, M., Gragnolati, M., & Ivaschenko, O. (2005). *Improving Child Nutrition Outcomes in India: Can the Integrated Child Development Services Be More Effective?* [Online]. Policy Research Working Paper Series No. 3647. Washington, DC: The World Bank. Retrieved from http://ideas.repec.org/p/wbk/wbrwps/3647.html.

Deaton, A., & Dreze, J. (2008). *Food and Nutrition in India: Facts and Interpretations* (pp. 42–65). XLIV: Economic and Political Weekly.

Docquier, F., Ozden, Ç., & Peri, G. (2014). The Labour Market Effects of Immigration and Emigration in OECD Countries. *The Economic Journal, 124*(579), 1106–1145.

Duraisamy, P. (2002). Changes in Returns to Education in India, 1983–94: By Gender, Age-Cohort, and Location. *Economics of Education Review, 21*(6).

FAO. (2009). *The State of Food Insecurity in the World (SOFI): The 2009 Hunger Report.* Rome: Food and Agriculture Organisation. http://www.fao.org.

Food and Agriculture Organization (FAO). (2010). *The State of Food Insecurity 2010.* Rome.

Ghani, E. (2010). *The Poor Half Billion in South Asia-What Is Holding Back Lagging Regions?* Oxford University Press.

Hanushek, E. A., & Woessmann, L. (2008). The Role of Cognitive Skills in Economic Development. *Journal of Economic Literature, 46*(3), 607–668.

Harding, K. L., Aguayo, V. M., & Webb, P. (2018). Hidden Hunger in South Asia: A Review of Recent Trends and Persistent Challenges. *Public Health Nutrition, 21*(4), 785–795. https://doi.org/10.1017/S1368980017003202.

Hassan, R. M., & Babu, S. C. (1991). Measurement and Determinants of Rural Poverty: Household Consumption Patterns and Food Poverty in Rural Sudan. *Food Policy, 6*(6), 451–60.

Headey, D., Hoddinott, J., & Park, S. (2016). Drivers of Nutritional Change in Four South Asian Countries: A Dynamic Observational Analysis. *Maternal and Child Nutrition, 12*(S1), 210–218.

Headey, D., Chiu, A., & Kadiyala, S. (2012). Agriculture's Role in the Indian Enigma: Help Or Hindrance to the Malnutrition Crisis? *Food Security, 4*, 87–102.

Herz, B., & Sperling, G. B. (2004). *What Works in Girls' Education: Evidence and Policies from the Developing World*. New York: Council on Foreign Relations.

Institute for Health Metrics and Evaluation Education. (2016). Retrieved November 11, 2016, from http://www.healthdata.org/education.

International Food Policy Research Institute. (n.d.). Retrieved August 21, 2019, from https://www.ifpri.org/division/south-asia-regional-office-sao.

Jayachandran, S., & Pande, R. (2013). *Why Are Indian Children Shorter Than African Children?* [Online]. Department of Economics Working Paper. Cambridge, MA: Harvard University. Retrieved from http://www.hks.harvard.edu/fs/rpande/papers/Indianchildrenheight.pdf.

Kanbur, R., & Squire, L. (2001). The Evaluation of Thinking about Poverty: Exploring the Interactions. In Miere, G., Stiglitz, J. E. (Eds), *Frontiers of Development Economics: The Future Perspective* (pp. 186–226). New York: Oxford University Press.

Malhotra, A., Grown, C., & Pande, R. (2003). *Impact of Investments in Female Education on Gender Inequality*. Draft in Mimeo, International Center for Research on Women.

Menon, P. (2012). Childhood Undernutrition in South Asia: Perspectives from the Field of Nutrition. *CESifo Economic Studies, 58*(2), 274–295.

Micklewright, J., & Stewart, K. (2001). Poverty and Social Exclusion in Europe: European Comparisons and Impacts on Enlargement. *New Economy, 8*(2), 104–09.

Moatsos, M. (2017). Global Absolute Poverty: Behind the Veil of Dollars. *Journal of Globalization and Development, 7*(2), 1–28. https://doi.org/10.1515/jgd-2016-0033.

Nubé, M. (2009). The Asian Enigma: Predisposition for Low Adult BMI Among People of South Asian Descent. *Public Health Nutrition, 12*, 507–516.

Olawumi, T. O., & Chan, D. W. M. (2018). A Scientometric Review of Global Research on Sustainability and Sustainable Development. *Journal of Cleaner Production, 183*, 231–250. https://doi.org/10.1016/j.jclepro.2018.02.162.

Ostry, J., et al. (2016). Neoliberalism: Oversold? *Finance & Development, 53*(2). International Monetary Fund. http://www.imf.org/external/pubs/ft/fandd/2016/06/ostry.htm.

Panagariya, A. (2009). *India: The Emerging Giant*. Oxford University Press.

Pande, R. (2003). Selective Gender Differences in Childhood Nutrition and Immunization in Rural India: The Role of Siblings. *Demography, 40*, 395–418.

Pogge, T. (2009). *Developing Morally Plausible Indices of Poverty and Gender Equity: A Research Program*. New York: New York University Institute of Public Knowledge.

Ravallion, M., Chen, S., & Sangraula, P. (2009). *Dollar a Day Revisited*. Policy Research Working Paper No. 4620. Washington, DC: World Bank.

San Juan, D. M. (2017). Why Marx Was Right: Third World Edition. *Journal of Developing Societies, 33*(1), 1–26. http://journals.sagepub.com/doi/pdf/10.1177/0169796X16653333.

Sanyal, P., & Babu, S. (2009). *Food Security, Poverty and Nutrition Policy Analysis: Statistical Methods and Applications*. New York: Academic Press.

Saunders, P. (2018). Monitoring and Addressing Global Poverty: A New Approach and Implications for Australia. *The Economic and Labour Relations Review, 29*(1), 9–23. http://journals.sagepub.com/doi/abs/10.1177/1035304618756208.

Sheikh, S. M., & Loney, T. (2018). Is Educating Girls the Best Investment for South Asia? Association Between Female Education and Fertility Choices in South Asia: A Systematic Review of the Literature. *Frontiers in public health, 6*, 172. https://doi.org/10.3389/fpubh.2018.00172.

Spears, D. (2013). *How Much International Variation in Child Height Can Sanitation Explain?* [Online] Working Paper. Princeton, NJ: Princeton University.

Spears, D., Ghosh, A., & Cumming, O. (2013). Open Defecation and Childhood Stunting in India: An Ecological Analysis of New Data from 112 Districts. *PLoS ONE, 8*, e73784.

The United Nations Children's Fund. (2015). World Health Organization. World Bank. UNICEF. *WHO-The World Bank Child Malnutrition*

Database: Estimates for 2015 and Launch of Interactive Data Dashboards. New York: The United Nations Children's Fund.

UNICEF (United Nations Children's Fund). (2006). Tracking Progress towards MDG1. http://www.childinfo.org/areas/malnutrition.

UNCTAD (United Nations Conference on Trade and Development). (2012). FDI Statistics Database; The United Nations Conference on Trade and Development: Geneva, Switzerland. Retrieved January 9, 2017, from https://unctad.org/en/Pages/DIAE/World%20Investment%20Report/Regional-Factsheets.aspx.

UNCTAD (United Nations Conference on Trade and Development). (2016). Investor Nationality: Policy Challenges; The United Nations Conference on Trade and Development: Geneva, Switzerland. Retrieved January 9, 2017, from https://unctad.org/en/PublicationsLibrary/wir2016_en.pdf.

UNDP. (2010). *Human Development Report.* New York: Palgrave Macmillan.

UNDP (United Nations Development Programme). (2015). The Millennium Development Goals Report 2015, s.l.: United Nations.

United Nations. (n.d.). Envision 2030 Goal 1: No Poverty. Retrieved April 11, 2019, from https://www.un.org/development/desa/disabilities/envision2030-goal1.htm.

UNFPA (United Nations Funds for Population). (2011). *World Population Report 2011.* New York: United Nations Funds for Population. http://www.unfpa.org.

United Nations General Assembly. Transforming Our World: The 2030 Agenda for Sustainable Development. Retrieved July 10, 2018, from https://sustainabledevelopment.un.org/post2015/transformingourworld.

World Bank. (2005). *Bangladesh Integrated Nutrition Project: Project Performance Assessment Report.* Report No. 32563. Washington, DC: World Bank.

World Bank. (2009). *Guidance for Responses from the Human Development Sectors to Rising Food Prices.* Washington, DC.

World Health Organization (WHO). (2013). *Guideline: Updates on the Management of Severe Acute Malnutrition in Infants and Children.* Geneva: World Health Organization.

16

Social Audit: A Tool to Access Social Justice

Nitin Dhaktode

Introduction

The ultimate goal of the executive, legislature, and judiciary system is to ensure justice for the citizens of India. Apart from this, there are other ways to access justice such as free legal aid, Public Interest Litigation (PIL), various efforts that have emerged out of the interventions of civil society organisations, such as Social Audit, Janta information system, Right to Information and so on. The idea of Justice (Social, Political, and Economic) introduced in the preamble of Indian constitution has roots due to thousands of years of evil practices and traditions of Indian society such as marginalization, exclusion (social and financial), vulnerability, oppression, discrimination, inequality, untouchability and so on.

The author would like to thank Ms Sowmya Kidambi, Director, Social Audit Unit, undivided Andhra Pradesh and Present Telangana, for a deputing author as Special Observer for Social Audit in Sillod, Aurangabad, and Mr Ujjwal Phurkar as Co-Observer, Ashwini Survase and Pradnyasurya Shende for reading drafts of the chapter and suggesting corrections in it.

N. Dhaktode (✉)
Tata Institute of Social Sciences, Mumbai, Maharashtra, India

© The Author(s) 2020
Md. N. Momen et al. (eds.), *Building Sustainable Communities*,
https://doi.org/10.1007/978-981-15-2393-9_16

The injustice to the marginalised caste communities has pushed society towards underdevelopment.

Development is the ultimate goal to get social justice for all. "Development is argued here, as a process of expanding the real freedom that people enjoy" (Sen 1999). Amartya Sen argues that freedom is more important for development. And therefore, he cited an example of the labourer from Bangladesh who lost his life in a communal riot due to lack of freedom of choice to not go in search of work, due to his poverty and dependency on a daily wage. It means freedom comes with comfort and security, which assures a secure life. The data used in this chapter suggests how the labourers from Buldhana district had to forcefully get migrated to Sillod block of Aurangabad district for the daily wage work under the Employment Guarantee Scheme amongst which five labourers lost their lives. Majority labourers belonged to Marginalised caste groups who do not have land or other resources to earn to meet their daily needs.

The mandatory social audit in the Mahatma Gandhi National Rural Employment Guarantee Act (NREGA) implementation provided an opportunity to access justice to the victims of structural violence by the system as Akhil Gupta has argued (2012). As the news of *Hindustan Times* dated 29 December 2013 suggests,[1] five NREGA labourers who had migrated from Buldhana district and worked in Sillod block committed suicides due to delayed payment. With the suo-moto intervention of the National Human Right Commission (NHRC), the Ministry of Rural Development (MoRD) had directed the government of Maharashtra to conduct the special social audit in Buldhana and Aurangabad district under the supervision of Committee of Experts on the Special Social Audit (COE henceforth). The COE combined members from Andhra Pradesh, Orissa, and Delhi. The Special Social Audit was conducted by the 'Sparsh' Non-Governmental Organisation (NGO) from Maharashtra. The special audit provided the platform for the labourers to raise their concerns and get justice.

[1] https://www.hindustantimes.com/india/delayed-nrega-payments-drive-workers-to-suicide/story-MlLZGwzDHkWE1ifOykxcrM.html.

This chapter is an attempt to present the idea of social audit as a tool to access justice with various theories and their critics. It further unfolds the process of social audit and its significance in the life of powerless labourers through the case of Maharashtra.

Theoretical Conceptualization of Problem

Before we discuss the key arguments, it is necessary to understand the concept of the problem concerning the marginalised section of the society through the lenses of different perspectives. The problem can be understood through the Hegelian perspective by the book *Elements of the philosophy of rights Edited* by Allen Wood (1991). Here, he argues that injustice begins with limiting the individual from accessing the rights and freedom. In fact, a similar argument is also made in the book called *Development as Freedom* by Sen (1999). Even the idea of liberty and equality with fraternity advocated by the key architect of the Indian constitution Dr. B.R Ambedakar was for the adequate freedom of individuals of social, political, and economic justice. Thus, contradiction with the relation of internal principles of human beings which come through the various experiences coming from the social, economic, legal institutions and events, including war.[2] The consciousness in the development of individuals plays a vital role that comes from socio-political-economic and cultural socialisation. For injustice practices, the most influential componants are social, economic and political. The unskilled labourers from Buldhana district who worked in Aurangabad district went through a similar process as those who did not have the right to access work and wages granted by the law, especially labour laws and social security benefits.

While problematising the society, the Marxist theory argues about the exploitation with reference to the class structure that occurs from the capitalist and proletariat society. The theory of class struggle was presented in the communist manifesto. Marx and Engels argue "the modern

---

[2] For more details, read a paper by Louis Althusser (1962) 'Contradiction and Over determination' published by Penguin Press.

bourgeois society that has sprouted from the ruins of feudal society has not done away with class antagonisms. It has been established new classes, new conditions of oppression, new forms of struggle in place of the old ones" (Marx and Engels 1848: 15) to historicise the problem; they believed the exploitation based on class is not new and the feudal system was practising it in those society where the poor were slaves. Today, the same rulers have established a capitalist society, based on the exploitation of the poor. By connecting the Hegelian perspective with the concept of the class as superstructure, they argue that the state and all the legal, political and ideological forms come together to oppress the poor (ibid.). The division of labour has just limited them only to get wages but capitalists earn more profits using workers' stable resources and labour.

Marx's writing leads to the conceptualization of injustice narrowing class perspective which is altogether different from India where caste plays an important role in a negative sense. Their arguments provide analysis of how the capitalist gets a hold on the production process on whom society became dependent for consumption and earning. Labourers became economically weak due to dependence on the bourgeois for work and wages. Whereas more profit earned by capitalists provided space to control state and create the laws in their favour. Such practices ultimately develop an unjust society. In the Indian society, specifically with reference to the Buldhana labourers' case, most of them were landless, belonging to Dalits and De-Notified-Tribes, and those who were in the power as Block Development Officers (BDOs) and engineers belonged to particular dominant caste that has political power backing. More than 1500 labourers were involved in the case presented below does not have either social or political capital. Therefore, they choose to end their lives. However, the social audit given by law came into the picture to rescue their families and provide justice.

Another perspective of Amartya Sen as a welfare approach invites attention. He argues about the comparative analysis based on the capabilities should be developed by individuals with the help of the state; also, he believes that development should consider the freedom to access various benefits or natural rights, rather than just giving importance to measure development based on Gross Domestic Product (GDP) or Gross National Product (GNP) by the state. Sen is one of the leading figures born in India to theorise the idea of justice who also criticised John Rowl and various other thinkers to expand his views on justice. His core argument around the 'freedom for development' leads to the justice of the individual. It is important to build the capabilities of individuals to access the freedom of choice. However, it was nothing new as Dr. Ambedkar has fought his entire life to annihilate the caste from the society. Caste as a horrific social institution has given birth to multiple socio-political and economic problems in India. "Ambedkar's perspective of social justice is based on social democracy which consists of three concepts of justice namely liberty, equality and fraternity" (Ranjithkumar 2011). The key concept of bringing liberty, equality, and fraternity in the Indian Constitution was to give social justice to the marginalised who are historically denied justice. The idea of socialist democracy given by Ambedkar is for justice. The nationalisation of the key resources such as land, industries, banks were considered as entry points for social justice through social democracy. However, to some extent, it could not happen. Access to social justice leads to positive changes in society, which brings ultimate development.

While conceptualizing the problem of Indian caste-based inequalities through 'Annihilation of Caste', Dr. Ambedkar unfolds the layers of the case and its process in India.[3] He argues, the caste system is the key problem in the society, which does not allow the equality, by presenting the fact from Karnataka's lower caste, he says that there was a rule made by upper caste people, and many restrictions were made for lower caste people with regard to their behaviour and living style (Ambedkar 1936). The text Manusmriti written by Manu did not allow the lower caste people to get educated or become rich; it explains that lower caste people were born from the god Vishnu's leg and they were considered Shudra and *atishudra.* Thus, they don't have any right to live the life the way the Kshatriya, Vaishya, and Brahmin live. They are the *Dasas,* that is, servants of these three Varnas and they have to serve by the division of work. There was no

[3] Annihilation of caste was an undelivered presidential speech written in 1936 by Dr. B. R. Ambedkar, which was supposed to deliver in Jat Pat Todak conference in Punjab but due to religious sentiment it got cancelled later it became an important text to understand the Indian caste system as exploitative and unjust.

opportunity for any lower caste individuals to change their caste-based occupation, which only serves the above three Varnas. Therefore, he argues the root cause of the injustice is inequality, restrictions based on the caste system, and untouchability practices which are key in the Indian society.

State Approach for Justice

To achieve the larger goal of development and provide social justice to the marginalized, the government of India has implemented many schemes and programmes to eradicate poverty and to provide employment, shelter, food and so on, but due to the corrupt practices, it has not reached the beneficiaries, and the goal set 70 years back still remains a dream. Since most of the beneficiaries come from the socio-economically and politically weak backgrounds they are not in a position to articulate and ask the question to the government. They have not been considered for their rights; for instance, most of the rural schemes are linked with the Gram Sabha that is considered the highest body of the village. However, Dalit, Tribal, women are not given space for their participation to raise their grievances.

At the beginning of the twenty-first century, the government of India gave two very important laws: first, the Right to Information Act 2005, and second, the National Rural Employment Guarantee Act (NREGA) 2005. In the History of India, for the first time, poor people got an opportunity to ask for the information they want through RTI and NREGA. Section 17 of NREGA makes it compulsory for all the states to conduct the social audit of schemes twice in a year. This was another important milestone for beneficiaries. The tool of social audit has become effective not only to curb the corruption but also to empower labourers to fight for their right, to challenge the system, to make the system accountable and transparent. States like Andhra Pradesh, Telangana, are doing their best to bring transparency and accountability in the system through the implementation of this tool and empower the people to access justice.

To conceptualize the contemporary issues in policy implementation by the government, it is necessary to understand the role of the system. For any kind of injustice in the social status is key responsible factor in a democratic country like India. Thus, the welfarist state acts as a caretaker of the marginalized to ensure justice. However, despite that, the injustice in the society has become sustainable rather than achieve the sustainable goals of development due to the corrupt practices. After the Millennium Development Goal (MDG), the Sustainable Development Goal (SDG) has become an important target for the state. Employment guarantee scheme is one of the key programmes to eradicate poverty in which the role of the state as implementing agency is crucial. Therefore, the proper implementation of such a programme by the state became the need of the hour.

After the 73rd and 74th Amendment of the Indian constitution, the decentralization of power has been initiated by the government of India to make sure that the people participate in the planning, implementation, and evaluation. The idea of participatory planning and governance presented by Robert Chambers in the 1980s are incorporated in the 73rd and 74th constitutional amendments. However, with that effort, corruption also got decentralized through traditional caste-based dominant stakeholders in the village. Corruption has been institutionalised in Indian bureaucracy and political system and has a direct impact on the poor and marginalized. The process of corruption begins with the recruitment of the employees and ends with their retirement (job or life) or in case of a politician with death. Politicians want their favoured top bureaucrats on posts; they lobby to recruit staff that support their party from village level to state and also take a certain amount of money from individuals for the post. Unfortunately, it has become an open secret which has been normalised in society. As Akhil Gupta argues, corruption became a culture to get done their work (Gupta 2012). In such corrupt practices, poor and socially marginalised suffer and face injustice. Thus, there is a need to make people aware and build their capabilities to raise the voice against such practices at democratic spaces such as Gram Sabhas.

Literature Review of Theory of Social Justice

Social justice is "justice in terms of the distribution of wealth, opportunities, and privileges within a society".[4] "'Social justice' is generally used to refer to a set of institutions which will enable people to lead a fulfilling life and be active contributors to their community" (Rawls 1971). "The term 'social justice' only became used explicitly from the 1840s. A Jesuit priest named Luigi Taparelli is typically credited with coining the term, and it spread during the revolutions of 1848 with the work of Antonio Rosmini-Serbati" (Zajda et al. 2006). The preamble to establish the International Labour Organization recalled that "universal and lasting peace can be established only if it is based upon social justice".[5] Later in the twentieth century, social justice was made central to the philosophy of social contract, primarily by John Rawls in *A Theory of Justice* (Rawls 1971). In 1993, the Vienna Declaration and Programme of Action treats social justice as a purpose of human rights education.[6]

Social justice is a property of social systems—or perhaps a "predicate of societies" (Jost and Kay 2014), as suggested also by Rawls (1971). A just social system is to be contrasted with those systems that foster arbitrary or unnecessary suffering, exploitation abuse tyranny oppression, prejudice, and discrimination (Jost and Kay 2014). "This concept of social justice emphasises individual liberties and equality of opportunity, rights and outcomes" (Occasion paper, 2011). J. S. Mill talks about the Utilitarian approach of social justice. As per his concept "the distribution of societal goods should be for the 'greatest net balance of satisfaction' for society" (Mill 1863). Mill argues, "utilitarianism was actually a 'standard of morality' which used happiness of the greater number of people as its ultimate goal" (ibid.). This view reflected an unequal distribution based on the status as it was often the disadvantaged whose rights were sacrificed for the good of the privileged classes of society.

Rawls' perspective on social justice was more social than economic. He criticizes the Utilitarian approach, Rawls argues, the conception of

[4] New Oxford American dictionary.

[5] International Labour organization preamble.

[6] Vienna Declaration and Programme of Action, Part II, D.

distributive justice provided that "all social values ... are to be distributed equally unless an unequal distribution of any or all of these values is to everyone's advantage" (Rawls 1971). The two fundamental principles of Rawls' original theory of social justice are "(1) Each person has an equal right to the most extensive system of personal liberty compatible with a system of total liberty for all. (2) Social and economic inequality are to be arranged so that they are both (a) To the greatest benefit to the least advantaged in society, (b) Attached to positions open to all under conditions of fair equality of opportunity" (ibid.). Basically, he argues more towards the liberty, freedom of the person along with the social values.

Miller's approach to social justice rests on the idea that the market is capable of giving individuals what they deserve. This theory treats individuals as responsible for their own actions and proportionately rewards (or punishes) them in accordance with their actions and efforts, insofar as the actions/efforts are the results of their individual choices (Miller 1999: 149–155). He sounds more a capitalist and market-oriented thinker who thinks that human beings are capable of coping up with the present situation. If they try to find out their own way of justice, they can get it because everything is available in the market.

Amartya Sen's *'The Idea of Justice'* gave a very important turning point for the Indians with the understanding of the Indian social system. He is more practical which provides an opportunity to take action on that to achieve. Therefore he argues, "Its aim is to clarify how we can proceed to address questions of enhancing justice and removing injustice, rather than to offer resolutions of questions about the nature of perfect justice" (Sen 1999). Sen's comparative approach explores social alternatives, ranking them based on the values and priorities of the community. It is clearly visible how Indian social structure reflects in Sen's social justice theory. He accepts that social arrangements have to make it possible for individuals to build their capabilities. For example, "a right to education concerns not simply an individual's access to appropriate educational material but the responsibility of the government to provide a stable presence of certain institutions and institutional frameworks" (ibid.). His key focus of social justice is to emphasize developing individual capabilities in many ways and that leads him to promote a welfare economy that gives space for the marginalized to build their capabilities.

However, the institutional reforms in social values, adequate freedom to achieve social justice is an important component of the process. Sen acknowledges that his economic master is Dr. B. R. Ambedkar, who had provided the Social Justice theory advocating liberty, equality, and fraternity as the centre of human life. He further made mandatory for the state to implement the programmes and policies with legal direction thorough Articles 14, 15, 17,19 and directive principles for the state. "The contents of Ambedkar's concept of Social Justice included unity and equality of all human beings, equal worth of men and women, respect for the weak and the lower caste, regards for human rights, benevolence, mutual love, sympathy, to learn and charity towards fellow being".[7]

Injustice with NREGA Workers

In the case of Buldhana district, five labourers who worked under Mahatma Gandhi National Rural Employment Guarantee Act 2005 (MG-NREGA) and committed suicide due to delayed payment and financial corruption by the implementing agency, the majority belonged to SC and ST.[8] Though the act says that there should be social audit after six months, it never happened until January 2014 in Buldhana, Aurangabad, Jalna, and Parbhani districts of Maharashtra and it is no different in other parts of the state. When the case was highlighted by the *Hindustan Times* a leading newspaper, the Ministry of Rural Development (MoRD) government of India appointed the Committee of Special Observers (CSO) and did the social audit by third-party organizations appointed by the government of Maharashtra. Unfortunately, the social audit was done as a formality, where oppressed voices were suppressed. The CSO submitted their report to the government of India in which they clearly mentioned that the process of the social audit was not done properly, five suicides took place due to government negligence and corruption made by the implementing agency.

[7] http://shodhganga.inflibnet.ac.in/bitstream/10603/10806/10/10_chapter%203.pdf.

[8] http://www.hindustantimes.com/india-news/delayed-nrega-payments-drive-workers-to-suicide/article1-1167345.aspx.

More than 1500 MG-NREGA workers who migrated from Buldhana district to Aurangabad district with the help of contractors worked in Sillod block on MGNREGA work for more than two years. During that period, they received weekly Rs. 300 per couple (Husband and wife). The MG-NREGA works were implemented by the Sillod Block Development Officer (BDO) and his staff on a contract basis with the help of a local contractor who would get a commission for as per number of labourers. Initially, labourers were promised that they will get Rs. 300 per couple for a weekly market and payment after completion of government process. The wages were decided based on the task which they were supposed to complete in a day; hence, they would work from 6 a.m. to 6 p.m. hoping that they would get good wages. However apart from the weekly amount, they did not receive any amount. To get the payment in the name of the labourers, the BDO and his staff made fake job cards with the help of the local Gram Panchayat and took the amount from the local post office. The total wage amount was more than four crores as per the calculation. The labourers who migrated from the Buldhana to Sillod had hopes and plans to fulfil from wages, for example, daughter's marriage, children's educational expenses, building home and so on. When they realised that their amount has been taken away by the government officials, five of them committed suicide.

Methodology

The researcher adopted a participatory method through a social audit that involves the mixed qualitative and quantitative data through in-depth interviews of the labourers and government officials. The data was collected through participation in the entire social audit process. The rationale of using this method is defined as "The oppressed suffer from the duality which has established itself in their innermost being. They discover that without freedom they cannot exist authentically. Yet, although they desire authentic existence, they fear it. They are at one and the same time themselves and the oppressor whose consciousness they have internalized. The conflict lies in the choice between being wholly themselves or being divided; between ejecting the oppressor within or not ejecting them; between human solidarity or alienation; between following prescriptions or having choices; between being spectators or actors; between acting or having the illusion of acting through the action of the oppressors; between speaking out or being silent, castrated in their power to create and recreate, in their power to transform the world. This is the tragic dilemma of the oppressed which their education must take into account" (Freire 2006). It needs to develop the capacities of the oppressed to fight for their rights by making them empowered enough. The empowerment can be possible only with their full participation in the process and bringing them into the mainstream flow of the development. The tool of social audit was the best method to access social justice for the MGNREGA labourers who were exploited in Buldhana and Jalana. Before going on to the practical methodology of the social audit, let's understand the concept of social audit.

Social Audit: A Tool to Access Social Justice

"The term audit originates from the Latin word '*audire*', which means '*to hear*'. Audire in ancient Greece referred to the 'hearing of accounts'" (Kidambi 2011). According to Charles Medawar, "The concept of social audit starts from the principle that in a democracy the decision-makers should account for the use of their powers and that their powers should be used as far as possible with the consent and understanding of all concerned".[9] Yamini Aiyar, who has done a study of social audit in Andhra Pradesh says, "This process of reviewing official records and determining whether state reported expenditures reflect the actual money spent on the grounds is referred to as the social audit".[10]

As per the Society for Social Audit Accountability and Transparency (SSAAT), a leading organization in MG-NREGA social audit, "Social audit is a process by which an attempt is made to find out whether the benefits of the project/activities reach the people for whom it meant. It is

[9] http://www.cgg.gov.in/publicationdownloads2a/Social%20Audit%20Toolkit%20Final.pdf also see the article by Dhaktode 2016.

[10] http://knowledge.nrega.net/193/1/transparency-accountability_Andhra.pdf also see the article by Dhaktode 2016.

a democratic process in which all stakeholders involved in a particular project take part. Social audit is not only an audit of expenses or decision but it also covers the issues of equity and quality in programme implementation. It is a verification of Government records with people and the works/projects executed at the field level in great detail" (Dhaktode 2016). "In a social audit, the people and the government jointly monitor the project. It brings on board the perceptions and knowledge of the people, involves people in the task of verification and also brings about much greater acceptability of the government".[11]

The concept of social audit is not new in India. It has been used in ancient times, also in kingdoms, but not under the name of "social audit". "Social Audit gained significance after the 73rd Amendment of the constitution relating to Panchayati Raj Institutions (PRI) in 1993".[12] The process of social audit in social policies was started with the Mazdoor Kisan Shakti Sangathan (MKSS), a people's organization that has fought for peasants' and labourers' rights since 1990. In June 1994, MKSS had first *Jan Sunwai*, the first public hearing in Kot Kirana (Pali district) in Rajasthan.[13] The process of social audit began in Rajasthan with the slogan '*Hamara Paisa Hamara Hisab* (Kidambi 2011; Dhaktode 2016).

Process of Social Audit

The SSAAT, a leading organization established by the state government of Andhra Pradesh in 2009 (including Telangana), has considered the ideal and acknowledged that the MoRD has issued social audit rule 2011 as a guidance to replicate the same model across the nation. To access the justice for the MGNREGA labourers, the following few states and methods have been used by the SSAAT.

Social audits are conducted for the most part by literate youth (men and women) from the labourers' families. The person begins with-

[11] http://125.17.121.162/SocialAudit/AboutUs.jsp.

[12] http://www.cgg.gov.in/publicationdownloads2a/Social%20Audit%20Toolkit%20Final.pdf.

[13] http://www.india-seminar.com/2001/500/500%20aruna%20roy%20et%20al.htm.

1. Drawing up the Social Audit Schedule
2. Intimation letters
3. Team Selection
4. Identification and training the Village Social Auditors
5. Training and consolidation of records
6. Formation of teams to do the Social Audit
7. The Social Audit process
8. Awareness building of the rights and entitlement
9. Social Audit Gram Sabha
10. Social Audit Public Hearing
11. Presiding Officers and Decisions Taken
12. Decisions Taken Report
13. Days Follow-up and Appraisals
14. Social Audit Report Public Disclosure
15. Action Taken Reports Public Disclosure.[14]

How It Links with Social Justice

The important steps in the above process to get justice are (1) auditors from the wage seekers' families from Village Resource Person (VRP) to State Resource Person (SRP) all belong to the same socio-political and economic process. They do their best to ensure justice for the wage seekers whose labourers face injustice, (2) the awareness built by the facilitators and auditors was most effective during the process. They provide information about the MGNREGA rights and entitlement along with their role in the entire process of implementation, monitoring, and evaluation of the scheme. Also, another important step is transparency and accountability with the use of RTI. They spread awareness about RTI. (3) Most important is an open forum where labourers raise the concerns and questions in Gram Sabha and taluka/block-level public hearing. Due to caste hierarchy and political pressure, they may not raise their voices in the Gram Sabha, but they have an opportunity at the Block level public

[14] http://125.17.121.162/SocialAudit/AboutUs.jsp.

hearing, in that they raise their issues. (4) Last but not least is the accountability ensured by the government: in the public hearing, the government officers, political leaders, media, labourers participate and the higher authority listen to both sides—labourers' issues and government employees' explanations on the same. Once it is proved that the functionaries are guilty, serious action is taken on them. In this particular process, the role of social auditors is very important in terms of presenting issues and evidence properly, facing political pressure, and so on.

With the above process, the Society for Social Audit Accountability And Transparency (SSAAT), government of Andhra Pradesh has conducted more than 9363 public hearings in 6–7 rounds in 22 districts until 30 April 2014 (SSAAT, Month of May state review meeting notes). In 2012–2013, SSAAT audited public funds worth Rs. 5000 crores. As of February 2014, misappropriated funds worth Rs. 93.2 crores have come to light, of which around Rs. 31.74 crores have been recovered, 64,907 field-level functionaries have been implicated; 5954 staff members have been dismissed; and 1741 suspended. Punishments have been imposed on 24,573 officials and 64,907 departmental enquiries have been initiated. All this has been possible due to the efforts of 86.612 Village Resource Persons (VRP) trained by District Resource Persons (DRPs) and State Resource Persons (SRPs).[15]

Justice for Buldhana NREGA Labourers Through Social Audit

As it has stated above, the social audit wasn't carried out as per the Social Audit Rule, 2011 in Sillod block. Nevertheless, the members of COE were from different places such as (1) Director of SSAAT Hyderabad, (2) Director of Orissa Society for Social Audit Accountability and Transparency (OSSAAT), (3) Secretary of 'PragatiAbhiyan' NGO, and other members two programme managers (researcher was one amongst) of SSAAT, and three members from Programme Advisory Group Secretariat (PAG), Ministry of Rural Development, New Delhi. The

[15] http://www.satyamevjayate.in/Kings-Every-Day/EPISODE-4Article.aspx?uid=E4-sAud-i1.

COE has submitted the report to MoRD and stated non-payment as the prime factor responsible for the suicides in the districts. The high amount of corruption had taken place in the implementation process of MG-NREGA, labourers were denied their rights and entitlement, even the serious issues were not raised by the social audit team, and the process was not followed up by the third-party organization.

Based on the report, government of India issued a guideline and has given compensation to the family members who had committed suicide, taken action on the government employees (they were in jail), recovery was done, and the payments were made to all the labourers. Government ensured that minimum 100 days work will be given to needy under MGNREGA at their village itself and check whether the payments are made properly. The labourers are getting a choice of freedom to work, getting right of work, and other entitlement including payment in 15 days, worksite facilities and so on. Today, labourers can ask the government about their rights and the demand for work. They could build their capabilities through this process. If the payment does not take place in a given time, the government employees are held responsible for this. Labourers participate in the work from planning to evaluation.

As Amartya Sen argues on the freedom of choice theory, the Buldhana labourers did not have a choice to get work at their village; therefore, they migrated and were exposed to exploitation. Similarly, they were unorganised and away from the mainstream village process. After social audit, they got a choice to get work at the village, they developed their capabilities to find out the ways to cope up with exploitation, the tool of social audit helped empower them. Paulo Freire, Brazilian Educator and Philosopher, argues that revolution can be possible only when the oppressed develop the capacity to fight against this. Here the social audit process has given a boost to the confidence of labourers to fight for rights. As a result of this, they got justice and created such a system where the process of injustice has gone down.

Conclusion

To achieve the goal of being a developed country, there is a need to achieve and ensure social justice for the citizens in the computation of becoming a developed country. Today, the meaning of development is more economy-oriented, where industrialization is playing a vital role. At the same time, promoting industrialization is pushing the state back from its key role. Once industrial development takes hold of the economy, its key objective is to earn profits. In this given situation, the principle of social justice gets manipulated and violated on various grounds. The budget allocation for health, education, and social security provide a depth of state seriousness towards these components. The people who get displaced from their land have no freedom of choice to work the way they want, since they have lost their land and natural resources. The policy of new liberalization adopted by the government is more money-oriented, which does not provide an opportunity for welfare-based policies.

The important social welfare policies such as MGNREGA, Mid-Day-Meal, Social Security Pension, Public Distribution System and so on are caught in the corruption and malpractices. The political and bureaucratic systems are becoming money-oriented. As a result of this, beneficiaries are suffering and exposed to various injustices. Though they have the right to get work in their village, they did not get that and hence migrated. In the case of Buldhana, it clearly shows this. Though, it is mandatory that their payment should be given in 15 days as per MGNREGA, they didn't get this for more than two years, since they migrated and were away from family and faced everyday life issues. Moreover, their children did not get an education. When the poor are not able to access the basic facilities, then absolute injustice happens to them on multiple fronts.

Until people become aware of injustice and get mobilised to fight against it, the problem is not going to be resolved. Paulo Freire, a Brazilian Educator and Philosopher, argues on the same grounds. He believes that there is a need to empower the oppressed, to revolt against the exploitative system. Nevertheless, the tool of social audit is all about this. It ensures people's participation through awareness about their rights and entitlement, while doing the social audit of particular schemes and programmes. It gives strength to the poor to access the rights and entitlement ensured by the Constitution of India.

References

Aiyar, Y., Kapoor, S., & Samji, S. *Guide to Conducting Social Audit Learning from the Experience of Andhra Pradesh* (p. 4). New Delhi: Accountability Initiative: Research and Innovation for Governance Accountability. Retrieved from http://www.accountabilityindia.in/sites/default/files/guidelines_-_le.pdf.

Ambedkar, B. (1936). *Annihilation of Caste*. Mumbai: Bharat Bhushan Printing Press.

Dhaktode, N. (2016). *Social Audit Through Community Organization: Case Study of United Andhra Pradesh*. Jaipur: Manas Publication.

Freire, P. (2006). *Pedagogy of the Oppressed* (30th Anniversary ed.). New York: Continuum.

Gupta, A. (2012). *Red Tape. Bureaucracy, Structural Violence, and Poverty in India*. Durham and London: Duke University Press.

Jost, J., & Kay, A. (2014). The history of psychology and Social justice (edt) Johnson, C., and Fridman, H. ABC Clio Publication.

Kidambi, S. (2011). *Termites, Earthworms, and Other Organic Gardeners*. New Delhi: Seminar wel-edition.

Marx, K., & Engels, F. (1848). *The Manifesto of the Community Party*. London: Communist League's Programme on the Instruction of Its Second Congress.

Mill, J. S. (1863). *Utilitarianism*. London: Parker, Son and Bourn.

Miller, D. (1999). *Principles of Social Justice*. Cambridge: Harvard University Press.

Ranjithkumar, A. (2011). Ambedkar's Notion of Social Justice a Different Perspective. *International Journal of Scientific & Engineering Research, 2*(12), 1–5.

Rawls, J. (1971). *A Theory of Justice*. Cambridge, MA: Harvard University Press.

Sen, A. (1999). *Development as Freedom* (1st ed.). Oxford: Oxford University Press.

Zajda, J., Majhanovich, S., & Rust, V. (2006). Education and Social Justice. *Review of Education, 52*(1), 9–22.

17

Social System Breakdown of Bangladesh Due to Rohingya Crisis: In Search for Possible Solution

Md. Zahir Ahmed

Introduction

Rohingya is the Indo-Aryan ethnic minority of Myanmar originated from South Asia. This indigenous community lives in the western Rakhine state of the country. Mostly, they are under the subordination of Arab, Mughal, and Portuguese rule and historically lived in the western part of the country for around 2500 years. They proclaim that they are the descendants of Arakan and colonial Arakan. Historically, this region was an independent state between Southeast Asia and the Indian subcontinent ("Rohingya people", Wikipedia Contributors 2020f). However, the Myanmar government has never recognized Rohingya ethnicity as their indigenous community. On the contrary, the national citizenship

Doctoral Candidate, College of Psychology, Northwest Normal University, China. Member—The Network of Asia Pacific Schools and Institutes of Public Administration and Governance (NAPSIPAG) & Jr. Research Coordinator, Policy Research Centre.bd (On Leave)

Md. Z. Ahmed (✉)
Northwest Normal University, Lanzhou, China

© The Author(s) 2020

Md. N. Momen et al. (eds.), *Building Sustainable Communities*,
https://doi.org/10.1007/978-981-15-2393-9_17

law passed in 1982 prohibits their citizenship, and the ruling party has always declared them as illegal immigrants from Bangladesh. Ever since the citizenship law came into force in 1982, Rohingyas became the largest outcaste ethnic group on earth, or in other words, refugees (Burmese Rohingya Organisation UK 2014). Rohingya ethnicity constitutes more than 10% of the world's total refugees and the number is increasing rapidly ("Rohingya people", Wikipedia Contributors 2020f; Wildman 2017). In 1978, first 20,000 Rohingya refugees fled to Bangladesh as an aftermath of Operation King Dragon conducted by Myanmar's Junta government four years prior to implementing that unscrupulous Citizenship Law ("Operation Dragon King", Wikipedia Contributors 2020b). Following 16 months' extended diplomatic efforts of the international community, mainly of The United Nations High Commissioner for Refugees (UNHCR), a Memorandum of Understanding was signed between Bangladesh and Myanmar and Rohingya refugees were repatriated subsequently (Kiragu et al. 2011). At that time, it was the second-largest repatriation in Asia. However, a number of Rohingya refugees remained in Bangladesh owing to the fear of future recurrent conflicts, and for the newly independent Bangladesh, it was an arduous task to identify all the illegal Rohingya refugees. Due to the administrative paucity of Bangladesh, many Rohingya refugees started residing in the country on a permanent basis (Abrar n.d.). When Aung San Suu Kyi was elected in Myanmar's general election of 1990, the Junta government launched military aggression on the local Muslims. The main reason for that action was the direct and enthusiastic support of the Muslims to the democratic movement. On that occasion, Aung San Suu Kyi was put under house arrest for the long term, and Muslims were tortured in every possible way by the Junta government and consequently, 2,50,000 refugees fled to Bangladesh for the second time (Burma: 20 Years After 1990 Elections, Democracy Still Denied 2010). During that influx, Bangladesh emphasized on peaceful resolution through strengthened border security and managing Rohingya refugees effectively.

Under the interposition of UNHCR the Rohingya repatriation was completed later on. During a massive administrative change in 1990, Myanmar's Junta government changed the name of Arakan state to Rohingya, which was the most frivolous example of a biased attitude towards the Rohingya community. Historically, this region was known as Arakan for thousands of years. This name alteration could be considered as the most effective measure to disband Arakani Muslims to acquire full citizenship of Myanmar. For over half of a century, former Arakani and present Rohingyas are being tortured by the combined military forces and Buddhist locals ("Arakan", Wikipedia Contributors 2020d). Through the State Law and Order Council (SLORC) in 1991–1992, the Myanmar government oppressed the Rohingya people ethnically labelling them as terrorists in the northern Rakhine state (United Nations 1998). Since then, Rohingya Muslims are being tortured by Myanmar soldiers and local Buddhists in all possible manners, that is, intention riot, extrajudicial killing, rape, destruction of property, kidnapping, mosque demolition, obstruction on religion, push-in to Bangladesh, and even forced labor in the military camps.

In 2012, Myanmar government wanted to instill the fear of being minorities among the Rakhine Buddhist community, which led to an internecine riot between local Muslims and Buddhists. During the riot, fatal incidents, including gang rape, killings, vandalism occurred. It is believed that during that occasion, the Myanmar government instigated the Buddhist community to curb Rohingya Muslims, and there was no justice restored to the side of Muslims, which was the clear indication of the perpetrating authorities' malevolence ("2012 Rakhine State riots", Wikipedia Contributors 2020a; Deadly Riots Raise Muslim-Buddhist Tensions in Myanmar 2012). Following the continuous authoritarian suppression, from 2012 to 2015, thousands of Rohingya have left Myanmar's borders by boat, risking their lives to escape such brutality and persecution ("They Tried to Kill Us All": Atrocity Crimes against Rohingya Muslims in Rakhine State, Myanmar, Fortify Rights and Simon-Skjodt Center for the Prevention of Genocide 2017; United Nations High Commissioner for Refugees 2017; C., 2015). Apart from their primary destination which was Bangladesh, the Rohingyas stated migrating to Malaysia, Indonesia, and Thailand (Fuller and Lyons 2015). On October 9, 2016, Rohingya armed insurgents attacked a border check post in the city of Maungdaw adjacent to the Myanmar–Bangladesh border. In that attack, rebels killed nine security personnel and looted firearms and ammunition ("Myanmar: A New Muslim Insurgency in

Rakhine State" 2016; Lewis 2016). Later, in the name of restraining suspected insurgents, the country's security forces launched the notorious Rohingya ethnic cleansing operation. Though the United Nations, Amnesty International, the US State Department, the government of Malaysia criticized the authoritarian suppression extensively, the Myanmar government and even Aung San Suu Kyi remained silent on this crucial humanitarian issue ("Myanmar: Security forces target Rohingya during vicious Rakhine scorched-earth campaign" 2016).

Despite the tremendous pressure from the international communities, the persecution forced 1.3 million of Rohingyas to take asylum in permanent and temporary camps of neighboring country, Bangladesh ("Bangladesh/Myanmar: Rakhine Conflict 2017" 2018). Meeting the basic needs of this huge number of refugees has really become a challenge for one of the world's densest countries. To mitigate this humanitarian crisis conjointly, the government of Bangladesh has been working with several local and international agencies since the very first day. Bangladesh government and international organizations are implementing various short- and medium-term solution projects for the management of Rohingya refugees, but since their basic needs are not being fulfilled and also because of their aggressive stance to spread throughout Bangladesh, the environment, internal security, and overall social system of Bangladesh have already been in extreme peril.

Objectives of the Study

This chapter focuses on the unforeseen Rohingya refugee crisis–induced adverse effect on Bangladeshi environment, wildlife, indigenous way of living, internal security, especially in the Cox's Bazar, Teknaf, Ukjiya and hill track areas.

Methodology

The core structure of this chapter is based on the qualitative research approach and data and other relevant information gathered from newspapers, journals, and relevant articles. Some information was gathered from the Bangladesh Statistical Bureau and the international community's viewpoints. A significant part of this chapter was taken from security experts and analysts as well. After gathering all possible information, the chapter remarks on the problems created in Bangladesh by the recent Rohingya crisis and concludes with possible solutions to mitigate this humanitarian crisis.

Endangered Indigenous Life in Hill Tracks

Although Rohingyas were granted asylum in Bangladesh on humanitarian ground, the uncertainty of repatriation is rising constantly due to the reluctance of the Myanmar government. Since the Rohingyas have been living in the hilly regions of Bangladesh, the indigenous way of life and living of the three hilly districts and Cox's Bazar are endangered. Currently, the cumulative number of Rohingya refugees in Bangladesh is over 1.3 million, while the total indigenous population of the three hilly regions of Bangladesh is 0.5 million ("Indigenous peoples in Bangladesh", Wikipedia Contributors 2020c). As the camps of this huge unwanted population are very proximal to the hill tracks, the social, cultural, and aggregated indigenous life is facing unnatural changes which is consequently accelerating the process of extinction. According to the 2001 census, the population of the three hilly districts of Bangladesh was 5,92,977 which reached 8,45,541 ("Bangladesh-indigenous/tribal population and access to secondary schools (draft): indigenous peoples plan: Indigenous peoples plan" 2008), so it is clear that apart from the Rohingya crisis, Bangladeshi people are migrating from other internal areas to the hilly region. As a result of increasing pressure, the indigenous people are often deprived of their rights to land and even forced to leave their homes and other property. In justification of this statement, within the 49 years since the Independence of Bangladesh, at least ten ethnic groups have completely disappeared from the hill tracks of Bangladesh (Cavallaro and Rahman 2009; "World Directory of Minorities and Indigenous Peoples—Bangladesh" 2018). Besides, if these 1.5 million Rohingyas remained here for a long period of time, their indigenous life of the hill tracks would certainly be seriously disrupted resulting in agitation (Fig. 17.1).

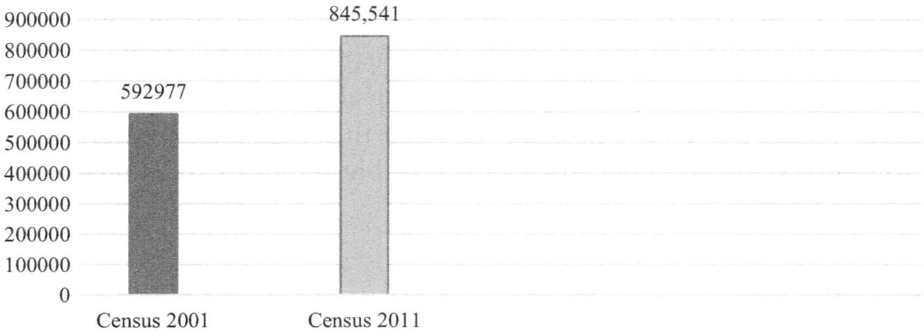

Fig. 17.1 Population of ethnic group of Bangladesh. (Source: Bangladesh Bureau of Statistics)

Anticipated Severe Environmental Degradation

Because of the abrupt Rohingya influx and acute shortage of Bangladesh's national resources, it was anticipated that there would be severe environmental degradation in the hill tracks. Seven reserved forests covering 2500 acres of land in Cox's Bazar district have already disappeared to accommodate this massive population and to supply their fuelwood within the very first two months of that influx ("Bangladesh to allocate 2,500 acres of land for fleeing Rohingyas" 2017; "Govt allocates 2,000 acres for Rohingyas" 2017). Today, Rohingya refugees are living in 1625 acres of natural forest and 875 acres of reserved forest in Ukhia and they are cutting millions of reserved trees every day. Within the first two months, 1500 acres of community and social forestry projects have been completely chopped down to supply the everyday fuelwood (Paul 2017). The age of the projects was around 15 years and it had the direct involvement of 2000 locals. These poor locals had hoped to get 45% of the benefits following the completion of the projects (Mahmud 2017). By the end of 2017, Rohingya refugees chopped down Bangladesh's reserve forests worth 18 million US dollars. The country's forest department corroborated that the Rohingyas have temporarily built their shelters in the reserved forests covering a total area of 6000 acres, and it was anticipated that the area would increase to 8000 acres soon. Besides Ukhia, Rohingyas are living in 450 acres forest land of Teknaf Range, 375 acres of Shilkhali Range, and 50 acres of Putibunia Range (Ahmed 2018).

The government of Bangladesh has already planned to allocate two more thousand acres of forests and hills for Rohingya refugees. Since 1978, the government of Bangladesh has acquired 4000 acres of forest for the accommodation of Rohingyas. However, because of the lack of proper monitoring and enforcement, only 40% of the refugees live within the assigned camp area, and unfortunately, the rest have chosen 200 hills as their habitat. Apart from chopping down trees belonging to the reserved forests, they have started to cut and raze the hills which could result in dangerous consequences such as landslides during the monsoon season. Besides, some unscrupulous locals of Bangladesh even destroy the forest and hills for their vested interest, and to some extent hire the Rohingya refugees to cut the hills to confuse the authority who believe them to be the acts of the Rohingyas ("Rohingya Influx Brings 'Environmental Catastrophe': Bangladesh Officials" 2017) (Fig. 17.2).

Fig. 17.2 Temporary shelters for Rohingya refugees in Cox's Bazar, Bangladesh. (Photo Credit @Author, 2018)

Rohingyas Dwelling on the Path of Elephants

According to the report titled "Status of Asian Elephants in Bangladesh" by International Union for Conservation of Nature (IUCN 2016), the number of elephants during the dry season in the southern region of Cox's Bazar (Ukhia-Teknaf) is 78 and the average number during the rainy season is 47. These elephants often come out of the sanctuary in search of food. There are 12 corridors for the elephants in Chattogram and 8 of those are in the Cox's Bazar area which is the main reason for their disoriented movement (Motaleb et al. 2016). Though there are warning signs on the path of elephants in the area from Kutupalong to Balukhali, thousands of Rohingyas have built temporary shelter there. The expected human–elephant conflict is already occurring in the Cox's Bazar area. According to the Forest Department of Bangladesh, at least ten Rohingya refugees including children have been trampled to death by wild elephants' attack at the Balukhali refugee camps ("Elephants kill 10 Rohingyas in Bangladesh: UN" 2018). Usually, elephants follow the same route for daily movement and they identify the path by trees around the vicinity. Indiscriminate destruction of forests disorients the elephants to trample people and destroy property. The elephants will be completely shifted to new areas and suspected to have more incidents of man-elephant conflict which ultimately might be the cause of their extinction if the Rohingya crisis remains unsolved and stays in the hill tracks of Bangladesh for a long time.

Internal Security Threats to Bangladesh

Bangladesh's internal security is threatened because of the Rohingya influx. In the meantime, there have been reports of a number of cases of forcible occupation of local's land and forest to accommodate thousands of Rohingya refugees. Since it is believed that some Rohingyas have strong links with armed insurgents, locals are in a state of panic. That is why to enforce their safety, locals along with the police guard their property at night. According to the local administration, there are already findings of incidents of looting and attacks . On October 27, 2017, several Rohingya refugees attacked four Bangladeshi labors with sharp weapons near the Balikhali Camp of Ukhia, Cox's Bazar; the local police have also confirmed the incident ("4 injured in Rohingya attack, arms recovered from attackers" 2017). Relatively wealthy Rohingyas residing in the refugee camps have been trying to enter Bangladesh employing illegal means. In the last couple of years, a number of Rohingyas attempted to acquire Bangladeshi passports with false birth registration certificates. On October 12, 2017, three Rohingya citizens were arrested in Bangladesh's Bogra district when collecting passports showing false papers ("3 Rohingyas detained from Bogra passport office", 2017). Besides, on July 4, 2019, two Rohingya youths were detained with Bangladeshi passports at Brahmanbaria check post while traveling to India ("2 Rohingyas held with Bangladeshi passport", 2019). Also on September 6, 2019, three Rohingyas were held with Bangladeshi passports in Chattogram ("3 Rohingyas held with Bangladeshi passport in Chattogram" 2019), and in all such cases, they have been forcefully sent back to their assigned camp. A group of unscrupulous people are taking advantage of helplessness Rohingyas and tempting them to own Bangladeshi passport and migrate illegally to Malaysia, Thailand, and Indonesia which is a blow on the sovereignty of Bangladesh. Apart from that, Myanmar's Rakhine state is the main manufacturing and trading hub of the deadly drug Yaba. It is believed that many of the refugees in the camps in Bangladesh are involved in drug dealing. It is apprehended that their residence within the boundary of Bangladesh would accelerate the availability of narcotics in Bangladesh. Moreover, Rohingyas' obvious connection with international militant groups could trigger criminal activities in Bangladesh. Several refugees have already been detained by the law enforcement agencies of Bangladesh for murder, assault, extortion, drug dealing, and possession of illegal weapons. ("Rohingya refugee arrested in Bangladesh, activists concerned" 2017).

Fears of Rising Militancy

Reports have confirmed clear links between international Islamist militant groups and Rohingya armed insurgents on Myanmar border police in 2016 (International Crisis Group 2016). This can be considered as the serious religious fallout of Myanmar's Buddhist–Muslim conflict. International Crisis Group has confirmed the link between the Rohingya insurgent group Harakah al-Yaqin with Saudi Arabia and Pakistan soon after the communal riot of 2012 ("Arakan Rohingya Salvation Army", Wikipedia Contributors 2020e). The ICG also says that some Afghan and Pakistani militants had been training the armed insurgents of Rohingya for over two years. Islamic States, Al-Qaeda, and the Taliban of Afghanistan repeatedly called for Jihad to protect the Rohingyas from Myanmar's Buddhist aggression. This illustrates that entire South Asia including Bangladesh would soon witness the rise of militancy if the humanitarian crisis remains unresolved as over 1.3 million Rohingya refugees could easily be encouraged by the militancy. The refugee crisis is not only confined to the economic aspect, but it is also now inlaid with political and security issues. If it is not possible to mitigate the Rohingya crisis soon, Rakhine will become a haven for militant groups. International militant networks could start operating there which would certainly risk the security of entire South Asia, including, Bangladesh.

Recommendations

To address the environmental degradation, sociocultural declension, and internal security issues of Bangladesh induced by the Rohingya refugee crisis, a collective endeavor from the government of Bangladesh, the international community, and the Rohingya themselves is absolutely imperative. It would be very difficult for any single party to resolve this crisis. This chapter recommends the following possible short-, mid-, and long-term solution schemes to address the Rohingya refugee crisis:

1. Imposition of a ban on new Rohingya refugees entering Bangladesh by the Bangladeshi government.
2. Create a digital database of Rohingya refugees living in Bangladesh and ensure rigorous surveillance so that they cannot relocate to any other place in Bangladesh outside their assigned camps, and prompt formulation of policies ensuring the Rohingyas are provided with proper facilities and their needs are met.

3. Assure safe and prompt repatriation with the active vigilance of international communities.
4. Use effective international pressure on the government of Myanmar in order to recognize Rohingyas as its citizens..
5. Ensure there are fair rehabilitation programs for Rohingyas in Rakhine state with close monitoring of the international community to avoid any untoward incidents in the future.
6. To enable the development of the living standards of backward Rohingyas ethnicity, Myanmar government and the international community need to jointly implement long-term projects..
7. Consider special provisions to prevent communal riots against Rohingyas.
8. Launch anti-militancy campaigns among the Rohingya people at the earliest.
9. Identify those Rohingyas who have already entered Bangladesh beyond the refugee camp boundary, and try and confiscate their Bangladeshi passports and arrange to send them back to the camp immediately.

Conclusion

Historically, the Rohingyas have been living in Arakan state for centuries, but due to religious bias, the Junta and the later quiescent government have been denying them the Citizenship Law of 1982. Since then, as a result of the authoritarian denial, the Rohingya people have been deprived of all civil and basic facilities. They do not possess the national identity card since the citizenship law extruded them, and without the card, they are unable to avail the basic facilities, that is, travel, education, medical facility, banking, and so on. This exclusion led to the Rohingyas lagging behind in all aspects and literally, more than 80% of Rohingyas are totally illiterate. Furthermore, under the land and property law of Myanmar, foreigners and excluded persons from national citizenship are not permitted to possess land and any other property in Myanmar, and since the Rohingyas are illegal as per the citizenship law, until today, they are disallowed to own any kind of property. Apart from depriving them of civic rights and basic needs for the past half a century, the government has made their lives unendurable by imposing restrictions on their religious activities. It is evident that the government has instigated communal riots against Rohingya Muslims by misguiding the peace-loving and non-violent Buddhists. Since this community struggles to meet basic needs and identity, it could create room to mislead them very easily, and undoubtedly, international militant organizations and extremist groups are looking for this opportunity.

As the living standard of 1.3 million Rohingya refugees in Bangladeshi camps is extremely low, the entire negative impact of this humanitarian crisis is threatening the social system of Bangladesh, especially the hill tracks. Moreover, owing to the constraint of national resources Bangladeshi government has failed to meet all the basic requirements of the Rohingyas which lead them for self-management and resulting in the subsequent destruction of hills and forests infatuatedly. They are even kingpins of drug dealings and involved in anti-social activities in the southern parts of Bangladesh. The overall situation has taken such a turn for the worse that by being generous on humanitarian grounds, Bangladesh's sovereignty is being threatened immensely. It should be noted that this negative impact on Bangladesh could entangle the entire South Asia. That is why to mitigate the crisis, the immediate and safe repatriation of Rohingya refugees to Rakhine is imperative.

References

2 Rohingyas Held with Bangladeshi Passport. (2019, July 5). *The Daily Star*. Retrieved from https://www.thedailystar.net/country/news/2-rohingyas-held-bangladeshi-passport-1767103.

3 Rohingyas Detained from Bogra Passport Office. (2017, October 12). *The Daily Star*. Retrieved from https://www.thedailystar.net/rohingya-crisis/3-rohingyas-detained-bogra-passport-office-1475545.

3 Rohingyas Held with Bangladeshi Passport in Chattogram. (2019, September 6). *The Daily Star*. Retrieved from https://www.thedailystar.net/rohingya-crisis/news/3-rohingyas-held-bangladeshi-passport-chattogram-1796398.

4 Injured in Rohingya Attack, Arms Recovered from Attackers. (2017, October 28). *Prothom Alo English*. Retrieved from https://en.prothomalo.com/bangladesh/4-injured-in-Rohingya-attack-arms-recovered-from.

Abrar, C. A. (n.d.). Repatriation of Rohingya Refugees. Retrieved from https://www.burmalibrary.org/docs21/Abrar-NM-Repatriation_of_Rohingya_refugees-en.pdf.

Ahmed, K. (2018, January 25). Bangladeshi Forests Stripped Bare as Rohingya Refugees Battle to Survive. *The New Arab*. Retrieved from https://www.alaraby.co.uk/english/indepth/2018/1/25/bangladeshi-forests-stripped-as-rohingya-refugees-battle-to-survive.

Bangladesh to Allocate 2500 Acres of Land for Fleeing Rohingyas. (2017, September 10). *Prothom Alo English*. Retrieved from https://en.prothomalo.com/bangladesh/Govt-to-allocate-2-500-acres-of-land-for-fleeing.

Bangladesh/Myanmar: Rakhine Conflict 2017. (2018, February 28). Retrieved from https://www.who.int/emergencies/mmr/en/.

Bangladesh—Indigenous/Tribal Population and Access to Secondary Schools (Draft): Indigenous Peoples Plan: Indigenous Peoples Plan (Rep.). (2008, April 4). Retrieved from http://documents.worldbank.org/curated/en/627571467997887013/Indigenous-peoples-plan.

Burma: 20 Years After 1990 Elections, Democracy Still Denied. (2010, May 26). Retrieved from https://www.hrw.org/news/2010/05/26/burma-20-years-after-1990-elections-democracy-still-denied.

Burmese Rohingya Organisation UK (2014). *Myanmar's 1982 Citizenship Law and Rohingya.*. Retrieved from http://burmacampaign.org.uk.

C., R. (2015, January 1). The Plight of the Rohingyas. *The Economist*. Retrieved from https://www.economist.com/the-economist-explains/2015/06/01/the-plight-of-the-rohingyas.

Cavallaro, F., & Rahman, T. (2009). The Santals of Bangladesh. *The Linguistics Journal-Special Edition*, pp. 187–199. Retrieved from https://www.ntu.edu.sg/home/CFCavallaro/Pdf%20files/Cavallaro%20and%20Rahman%20 2009.pdf.

Deadly Riots Raise Muslim-Buddhist Tensions in Myanmar. (2012, June 9). *The New York Times*. Retrieved from https://www.nytimes.com/2012/06/10/world/asia/7-die-in-sectarian-riots-in-myanmar.html.

Elephants Kill 10 Rohingyas in Bangladesh: UN. (2018, March 6). *Press TV*. Retrieved from https://www.presstv.com/Detail/2018/3/6/554587/Bangladesh%20Rohingya%20Refugees%20Elephants.

Fortify Rights & Simon-Skjodt Center for the Prevention of Genocide. (2017). "They Tried to Kill Us All": Atrocity Crimes against Rohingya Muslims in Rakhine State, Myanmar. Retrieved from https://www.fortifyrights.org/downloads/THEY_TRIED_TO_KILL_US_ALL_Atrocity_Crimes_against_Rohingya_Muslims_Nov_2017.pdf.

Fuller, T., & Lyons, P. J. (2015, May 14). How Myanmar and Its Neighbors Are Responding to the Rohingya Crisis. *The New York Times*. Retrieved from https://www.nytimes.com/interactive/2015/05/13/world/asia/15rohingya-explainer.html?mtrref=www.google.com&gwh=89E4F5C3D554E8033AADCA884E435B1C&gwt=pay&assetType=REGIWALL.

Govt Allocates 2,000 Acres for Rohingyas. (2017, September 11). *The Daily Star*. Retrieved from https://www.thedailystar.net/frontpage/govt-allocates-2000-acres-rohingyas-1460356.

IUCN Bangladesh. (2016). Status of Asian Elephants in Bangladesh. IUCN, International Union for Conservation of Nature, Bangladesh Country Office, Dhaka, Bangladesh. pp. xii + 102.

Kiragu, E., Rosi, A. L., & Morris, T. (2011). States of Denial a Review of UNHCR's Response to the Protracted Situation of Stateless Rohingya Refugees in Bangladesh. Retrieved from United Nations High Commissioner for Refugees Website. http://www.unhcr.org/4ee754c19.pdf.

Lewis, S. (2016, December 16). Myanmar's Rohingya Insurgency Has Links to Saudi, Pakistan—Report. *Reuters*. Retrieved from https://www.reuters.com/article/us-myanmar-rohingya/myanmars-rohingya-insurgency-has-links-to-saudi-pakistan-report-idUSKBN1450Y7.

Mahmud, T. (2017, October 21). Rohingya Influx: 15-Year-Old Forestation Project Destroyed in 57 Days. *Dhaka Tribune*. Retrieved from https://www.dhakatribune.com/bangladesh/2017/10/21/rohingya-influx-15-year-old-forestation-project-destroyed-57-days/.

Motaleb, M. A., Ahmad, M. S., Islam, H., & Haque, M. A. (2016). *Atlas: Elephant Routes and Corridors in Bangladesh*. IUCN, International Union for Conservation of Nature, Bangladesh Country Office, Dhaka, Bangladesh. PP. xii + 104 (pp. 11–32).

Myanmar: A New Muslim Insurgency in Rakhine State. (2016, December 15). Retrieved from International Crisis Group Website. https://d2071andvip0wj.cloudfront.net/283-myanmar-a-new-muslim-insurgency-in-rakhine-state.pdf.

Myanmar: Security Forces Target Rohingya during Vicious Rakhine Scorched-Earth Campaign. (2016, December 19). Retrieved March 19, 2020, from https://www.amnesty.org/en/latest/news/2016/12/myanmar-security-forces-target-rohingya-viscious-scorched-earth-campaign/.

Paul, R. (2017, October 5). Bangladesh Carving Out Forest Land to Shelter Desperate Rohingya. *Reuters*. Retrieved from https://www.reuters.com/article/us-myanmar-rohingya/bangladesh-carving-out-forest-land-to-shelter-desperate-rohingya-idUSKBN1CA0ZN.

Rohingya Influx Brings 'Environmental Catastrophe': Bangladesh Officials. (2017, October 30). *Radio Free Asia*. Retrieved from https://www.rfa.org/english/news/myanmar/refugees-environment-10302017171539.html.

Rohingya Refugee Arrested in Bangladesh, Activists Concerned. (2017, July 13). *UCA News*. Retrieved from https://www.ucanews.com/news/rohingya-refugee-arrested-in-bangladesh-activists-concerned/79750.

United Nations. (1998, September 17). Burma [Myanmar]: State Law and Order Restoration Council (SLORC)/State Peace and Development Council (SPDC)/National League for Democracy (NLD)/Burmese Dissidents. Retrieved March 15, 2020, from https://www.refworld.org/docid/3df097254.html.

United Nations High Commissioner for Refugees. (2017, September 18). Rohingya Emergency. Retrieved March 16, 2020, from https://www.unhcr.org/uk/rohingya-emergency.html.

Wikipedia Contributors. (2020a, February 20). 2012 Rakhine State Riots. In *Wikipedia, The Free Encyclopedia*. Retrieved 12:55, March 15, 2020, from https://en.wikipedia.org/w/index.php?title=2012_Rakhine_State_riots&oldid=941716628.

Wikipedia Contributors. (2020b, February 20). Operation Dragon King. In *Wikipedia, The Free Encyclopedia*. Retrieved 20:13, March 12, 2020, from https://en.wikipedia.org/w/index.php?title=Operation_Dragon_King&oldid=941811653.

Wikipedia Contributors. (2020c, February 26). Indigenous Peoples in Bangladesh. In *Wikipedia, The Free Encyclopedia*. Retrieved 07:28, March 19, 2020, from https://en.wikipedia.org/w/index.php?title=Indigenous_peoples_in_Bangladesh&oldid=942753916.

Wikipedia Contributors. (2020d, January 15). Arakan. In *Wikipedia, The Free Encyclopedia*. Retrieved 18:33, March 14, 2020, from https://en.wikipedia.org/w/index.php?title=Arakan&oldid=935908164.

Wikipedia Contributors. (2020e, March 18). Arakan Rohingya Salvation Army. In *Wikipedia, The Free Encyclopedia*. Retrieved 08:41, March 19, 2020, from https://en.wikipedia.org/w/index.php?title=Arakan_Rohingya_Salvation_Army&oldid=946123826.

Wikipedia Contributors. (2020f, March 3). Rohingya People. In *Wikipedia, The Free Encyclopedia*. Retrieved 18:39, March 12, 2020, from https://en.wikipedia.org/w/index.php?title=Rohingya_people&oldid=943704188.

Wildman, S. (2017, March 31). The World's Fastest-Growing Refugee Crisis Is Taking Place in Myanmar. Here's Why. *Vox Media*. Retrieved from https://www.vox.com/.

World Directory of Minorities and Indigenous Peoples—Bangladesh. (2018, July). Retrieved from https://www.refworld.org/docid/4954ce6519.html.

18

Impact of Organic Farming on Sustainable Livelihood of Farmers

Ramesha Naika, C. Sivapragasam, and Sri Krishan Sudheer Patoju

Introduction

Agriculture sector is the backbone of the Indian economy, contributing significantly to the national GDP. India is predominately rural, and a large number of people live in the villages. According to the 2011 census report, this number accounts for almost 75 percent of the total population. Out of these, a large number of people, nearly 68 percent, depend on agriculture for their livelihood (Central Statistics Organisation 2014–2015). There are different kinds of farming system practices across India. The major crops in the southern part are rice and wheat, while in the states of Maharashtra, Tamil Nadu, Karnataka and Andhra Pradesh, farmers are cultivating sugar cane, fruits and cash crops. Today, India is

R. Naika (✉) • S. K. S. Patoju
Tata Institute of Social Sciences, Tuljapur, India
e-mail: sudheer.patoju@tiss.edu

C. Sivapragasam
Department of Applied Research, The Gandhigram Rural Institute, Deemed University Gandhigram, Chinnalapatti, India

© The Author(s) 2020
Md. N. Momen et al. (eds.), *Building Sustainable Communities*,
https://doi.org/10.1007/978-981-15-2393-9_18

the world's largest producer of milk and the second-largest producer of wheat, rice, fruits, sugar and groundnuts. However, the gross domestic product (GDP) from the contribution of agriculture has declined in the last few decades (Bhattacharyya 2005).

Before independence, traditional farming was organic by default, because most of the farmers were not using chemical fertilizers and pesticides for cultivation. After the green revolution, farmers started doing organic cultivation and increasing the production of high yield varieties. But in the present scenario, most of the farmers are aware of the basic actualities of organic farming, such as environmental protection, health benefits conservation and improved food quality for sustainable livelihood. Some people believe that organic farming is the 'right' way to practice farming, and it is better for the long-term sustainable future of the earth. Farmers who sustain organic practices are economically well off, and their lifestyle has changed (Reddy 2010). Reddy's (2010) study found that most organic farmers never use chemical fertilizers and pesticides, because the farmers want their food to be naturally pure, safe, healthy, environmentally friendly and protected from diseases. Future studies also documented people's awareness about the government's support in promoting organic farming through various methods in order to achieve a sustainable livelihood for organic farmers (Roy 2014). In Karnataka, there were about 35,000 certified organic farmers and 60,000 ha (2010–2011) of land under organic certification. Although there are many more organic farmers producing a variety of commodities, their area and products are yet to be assessed and certified (Lukas 2008). There is a great need to educate organic farmers and facilitate the process of certification for organic products. In India, many programmes have been introduced by the government to eradicate poverty in rural areas. Further, major benefit of organic farming is it provides a sustainable livelihood for marginal organic farmers (Udin 2014).

Nemes (2009) in his study found that the profitability of organic farming must be distinguished from relative profitability and income of organic farms and the agricultural income of non-organic farms in terms of prices, yield and production cost. Further availability of price premiums seems to be a crucial factor in good economic performances of organic systems and in most cases, studies have found that organic agriculture is economically more profitable (Nemes 2009).

Organic farming has been a necessary part of agrarian practices worldwide for quite a long time now. In the present scenario, the main focus is on the impact of organic farming on humans and the environment. Organic farming is the natural way of farming, in which no chemical fertilizers or pesticides are used, only organic waste, animal waste or compost (Chandrashekar 2010). Organic farming protects human health and conserves, maintains or improves natural resources, ensuring the quality of the environment for future generations for eco-friendly and economically sustainable agriculture. In recent years, organic farming has expanded rapidly in the world because of technological advancements and improvements in knowledge. Ultimately, most farmers are economically empowered in their livelihood activities. Chhetri (2015) observed that organic farming benefits range from reduced environmental pollution to increased soil quality.

Organic farming is a method that uses a different method to cultivate the land and plant crops, thereby sustaining the earth for future generations. This could be achieved through the use of biological materials and organic wastes along with beneficial microbes to release more nutrients to crops for increased sustainable production. Organic agriculture is a holistic production managing system which promotes the agro-ecosystem with biological cycles, biodiversity and soil systems (Altenbuchner et al. 2014).

Need for Organic Farming

Agriculture and related activities are the basic life of poor people in India; most organic farmers are small and marginal. In the present situation, due to rapid increases in population, our daily needs depend not only on agriculture production and related activities but also on sustainable production. The people know that the green revolution will contribute more to the farming system and to reach a plateau. Organic farming contributes to the health of the environment by controlling pollution (Shetty 2013).

Why Organic Farming

There are a number of reasons for farmers to adopt organic farming practices, such as economy, beliefs and lifestyle. But in the present scenario, there has been an increase in market demand for Indian organic products at the international level. As a result, farmers are getting good market prices for their products. In rural areas, farmers believe that organic farming is a 'good' way to do farming and offers a long-term sustainable livelihood. Most of the people are aware that organic farming is a better way of life; we can see that farmers are changing their lifestyle because they are using good organic manure in the field and are not using chemicals on their farms. They are the most enjoyable and environmentally sustainable lifestyle (Dahama 2003).

Status of Organic Agriculture in the World

According to the latest FiBL (Research Institute of Organic Agriculture) survey on organic agriculture certified worldwide, as of the end of 2015, there is data available on organic agriculture for a total of 179 countries. In 2015, organic agricultural land is 50.0 million hectares as a conversion area. Oceania has the most organic agricultural land (22.8 million hectares) in the world.

Asia is home to 35 percent of the world's organic producers, with almost 2.4 million producers in 2015. Africa and America are the other two countries with the highest organic farmland. India (585,200) is one of the most productive countries in the world. There has been an increase in the number of producers of organic farming since 2014.

Status of Organic Agriculture in India

In India, organic agriculture sectors have made significant growth in the past decade from nearly 528,171 hectares in 2007–2008 to1.18 million hectares in 2014–2015. There was almost threefold increase in cultivable land and remaining land of 3.71 million hectares of forest and wild areas. In 2016, Sikkim became the first fully organic state. In this state, 700,000 hectares' land cultivated as organic farming and there was the prohibition for the use of chemical fertilizer and pesticides. The other states replicating the same model and following organic farming are Nagaland, Himachal Pradesh, and so on. In India, organic food increased to about 25 percent annually from 0.36 billion US dollars in 2014 (Narayanan 2005).

Organic Agriculture in Udupi District

The coastal part of Udupi district of Karnataka is primarily an agrarian district, the majority of the population involved in agriculture. The major crop production in the district is paddy. The total area under organic farming was 4323 ha. In this district, the major foods cultivated are rice, peppers, banana plants and some other plantation crops.

Organic farming in this district was less common in the past, but today most NGOs are working to promote organic farming through government support to the farmers who adopt organic practices.

Table 18.1 Varities of Organic Crops in Udupi district, paddy production is the major crop of organic farming (40.32 percent). It is grown during Kharif, Rabi and the summer seasons. Other food crops grown include pulses (12.35 percent). Plantation crops grown include coconuts, areca nuts, cashew nuts and rubber (24.25 percent). Groundnuts are the

Table 18.1 Varieties of organic crops in Udupi district

Crops grown	Area under crop (ha)	Percentage
Paddy	1743	40.32
Pulses	534	12.35
Oilseeds	216	4.95
Fruits	369	8.54
Vegetables	141	3.26
Plantation crop	1048	24.25
Flowers, spices and another crop	272	6.31
Total	4323	100

Source: Agriculture department ZP, Udupi, 2011

main oilseeds grown in the district. The major fruit crops grown are mango, banana, pineapple, sapota and lemon (8.54 percent), cropping in organic farming in Udupi district.

Organic Farming Through Integrated Farming System

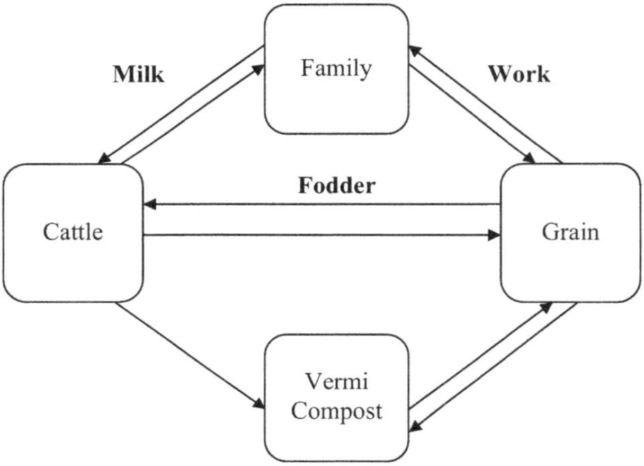

Organic farming through the integrated farming system eliminates organic waste; it will be interlinked with all processes. The farmers are using chemical fertilizer for the integrated farming system; they can use bio-fertilizer and composting culture in the recycling process. Organic agriculture is one approach to sustainable farming, in that many other goods and techniques are used in farming, such as crop rotation, inter-cropping and mulching, in addition to different crops and livestock (Hsieh 2008). Organic farming can, therefore, be achieved only through an integrated farming system, in which one can achieve economic and sustainable agricultural production to meet the diverse requirement of the farm household and maintain high environmental quality. In this integrated farming system, organic farming is very low cost, the products are good quality and quantity, and the environmental quality and soil degradation are improved. When composting the recycled waste products of organic farming, pest and diseases should be utilized through the integrated organic farming system in order to achieve the main purposes of organic farming. By adopting organic farming based on an integrated approach, all farmers are encouraged to be sustainable in the agricultural sectors.

Materials and Methods

Statement of the Problem

In Karnataka State, organic farming is practiced in a different place and to date, several training programmes have been conducted for these farmers. Organic farming can maintain good soil health, protect the environment and sustain crop production. Further, it maintains ecological balance in addition to ensuring sustainability in terms of food production, human health and safety.

A number of studies of organic farming show that the farmers who are presently practicing organic farming were not always aware of organic farming and they were basically depending on non-organic farming. The limited knowledge and lack of awareness regarding organic farming was an important factor in the farmers' non-use of organic methods. In addition, they were also using a huge amount of chemical fertilizer in their field, which benefits farmers with more yield. In general, farmers today are not aware of things such as proper compost culture, vermicompost and cow dung and how to utilize them in their cultivation of land. However, the effects of the green revolution encouraged farmers to take up organic farming, and today various government programmes and schemes are also focusing on organic farming. In addition, NGOs are also focusing on organic farmers, and by practicing organic farming, most of the farmers' lifestyles have been changed positively. Their standard of living has improved. At the same time, farmers are facing difficulties with the cultivation practices of organic farming. The present study is an attempt to study the organic farmers with respect to the impact of organic farming on the lifestyle of farmers and to analyse the difficulties they face.

Objectives of the Study

- To analyse the impacts of organic farming on the life of farmers in Udupi district
- To analyse yield level and cultivation practices among the organic farmers

Selection of Respondents

Udupi district was purposively selected for this study for many reasons. This district depends on monsoon rain for agriculture. The district is primarily irrigated by rivers rising from the Western Ghats. The Seetha River affords perennial irrigation to farmers which are normally used for two crops' yield. There are also tube wells and open wells in different parts of the district, augmenting the source of irrigation. In Udupi district, farming is carried out in two cropping seasons; the first takes place from June to September and the second from October to February. In view of the district's environmental potential, the researcher desired to study the impact of organic farming on the livelihood of the farmers.

In this district, there are three blocks, namely, Udupi, Kundapur and Karkala. From each selected block two villages Panchayath were selected randomly, and from each village, 25 members who are practicing organic farming were picked randomly. A total of 150 organic farmers from the district were randomly chosen to constitute the study sample.

Tool for Data Collection

The present study is mainly based on primary data. A well-structured interview schedule covering the study aspects were used to collect primary data from the study respondents. In addition, observations at visits to organic farmers were made in order to get more information about their farming activities and assess the impact of organic farming on the life of the respondent. The necessary secondary data was collected from books, journals, magazines, reports and government records.

Study Design

The study used a descriptive design. The researcher used this design to describe the socio-economic conditions of the respondents, income from farming and the impact on the life of the farmers involved.

Universe of the Study

Organic farmers in Udupi district of Karnataka constitute the universe of the study. A sampling frame was prepared and used for relating the sample respondents accordingly; 150 organic farmers were randomly selected as respondents for the present study in the selected blocks in the Udupi district.

Statistical Analysis of the Data

The data collected were checked, edited, coded, classified and tabulated before they were subjected to further statistical analysis. The data were entered into SPSS-20. Tables were prepared for analysis, simple descriptive statistics were calculated, including percentage, and analysis was carried out.

Results and Discussion

Table 18.2 describes the agricultural land holdings of the respondents and the major crops cultivated on their own land. Nearly all (96.7 percent) of the respondents own agricultural land 58.7 percent of respondents own 3–6 acres of land, 20.7 percent of the respondents own 0–3 acres of land and 17.3 percent of the respondents own 6–9 acres of land.

Table 18.2 Agriculture landholding

Particulars	Number of respondents (N = 150)	Percentage
Own agriculture land		
(a) Yes	145	96.7
(b) No	5	3.3
If Yes, how many acres		
(a) 0–3 acres	31	20.7
(b) 3–6 acres	88	58.7
(c) 6–9 acres	26	17.3

Table 18.2 shows that the majority of the farmers were agricultural landholders in the district, and other farmers have cultivated crops on other agricultural lands.

Figure 18.1 describes the area of cultivation in organic and non-organic land; 94.7 percent of the respondents cultivating organic land at 0–3 acres because most of the farmers started cultivating in small areas and yielded more at further levels, while 5.3 percent of the respondents have cultivated 3–6 acres of organic land, 59.3 percent of the respondents have cultivated 0–3 acres of non-organic land, 36 percent of them have cultivated 3–6 acres of non-organic land and the remaining 4.7 percent of the respondents have cultivated 6–9 acres of non-organic land. The figure shows that most of the farmers have also engaged in non-organic farming. For the first two years, farmers were not able to get good crop yields on their land.

Figure 18.2 explains the respondents' employment position before starting organic farming. Sixty-one percent of respondents were farmers, 21 percent of respondents were unemployed, 11 percent of them were agricultural labourers and 7 percent of them engaged in business. These numbers show that most of the respondents were farmers before starting organic farming. But in this figure, 61 percent of farmers have a background in agriculture; they started cultivating their land in different ways a year ago. Further, it is also found that farmers are more knowledgeable in terms of seasonal crops. Farmers are very well acquainted with the sustainable eco-friendly environment in organic farming.

Table 18.3 shows the respondents' income from organic farming. Eighty percent of the respondents were earning a good income from organic farming because farmers received good yield in their organic

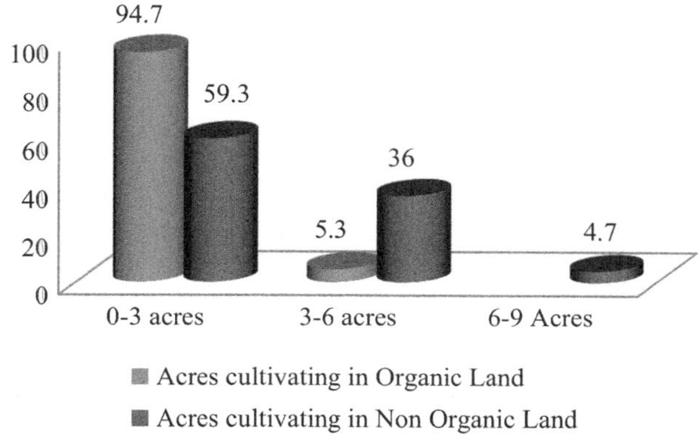

Fig. 18.1 Area of cultivation (in acres)

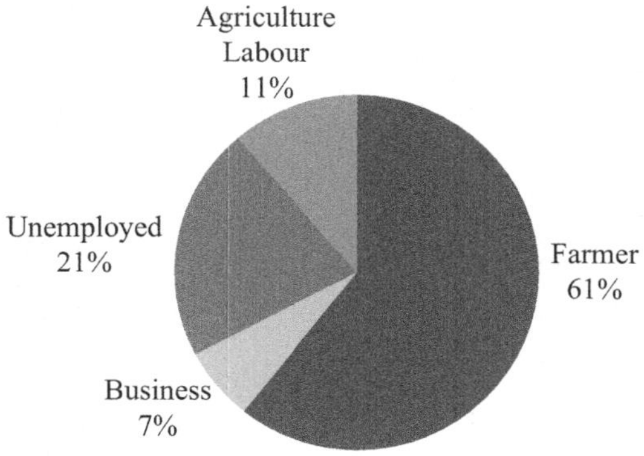

Fig. 18.2 Before starting organic farming

farming and good market rates for their products. Twenty percent of the respondents were not earning a good income from organic farming.

Regarding Annual income from organic farming, 28 percent of respondents earned Rs. 15,000–20,000, 21 percent of respondents earned above Rs. 20,000, 19.3 percent of respondents earned Rs. 10,000–15,000, 8 percent and 3.3 percent of the respondents earned, respectively, Rs. 5000–10,000 and below 5000 from organic farming.

Table 18.4 shows the respondents yield in quintal per acre, prior to organic farming and an increase in the yields after adopting to organic

Table 18.3 Annual Income from organic farming

Particulars	Number of respondents (N = 150)	Percentage
Annual Income from organic farming		
(a) Yes	120	80.0
(b) No	30	20.0
If Yes, how much per acre		
(a) Below 5000	05	3.3
(b) 5000–10,000	12	8.0
(c) 10,000–15,000	29	19.3
(d) 15,000–20,000	42	28.0
(e) Above 20,000	32	21.0

Table 18.4 Crop yield in organic and non-organic

Crops	Quintal per acre	Yield prior to organic farming (N = 150)	Yield in organic farming (N = 150)
Paddy	1–5	22 (14.6)	0
	5–10	54 (36)	26 (17.3)
	10–15	72 (48)	62 (41.3)
	Above 15	2 (1.3)	62 (41.3)
Areca nut	1–3	7 (4.6)	0
	3–6	31 (20.6)	9 (6)
	6–9	10 (6.6)	31 (20.6)
	Above 9	0	8 (5.3)
Coconut	1–3	17 (11.3)	0
	3–6	5 (3.3)	20 (13.3)
	Above 6	0	2 (1.3)
Vegetables	1–3	12 (8)	0
	3–6	0	12 (8)

Figures in brackets are percentage.

farming. Accordingly, a total of 48 percent of the respondent yielded 10–15 quintal paddy before organic farming, whereas the yield was 41.3 percent in organic farming. Further only 1.3 percent of the respondent reported a yield above 15 quintals prior, whereas the number of respondents with increased yield was 41.3 percent in organic farming. As shown in the above figures, the non-organic agriculture system followed by the farmers was resulting in good yield in many instances and this yield was resulting from the use of pesticides and fertilizer.

Areca nut was cultivated more in organic farming: 4.6 percent of the respondents yielded 1–3 quintal per acre areca nut in yield prior in organic farming, 20.6 percent of the respondents yielded 3–6 quintal per acre in yield prior and 6 percent of the respondents yielded areca nut in organic farming, 6.6 percent of the respondents yielded 6–9 quintal areca nut prior in organic, 20.6 percent of the respondents yielded in organic farming and also 5.3 percent of the respondents yielded above 9 quintals per acre in areca nut in organic farming.

Coconut is the main yield in this district; most of the farmers are cultivating coconut for their daily purposes, such as cooking, oil, and so on. In this, 11.3 percent of the respondents yielded 1–3 quintal per acre prior in organic farming, 3.3 percent of the respondents yielded 3–6 quintal per acre prior to organic farming and also 13.3 percent of the respondents yielded in organic farming. While 1.3 percent of the respondents yielded above 6 quintals per acre in organic farming.

Farmers were getting more yields of vegetables by using organic manure and bio-fertilizer and selling their products in the daily market. Eight percent of the respondents yielded 1–3 and 3–6 quintals per acre prior to adopting organic and non-organic farming respectively. They are sustaining their crop productivity through organic farming.

Suggestions

- Compost manure and vermicompost preparation for use in organic farms should be ensured.
- Affordable prices for green manure, bio-pesticide, bio-fertilizer need to be maintained.
- Agricultural universities and research institutions should be encouraged and adequately funded to forward research in organic farming and disseminate the research findings to the farmers.
- Organic agriculture at a grassroots level should be encouraged through comprehensive planning.
- In the study area, separate markets for organic farming products, paddy and other crops should be established.
- NGOs should encourage organic farming as one of their important roles in rural development.

- The government should be withdrawing their subsidies to chemical fertilizer and pesticide companies.
- Organic farmers should be given incentives for motivation.

Conclusion

Agriculture is the main livelihood of rural people. The prosperity and quality of people's life in Indian villages depends on the productivity in agriculture. Indian agriculture has been facing numerous challenges for several decades. However, there has been a huge stride in the farming system in the past few decades, characterized primarily by using machinery and chemical technology which has come to replace traditional agricultural methods, namely, organic farming.

This study concludes that organic farming sustains a better and more balanced environment and provides better food and a much better standard of living to farmers. This study shows that 61 percent of farmers achieved good income within a year of adopting organic farming. This study shows that organic farming in the Indian context is much helpful in terms of marketability and profitability in contrast to no-organic farming. Organic farmers' use of vermicompost, bio-fertilizer and compost manure in farmland has led to the maintenance of soil fertility levels and stayed eco-friendly to sustain a future for organic farming. A consistent and continuous effort should be made by both the government and NGOs located in the study area to motivate farmers to switch to organic farming.

References

Altenbuchner, C., Larcher, M., & Vogel, S. (2014). The Impact of Organic Cotton Cultivation on the Livelihood of Smallholder Farmers in Meatu District, Tanzania. *Renewable Agriculture and Food Systems, 31*(1), 22–36.

Bhattacharyya, P. (2005). Current Status of Organic Farming in India and Other Countries. *Indian Journal of Fertilisers, 1*(9), 111–123.

Chandrashekar, H. M. (2010). Changing Scenario of Organic Farming in India: An Overview. *International NGI Journal, 5*(1), 34–39.

Chhetri, M. (2015). Empowering Farmers Through Organic Farming in the Himalayan State of Sikkim. *American International Journal of Research in Humanities, Arts and Social Sciences, 15*, 2328–3696.

Dahama, A. K. (2003). *Organic Farming for Sustainable Agriculture*. Jodhpur: Agrobios.

Hsieh, S. C. (2008). Organic Farming for Sustainable Agriculture in Asia with Special Reference to Taiwan Experience. *Research Institute of Tropical Agriculture*, 19–26.

Lukas, M. C. (2008, June). Organic Agriculture and Rural Livelihoods in Karnataka, India. *IFOAM Organic World Congress*, 16–20.

Narayanan, S. (2005). *Organic Farming in India: Relevance, Problems, and Constraints*. Mumbai: Nabard, Department of Economic Analysis and Research.

Nemes, N. (2009, June). Comparative Analysis of Organic and Non-organic Farming Systems: A Critical Assessment of Farm Profitability. *Food and Agriculture Organisation of the United Nations*, 18–34.

Reddy, B. (2010). Organic Farming: Status, Issues, and Prospects – A Review. *Agricultural Economics Research Review, 23*, 343–358.

Roy, T. (2014). A Critical Appraisal of Organic Food Market in India. *Global Journal of Enterprise Information System, 6*(2), 96–100.

Shetty, P. K. (2013). Status of Organic Farming in Agro-ecosystems in India. *Indian Journal of Science and Technology, 6*(8), 5084–5088.

Udin, N. (2014). Organic Farming Impact on Sustainable Livelihoods of Marginal Farmers in Shimoga District of Karnataka. *American Journal of Rural Development, 2*(4), 81–88.

19

Managerial Efficacy of Jagratha Samithi cum Family Empowerment Forum: A Case Study of Panancherry Panchayat

Neena Joseph

Introduction

Many social projects may be built upon a strong bedrock of ideology and launched strategically. The project may be addressing a felt need of the community. In spite of these factors, the sustainability may be at stake due to managerial non-viability. Violence against women[1] is rampant in Kerala[2] and the root cause can be traced to gender power asymmetry

[1] Kerala Police Crimes Records Bureau reported 32 dowry deaths in Kerala and 5216 cases of cruelty against women by husband and relatives in 2012. INCLEN PROWID Survey shows that, in Kerala, 46% of rural women face physical violence from husbands (against the national figure of 44%) and 69% of women face physical violence (against the corresponding national-level percentage of 49%). See also Annexure 1 for the situation at Panancherry.

[2] Kerala is the southernmost state of India which is composed of 29 states and 7 union territories. It has an area of 38,863 km^2 (1.2% of India). Kerala is divided into 14 districts and Thrissur is one of them.

N. Joseph (✉)
Institute of Management in Government, Kochi, Kerala, India

© The Author(s) 2020
Md. N. Momen et al. (eds.), *Building Sustainable Communities*,
https://doi.org/10.1007/978-981-15-2393-9_19

generated and perpetuated through patriarchy. Jagratha Samithi cum Family Empowerment Forum (JScFEF) which is a project of the Self Employed Women's Association (SEWA) envisaged to empower families in Panancherry panchayat[3] by democratizing them and thus eliminating violence from the community through legal and psychological counseling center, gender desk in schools, library cum training center and short-stay homes. SEWA conceptualized, formulated and implemented JScFEF through panchayat, ran it for 20 months and then handed it over to the panchayat on February 2006. The evolution of SEWA is traced through five phases. Initially, the counseling activities of the forum were intended to settle family issues and issues of violence against women. But eventually in response to demand, it also started to mediate in all sorts of disputes including boundary disputes and finance-related issues as well. The forum was a great relief to the community because the dispute-settling mechanism resulted in quick justice delivery at the doorstep and saved them from protracted legal battles. The forum stands closed as on date. The objective of the paper is to analyze the managerial efficacy of the project from the genesis to the present situation. The specific research question addressed in this paper is (i) was the management of the processes during the different phases effective? The framework used is the McKinsey 7S model. The methodology employed was focus group discussion with women in the community and also interviews with key informants. Records of SEWA rendered useful basic information. The chapter is divided into five sections. The introduction comprises the background, the research objectives, the research questions and the structure of the paper. Literature review is given in the second section. The third section gives methodology and conceptual framework. Section four discusses the project in five phases starting from genesis to the present state of affairs along with the respective findings. Section five gives conclusions and recommendations.

[3] Panchayats are the local self-governments in India. A three-tier system of panchayats exists in Kerala—gram panchayats in villages and block and district panchayats, respectively, at blocks and districts. There are 978 gram panchayats in Kerala.

Literature Review

Mediation is the process through which the mediator facilitates the parties to invoke the inherent power of the parties for effecting the resolution of the current conflict and thus empowering them to manage future conflicts (Bagham 1995). An ADR system "stands a better chance of moulding consensus because it operates in an everyday manner as well as milieu" (Danzig 1973) The factors contributing to ADR's success are the fairly equal balance of power, voluntariness of both parties to undergo the mediation process, significance of ongoing relationship between both parties and perception about the mediator's impartiality (Bagham 1995). The manipulative potential of "informal" justice (Rifkin 1984) is a concern. When it comes to the issues of women, children or other vulnerable sections, informal jury may be swayed by gender biases and class considerations. A case in point is khaps.[4] (Yadav 2009). Freshly introducing a mediation project into a patriarchal community and that too with a feministic underpinning throws up formidable cultural challenges to the leader along with the issues of resource crunch and volatility of political milieu. NGOs usually have social change mission. Vision, convictions and commitment of the leader are vital. The organization carved out for bringing about the change needs to be managed robustly. Successful performance of non-governmental organizations depends on management capability, strategic leadership, networking and linkages, financial resources, physical resources and human resources, respectively (Nanthagopan 2011). In this context, the McKinsey 7S framework assumes relevance. Strategy implementation is concerned with the building of a capable organization. Strategy will have to be executed well if the firm is to obtain success in its operations (Mutisya 2013). In this context, the style of leadership assumes importance. Participation has become important in the struggle to improve the effectiveness of both the "management of organizations" and the "management of development." Participatory management for NGOs can help to improve organizational

[4] Khaps are clans and they correspond to gotra (lineage segment) within caste groups.

impact on beneficiaries through the greater involvement of staff in decision-making (Sheehan 1998). In spite of the exceptional qualities of many NGO leaders, there has been no systematic approach to the development of successful second-line leadership (Siddiqi 2001). To ensure the sustainability of the project systems, staff and skills also need to be aligned with the vision along with the appropriate leadership style, strategy and structure.

Methodology and Conceptual Framework

This is an exploratory research. Focus group discussion was held with ten women from Kudumbashree.[5] Interviews with key informants and secondary data from the records of panchayat were the source of information. Twenty-four informants were interviewed during September and October 2014. The framework used to analyze the information is the McKinsey 7S framework which hinges around the 7 Ss: shared values, strategy, systems, structure, staff, skill and style (see Diagram 19.1). Strategy is the broad plan. Systems are the daily activities and procedures to get the work done. Shared values are the core values evidenced in the culture and general work ethic. Style is the type of leadership. Staff denotes the employees and their general capabilities. Skill denotes the actual skills and competencies. The 7S model can be used to analyze a project or parts of a project for its managerial effectiveness. McKinsey's framework is furnished below. The diagram encapsulates the interdependence between the elements with shared values at the center, emphasizing that this is central to the development of all other elements which in turn stems from why the organization was originally created. When values change, all other elements change. The framework depicts that the seven elements are to be aligned in a mutually reinforcing fashion and in consistency with one another (Waterman et al. 1980).

---

[5] Kudumbashree is the network of neighborhood groups of women in Kerala which is managed by Kudumbashree Mission of Kerala Government with the vision to eradicate poverty in all its dimensions under the leadership of local bodies by empowering families through the empowerment of women. Now in Kerala, 4,100,000 women members are federated into 1072 CDS which is the federation of over 187,000 neighborhood groups at gram panchayat level. In PGP, about 7500 women are organized into about 600 neighborhood groups.

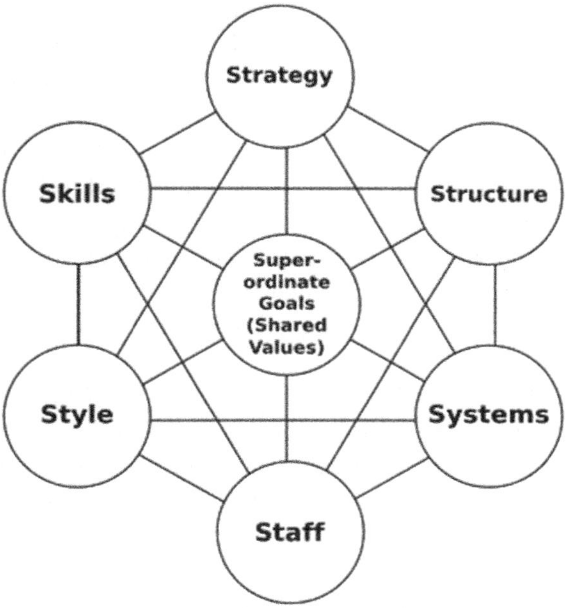

Diagram 19.1 Source of the framework: journal article "Structure Is Not Organization," Business Horizons, 1980

Managerial Effectiveness of the Three Phases of the Project: Discussion and Findings

Phase I Genesis

With the 73rd and 74th Amendments of the Indian Constitution in 1994, local bodies[6] became self-governments with the devolution of functions, functionaries and funds. During 2004, Programme on Capacity Development for Decentralization in Kerala invited selected NGOs including SEWA who work at grassroots level for a discussion

[6] Governance structures set up at local level which are governed by elected representatives. These local bodies in rural areas are known as panchayats. The panchayat system has three tiers: the grassroots gram panchayats, block-level panchayats and district-level panchayats. Gram panchayats are further divided into wards.

meeting on the possible projects for empowerment of local self-governments. As suggested by Nalini Nayak, Gracy represented SEWA Kerala in the meeting. During the meeting, it was learned that projects could be submitted under the broad categories of environment, energy, sustainable development, water harvesting, gender and so on. The partnering NGO could conceive and implement the project and CapDecK will have only an overall control. Thanks to Gracy's relevant background and experience in gender issues,[7] she conceived and formulated the project JScFEF for Panancherry gram panchayat which was fine-tuned and forwarded to CapDecK by Nalini Nayak. The project was approved by CapDecK in 2004.

The project had high relevance because, right from 1997, Government Order (GOs) had been issued by the government regarding the constitution and functioning of Jagratha Samithis at district- and local-level panchayats.[8] The secretary of the Kerala Women's Commission[9] had also revived the formation of JScFEF during January 2004. But Jagratha Samithis had not taken off. Arguably, lack of political will, perceived disutility of such projects for electoral politics, lack of clarity regarding its social need and significance and paucity of funds were the reasons. In the GO of 2004, the scope for utilizing foreign funds is open.

JScFEF of Panancherry envisioned by Gracy had a much more comprehensive scope than the Samithis as per the government order of 2004.[10] The then GO envisaged it only as a forum to settle women-specific cases at panchayat level. But the Panancherry project had the additional component of family empowerment which targeted the empowerment of families by transforming them into egalitarian spaces based on mutual respect. JScFEF ultimately targeted to have a panchayat sans violence so that the entire attention could be focused on human development by intensifying the activities of gender desks, library cum training center and so on. Hence, in addition to case hearing, the project

[7] Refer to Annexure Background of Gracy.

[8] See Annexure Reference of Government Orders.

[9] Kerala Women's Commission is a statutory body which came into force in 1996 to imiprove the status of women in Kerala and to inquire into unfair practices and the pertinent matters affecting women.

[10] The Jagratha Samithi cum Family Empowerment Rules Approved by Panchayat Committee, 2008

included library cum training center, legal and non-legal training center, special counseling team, gender desk at schools, short-stay home, pre-marital training for youth and so on. Jagratha Samithi conceived by the government had only one woman ward member. But in the Panancherry model, all women members of the panchayat are incorporated in the JScFEF. Also there are three women social workers, out of which one person shall be an SC/ST. Nominations will be made by the members which shall later be approved by the panchayat committee and then by the district Jagratha Samithi. Three social workers are included because proper persons with the appropriate skill set are required for handling the multifaceted responsibilities. All women members are included with the objective of giving a chance for all women elected members to get involved in the wide gamut of activities irrespective of political affiliations. Further, JScFEF had vigilance cells at the ward level constituted by carefully selected personalities with general public acceptance and who are reputed for their integrity and impartiality. The structure and functioning were governed by a bye law which was hurriedly handwritten by Gracy prior to the inauguration in the same week. Bye law was taken for discussion individually with Ms. Sakunthala Unnikrishnan who had recently become the panchayat president. It was also discussed with many ward members prior being presented before the panchayat committee. This bylaw was unanimously approved by the panchayat. The JScFEF in Panancherry is functioning according to this bye law.

Discussion

It can be seen that when SEWA Kerala received the invitation to attend the meeting, there was not much clarity regarding what the meeting was about. In fact the meeting was attended in an exploratory attitude. This is the entrepreneurial spirit contrary to the bureaucratic style marked by intolerance to ambiguity regarding each step taken. It was at the meeting convened by CapDecK that Gracy learned that opportunity was for proposing any project to empower local bodies and also for managing the project independently with the least interference from CapDecK. If the meeting was missed, JScFEF would not have happened. SEWA was able to capture the opportunity of the GOs regarding the constitution of Jagratha Samithi and the funds availability with CapDecK. Again, the Panancherry model acknowledges the fact that unless there is gender parity, the issue of violence cannot be tackled. Further there was understanding that unless there is people's participation and ownership, the activities of the forum cannot be sustained. Hence there are enlarged and enhanced structure and functions reaching out to grassroots, vast array of activities and the concomitant structure. A Jagratha Samithi constituted merely to comply with the GO cannot be expected to have this vision-based ideological underpinning which is the soul of any Jagratha Samithi. All activities of JScFEF were tied up with the vision. Gracy is an NGO member with the necessary academic, training and activist background. This points out to the necessity of identifying relevant NGOs who can fruitfully partner with such social projects of the government. It is concluded that the structure dovetailed with the overall vision.

Phase II from the Approval by Panchayat Committee Till Inauguration

Gracy started the activities of JScFEF in July 2004. On August 12, 2004, a house was taken on rent at Panancherry from where Gracy operated to build up JScFEF. An amount of Rs 7 lakhs[11] was received from CapDecK. In spite of having funds, extravagance was avoided in food and choice of venue for training programs, because the much needed voluntarism might get diluted in the place of luxury. Gracy was the project director of JScFEF. A coordination committee was constituted to assist the project director to implement the vision of the project. Mr. Rajaji, who was active in party politics at national level, and Sr. Lissy, who was a lawyer by qualification and who was also qualified as counselor, were handpicked by Gracy to be the members of the coordination committee to assist Gracy in advisory capacity. CapDecK was also helping. Rajaji was expected to liaise with the politicians and Lissy was planned to be retained as legal expert after SEWA quits. The coordination committee

[11] Rs 7lakhs =Rs 700,000=$11,667@$1=Rs 60

members were given an honorarium of Rs 3000 per month. Adv. RK Asha was appointed as the full-timer who was supposed to supervise the field staff and the office staff and her remuneration was Rs 5000 per month. Five persons had worked as field staff—Libish, Ramesh, Prince, Lilly Francis and Jijo. Adv. Asha could not be retained for long. Sr. Lissy was given temporary charge as full-timer. Lilly Francis who impressed Gracy as a vibrant participant in one of the training programmes conducted by JScFEF. Later she was replaced by Beena who had experience in decentralized planning project of Kalliasseri[12] fame. The structure is given in Annexure 5. Lilly Francis continued as field staff. Under the full-timer, there was one office staff. The last office staff was Saramma. Finally, when Jagratha Samithi was handed over to the panchayat on February 2006, only Sr. Lissy and Jijo were retained.

The duty of the office staff consisted of going to the ward, meeting people, visiting houses and briefing them about JScFEF, wooing people to participate in classes, reporting to project director, assisting in the implementation of training programs, identifying respected and well-accepted persons who could be members of ward-level vigilance committees, making field enquiries, visiting sites and so on. The office staff would look after the office matters including correspondence, receiving complaints and issuing receipts for the same, maintaining records and so on. The project director used to hold meetings with the field staff, office staff and full-timer. The full-timer reports to the project director. The full-timer's job was to supervise the work of field staff and office staff so that they implement the decisions of the project director. The working time was between 9 AM and 5 PM. The full-timer used to be invited to the meetings of the coordination committee.

At field level, ward vigilance cell was formed to be the ears and eyes of JScFEF. There were gender desks both at panchayat level and in the two high schools in the panchayat. Gender desks had representatives of panchayat members, parents and teachers. Classes for teachers, students and parents and counseling were the main activities of the gender desks.

Within 7 months before the inauguration, attempts for consciousness raising were done as follows:

---

[12] Kalliasseri in Kannur, North Kerala, was the cradle of people's planning campaign.

Trainings

1. Five 2-day gender sensitization workshops were conducted for women. During the first stage, a preliminary list of participants was prepared through Kudumbashree members who were instructed to select impartial persons with leadership skills from religious and non-religious organizations, temples, clubs, youth clubs, political parties and so on. This list was discussed with the SEWA team. Preliminary enquiries were made about the persons through vigilance cells. The field staff visited each person at their home and invited them directly for the program with the invitation letter signed by the panchayat president. During the first day, discussion-oriented gender sensitization classes were held which attempted to explore some popular gender-related myths followed by classes on structural analysis of society and women's subordination. In the afternoon, the participants had to fill up a questionnaire designed to assess their status and also their level of safety and security. The highlights of the survey are given in Annexure 1.
2. Two of such 2-day workshops were conducted for men and male youth. In spite of boycotts and protests, Gracy continued the classes and those who boycotted returned with curiosity to listen to the contents. They actively participated in the meeting.[13]
3. Conducted workshops in all wards and also for young women through adolescent clubs of ICDS.
4. Conducted centralized training for selected young girls.
5. Ward-wise selection was made from the above trainees on the basis of their initiative, interest in women's issues, leadership quality, impartiality and so on. These persons were given joint training.
6. Prior to inauguration, corner meetings were conducted during evening time in important junctions with mike announcements. The highlights of the survey were used for establishing the need of JScFEF.

[13] A 80-year-old farmer advised a young man who boycotted the class to first listen to the resource person, understand the points and then respond. The elder generation were more receptive while youngsters were defensive.

7. Bit notices were distributed among the citizens highlighting the findings of the survey.
8. The second phase of trainings for selected persons included topics like legal awareness, health education, sex and sexuality and so on.

Before formal inauguration, Gracy began to hear cases at SEWA and had indulged in activism in her capacity as a member of Streevedi. By the time of inauguration, there was awareness among citizens about JScFEF and its scope. Nearly 5000 persons underwent this education process. Library was formed. Books were purchased. Gender desks were constituted at panchayat and schools. Classes were commenced. At the end of the classes, children could stay back for clearing their doubts. Due to paucity of time, the other envisaged activities like setting up a counseling team and training them and constituting a training centre to impart training on legal and non legal matters could not be materialized.

Discussion

It can be seen that there were clear vision-linked objectives for the endeavor. There were many meetings within SEWA during this phase. These meetings were excellent fora to contemplate the vision and to clearly educate the team regarding how the planned-out activities were connected to the vision. This phase lasted for only about 7 months within which the constitution of structure, selection of staff, conduct of classes and inauguration happened.

There was mass mobilization. Through the classes the mass was educated. About 5000 persons underwent these trainings. The penetration was high.[14] The attempts to share the vision are praiseworthy.

Since vision was pivoted on a gender-just community, the structures created though elaborate are very compatible with the vision. But the planning for structures was a bit too ambitious and hence all structures could not be constituted including short-stay home. Nevertheless, the envisaged structures remaining in the records could be considered in the

[14] If 5000 persons pass on the ideas of the classes to about 4 persons each, the coverage is 200,000 vs. the current population of 42,418. The population at that time might have been even less.

future. One effective step taken was instituting ward vigilance cell and ward committee at the ward level. The former was for doing field investigation of cases, assisting the field staff, organizing the classes and so on. Most of them were selected very carefully based on identified talents and public opinion. It is notable that right from the inception, strategic moves were planned. The ownership of panchayat and the community was ensured by involving both in each and every step of implementation. The selection of trainees was done by Kudumbashree, but was cross-checked by field staff and finalized in SEWA meetings. The invitation letters though drafted at SEWA was signed by the president and sent. The questionnaire was a major strategic tool to make the participants actively aware of their condition and position during the very process of getting them filled up. It was an excellent tool to gain an entry into the community. The major findings were used as tools to establish the need and to gain acceptance of JScFEF among the panchayat members and the general community. The statistics revealed through the survey was quoted during the meetings. Strategic moves were at their best in the context of inaugural function. The **system** of meetings and the liaison with panchayat through meetings was a very beneficial system for accomplishing the objectives. The system of placing the panchayat in the lead position yielded good results by way of participation. The system of staff selection was excellent considering the fact that time available was very less and much had to be accomplished within the time constraints. But correct persons fit for the jobs were selected by Gracy herself. The founder could exercise her autonomy to decide which of the persons would be able to deliver results. The system of field staff who make enquiries and shortlist eligible persons at ward level for training and future association is very effective. The decision to involve all women panchayat members as the members of Jagratha Samithi is welcome, because this ensures their involvement in spite of party affiliations. Till the project ended, there were funds to pay salary to the staff and honorarium to the members of the coordination team, TA to ward-level committee members and so on. When the project ended, these funds were not available. But the frequent change of staff due to administrative reasons was not conducive to the morale of the staff and might have to some extent contributed to the disconnect between the founder and JScFEF in the subsequent phases.

Phase III Inauguration

Inauguration was planned on March 8, 2005. The above activities facilitated the inaugural function. Mass mobilization became easily possible through the involvement of trainees, panchayat members and so on. Deliberate lobbying was done by SEWA to make the inauguration of JS of the panchayat as the Women's Day celebration program of Kerala Women's Commission for the year 2005 especially through Kulsu, one member of the commission. The SEWA team leader succeeded in convincing the chairperson that through their presence in the inauguration, the significance of Jagratha Samithis will be boosted across the state. Prior to inaugural function and notice distribution, mobile loud speaker announcements in each nook and corner of panchayats were done. This helped mass mobilization. It was conducted as a rally with loudspeakers, placard, banner and band set. It was a rally cum indoor convention. Notice for the meeting was printed by government press, and since the Women's Commission was present for the inauguration, their directions had to be taken regarding the function. About 3000 persons participated. In the public meeting, all the Jagratha Samithi members took an oath in front of the congregated mass. At the time of inauguration, Mr. MK Sivaraman was the panchayat president. The vigilance cell members also assumed responsibility after taking an oath at gram sabhas.

Discussion

The presence of all the Women's Commission members on Women's Day of March 8 at the inaugural function of JScFEF was of great strategic significance. The fact that it was inaugurated by the chairperson of Kerala Women's Commission Ms. Kamalam adds to this significance. The entire team of the Women's Commission was present with the exception of one. The publicity gave a boost to the concept of Jagratha Samithi in general and to the image of JScFEF in particular. Again, the oath taken by the Jagratha Samithi members in front of the congregated mass made them accountable to the public for sustaining JScFEF. By the time the formal inauguration was done, there were a mass of trainees in the panchayat who were already exposed to the concept and functions of JScFEF through trainings and a few who had benefited by way of counseling and mediated settlement done at SEWA. JScFEF acquired an elevated status and legitimacy consequent to the function. The seriousness given to JScFEF motivated the police to be readily available for the services of JScFEF. The presence of the Women's Commission was a deliberately adopted strategy to project JScFEF at state level so as to enhance its chances of replicability. Intense interaction with the public, involvement and solutions offered for their problems through the platform of JScFEF resulted in the electoral victories of the persons involved—one person won election as MLA (Member of Legislative Assembly) and six persons as members of panchayat.

Phase IV After Inauguration Till the Handing Over to Panchayat in July 2006

Originally, the project duration was only 1 year. After completion of 1 year, SEWA got an extension for further 8 months. Now SEWA began to refer cases to JScFEF and also to police and members of panchayat. Gracy wanted to retain Sr. Lissy as a legal expert.

After the inauguration, JScFEF shifted to a building within the premises of panchayat office and with the permission of the panchayat did some maintenance work utilizing the project fund. A phone in the name of SEWA was shifted to the name of JScFEF. Till date panchayat is meeting the telephone expenses.

Mediated settlement was the modus operandi. When a complaint is received, it will be entered into the register, and a receipt will be issued to the party. JScFEF field staff would go to the field and with the help of vigilance cell members make enquiries and visit the site as the particular case warrants. Report is submitted to JScFEF. Then a date is fixed for the sitting and the parties are intimated. In the instances where the respondent refuses to appear, the help of the police is sought and JScFEF had always got the help from the police. During the hearing, Gracy, the advocate and some JScFEF members would be present. In the instances of cases in a ward, sometimes the respective ward member insists to be present. This was not allowed. The parties would be heard separately and then jointly. Then they will be facilitated to reach an amicable settlement. The emphasis is not on law points, but the humanitarian considerations and the reality of undue delay in justice delivery if one resorts to litigation. Everyone wants to have some settlement of cases instead of chasing elusive justice through protracted legal battle. Even cases not settled at courts in spite of more than a decade of legal battle had been settled here.

Once the agreement is reached, both parties sign on the settlement. Follow-up of the case is also done through the field staff with the help of vigilance cell members. The settlements happened in cordial atmosphere.

Discussion

Sr. Lissy was groomed for assuming the position of legal expert. Gracy liaised with panchayat to appoint Sr. Lissy as the legal expert so as to ensure continuity of vision and activities. Succession planning was done consciously and deliberately.

For the settlement to become an executable decree, three conditions are to be satisfied—both parties should have signed a document expressing their willingness for entering into a mediated settlement, KELSA (Kerala Legal Services Authority) has to be involved in the settlement and both parties should sign the settlement reached. In the absence of these conditions, the settlement does not become an executable decree and hence can be challenged in a court.[15] The question is whether the settlements reached in such a forum should have the status of an executable decree. If yes, serious thinking has to be done regarding the selection of the lawyer and regarding how the network of TLSC (Taluk Legal Services Committee) of KELSA can be roped in. According to the bye law of JScFEF, the lawyer is to be selected by the panchayat committee and approved by the district panchayat committee. Again the question is: Is it enough to have one lawyer alone? Can we not think of a jury system wherein government officials including the retired hands can be included as in countries such as Belgium? Then we also will have to think about a

[15] Adv. Sandhya Raju, director, Human Rights Law Network, Ernakulam

minimum qualification and the necessary attitude and values of the selected persons and also about a system of approval by a competent panel of say retired judges, eminent social workers and psychologists, who can make their choice through an interview. The preliminary panel may be passed by the panchayat samiti. As long as such foolproof procedures are absent, is it not safe (from the point of justice delivery) if the settlements remain as non-executable settlement? Anyhow the answers to these questions are beyond the scope of this chapter, but they need to be mentioned because assessment of systems from the managerial angle alone will be incomplete. The legal viability of the system has inadequate scope for independent research.[16] Jagratha Samithi has a great role to play as a vigilante set-up so that it can alert the appropriate authority. Jagratha Samithis can link up with TLSC (Taluk Legal Services Committee) wherein civil cases and compoundable criminal cases can be tried in Lokayukta so that parties will get executable decrees. Non-compoundable criminal cases should not be taken up by JScFEF.

In retrospect, considering the success of the settlements and considering the fact that it was possible to reach harmonious solutions, one is tempted to opt for informal settlement. Panancherry panchayat is a hilly area remotely situated with 11% SC/ST population and everything worked fine.[17] But there are caveats when the model is attempted to be replicated in urban areas where the education and legal awareness status of people can be expected to be higher.

The system of field investigation coupled with counseling and mediated settlement had provided solace to many persons.

JScFEF functionaries need to understand the other systems[18] existing in the locality, adopt case specific, customised services and network with

---

[16] Mr. U Saratchandran, member, CAT, and former Member Secretary, National Legal Services Authority

[17] See Annexure for the profile of the panchayat.

[18] Protection officer: District-level officer for protecting women from domestic violence. Protection officer is networked with NGOs, police, magistrate, shelter homes and so on.

Child Welfare Committee: District-level committee comprising of selected jury who oversees the welfare of children

Child protection officer: District-level officer in charge of the protection of children

RDO government officer: One who is authorized to take action against dowry offenses and offenses against senior citizens them eclectically and leverage their strengths while maintaining one's independence.

The autonomy of JScFEF was visibly established through a separate building and phone connection.

Phase V After Being Handed Over to Panchayat Till Date

When JScFEF was handed over to panchayat in February 2006, Sr. Lissy and Jijo were the only persons left. Now there was no more salary from the project. Sr. Lissy with her background of law and counseling was working tirelessly, but the absence of remuneration did not deter her. It is also same with Jijo who continued the work with missionary zeal. In fact she had a very supportive husband who had even purchased a two-wheeler for her to facilitate her access to the nook and corner of the panchayat in connection with enquiry and field work. Lilly Francis was the panchayat president from October 2, 2005 to September 30, 2010. The committee was stable. Sr. Lissy, Jijo and the panchayat worked in tandem. JScFEF was vibrant. Mediated settlements were done twice a week. Through gender desks, the schoolchildren, parents and teachers were educated. During the period 371 cases were handled, out of which 123 were of civil nature and 117 of criminal nature. Family cases constituted 126. One case was taken up suo motu. 269 (73%) cases were settled in JScFEF.[19] Seven were transferred to courts and 63 were transferred to police and 1 to family court. Gender desk interventions resulted in the handling of problems of children. JScFEF used to be run with the help of

Bhoomika: An initiative of the Health Department where in each district there is a one-stop crisis intervention center for gender-based violence with doctors, psychological counselors, lawyers and so on

Seethalaya: District-wise arrangement of the Homeopathy Department where the services of doctors, psychologists and psychiatrists are available

Snehitha: Is the short-stay home of Kudumbashree which has counseling facilities and linkages with

district police cell

[19] "Jagratha Samithi Experiences: Panancherry Panchayat" published by Kerala Vanitha Commission during 2008

planned funds garnered through the regular procedures of project formulation, submission and approval. During the tenure of Lilly Francis, functions of the forum could be conducted smoothly with this fund.

Sr. Lissy managed to continue till the end of the tenure of Smt. Lilly Francis. But when the next Samithi took over in November 1, 2010, she prepared to quit. In fact her exodus was abrupt. Jijo was left alone. It was by the time JScFEF was acknowledged to be a necessary institution and Jijo had found it a meaningful occupation. She was desperately searching for a legal expert and finally identified Adv. Willy and entreated her to join JScFEF. Her appointment was regularized and a payment of Rs 10,000 per year was fixed as honorarium. Sr. Lissy was present on all days, but Adv. Willy used to render her services for a few days per week.

Eventually, the functions of JScFEF became progressively narrowed down to case hearing and settlement. The subsequent panchayat samiti did not endeavor to procure government funds for the running of JScFEF. During the time of the project, TA could be disbursed to the members of the vigilance cells at ward level for coming to meetings. But post-project, funds were not available for this purpose. So there was reluctance for vigilance cell persons to participate in the activities. Another reason quoted is the unwillingness of the people to accept a vigilance cell person as a mediator due to too much familiarity with the vigilance cell member being a next-door person.[20] The ward-level vigilance cells weakened and became almost dysfunctional.

Jijo doubled up as field staff and a member in the case hearing. Popularity of Jagratha Samithi was escalating and so also that of Jijo.

Jijo was functioning without salary. The panchayat used to allot some clerical works of the panchayat to Jijo and in return made some payments. In the meanwhile Kerala Institue of Local Administration (KILA)[21] had the program of Suchitwa Mission where in each gram sabha classes have to be conducted. The classes have to be coordinated in all the 23 wards.[22] In the absence of other persons to take up the coordination, the panchayat presi-

[20] The average population per ward is 1844 in Panancherry panchayat: Annual Plan 2013–2014.

[21] KILA is an autonomous institution in the state constituted in 1990 to empower the local self-governments through research, training, consultancy and publications.

[22] Currently, the number of wards is 23.

dent entrusted the coordination work to Jijo. Her work consisted of disbursing money to the ward members who in turn conducted the function and handed over the vouchers and the accounts back to Jijo. In the meanwhile, one member in the committee questioned Jijo as to why she did not consult the member before taking up the task to which she replied "How can I refuse a work entrusted to me by the panchayat president?" A former member and this particular member went to KILA after collecting the documents related to the conduct of the classes in gram sabhas showing that most of the attendees were fictitious as revealed from the obviously (because many signatures had the same handwriting) forged signatures. In panchayats it is a prevalent practice to "create" fictitious attendees to compensate for the loss to the organizers for the expenses of the extra refreshments ordered for the persons over above those who had actually turned up. The contention of the particular member and the ex-member is that the number of fictitious persons are more than reasonable and a person with due diligence will not plan for an unduly large number of participants. The ex-member has evening paper and he flashed the news that JScFEF operations are corrupt. This incident happened in July 2014. A committee was constituted to investigate the issue and the report of the committee is not yet tabled at the panchayat samiti. In the newspaper, the name of Jijo was not mentioned, but it was very obvious against whom it was targeted. With this incident, Jijo tendered her resignation to the panchayat secretary. Presently, the Jagratha Samithi office stands locked up.

Discussion

When SEWA left handing over JScFEF to panchayat, it seemed that the structures and processes were handed over, but not fully the vision of the all-encompassing preventive, promotive and ameliorative aspects of family empowerment through gender desks, library, premarital courses and so on with the assistance of ward vigilance cells. To some extent those missing can be attributed to transmission loss of vision to the successor and to the paucity of funds resulting from the lack of efforts to raise funds for functioning by preparing budgets and procuring due sanction. A disconnect could be sensed with the conspicuous absence of Gracy from certain occasions like showcasing the model of Panancherry to other panchayats in important seminars. But since Sr. Lissy and Jijo were present from the inception, the continuity aspect was not that dismal. But subsequently the structure of settlement forum alone remained whereas no other structures did survive. The situation was exacerbated with the abrupt exodus of Sr. Lissy. There was no scope left for identifying a suitable person and completing the due processes. Jijo did the enquiry and also sat for hearing along with the lawyer and police. Technically, the person who makes enquiries and the one who sits for the hearing need to be different (to technically ensure impartiality). The structure and system according to the founder's vision need funds. Not providing sufficient funds for functioning has weakened JScFEF and thwarted the sustainability. Procedures and systems have to be in place for employing staff and the rules need to be amended so that panchayat members need not make adjustments for actual expenses incurred by them. Panchayat did not have provision to pay Jijo. But panchayat managed to remunerate her through entrusting certain works to her. For the training program under discussion also, Jijo was paid a coordination fee. In that particular case also adjustments might have been made according to the usual practice. Adjustments made even in good faith can create problems. Hence what need to be put in place are systems which are compatible to the smooth functioning related to vision fulfillment. Instead of running JScFEF with an office saff who is occasionally paid on a sort of piecework basis, panchayat could employ one of their staff or give additional duty. Coming to the structure, there are possible linkages which can be forged with the agencies as mentioned earlier in footnote 17. Linking with the most appropriate agency is crucial to solve the problems of the client. But the programs of JScFEF started to narrow down to case hearing and settlement alone. During interviews and the FGD, each and every one believed and desired that JScFEF needs to be revived. During FGDs, there were no adverse remarks about the impartiality and confidentiality of the JScFEF proceedings. Lack of staff with predictable funds to run JScFEF was a major flaw. Price had to be paid for expediency. In spite of having no such systems, JScFEF was alive till 2010. There was continuity of tenure of panchayat president between 2005 and 2010. From 2010 till date the panchayat already saw three presidents and the current president has tendered his resignation informally. Only a stable government can focus on social projects like this which has little value in electoral politics. Political will is the primary requirement for sustaining such projects.

Conclusion and Recommendation

When the upscaling and replication of such social change projects are contemplated, the vision and passion of the founder is of primary importance. When a cultural change is contemplated, one has to primarily work with the mindset changes through trainings and campaigns. Structure, systems, strategy, staff, skill and style have to be consistent with and built upon the central vision. The values which drive the vision need to be shared horizontally among the team and vertically down to the successor. Strategies need to be employed to realize the vision. Strategic moves are required at all stages. Sufficient staff required to realize the vision need to be employed. Lack of political will and political stability can destroy the project if the community does not take the ownership. Grassroots ramification of the project is required to be laid out as in the case of ward vigilance cells. In this particular project, forging linkages with agencies working on similar lines is required to provide the most appropriate services to the community members. For this particular project, replication among a more urbanized society can raise new challenges like questions of legal validity of such informal mediated settlements. When informal arrangements are thought of, there has to be awareness on the legal validity and the limits (in this case study, avoiding to take up non-compoundable cases) while simultaneously leveraging the benefits of informality.

Annexure 1: Highlights of the Survey

Questionnaires were distributed to 378 women in the panchayats. 87% were married. 4% were uneducated and 49% were above 10th standard. 76% of women said that there is no gender equality. 86% responded that women are subjected to physical and mental violence within homes. 79% women said that, in their knowledge, there were suicides and attempts of suicide by women. Out of these, in 43% of cases, there was violence by husbands. 91% of respondents told that domestic violence is kept as secret by women. 93% of women responded that Jagratha Samithi at panchayat level was required to address the problems of women.

Annexure 2: Background of Panancherry Panchayat[23]

The panchayat is situated in Thrissur district. The area is 141.7 km^2 and it has 23 wards.[24] There are 12,412 houses. The population is 42,418 with a sex ratio of 1028. According to the 2011 census, the sex ratio is 1058 for Kerala. Tribal and scheduled caste then constituted 11% of the population. There are 15 schools in the panchayat. 70% of the population still depend directly or indirectly on agriculture.

Annexure 3: Background of MM Gracy and SEWA Thrissur

SEWA Thrissur is a non-governmental organization (NGO). Gracy belongs to Panancherry gram panchayat of Thrissur district, Kerala, India. She is a post-graduate of economics and a graduate of law. She was a member of PCO (Programme for Community Organisation), an NGO which is actively involved in fisher people's movement: research, training and activism. PCO was working for the upliftment of fisher people and gender equality was one of its major agendas. Her career and learning trajectory are furnished below.

1. Undergone 31/32-month-long course on social analysis and legal aid at ISI (Indian Social Institute at Bangalore) and also half-month-long research methodology program
2. Program coordinator of PCO center, Cochin

[23] Panancherry gram panchayat, 12th Five Year Plan, Annual Plan 2013–2014

[24] Each local self-government at grassroots level is divided into wards.

3. Participant of 1-month-long South Asian Workshop on Women and Development organized by FAO
4. Half-month-long study program at Norway organized by Norwegian Society for the Conservation of Nature and CDS jointly
5. Participated in the study on sexual harassment at workplace for Sakhi, Delhi
6. Women's program coordinator of PLDP (Participatory Local Development Programme) by IRTC (Integrated Rural Technology Centre), Mundoor
7. Post-graduate diploma in Universalising Socioeconomic Security for the Poor from the International Institute of Social Studies, Hague, Holland
8. Member, IAWS (Indian Association for Women's Studies)
9. One of the founding members of Kerala Streevedi
10. Member, NAPM (National Alliance of People's Movements)
11. Presently secretary, SEWA Thrissur

Annexure 4: Government Orders (GOs) Regarding Jagratha Samithis

GO (Ms) No 39/2004/SWD dated May 28, 2004

This GO quotes previous five GOs related to Jagratha Samithis issued during 1997 and 1999. The above GO also quotes a letter from the secretary of the Women's Commission in which paucity of funds was mentioned to be the reason for the Samithis not to function properly. The GO upheld the need for the qualitative strengthening of the status of women in society under the program for capacity building for decentralization. The program has the scope for initiating and strengthening Jagratha Samithis under the leadership of panchayat. The program will render financial support through selected NGOs preferably those having FCRA registration. For gram panchayat, the panchayat president will be the chairman, and one person under ICDS (Integrated Child Development Services Scheme) selected by the district social justice officer will be the convener. Other members are circle inspector/sub-inspector, one woman panchayat member, one lady lawyer and one lady social worker or councilor belonging to SC/ST community. The latter two are nominated by the chairman of the district samiti in consultation with the Kerala Women's Commission. The panchayat president can nominate the other members.

Annexure 5: Structure Before Handing Over to Panchayat

Project Director cum Co ordinator (Gracy)

↓

Full Timer (Adv Asha, Sr Lissy for a short term, Lilly Francis for a short while)

↓

Field Saff (Jijo for some time , Ramesh, Libish, Prince Lilly Francis for a short while) Office Staff (Greeshma, Suma,Saramma, Jijo for some time)

References

Annual Plan (2013–2014), Panancherry Panchayat.

Bagham. (1995). Mediating Family Disputes in Statutory Settings. *Australian Social Work,* Vol. 48, no. 4.

Danzig. (1973). Towards the Creation of Complementary, Decentralized System of Justice.26 *STAN.L. Rev*. 1.

David, B. (2005). *Succession Planning and Management: A Guide to Organizational Systems and Practices*. North Carolina: Centre for Creative Leadership.

GO (Ms) No 39/2004/SWD dated 28 May 2004.

GO(Ms)8/97.

Gracy, M. Towards Creating a Women Vigilance Committee and Generation of Awareness in Panancherry Grama Panchayat. Thrissur.

Hailey, J., & James, R. (2004). "Trees Die from the Top": International Perspectives on NGO Leadership and Development. *Voluntas: International Journal of Voluntary and Nonprofit Organization, 15*(4), 343–353.

Mutisya, P. M. (2013). Strategy Implementation by Milk Processors in Kenya. Accessed on 23 July 2020, from http://erepository.uonbi.ac.ke/handle/11295/59264.

Nanthagopan, Y. (2011).*Critical Dimensions of Organizational Capacity: An Empirical Study of NGOs in Vavuniya District of Sri Lanka.* International Conference on Business Management September 8, 2011.

Pascale, R., & Athos, A. (1981). *The Art of Japanese Management.* London: Penguin Books.

Peters, T., & Waterman, R. (1982). *In Search of Excellence.* New York; London: Harper & Row.

Retrieved November 10, 2014, from http://en.wikipedia.org/wiki/Kudumbashree.

Retrieved November 10, 2014, from http://www.kudumbashree.org/?q=homepage.

Retrieved November 10, 2014, from http://en.wikipedia.org/wiki/Local_governance_in_Kerala.

Retrieved November 10, 2014, from https://www.google.co.in/?gfe_rd=cr&ei=1Em-U_ntBenV8ge98YF4&gws_rd=ssl#q=area+of+india.

Rifkin, J. (1984). Mediation from a Feminist Perspective: Promise and Problems. *Law & Inequality*, 21.

Sheehan, J.(1998). *NGOs and Participatory Management Styles: A Case Study of CONCERN Worldwide, Mozambique.* International Working Paper Series, 2. Centre for Civil Society, London School of Economics and Political Science, London, UK. ISBN:075301257X.

Siddiqi. (2001).*Who Will Bear the Torch Tomorrow? Charismatic Leadership and Second-Line Leaders in Development NGOs.* International Working Paper Series, 9. Centre for Civil Society, London School of Economics and Political Science, London, UK. ISBN:0753019000.

The Jagratha Samithi cum Family Empowerment Rules Approved by Panchayat Committee, 2008.

Waterman, R., Jr., Peters, T., & Phillips, J. R. (June 1980). Structure Is Not Organisation. *Business Horizons, 23*(3), 14–26.

Yadav, B. (2009). Khap Panchayat Freedom*? Economic and Political Weekly*, Vol. 44, No. 52.

Personal Interviews

Adv. Reena John, Legal Expert, JScFEF.
Adv. Sandhya, Director, Human Rights Law Network.
Adv. U Saratchandran, Member, CAT, Ernakulam, and Former Member Secretary, NALSA.
Adv. Willy James, Counsellor, Family Court, Thrissur, and Former Legal Expert.
Mr. Chackochan, Member, PGP.
Mr. Damodaran, Former PGP during Whose Tenure JScFEF Was Inaugurated.
Mr. Damodaran, Member, PGP.
Mr. EM Varghese, Former Member, PGP.
Mr. Jofy, Former Member, JScFEF.
Mr. Joy Elamon, Former Director, CapDecK—When the Project Was Being Conceived and Implemented.
Mr. Libish, Former Field Staff, JScFEF (Telephone Interview).
Mr. Madan Mohan, coordinator, Gram Panchayat Association.
Mr. Rajaji Mathew Thomas, Former Member, Coordination Team in the Pre-transference Phase.
Mr. Ramesh, Former Field Staff, JScFEF.
Mr. Roy K Devassy, President, PGP (Panancherry Grama Panchayat, Thrissur District).
Mr. Thomas Samuel, Former Member, PGP.
Ms. Aleyamma Vijayan, Women's Resource Centre, Sakhi, Trivandrum (Telephone Interview).
Ms. Jijo, Staff, JScFEF.
Ms. Lilly Francis, Former Full Timer and Subsequently Field Staff in the Pre-transference Phase and Subsequently President, PGP, Who Completed a Tenure of 5 Years.
Ms. MM Gracy, Secretary, SEWA Thrissur.
Ms. Sakunthala Unnikrishnan, Vice President, PGP, and Former GPP during Whose Tenure JScFEF Was Sanctioned.
Ms. Savithri Sadanandan, Member, PGP.
Ms. Sheela, CDS (Community Development Society) of Kudumbashree23, Chairperson, Panancherry Panchayat.
Sr. Lissy Chacko, Former Coordination Committee Member during the Pre-transference Period and Legal Expert in the Post-transference Phase, JScFEF (Telephone Interview).

20

People with Disabilities: The Role of Social Workers for Rehabilitation in Bangladesh

Isahaque Ali, Azlinda Azman, Paramjit Singh Jamir Singh, Zulkarnain A. Hatta, Ndungi Wa Mungai, Muhammud Shariful Islam, and Tahmina Akhtar

Introduction

Bangladesh is a low-income country located in South Asia. Form the country's 170 million population, a reported 45% are living below the poverty line and earn under USD 1 a day (World Bank 2015). The country is facing severe poverty, where people have minimal access to education

I. Ali • A. Azman (✉) • P. S. Jamir Singh
Social Work Section, School of Social Sciences, Universiti Sains Malaysia, Penang, Malaysia
e-mail: azlindaa@usm.my

Z. A. Hatta
Faculty of Social Science, Arts & Humanities, Lincoln University College, Petaling Jaya, Malaysia

© The Author(s) 2020
Md. N. Momen et al. (eds.), *Building Sustainable Communities*,
https://doi.org/10.1007/978-981-15-2393-9_20

and medical care. Furthermore, Bangladesh is facing an unemployment crisis, lack of social security systems and is highly exposed to natural disasters (World Bank 2015). In the meantime, persons living with some form of disability account for about 20% of the world's population (WHO and World Bank 2015). "Persons with Disability" (PWD) in Bangladesh comprises people living with physical disabilities, visual impairment, hearing impairment, speech impairment, mental disability (cognitive disability or mental retardation) and mental illness, characterised by either part or full injury or loss of intellectual poise. The Disability Welfare Act states that PWD comprises of persons with multiple disabilities, whether the reason for the infirmity is inherited or a result of abuse, illness, accident or additional causes. The Act also classifies that disabled persons have a disability and are incapable of leading a normal life either partly or wholly due to the infirmity or psychological deficiency (Disability Welfare Act 2001).

An estimated 10% of Bangladesh's population, or approximately 17 million people, are living with disability (CDD 2015). Out of this number, a majority (41.5%) have physical disabilities, 19.6% have speech and hearing, 19.7% have visual disabilities, 7% have cerebral palsy, 7.4% have intellectual disabilities, 4.4% have multiple disabilities and 2.4% have a mental illness (World Bank 2015). Meanwhile, 70% of the PWD are illiterate, 68.9% of PWD have no access to medical assistance or rehabilitation due to economic hardship and 96.7% of PWD do not receive help from any organisation (Mostafa 2013).

N. W. Mungai
Social Work Department, School of Humanities & Social Sciences, Charles Stuart University, Bathurst, NSW, Australia

M. S. Islam
Social Work Department, University of Rajshahi, Rajshahi, Bangladesh

T. Akhtar
Institute of Social Welfare and Research, University of Dhaka, Dhaka, Bangladesh

PWD in Bangladesh are regularly ignored, deprived of basic needs, excluded from most public development policy initiatives and have difficulties accessing education, housing, medical care and basic living facilities (Park et al. 2002; Bowe 2006; Parish and Cloud 2006).

Global political instability, in recent years, has threatened peace and security, weakened economic recovery, delayed economic growth and poverty eradication. It has resulted in a lack of political will, poor governance, poor human rights, and lack of priority towards other social issues among women and children in difficult circumstances, people with AIDS, the elderly and drug users. These issues have contributed to the marginalisation of PWD. In this light, public development agencies and service providers have frequently ignored PWD. There are also cases where "special" programmes were implemented to effectively exclude PWD from mainstream society (Yeo and Moore 2003).

Social work is the scientific discipline that focuses on addressing problems and bringing about the desired changes at the individual, family, society, organisational and international levels (Ali and Hatta 2014). Since its inception, social work in infirmity services has transformed several of its important functions. Standard social services are focused on providing psychotherapy, information and referral, resource acquirement and case support (Beaulaurier and Taylor 2001). However, some PWD question the efficacy of social work based on their expectations of the role of social workers. These attitudes emerged in the 1970s–1980s where changes in numerous social work aspects were conducted in large measures to support the efforts to protect the civil rights of PWD (Beaulaurier and Taylor 2001). Other initiatives were campaigned to increase facilities for PWD so that they are capable of living in the society without institutional help and of creating positive attitudes and acknowledgement of their rights. The scope of disabilities creates an exclusive opportunity for social workers to collaborate with others to develop understanding and share expertise (CASW 2015). The PWD is known to experience higher living risk, thus increasing the need for advocacy, services and supports that can be provided by a social worker (Chaplin 2004). Social workers can undertake significant roles in resource relocation to assist and enable the disabled to be more autonomous and to create earnings opportunity in Bangladesh.

The Scenario of People with Disabilities in Bangladesh

The rights of PWD is one of the least understood issues in the area of development in Bangladesh. Issues concerning PWD have not been addressed widely in mainstream human rights, women's rights, child rights, religious minorities' rights or indigenous peoples' rights organisations (Ackerman et al. 2005). It is also historically evident that disability rights are one of the most neglected and forgotten development areas by the governments and NGOs. The government and other development agencies have not realised that a close relationship exists between disability and development and believed that disability is a welfare issue. PWD are considered recipients of charity and welfare in Bangladesh. Most parents with disabled children in Bangladesh are not ready to address the issue faced by their disabled children in front of the public due to negative perceptions of having a disabled child as a curse and sentence from the Creator (God) (Ackerman et al. 2005).

The PWD in rural Bangladesh are perceived negatively and experience wide-spread misconception and mistreatment making their lives miserable. The dependence and inability contribute to few facilities open to PWD and decrease their role in their family, society and nation (Morales and Sheafor 2004). The mistreatment of PWD is a culturally embedded and socially accepted form of oppression against PWD (Ackerman et al. 2005; Chiterika 2010). Moreover, the lack of experience in addressing the needs of this cohort may result in negative attitudes towards PWD from healthcare professionals who may struggle with communicating at appropriate developmental levels in Bangladesh (Ward et al. 2010). The PWD seems to be sidelined from mainstream national policies due to the public's negative impression of them, inadequate knowledge of policy-makers and planners, shortage of resources, insufficient knowledge and skills to address the needs of PWD in Bangladesh (Alam et al. 2005).

Bangladesh has ratified the United Nations Convention on the Rights of Persons with Disabilities (UNCRPD) and has a Disability Welfare Act-2001. The Government of Bangladesh (GB) and its concerned ministries, directorates, national and international agencies are aware of the situation of PWD. These agencies are working together for the development of PWD in the country. While the Ministry of Social Welfare (MSW) has provided a special allocation for PWD in its yearly budget, the allocation is undoubtedly inadequate to address the immediate needs of millions of PWD in the country (DRRA 2012). Nonetheless, all of these efforts could cover only 20% of the total disability in Bangladesh (DRRA 2012). The programme covered a few PWD in the country, but a major part of the PWD population cannot access these services in Bangladesh. Despite the Welfare Act 2001, no other laws allow PWD to claim and establish their rights in society (DRRA 2012).

The government provides monthly financial support to registered PWD. However, the cash grant amount is only sufficient to cover the expenses for a week, making them dependent on their families and other cash handouts from charities. Limited resources have restrained agencies from initiating and expanding rehabilitation and other services for PWD. Only a handful of PWD are involved in mainstream employment activities due to the lack of necessary expertise (Alam et al. 2005). Many international organisations are involved in disability rehabilitation in Bangladesh. Nevertheless, PWD is hardly engaged in the formulation of policy for these institutions, which makes it very hard to form actual policies for full inclusion of PWD (Yeo and Moore 2003).

There is also a lack of understanding and skills among development institutions to involve PWD with their policy and planning. Moreover, training organisations have inadequate capacity to train rehabilitation workers in Bangladesh, as well as inadequate professional facilities for sign language, braille and speech therapy that restrict the inclusion of PWD. The majority of the organisations are unwilling to employ PWD due to inexperience and lack of knowledge and expertise on the needs of PWD, negative perceptions, and the lack of intention to adopt a PWD-friendly working atmosphere in Bangladesh. Moreover, the PWD in most communities around the world are given little opportunity to become involved as a development partner due to the common perception that they are incapable of performing duties (Mostafa 2013).

There is still a general lack of acceptance of PWD in Bangladesh. The PWD is still frequently pitied, socially shunned, humiliated and discriminated against (Zastrow 2004). This vulnerable condition makes it more challenging and occasionally difficult for them to claim their political, social and financial rights. The voices of the disabled are not heard in Bangladesh. They are a disadvantaged group in society and live in poverty. In this regard, along with ethnic minorities, the elderly and women, they are facing constant discrimination (Brantlinger 2001; Das and Addlakha 2001; Larkin et al. 2003). Studies have argued that depriving policies, legislations, programmes and cultural rights contribute to a situation that denies people control over their lives (Chiterika 2010).

The MSW, Department of Social Services (DSW) and Foundation for the Development of PWD are working with other ministries and agencies to support the implementation of the laws and provisions. Hundreds of non-government agencies are working to address disability issues and encourage their inclusion in mainstream development. There is an apparent goal among the stakeholders working with PWD. However, many initiatives benefited only a small number of PWD in remote and urban areas, and there is a lack of cooperation between agencies (Mostafa 2013). In this light, PWD are vulnerable due to the lack of self-organisation, limited knowledge of their basic rights, low confidence, lack of respect from the society, lack of education facilities, limited social contact, weak leadership and limited advocacy in Bangladesh (Chiterika 2010) (Fig. 20.1).

Legislation and Institutional Frameworks on Disability Issues in Bangladesh

In the 1970s, the rights of PWD were recognised as the main agenda of the United Nations (UN) Agreement on the Civil Rights of the Mentally Retarded in 1975 (IFSW 2012). The government of Bangladesh accepted the rights of the PWD in 1993, with the establishment of the National Coordination Committee on Disability under the Social Welfare Ministry. The Constitution of Bangladesh specifically emphasised the equality, human dignity and social justice for the people of Bangladesh as stated in Articles 10, 11, 15, 17, 19, 20, 21, 27, 28, 29, 31, 32, 36. In this regard, the government of Bangladesh has guaranteed the rights and

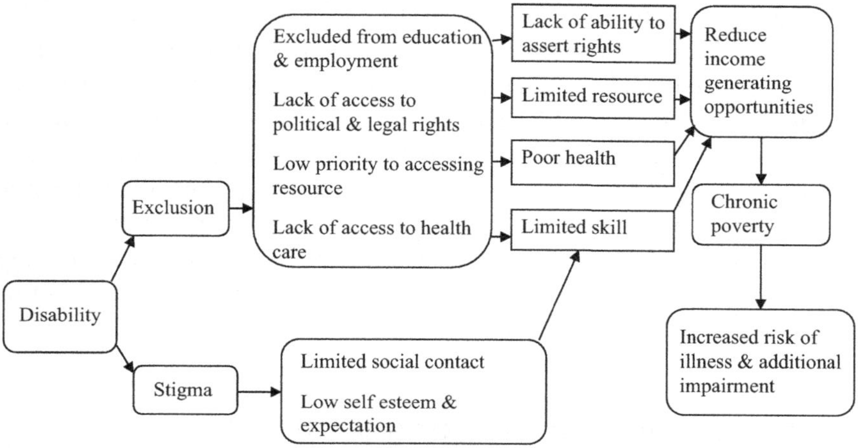

Fig. 20.1 The vicious circle of poverty and disability. (Source: Yeo and Moore 2003)

self-esteem of all citizens of Bangladesh, equally and without any prejudice (DRRA 2012).

The GB has passed several Acts for PWD to improve their living standards, including the Disability Welfare Acts 2001 and 2012. The Disability Welfare Act 2001 is based on a medical, welfare and charity-based approach (Mostafa 2013). During the last decade, government agencies have also introduced legislation against disability discrimination (Ackerman et al. 2005). However, the implementations and effects of such legislations have been limited (Yeo and Moore 2003). While the national constitution prohibits any form of discrimination in the employment for the disabled, the full inclusion of PWD in employment or income generation activities in Bangladesh has yet to materialise (DRRA 2012). It ignores the development and rights components of development cooperation and regards the affair as a "welfare" issue (DRRA 2012). The Disability Welfare Act 2012 has failed to meet the goals due to the prime fact that this was anti-UNCRPD legislation. Hence, it failed to uphold the universal standards applicable in this field (Mizan and Sarwar 2012). In this light, the act is grounded on the Disability Welfare

Act 2001 action towards the universal modification of the education of disabled children. However, no financial costs were included in this plan.

The discussion above shows that the present policies and laws in Bangladesh concerning the rights of PWD do not effectively defend the rights to equality and non-discrimination. The majority of the programmes, legislations, practices and policies aim to address PWD are not suitable for community, are functionally inadequate and ineffective for the welfare of PWD in Bangladesh (Mostafa 2013). More comprehensive legislation is required in Bangladesh to guarantee the rights of PWD in all phases—civil, financial, social, cultural and political.

Organisations for Working with PWD in Bangladesh

The Ministry of Social Welfare (MSW), Department of Social Welfare (DSW) and National Foundation for Development of the Disabled Persons are the agencies responsible for fulfilling the needs of PWD in Bangladesh. Besides government agencies, many NGOs are working for PWD in Bangladesh. These NGOs include Action on Disability and Development (ADD) and Bangladesh Disability Welfare Association, Centre for Services and Information on Disability (CSID), Centre for the Rehabilitation of the Paralysed (CRP), Young Power in Social Action (YPSA), The National Federation of the Blind and National Federation of the Deaf Disabled Rehabilitation and Research Association (DRRA). Moreover, some international NGOs such as Christian Blind Mission (CBM), Sightsavers, Helen Keller International, Oxfam, Action Aid, ORBIS, Lillian Fonds, Handicap International and Leonard Cheshire Disability—International, Rotary International, Niktan, Eblis Foundation, The Canadian International Development Agency (CIDA) and other agencies are among the frontline NGOs that support the development of persons with disabilities in the country. Addressing disability issues is important in all facets of development: equality, empowerment, human rights, poverty and marginalisation (Lee 1999). There are several limitations in relying on NGOs to meet the needs of PWD. This includes low financial assistance, disorganised service delivery, too flexible and unregulated (Harriss-White 1999; Yeo and Moore 2003). Moreover, many mainstream development NGOs continue to claim that they are not "specialists" and therefore, do not want to consider disability issues (Yeo and Moore 2003). It is clear that most mainstream development NGOs do not fully include disabled people in their work in Bangladesh.

Social Work and Disability

In caring for people with disabilities, social workers are generally responsible for working together with persons, groups and families who have several, difficult and co-existing personal and societal complications.

The main goal of social work is to enable PWD and their families to identify their needs, anxieties and aspirations, provide solutions and help them take control of their lives, avoid and cope with grief and catastrophe and ensure public involvement (Wilkinson 2015). Social work has existed as a discipline for over 100 years (Weaver 2006). During the Second World War, social casework has provided healthcare rehabilitation and social workers carried on more frequent interaction with PWD. In the golden age of social work in the 1970s and early the 1980s, the principal form of social work exercise has focused on personal casework with PWD (Langan 1992; Leece and Leece 2011). Besides that, the social casework methods were previously used for preparing individuals, family and society with a new disability outside of the hospital setting (Beaulaurier and Taylor 2001). The health rehabilitation task largely used a collective method that solely depends on the assistance of social workers (Berkowitz 1980).

The National Association of Social Workers (NASW) mentioned that social workers provide education, employment and skills development training, healthcare, affordable housing and mental support for PWD (NASW 2006a; Bean and Krcek 2012). Social workers help displaced, vulnerable people, elderly, children and PWD (Weaver 2006; NASW 2006b). In this light, the social workers as specialists could strengthen a person's skills in adjusting to the environment, to rebuild a person's relations with the community, while helping them to integrate into society, and promoting social functioning (Laima et al. 2003). Social workers have a responsibility to prepare future practitioners with the skills necessary to meet the needs of PWD (Robinson et al. 2012).

Social workers are responsible for meeting the needs of this population and can be integral to providing solutions (Robinson et al. 2012). Social workers are in an ideal position to work with PWD because they are highly trained to address the needs of people with disability (Hanley and Parkinson 1994; Bean and Krcek 2012). Furthermore, social workers value rights, dignity, individual uniqueness, self-determination and clients' access to resources (NASW 2006a). At the micro and macro levels, social workers are focused primarily on the social well-being of individual clients and their families, equally valued with the importance of their physical, mental and spiritual well-being. At the macro level, they establish a better capability to look beyond the ill health and medical care, to treat them as human beings and to take care of the political and social concerns of PWD (CASW 2015). Social workers assisting PWD at individual levels, assist families who have family members or children with infirmity, as well as with societies in local and international levels. The PWD and their families are served by social work providers in traditional service delivery systems and in social service agencies (Parish and Lutwick 2005). These scopes include research, advocacy and counselling, policy practices, casework, community development and direct exercise. Social workers play important roles in the field of disability to develop policies, programmes and legislation to extend support to PWD in Bangladesh. Besides, social workers assisting PWD and families often encourage social inclusion, working opportunity, society living, rehabilitation and family care (IFSW 2012). Consequently, social workers help to improve the human function and efficacy of the social structure by offering wealth and facilities for PWD, clients and service recipients (DuBois and Miley 2005). Social work also addresses the issue of grief, loss and bereavement associated with a mental and physical disability.

The targeted outcomes of social work intervention include accepting and recognising PWD. Social workers are the change agents concentrating their involvements on the social connections between persons and community, as social work is about strategic change (Tan and Envall 2004). Social work is a scientific profession that helps clients to change them, their environment or accepts aspects of themselves (Adams et al. 2002; Leece and Leece 2011). Social workers can build on their connection with users and develop roles through advocacy, group work and counselling with PWD (Kirkpatrick et al. 1998; Leece and Leece 2011). In working with PWD, social workers' are required to make appropriate referrals, provide supportive counselling, offer advocacy and to engage in group work, service coordination, team building and direct services (Andrews and Wikler 1981).

Role of Social Workers for Rehabilitation in Bangladesh

Social workers play a vital role in disability services offered by NGOs and government agencies. This is because they have vast knowledge in identifying both personal and logical dimensions of support and motivation, play a significant role in evaluation, case management and planning across all kind of service delivery (Wilkinson 2015). The principal mission of social work is to enhance their full potential, improve their living standard and avoid malfunctions in the communities (Hossain and Mathbor 2014). The role of social workers include:

Advocacy Services

Disability does not just affect individuals, but also the family and community surrounding them. Social workers play an important role in meeting the needs of PWD in Bangladesh, as they provide assessment, intervention and advocacy for people with PWD and their families. However, there is a serious shortage of social workers caring for PWD, as well as other disadvantaged groups, such as the elderly (Robinson et al. 2012). Social workers are involved in social planning, policy development, inspire and encourage community to engage in social action in order to influence social policy and financial improvement of PWD in Bangladesh (IFSW 2012; Hossain and Mathbor 2014). Social workers should also be involved with social justice to advocate for people who are facing various injustices in Bangladesh. Furthermore, they need to ensure that the engagement of service recipients and develop roles of advocacy, counselling and collective work in Bangladesh (Kirkpatrick et al. 1998).

Medical Social Work

The DSW operates hospital social service programmes in Bangladesh. Medical social workers in Bangladesh are responsible for supplying medicine, familiarising the patients with the hospital environment, provide healthy food and other healthcare provisions for the poor, rehabilitation of PWD and new patients. Medical and psychiatric social workers in hospitals in Bangladesh collaborate with doctors, nurses and other therapists in the diagnosis and treatment planning process of disabled people (Hossain and Mathbor 2014). As a member of in-patient or out-patient interdisciplinary team in the hospital, social workers are responsible for supporting the rehabilitation process so that PWD have the highest freedom and self-sufficiency in various areas of his or her functioning, (CASW 2015). Social workers encourage PWD to know their status and seek appropriate treatment from hospitals in Bangladesh. In these settings, medical social workers might visit the PWD and their families in their homes and are responsible for providing proper information, treatment or training, and provide adequate support for PWD, family and community members in Bangladesh (Chiterika 2010).

Counselling Services

Social workers provide counselling to PWD to support and encourage the inclusion of PWD in the community in Bangladesh. The counselling sessions could address individuals, family and interpersonal issues to empower the education of PWD and increase their economic opportunities. In rehabilitation settings, social workers may serve as a mediator between family and the agency to discuss issues pertaining to rehabilitation and other essential services and plans (Zastrow 2009). This procedure involves the family, clients and society. The social worker assists the clients and their family to cope with vulnerable situations, concerns and reliance to one of increased self-reliance and hope, through diagnosis and interventions in the areas of psychosocial functioning (CASW 2015). This can also be linked to the advocacy role whereby social workers can lobby the government to implement legislation specifically targeting PWD in Bangladesh properly. The social workers could effectively guide the government concerning the inclusion of disability in all legislation, policies and programmes in Bangladesh (Yeo and Moore 2003). In this regard, the Code of Ethics calls for social workers at the macro level to advocate for changes in policy and law to improve social settings to meet social needs and enhance social justice for PWD (NASW 2006a).

Financial and Equipment Support

Bangladesh is a low-income country where 49% of people live below the poverty line and earn less than US$1 a day. People are living with inadequate health, education and social security services, low employment and exposed to natural disasters (World Bank 2015). Poverty hinders parents from getting assistance and therapy that could ensure the proper growth and development of PWD. Moreover, poor parents are not in the position to purchase assistive devices for their disabled children. Social workers assist by providing their clients with various social service offices/agencies for welfare benefits, and assistive devices like wheelchairs, special seats, standing and walking devices and orthotics and prosthetics, as well as hearing aids, toys for activities of daily living, tricycles, tools and equipment for work, protective footwear, simple devices for braille reading. They also provide home services to increase accessibility.

Vocational Training and Empowerment

Currently, many stakeholders are working on behalf of disabled people. However, they have been working through exclusion, isolation and demeaning welfare programmes (Yeo and Moore 2003). In this light, providing charity becomes insufficient to address the issues of PWD and change the notion that PWD is inferior. Charity is not part of the development process, and PWD have not benefited from it. This is because the PWD desire to be treated equally, participate as equal citizens in their own communities with ordinary citizens with equal rights (Chiterika 2010; Suharto 2006). Thus, the empowerment perspective has encouraged social workers to develop collaborations with oppressed PWD and persons living in poverty (May 2005) as individuals with disabilities are becoming increasingly interested in gaining empowerment (Mackelprang and Salsgiver 1996; Beaulaurier and Taylor 2001).

Therefore, rehabilitation programmes need to empower the PWD to enhance their confidence and self-motivation by mutual discussion as a support group (Suharto 2006). The principal objectives of the rehabilitation programmes should aim to remove the barrier of disability while enhancing employment opportunities for self-reliance and freedom (Suharto 2006). PWD is one of the most disadvantaged and helpless groups in Bangladesh, as they have no or limited access to education, basic needs, employment opportunities, skills or vocational training. This has worsened their situation further. In this regard, social workers might provide training for PWD who lack the skills and qualification to venture into employment and provide employment opportunities for qualified PWD (IFSW 2012).

Professional Schools

Children and youth with disabilities have limited access to education in Bangladesh. While providing educational opportunities for PWD is central to economic empowerment, however, access to education for PWD is very restricted due to the negative attitudes of community leaders, school authorities and village leaders who do not admit children with disabilities in Bangladesh. They do not include adequate hygiene (toilets), suitable infrastructure, educational materials, assistive devices like hearing and braille in Bangladeshi schools (Ackerman et al. 2005). School workers can directly assist individual children, groups or entire classrooms. Additionally, they might assist families to receive services that are required for them, consult with parents and teachers and others in the issue of children's lives (NASW 2006b). In Bangladesh, the education of children with disabilities is dependent on NGOs due to minimal government programmes. Consequently, social workers may influence political, administrative and legislative changes through relevant proposals and submissions to government bodies for research, continuous professional development, education and training and promoting and facilitating the inclusion or integration of PWD within the broader community in Bangladesh.

At the same time, medical social workers might play a significant role in providing information on the available therapies, discuss treatment options with nurses and doctors, teaching health education and help create rapport with doctors and patients. The scope of social workers' roles includes cooperating with other professionals, collaborating with community members, supporting PWD, developing healthcare model and resources, consulting for adequate services, changing policy to eradicate poverty, housing, employment and social justice and assisting the development of preventive programmes for PWD. Furthermore, social workers could facilitate the interference, support and increase access to services, information and individual and public teaching for PWD (CASW 2015). PWD are most likely to run their small businesses with profits to improve the living standard (Ackerman et al. 2005). Social workers play an increasingly prominent role in employee assistance programmes in addressing issues affecting PWD in the workplace in Bangladesh. The majority of PWD are mostly uneducated and receive no training. Hence, they are eventually unable to run their business with profits. In response, social workers could arrange need-based training on income generation activity (IGA) to enhance skills in the market promotion.

Lobbyists

Social workers could conduct collaborative, strengths-cantered activities to promote, support and advocate awareness of the rights and benefits of adaptive technology for PWD (DuBois and Miley 2005). There is a lack of self-organisation among PWD (Alam et al. 2005). Disabled people are socially excluded, as they often have little interaction with other disabled people in their home country and other parts of the world (Yeo and Moore 2003). Social workers can play the role as a lobbyist to strengthen self-help disability organisations particularly in the area of identifying and training of leaders, linking self-help groups with potential funding partners, training on operation and management of cooperatives, seminars on small business, development and management and training on advocacy and awareness-raising in Bangladesh. Disabled people are rarely involved in the policy-making at the national and international levels (Yeo and Moore 2003). Social workers can build collaboration and coordination as well as exchange information, knowledge, resource and services among the governmental and NGOs at both national and international levels in Bangladesh. The social workers could also help empower PWD through the formation of self-help groups and in rural areas and urban slums (Laima et al. 2003).

Increase Awareness

Illiteracy is a major problem in Bangladesh, and illiterate parents mostly live in poverty. Subsequently, they are not aware of the constitutional facilities and basic rights of their disabled children and the education options available to them. Children with disabilities are often hidden as victims due to humiliation and potential teasing and ridicule from others (Alam et al. 2005). The print and electronic media have also given minimal attention to PWD. Therefore, social workers could work collaboratively to fulfil the requirements of PWD and increase the level of public awareness regarding their fundamental rights, and to create new levels of coordination into social life in Bangladesh. Social workers in their organisations work on behalf of their clients, and this includes efforts to influence other professionals to improve the living standard of PWD. Social workers could also provide counselling campaign for the provision of service, and establish social or public policies that benefit PWD and their caregivers (Parish and Lutwick 2005; Heller et al. 2007; Robinson et al. 2012).

Provide Health Education

Medical social workers are responsible for delivering education, knowledge, family planning and counselling, immunisation, food nutrition, maternal illness and healthcare for children in Bangladesh. However, most PWD is living in poverty and in rural areas without access to welfare programmes and services in Bangladesh. Healthcare services are scarce for disadvantaged people in Bangladesh, including PWD. The lack of nutritious food, hazardous working and housing conditions, lack of access to vaccination, poor sanitation and hygiene, insufficient information, and risk of war, conflict and natural disaster are significant issues for PWD in Bangladesh. This has generated a negative impact on family care, and primary interference is an important instrument to help people with disability. Hence, social workers need to empower their parents and family members with the knowledge and skills to understand and fulfil the requirement of PWD in Bangladesh (Ackerman et al. 2005).

PWD continue to be isolated from mainstream society due to their physical inability and negative stigmas regarding their competence in Bangladesh. While there are no laws that restrict PWD from participating in the society, social works in Bangladesh are still hindered by physiological obstacles and deprivation which made it hard for the disabled to be accepted by the community in Bangladesh (Beaulaurier and Taylor 2001). Negatives attitudes, prejudice, and employers' reluctance to hire and accommodate PWD have somewhat significantly impacted their employment opportunities (Schartz et al. 2006a; Schartz et al. 2006b; Samant et al. 2009).

Professional social workers act as trainers, consultants and programme designers to assist engagement, maintenance, accommodation and training workers with disabilities (Straussner 1990). As counsellors and mentors, social workers could assist organisations to gain a better understanding of disability matters for employment, legislation and accommodating PWD (Samant et al. 2009). Therefore, social workers could develop a relationship with international development agencies to encourage the participation of PWD in their programmes and services and spread awareness so that the future generation accepts PWD as equal members of the community in Bangladesh.

Conclusion

Undoubtedly, PWD is one of the most discriminated minority groups in the world and have received constant systematic violations of their human rights. These violations result in worsening living conditions, degrading human treatment, lack of access to housing, healthcare and employment, social exclusion and even death. PWD are seen as perpetually dependent and unproductive. Some families tend to neglect their developmental needs. With a common perception that nothing can be expected from PWD, families do not care to invest for the education, healthcare and other needs of their family members with disabilities in Bangladesh. Inadequate and ineffective execution of the current laws and policies in the lack of attentive monitoring system and apathy of GOs and NGOs in imposing them due to the lower priority assigned to programmes regarding disability. Generally excluded from social, economic and political institutions and interactions, disabled people lack the power to exert an influence on policymakers in Bangladesh.

Advocacy of PWD is crucial to generate awareness of disability issues. Providing equal rights and chances for the disabled to get involved in finance and politics, as well as creating more social connections will contribute to the increase of living standards in Bangladesh (Yeo and Moore 2003). There is a dearth of professional social workers with the capability to work with PWD to address issues related to people with physical, visual, hearing and intellectual disability. Social workers can be the catalyst for change, as they are responsible for training, education, advocacy, and identification of needs on individual, community and state. In this regard, they can create and sustain the necessary partnerships to reduce the disparities experienced by PWD (Robinson et al. 2012). However, there are still many mountains to climb before equity and social inclusion can be a reality for PWD. Multi-sectoral collaboration efforts are needed to ensure the full integration of PWD. With unprecedented global efforts to improve living standard, PWD should be fully incorporated as a major target for substantial improvement to PWD living standards to become a reality.

References

Ackerman, P., Thormann, M. S., & Huq, S. (2005). *Assessment of Educational Needs of Disabled Children in Bangladesh*. Washington, DC: United States Agency for International Development (USAID).

Adams, R., Dominelli, L., & Payne, M. (2002). *Social Work: Themes, Issues and Critical Debates* (2nd ed.). Basingstoke: Palgrave Macmillan.

Alam, K. J., Bari, N., & Khan, M. A. (2005). *Community Based Rehabilitation Practices and Alleviation of Poverty of People with Disabilities in Bangladesh*. Retrieved from http://www.worldenable.net/cbr2005/paperbangladesh.htm.

Ali, I., & Hatta, Z. A. (2014). *Zakat* as a Poverty Reduction Mechanism Among the Muslim Community: Case Study of Bangladesh, Malaysia, and Indonesia. *Asian Social Work and Policy Review, 8*(1), 59–70.

Andrews, S., & Wikler, L. (1981). Developmental Disabilities. *Health and Social Work, 6*(1), 62–68.

Bean, K. F., & Krcek, T. E. (2012). The Integration of Disability Content into Social Work Education: An Examination of Infused and Dedicated Models. *Advances in Social Work, 13*(3), 633–647.

Beaulaurier, R. L., & Taylor, S. H. (2001). Social Work Practice with People with Disabilities in the Era of Disability Rights. *Social Work in Health Care, 32*(4), 66–91.

Berkowitz, E. D. (1980). *Rehabilitation: The Federal Government's Response to Disability 1935–1954*. New York: Arno Press.

Bowe, F. (2006). *Disability in America*. New York: Hofstra University.

Brantlinger, E. (2001). Poverty Class and Disability: A Historical, Social and Political Perspective. *Focus on Exceptional Children, 33*(7), 1–15.

Canadian Association of Social Work (CASW). (2015). *The Role of Social Work in Mental Health*. Retrieved November 1, 2015, from http://www.casw-acts.ca.

Centre of Disability in Development (CDD). (2015). *Disability Situation in Bangladesh*. Retrieved October 24, 2015, from http://www.cdd.org.bd.

Chaplin, R. (2004). General Psychiatric Services for Adults with Intellectual Disability and Mental Illness. *Journal of Intellectual Disability Research, 48*(1), 1–10.

Chiterika, C. (2010). People with Disabilities and the Role of Social Workers in Lesotho. *International Journal of Social Work and Society, 8*(1), 82–93.

Das, V., & Addlakha, R. (2001). Disability and Domestic Citizenship: Voice, Gender, and Making of the Subject. *Public Culture, 13*(3), 511–531.

Disability Welfare Act. (2001). Govt of Bangladesh.

Disabled Rehabilitation & Research Association (DRRA). (2012). *Contextual Analysis of Socio-economic Situation of Persons with Disabilities in Bangladesh*. Dhaka: Bangladesh.

DuBois, B., & Miley, K. (2005). *Social Work: An Empowering Profession*. Boston: Pearson.

Hanley, B., & Parkinson, C. B. (1994). Position Paper on Social Work Values: Practice with Individuals Who Have Developmental Disabilities. *Mental Retardation, 32*(6), 426–431.

Harriss-White, B. (1999). Onto a Loser: Disability in India. In B. Harriss-White & S. Subramanian (Eds.), *Ill Fare in India: Essays on India's Social Sector in Honor of S. Guhan* (pp. 135–158). New Delhi: Sage Publications.

Heller, T., Caldwell, J., & Factor, A. (2007). Aging Family Caregivers: Policies and Practices. *Mental Retardation and Developmental Disabilities Research Review, 13*(2), 136–142.

Hossain, M. I., & Mathbor, G. M. (2014). Social Work Practice for Social Development in Bangladesh: Issues and Challenges. *Asian Social Work and Policy Review, 8*(2), 123–137.

International Federation of Social Work (IFSW). (2012). *People with Disabilities*. Retrieved from http://www.ifsw.org.

Kirkpatrick, I., Ackroyd, S., & Walker, R. (1998). *The New Managerialism and Public Service Professions: Change in Health, Social Services and Housing*. London: Palgrave Macmillan.

Laima, S., Ligija, S., & Jurate, G. (2003). The Role of Social Worker in Team of Rehabilitation: Methodological Approach. *MEDICINA, 39*(9), 889–833.

Langan, M. (1992). The Rise and Fall of Social Work. In J. Clark (Ed.), *A Crisis in Care*. London: Sage.

Larkin, V. M., Alston, R. J., Middleton, R. A., & Willson, K. B. (2003). Underrepresented Ethnically and Racially Diverse Ageing Populations with Disabilities: Trends and Recommendation. *Journal of Rehabilitation, 69*(2), 26–31.

Lee, H. (1999). *Discussion Paper for Oxfam: Disability as a Development Issue and How to Integrate a Disability Perspective into the SCO*. Oxford: Oxfam.

Leece, J., & Leece, D. (2011). Personalisation: Perceptions of the Role of Social Work in a World of Brokers and Budgets. *British Journal of Social Work, 41*(2), 204–223.

Mackelprang, R. W., & Salsgiver, R. O. (1996). People with Disabilities and Social Work: Historical and Contemporary Issues. *Social Work, 41*(1), 7–14.

May, G. E. (2005). Changing the Future of Disability: The Disability Discrimination Model. In G. E. May & M. B. Raske (Eds.), *Ending Disability Discrimination: Strategies for Social Workers* (pp. 82–98). Boston: Pearson Education, Allyn & Bacon.

Mizan, A. S., & Sarwar, M. G. (2012). *In Pursuit of Deliverance: Critical Analysis of the Draft Rights of the Persons with Disabilities Act 2012*. Retrieved from https://www.academia.edu/6615613/An_Overview_of_the_Rights_of_Persons_with_Disabilities_Act_2013.

Morales, A. T., & Sheafor, B. W. (2004). *Social Work: A Profession of Many Faces*. Boston, MA: Allyn & Bacon.

Mostafa, M. G. (2013). *The Convention on the Rights of Persons with Disabilities: A Study on Bangladesh Compliance*. National Human Rights Commission, Bangladesh (NHRCB). Dhaka: Bangladesh.

National Association of Social Workers (NASW). (2006a). *How Do Social Workers Help the Families of Children with Disabilities*. Retrieved from http://www.nasw.org.

National Association of Social Workers (NASW). (2006b). *Assuring the Sufficiency of a Frontline Workforce: A National Study of Licensed Social Workers*. Retrieved from http://www.workforce.socialworkers.org/studies/nasw_06_execsummary.pdf.

Parish, S. L., & Cloud, J. M. (2006). Financial Well-Being of Young Children with Disabilities and Their Families. *Social Work, 51*, 223–232.

Parish, S. L., & Lutwick, Z. (2005). A Critical Analysis of the Emerging Crisis in Long Term Care for People with Developmental Disabilities. *Social Work, 50*(4), 345–254.

Park, J., Turnbull, A. P., & Turnbull, H. R. (2002). Impacts of Poverty on Quality of Life in Families of Children with Disabilities. *Exceptional Children, 68*(2), 151–170.

Robinson, L. M., Dauenhauer, J., Bishop, K. M., & Baxter, J. (2012). Growing Health Disparities for Persons Who Are Aging with Intellectual and Developmental Disabilities: The Social Work Linchpin. *Journal of Gerontological Social Work, 55*(2), 175–190.

Samant, D., Soffer, M., Hernandez, B., Adya, M., Akinpelu, O., Levy, J. M., Repoli, E., Kramer, M., & Blanck, P. (2009). Corporate Culture and Employment of People with Disabilities: Role of Social Workers and Service Provider Organizations. *Journal of Social Work in Disability & Rehabilitation, 8*(3), 171–188.

Schartz, H., Hendricks, D. J., & Blanck, P. (2006a). Workplace Accommodations: Evidence-Based Outcomes. *Work, 27*, 345–354.

Schartz, H. A., Schartz, K. M., Hendricks, D. J., & Blanck, P. (2006b). Workplace Accommodations: Empirical Study of Current Employees. *Mississippi Law Journal, 75*, 917.

Straussner, S. L. A. (1990). *Occupational Social Work Today*. New York: Haworth.

Suharto, E. (2006). Role of Social Workers in Indonesia: Issues and Challenges in Rehabilitation People with Disabilities. Retrieved July 16, 2019, from http://www.policy.hu/suharto/Naskah%20PDF/CibinongDifableSocialWork.pdf.

Tan, N. T., & Envall, E. (2004). Social Work; Challenges in the New Millennium. In N. T. Tan & E. Envall (Eds.), *Social Work Around the World*. Bern, Switzerland: IFSW.

Ward, R. L., Nichols, A., & Freedman, R. (2010). Uncovering Health Care Inequalities Among Adults with Intellectual and Developmental Disabilities. *Health and Social Work, 35*(4), 280–290.

Weaver, H. N. (2006). Social Work Through an Indigenous Lens: Reflections on the State of Our Profession. In N. Hall (Ed.), *Social Work: Making a World of Difference, Social Work Around the World 1V in the Year of IFSW's 50th Jubilee* (pp. 37–51). Berne: International Federation of Social Workers and Fafo.

Wilkinson, G. (2015). *Social Work Practice Statement: The Contribution Social Work Can Make to the National Disability Insurance Scheme (NDIS)*. Retrieved from https://www.aasw.asn.au/document/item/7420.

World Bank. (2015). *Bangladesh Country Profile*. Retrieved from http://www.wb.org.

Yeo, R., & Moore, K. (2003). Including Disabled People in Poverty Reduction Work: "Nothing About Us, Without Us". *World Development, 31*(3), 571–590.

Zastrow, C. (2004). *Introduction to Social Work and Social Welfare: Empowering People* (8th ed.). Belmont, CA: Brooks/Cole.

Zastrow, C. (2009). *Introduction to Social Work and Social Welfare: Empowering People* (10th ed.). Baltimore, MD: CENGAGE Learning.

21

Poverty and Social Inequality: Bangladesh Experience

M. Rezaul Islam, Niaz Ahmed Khan, Adi Fahrudin, Md. Rabiul Islam, and A. K. M. Monirul Islam

Introduction

Notwithstanding the gradual recognition of the intertwined nature of poverty and social inequality in the development discourse, there has been strikingly limited research on the subject, especially in the context of Bangladesh. The conceptual evolution of the two as developmental

---

M. Rezaul Islam (✉) • Md. R. Islam
Institute of Social Welfare and Research, University of Dhaka, Dhaka, Bangladesh

N. A. Khan
Department of Development Studies, University of Dhaka, Dhaka, Bangladesh

A. Fahrudin
Department of Social Welfare, University of Mohammadiyah, Jakarta, Indonesia

A. K. M. M. Islam
Bangladesh National Commission for UNESCO, Ministry of Education, People's Republic of Bangladesh, Dhaka, Bangladesh

© The Author(s) 2020

Md. N. Momen et al. (eds.), *Building Sustainable Communities*,
https://doi.org/10.1007/978-981-15-2393-9_21

issues has been complex, and each presents particular features (Villar 2017). Sen (1976) suggested a way of introducing inequality into poverty measurement by substituting the mean income of the poor in equation by the mean income deflated by Gini inequality index applied to the income distribution of the poor. Of late, there has been a growing concern amongst policy makers and development practitioners regarding the current trend of poverty and social inequality, and its widespread negative implications and ramifications over communities and nations. The Research Institute of the Credit Suisse, among many others, reported that Europe holds about one-third of the world's total wealth, North America owns another third and the rest of the world the remaining one. Africa, China, India and South America only account for about half of that last third, which implies that more than half of the world's population only gets approximately 15% of the total wealth (Davies et al. 2014). The report further notes that the whole continent of Africa holds just 1% of the total wealth. Such unequal distribution of resources results in high levels of inequality and poverty in a large fraction and unfair distribution of opportunities (Villar 2017). The Sustainable Development Goals (SDGs) include and treat both these concepts separately and assert that growth without redistribution cannot eliminate poverty or boost-up sustainable development.

Reduction in poverty and inequality poses a fundamental challenge for Bangladesh. The government of Bangladesh has subscribed to the SDGs as part of its international commitment to eradicate extreme poverty and hunger and foster global collaboration for development by 2030. Complementarily, the nation has also adopted its own 'Vision-2021' with the commitment to reduce the rate of poverty from 25% to 15% by 2021. The current (7th) Five Year Plan 2016–2022 also aims to attain average real GDP growth of 7.4% per year, reduce the head-count poverty ratio by 6.2% point and extreme poverty by about 4.0% point and create good jobs over the period. Despite country's policy obligation, the trend of poverty and social inequality in Bangladesh is increasing over time. Afsar (2010) argued that weak governance as a result of clientelist practice is often considered as the single-most important factor behind the growing inequality and persistent poverty in Bangladesh. Khan et al. (2011) reported that in Bangladesh, the overall development process has favoured the rich and powerful and has generated gross inequity across socio-economic groups. They added that there are huge disparities between the poorest and the richest quintiles. They showed that out of 49 factors, 16 show increasing inequity, whereas 22 show a decrease in the rich–poor gap. The BBS (2016) reported that currently, 10% of the rich people hold 38% of the total national income. The economists of the country state that the unequal distribution of wealth affects the overall progress of the nation in many ways, and all of them produce intense negative effects.

In the above backdrop, this chapter aims to critically review the nature and processes of poverty and inequality in Bangladesh. The study explores and examines poverty and inequality from diverse angles including region, income, gender, education, land ownership and social opportunities. After revealing the intertwined dynamics and implications of poverty and inequality, the chapter offers some relevant policy options that may illuminate strategic decisions towards reducing poverty and social inequality in Bangladesh.

The Key Concepts

Poverty

Poverty is a complex phenomenon that refers to the difficulty of having access to those goods and services that ensure living with dignity and developing a satisfactory social and personal life (Villar 2017). Poverty is perceived as economic, social and psychological deprivation, occurring either among people or countries that lack resources to maintain or provide either individual or collective minimum levels of living. It is also described as something that impairs the ability to provide for minimum nutrition, health, shelter, education, security, leisure or other aspects considered necessary for life. Poverty may also be represented as an exclusionary relationship, including exclusion from an institutional network sufficient to maintain one's survival (McCarthy and Feldman 1988; Kamruzzaman 2015). According to Green and Hulme (2005), poverty is not a natural fact, but a social experience. Poverty is not a characteristic of a certain group of people; rather, it characterizes a particular situation in which people may find themselves at a given point in time (Bastiaensen et al. 2005). According to the World Bank, in estimating poverty worldwide, the same reference poverty line has to be used, which is expressed in a common unit across countries. The bank uses reference lines set at US$1 and US$2 per day (more precisely US$1.08 and US$2.15 in 1993). It is not very clear, though, why the same poverty line should be used for all poor countries. Moreover, the World Bank's heavy reliance on per capita income to categorize countries, and its use of US$1 or US$2 a day as its most widely cited benchmarks, fuels the view that income must be the main criterion for assessing poverty (Nguyen-Marshall 2008).

Besides the monetary features, there are non-monetary aspects of poverty. For example, Sen (1981) emphasizes one's capability, entitlement and deprivation response in this regard. He illustrates how a peasant and a landless labourer may both be poor, but their fortunes are not identical. What emerges is an alternative conceptualization of poverty as a combination of quantitative and qualitative factors, which recognize that poverty arises from various forms of deprivation, including social and psychological as well as economic dimensions. The recognition of non-income dimensions reflects a greater acceptance that well-being and poverty are multidimensional and, in particular, that no single one-dimensional measure adequately captures the full gamut of well-being achievement. Yunus (2002) reminds us that the first thing to remember is that poverty is not created by poor people. It is created by social, political and financial institutions, and the designers and managers of these institutions. However, poverty can be perceived through a range of contested definitions, which overlap and sometimes contradict each other (Kamruzzaman 2015). In contrast to a universally applicable definition of poverty (for *all* poor countries), it is proposed that local understanding of poverty may be more effective for ameliorating local situations.

In Bangladesh, poverty has been viewed along 'food' and 'non-food' lines. In the food poverty line, three criteria are selected: a basic food basket (eleven food items), the quantities in the basket are scaled according to the nutritional requirement of 2122 kcal per person per day, and the cost of acquiring the basket is calculated. This estimated cost is taken as the Food Poverty Line (FPL). The non-food poverty category is calculated by estimating the cost of consuming non-food items by the households close to the food poverty line. Another two criteria—'lower' and 'upper' poverty lines are also used to calculate poverty in Bangladesh. In lower poverty line, the extremely poor households are those households whose total expenditures on food and non-food combined are equal to or less than the food poverty line. The upper poverty line is estimated by adding together the food and non-food poverty lines. The moderately poor households are those households whose total expenditures are equal to or less than the upper poverty line. Despite the large numbers of people living in poverty, the definition of poverty has been the subject of debate. The mainstream emerging sees poverty as generally being characterized by the inability of individuals, households or entire communities, to command sufficient resources to satisfy a socially acceptable minimum standard of living. The alternative view understands poverty as a part of social property relations.

Social Inequality

There are wide variations in the meaning and understanding of 'social inequality'. It may refer to the graduated dimensions (Blau 1977), vertical classifications (Ossowski 1963: Schwartz and Winship 1980) and bounded categories (Tilly 1998), or hierarchical relations (Burt 1982) by which human populations at varying levels of aggregation are differentiated. The concept is dated back Plato's conception of the republic and developed subsequently in the social theories of Marx in 1859, Moscain in 1939, Weber in 1947, Simmelin 1896, Sorokin in 1941, Eisenstadtin 1971 and Merton in 1968. The idea of social inequality has been muddled with some other concepts such as social class, social stratification, socio-economic status, power, privilege, cumulative advantage, dependence, and dominance (Encyclopaedia of Sociology 2001). Social inequality ensues in a society when resources are distributed unevenly among the socially defined categories of persons. Wade (2014) argues that this uneven distribution occurs in the society when preference of access of social goods are considered based on power, religion, kinship, prestige, race, ethnicity, gender, age, sexual orientation and class. Some other social conditions and opportunities are also related with this, such as labour market, the source of income, health care, and freedom of speech, education, political representation and participation (Kerbo 2003; Blackburn 2008). Hradil (2001) called these 'valuable goods' that a person earns through social network of relationships. This definition implies that differences in eye colour, body height, physical handicap, and so on cannot be called social differences or social inequality because they are not the result of a social process (Hoffmann 2008). In discussing social inequalities, it is important to make two distinctions. First is the difference between the unequal distribution of desirable life outcomes such as health, happiness, educational success, or material possessions, and the unequal distribution of opportunities (access to power and life chances that facilitate attainment of desirable outcomes). Second is the distinction between the unequal distribution of opportunities and outcomes among individuals and between groups (Carter and Reardon 2014).

The Methodology

Research Method and Tools

This study primarily bases itself on a Qualitative Interpretative Meta-Synthesis (QIMS). QIMS is a nonlinear conceptualization of a cross-study data collection tool that intends to merge themes from a collection of related studies that ultimately result in a holistic understanding (Schuman 2016; Ruiz and Praetorius 2016; Islam 2016, 2017; Reza et al. 2018; Chowdhury et al. 2018). This approach is consistent with previous papers similar to those of Joffres et al. (2008), Ruiz and Praetorius (2016), Schuman (2016), Islam (2016, 2017), Islam and Mungai (2016), Reza, Subramanian and Islam (2018), and Chowdhury, Wahab and Islam (2018).

Sampling

The data for QIMS were gathered through using computer databases and reference lists in the English-language literature, broadly in social sciences in both Scopus and Web of Science (Thomson Reuters) using keyword searches—namely 'poverty and social inequality' and 'poverty and social inequality in Bangladesh'. The duration of the search ranged from January 2008 to December 2017 in both Scopus and Web of Science. Altogether we found 1964 publications from Scopus and 2416 from Web of Science on the keywords 'poverty and social inequality'. We observed a decreasing, and somewhat fluctuating, trend in the number of publications on both sources (Fig. 21.1). We also found only nine publications from Scopus and 11 from Web of Sciences on that time (Fig. 21.2). It is imperative to note here that there was considerable overlap in the two sources as both sources recorded the same publications. Under the circumstances, we additionally considered some relatively credible unpublished and published national and international reports and documents including reports published by World Bank, United Nations Development

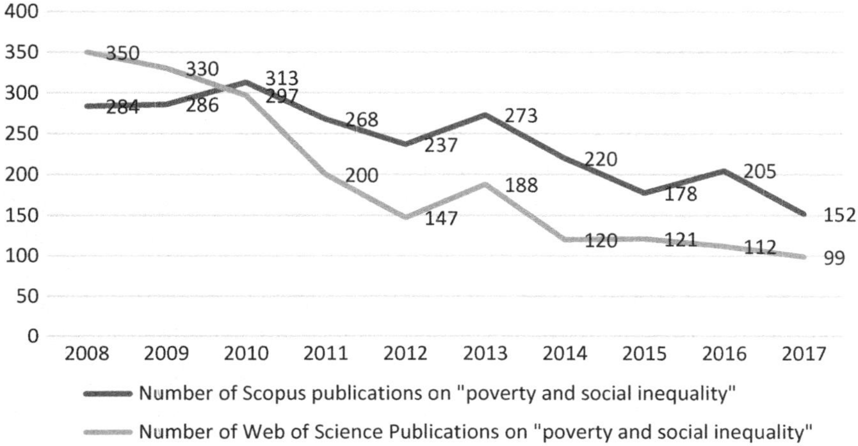

Fig. 21.1 Number of publication of Scopus and Web of Science on 'poverty and social inequality in Bangladesh'. (Sources: Scopus 2018; Web of Science 2018)

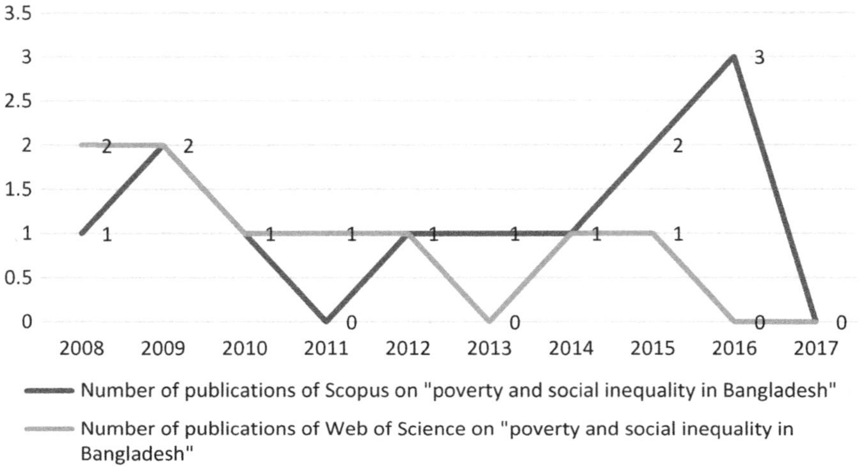

Fig. 21.2 Number of publications of Scopus and Web of Science on 'poverty and social inequality'. (Sources: Scopus 2018; Web of Science 2018)

Programme (UNDP) and Bangladesh Bureau of Statistics (BBS). Finally, we considered 42 publications which were immediately relevant to the objective of this study.

Data Analysis

We used a thematic approach for the data analysis. The major themes explored include rural–urban disparity, income distribution and income inequality, access to education, food intake and access to health, access to land and social opportunities and poverty and inequality within gender dimension. The original themes were examined within each literature to maintain the integrity of each author's work. Common factors were identified as themes, harvested across studies and combined to form a synergistic understanding, resulting in a number of discrete categories. A structural and contextual understanding emerged from the process that the studies and reports described 'poverty and social inequality' and 'poverty and social inequality in Bangladesh'. Throughout the theme extraction process, the authors triangulated with a specialist panel to ensure validity of evaluation and accuracy in theme selection (Reza et al. 2018; Chowdhury et al. 2018).

The Review Findings: Results and Discussion

Rural-Urban Disparity

Although some 65% of the population still live in the rural areas (World Bank 2016), in terms of education, income, housing, and water and sanitation, these people get lower facilities and opportunities compared to their urban counterparts. According to the Household Income and Expenditure Survey (2016), the literacy rate in rural area is 63% while it is 72% in urban areas and the expenditure on education in the two areas are Taka 604 (US$7) and Taka 1796 (US$22)—marking a striking disparity (Table 21.1). The income per household in rural areas is monthly Taka 13,353 (US$163); it is nearly 80% higher (Taka 22,565/US$275) in urban areas. Around 26% of the urban people have brick/cement roof house which is five times less (5.32%) in rural area. More than 37% of the urban people get supply water, which is only 2% in rural area; 94%

Table 21.1 Rural–urban disparity on some selected components

Components	National	Rural	Urban
Population (%) in 2016	100	64.96	35.04
Literacy rate (%)	65.6	63.3	71.6
Expenditure on education (in Taka)	925	604	1796
Income per household per month (in Taka)	15,945	13,353	22,565
Income per capita per month (in. Taka)	3936	3256	5748
Expenditure (in Taka)	15,715	14,156	19,697
Brick/cement house (%)	11.06	5.32	25.73
Supply water (%)	12.01	2.14	37.28
Electricity (%)	75.92	68.85	94.01
Sanitary latrine use (%)	61.37	53.27	82.12
Lower poverty line head count (%)	12.9	14.9	7.6
Upper poverty line head count (%)	24.3	26.4	18.9
Dependency	65.3	78.1	60.3

Source: Bangladesh Bureau of Statistics (BBS 2016) and World Bank (2016)

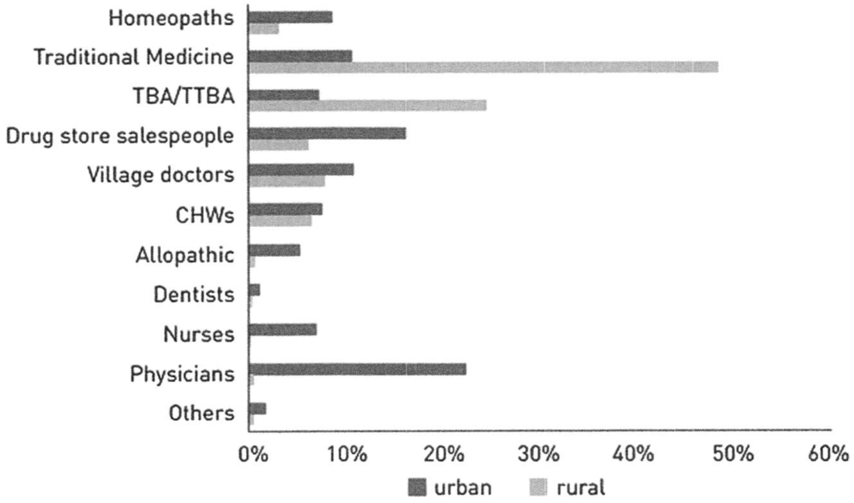

Fig. 21.3 Rural–urban distribution of health care providers by type (per 10,000 populations). (Source: Ahmed et al. 2015)

of the urban people get electricity, 82% use sanitary latrine which is 69% and 53% in rural areas, respectively. The head count rural poverty rate in rural area is 15% and upper poverty line 26%, which is 8% and 19% in urban areas. The dependency rate is 60% in urban areas and 78% in rural areas. According to the Fig. 21.3, the household monthly income, expenditure and food expenditure has increased in both rural and urban areas over time, but still there is considerable variation between the areas.

Income Distribution and Income Inequality

The pattern of income distribution can be a key driver to poverty and economic inequality. The growing income inequality has become a major concern for Bangladesh. According to the last Household Income and Expenditure Survey by BBS (2017), there is a significant level of differences in 2010 and 2016 in both income and expenditure in the rural and urban areas. Although the income and expenditure levels in both rural and urban areas have increased, the gap between the two areas has

Table 21.2 Income and expenditure in rural and urban areas in Bangladesh (in Taka)

Indicator	Household income and expenditure survey 2016			Household income and expenditure survey 2010		
	Total	Rural	Urban	Total	Rural	Urban
Income						
Income per household	15,945	13,353	22,565	11,479	9648	16,475
Income per capita	3936	3256	5748	2553	2130	3741
Expenditure						
Total expenditure per household	15,715	14,156	19,697	11,200	9612	15,531
Consumption per household	15,420	13,868	19,383	11,003	9436	15,276

Source: Based on BBS (2017)

Table 21.3 Income levels, 2015

Areas	Average monthly household income	Average income per capita per annum
	(taka)	(US dollars)
Dhaka City	55,086	2117
Other urban	24,031	769
Rural	18,349	580
All Bangladesh	31,883	1073

Source: PPRC and UNDP (2016)

widened more starkly. According to the data in 2010, the urban–rural household income difference per household was Taka. 4916, which doubled to Taka. 9212 in 2016. The difference of income per capita was Taka. 1611, which became Taka. 2492 in 2016. The difference of the expenditure level, however, decreased slightly between urban and rural in 2010 and 2016 respectively (Table 21.2).

According to the income level data in 2015, the average monthly household rural income was nearly one-third of that experienced in the Dhaka City: approximately Taka 6000 less than other urban areas and more than Taka. 13,000 less than the national monthly household income (Table 21.3). Income data also showed considerable variations across major occupational groups (Table 21.4). The highest income

Table 21.4 Income variation across occupational groups, 2015

Major income	Dhaka City		Other urban		Rural	
	Average monthly income in Taka	% of change	Average monthly income in Taka	% of change	Average monthly income in Taka	% of change
Business	121,902	6.2	97,363	32.9	60,375	n.a.
Petty/small business	19,556	20.6	20,637	8.9	22,774	28.9
Labour	18,024	32.7	16,499	37.7	10,992	31.3
Service	35,799	35.7	23,130	22.8	21,179	42.4
Professional	109,150	34.3	80,000	33.3	16,000	30.2
Agriculture	22,470	35.3	13,692	8.9	15,311	24.5
Non-earner (rental, remittance, savings	42,322	12.7	38,264	31.9	35,156	28.4
Others	28,825	17.2	18,000	50.0	16,167	29.3
All	55,086	16.4	24,031	23.4	18,349	10.3

Source: PPRC and UNDP (2016)

Table 21.5 The income dynamics for the period 2012–2015

Areas	% rise in average monthly household income over 2012–2015	% rise in average monthly household income over 2012–2015 adjusted for inflation
Dhaka City	16.4	−4.8
Other urban	23.4	0.9
Rural	10.3	−9.73
All Bangladesh	15.9	−5.2

Source: PPRC and UNDP (2016)

groups in urban areas are 'business', 'professional' and 'non-earners' (rental, remittance and savings), whereas these are business and non-earners in rural areas. The poorest groups recorded are labour 'and' petty/small businesses. All income groups have experienced a rise in their nominal income in 2015, although there are variations in the rate of change amongst the groups.

Table 21.5 shows the changes in the income levels over the three foregoing years. Data show that the nominal income increased for all regions.

Table 21.6 Income disparities

Income deciles	Dhaka city	Other urban	Rural
	Average monthly household income (Taka)		
Decile 1 (lowest)	9781	5570	5851
Decile 2	14,277	8485	8368
Decile 3	16,384	9976	9948
Decile 4	19,220	11,900	11,826
Decile 5	21,940	14,415	14,357
Decile 6	25,380	16,357	16,293
Decile 7	35,410	19,287	19,287
Decile 8	47,381	23,900	23,353
Decile 9	64,083	30,770	31,314
Decile 10	311,190	75,040	68,391
All	55,086 (*N* = 500)	24,031 (*N* = 340)	18,349 (*N* = 660)

Source: PPRC and UNDP (2016)

However, when the rise in nominal income is adjusted for inflation, income levels were essentially static over the recent past (2012–2015). Average income masks significant disparities in three regions. From Table 21.6, we see that all income deciles have moderately higher average income in Dhaka City compared to both 'other urban' and 'rural'. However, disparity within three areas is remarkable. The severe nature of such disparities is brought out further in Table 21.7, which groups the sample into 'bottom 40%', 'Middle 50%' and 'Top 10%'. A disparity in income is prominent for these three groups, particularly in Dhaka City. The top 10% of Dhaka City households enjoyed a per capita annual income of US$ 11,791 in 2015 (Table 21.6). On the other hand, the bottom 40% had a per capita annual income only 555 US$. The difference is less harsh for 'other urban' and 'rural', but still quite prominent per capita annual income of US$ 2770 and US$ 1836 respectively for the top 10%—compared to US$ 308 and US$ 264 for the bottom 40%. As depicted in Table 21.7, the per capita annual income for the bottom 40% was US$ 359 in 2015, while the income was US$ 867 for the middle 50% and US$ 4962 for the top 10%.

Table 21.7 Income share and average income groups, 2015

Income group	Variable	Dhaka City	Other urban	Rural	All Bangladesh
Bottom 40%	Average monthly income (taka)	14,421	9615	8342	10,657
	Annual income per capita (US$)	555	308	264	359
	% of share of total income	10.5	16.0	18.2	13.4
Middle 50%	Average monthly income (taka)	37,323	23,047	18,404	25,763
	Annual income per capita (US$)	1436	737	581	867
	% of share of total income	33.9	48.0	50.1	40.4
Top 10%	Average monthly income (taka)	306,567	86,612	58,106	147,388
	Annual income per capita (US$)	11,791	2770	1836	4962
	% of share of total income	55.7	36.0	31.7	46.2
All		*N* = 500	*N* = 340	*N* = 660	*N* = 1500

Source: PPRC and UNDP (2016)

Access to Education

Education is possibly the best indicator to measure social poverty and inequality because it has a tendency to be more unfavourable to the poor than the non-poor. The poor education generally leads to lower income, and lower income, in turn, leads to poor education of children. In 2010, the literacy rate for population 7 years and over was 57.91% at national level, which was 61.12% for male and 54.80% for female. This rate for both genders in the rural areas was 54.19% compared to 68.40% in the urban areas and male and female literacy rates in the rural areas were 56.90% and 51.39%, respectively, compared to 70.84% and 65.95% for urban areas in 2011 (BBS 2010).

Table 21.8 shows that in 2005 the primary net enrolment rate was 14% points lower for the poor compared to the non-poor in Bangladesh. The data also illustrate that inequality in enrolment broadens as children

Table 21.8 Education indicators in Bangladesh 2005

	Poor	Non-poor	Total	% difference between poor and non-poor	% point difference between poor and non-poor
Primary net enrolment rate (%)	61	75	68	23	14
Secondary net enrolment (%)	28	56	45	100	28
Higher secondary net enrolment (%)	3	12	8	300	9
Tertiary net enrolment (%)	1	7	5	600	6
% of 15–19 year olds completing primary	58	81	74	40	23
% of 20–24 year olds completing secondary	8	33	25	313	25

Source: Al-Samarrai (2008)

Note: Children are defined as poor if their household's per capita falls below the 2005 poverty line

grow up, with children in non-poor households twice as likely to be enrolled in secondary school than their poor counterparts. This is partly due to higher primary school completion rates among non-poor children. Access to tertiary education is profoundly limited; here too inequalities continue to broaden; children from non-poor households are 6 times more likely to be enrolled in post-secondary than the children from poor households. Given the strong link between livelihood opportunities and levels of education, these widening gaps in education are likely to make worse broader inequalities in society.

Al-Samarai (2008) argued that education inequality, particularly at the primary level, has been the result of poor governance; biases in the decision-making process and weaknesses in accountability institutions result in resource allocations favouring the non-poor. Goetz and Jenkins (2005) identify three main areas that lead to widening inequality. Firstly, where corrupt practices lead to the diversion of resources, the poor are less likely to be in a position to substitute their own resources for misappropriated government resources. For example, non-poor households are more likely to be able to afford private tuition to substitute for unauthorized teacher absence compared to non-poor households. Primary school students from the wealthiest households are 2–4 times more likely to pay for extra tuition compared to the poorest students. Poor households also have fewer exit options from the government-supported education system and will therefore be more affected by declines in access and quality associated with corruption. Secondly, the poor are in a weaker position in society as a whole compared to wealthier households and this makes it less likely that they will resist and expose corruption in the education sector. This lack of power also means they are more likely to face informal payments to access education services and finally, capture strengthens the position of local elites because they are often the direct beneficiaries and further reinforce the wide differences in education and livelihood opportunities of the poor compared to the non-poor (Al-Samarai 2008).

Food Intake and Access to Health

Food intake and access to health services are two important indicators for measuring poverty and social inequality. Regarding food intake, the data show that the calorie intake among the rural and urban or male and female does not differ significantly, but the difference is in the quality of food consumed, such as meat, egg, milk, and fruits, particularly those consumed on a daily basis. For example, only 2.3% of the households are able to afford meat, 14.4% egg, and 4.6% fruits on a daily basis, which suggests a striking difference when compared to the urban people (Table 21.9).

In Bangladesh, due to a wide gap between the rich and poor, the poor people can secure very few of the benefits of development. Khan et al. (2011) mentioned that the poor have less access to health care than the rich. They reported that the poorest 20% of women are nearly three times less likely to have received antinatal care (ANC) and nearly six times less likely to have a skilled attendant at birth (Fig. 21.3). The rate of women who delivered under a skilled attendant is very low (15% in 2001). Nearly

Table 21.9 Dietary adequacy: Food intake behaviour

Food item	Dhaka City		Other urban		Rural	
	% of Households able to have daily	% of Households able to have weekly	% of Households able to have daily	% of Households able to have weekly	% of Households able to have daily	% of Households able to have weekly
Fish	21.2	78.8	28.0	70.8	27.3	70.9
Meat	5.6	69.4	2.7	56.5	2.3	47.4
Egg	21.8	73.0	23.4	70.7	14.4	77.1
Milk	19.0	32.6	29.8	31.2	27.5	27.2
Fruits	9.2	50.6	6.5	46.9	4.6	37.9

Source: PPRC and UNDP (2016)

50% of the rural people are still taking medical services from the traditional health care, which is five times less (10%) in urban areas. The urban people have access to drug stores, nurses and physicians unlike those in the rural areas.

Ahmed et al. (2015) argued that the existing health workforce in Bangladesh is inequitably distributed. All significant health institutions (secondary and tertiary hospitals), pharmacies and health workers are greatly centralized in the capital city and big urban areas in the country, although 65% of the population lives in rural areas. The overwhelming urban bias of the distribution of qualified professionals remains a continual phenomenon in Bangladesh. On the other hand, the unqualified/semi-qualified allopathic practitioners such as village doctors and Community Health Workers (CHWs) are mainly concentrated in rural areas. According to the UNICEF (2018), the antenatal care coverage at least 4 times among the rural area was 18%, skilled attendant at birth 36%, institutional delivery 31%, and postnatal of mothers within 2 days 52%, which were 43%, 61%, 57%, and 75% respectively in urban area. According to the household wealth, the rates were 9%, 18%, 15% and 41% respectively among the poorest, which were 53%, 74%, 70% and 81% respectively among the richest. However, data at both residence and

Table 21.10 Disparity of maternal and newborn health intervention, 2014

		Demand for family planning satisfied by modern methods	Antenatal care coverage at least 4 times (%)	Skilled attendant at birth (%)	Institutional delivery (%)	Delivered by caesarean section (%)	Postnatal of mothers within 2 days
Residence	Urban	74.5	43.2	60.5	56.8	38.1	74.7
	Rural	71.5	17.9	35.6	30.6	17.6	52.3
Residence ratio (urban to rural)		1.0	2.4	1.7	1.9	2.2	1.4
Household wealth	Richest	71.4	52.9	74.4	70.2	51.4	81.2
	Poorest	72.9	9.0	17.9	14.9	6.7	40.6
Household wealth ratio (richest to poorest		1.0	5.9	4.2	4.7	7.7	2.0

Source: UNICEF (2018)

household wealth ratios show major disparity between rural and urban regions, and amongst the poorest and the richest (Table 21.10).

Access to Land and Social Opportunities

Land and local social opportunities (e.g., *khas* land, char lands, local market, access to local government properties and services) are important sources for the poor people, but their access is very limited there. Barkat (2004) argued that Bangladesh seems to be caught in the trap of economic criminalization. This trap is attributable to the legacy of anti-poor political economy and crisis in governance. He mentioned that the total *khas* land (agricultural and non-agricultural) and water bodies are of 3.3 million acres, but most of this land is captured by land grabbers and politically influential people. However, the access to the *khas* land is very limited among the landless and poor people. Additionally, there are some other issues, such as limited land rights of the religious and ethnic minorities, women's lack of access to land, fishing community's constrained access to water bodies, that are very common in Bangladesh. All categories of landless rural households are living below poverty. The average net annual income of a landless household is five times less than that of large landowner household and the valuation of capital assets of a landless household is 15.8 times less than that of a large landowner. Afsar (2010) found that the rising inter- and intra-regional inequality is an outcome of inequitable distribution of public goods and services confirming the clientelist syndrome of the government. The land grabbing among the ethnic minorities and ex-enclaves are very common in Bangladesh (Abusaleh and Islam 2019). Islam and Morgan (2012), Islam (2016), and Islam (2017) noted that the rural poor people have a lower level of access to the local market and community resources. According to the Sharia, a Muslim woman has some limited right to inherit property, while Daibhag does not have any provision for a Hindu woman to inherit.

Poverty and Inequality Within Gender Dimension

Gender also has serious implications for the inequality between male and female. Ferdaush and Rahman (2011) argued that such causes as the patriarchal structural and social institutions, lack of explicit policy initiatives, certain deeply embedded cultural values and mindset and early marriage trigger gender inequality in Bangladesh. Despite recent initiatives to enhance the female literacy in Bangladesh, there remains a considerable gap in enrolment between male and female. Literacy as well as the significantly higher proportion of female dropout from the system is still a major concern. There are also high rural–urban variations in case of the literacy rate by gender where the rural women are far behind than their urban counterparts, and male counterparts as well. Gender disparity is significantly high in tertiary-level education (university level). In 2001, among the total students in the public universities, only 24.3% were female students whereas the male enrolment was almost three times higher (75.7%) than that of the female (Ferdaush and Rahman 2011).

There exists a massive inequality between males and females regarding employment status. Figure 21.4 shows that women employment has

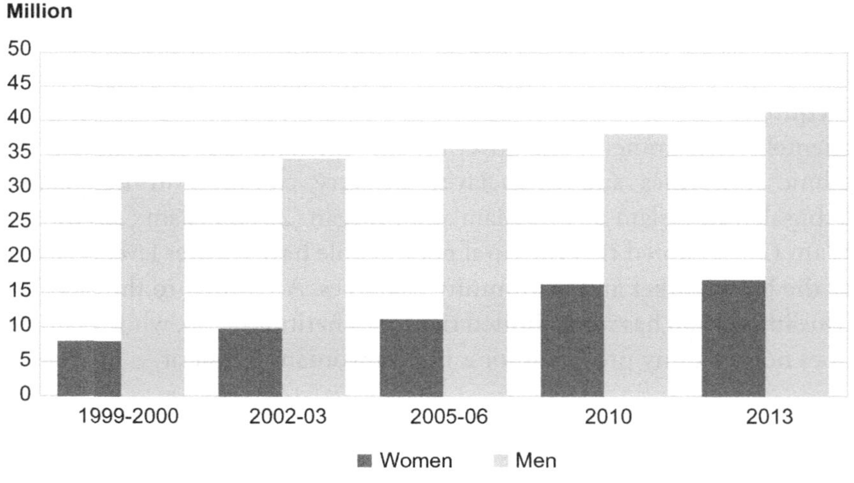

Fig. 21.4 Employed populations in Bangladesh (in million). (Source: BBS 2015)

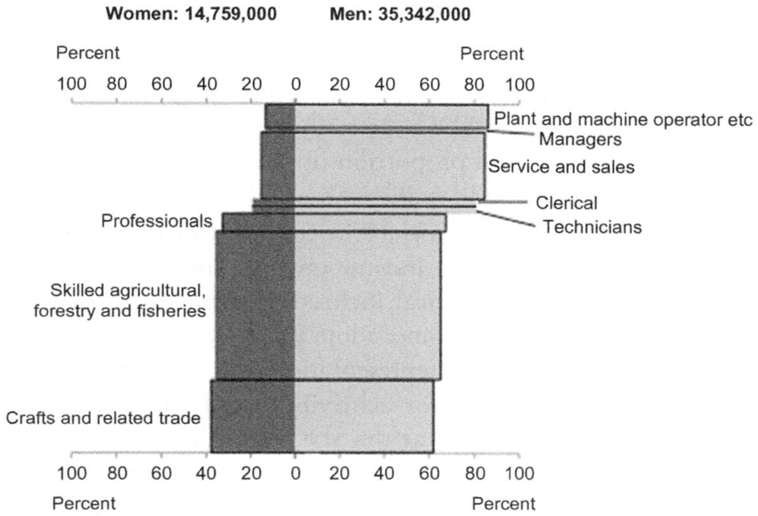

Fig. 21.5 Employed population aged 15 and above, by occupation, year 2013. (Source: BBS 2013)

increased over time but the rate is less than half of their male counterparts. Data from the BBS (2013) showed that the women employment rate was 17%, which was 42% among men (Fig. 21.5). At all occupational levels, men are dominating by 24% to 72% points with largest share (86%) in plants and machine-operating occupations and managerial positions (BBS 2013). Compared to men, women are mostly engaged in craft (38%) and in agriculture, forestry and fisheries (35%).Child marriage rate is overall quite high (more than 60%) in four quintiles (poorest to the fourth quintile); with 70.9% at the second quintile. The rate decreases quite notably to 49.4% when girls belong to the richest families. Thus the rate of child marriage among wealthy quintile decreases by 21.5% points when they are at the richest quintile. However, there is negative relationship between child marriage and wealth status and girls from better economic condition are less prone to child marriage. The difference in child mortality rate of males and females is very high, where female child mortality is higher than that of their male counterparts.

Epilogue: Selected Policy Implications

This chapter has explored the poverty and inequality scenarios in Bangladesh. Goal 1 of the SDGs prescribes such means to achieve 'no poverty' as reduction of the proportion of rural poor population, social protection, equal rights, build resilience among poor and resource mobilization. On the other hand, Goal 10 (social inequality) has highlighted the following remedial aspects: income growth; empower and promote the social, economic and political inclusion; ensure equal opportunity and reduce inequalities of outcome; adopt policies, especially fiscal, wage and social protection; ensure representation and voice for developing countries in decision-making for achieving social equality among the nations. If we reckon and examine the SDGs in light of the current poverty and social inequality in Bangladesh (as discussed above), Bangladesh has done better over time in terms of reducing poverty, but there still remain significant differences between poor and non-poor people, rural and urban people, male and female, and race and within religions in terms of income, employment, land ownership, education, food intake and health, and local resources and social opportunities.

A careful review of the report of the 7th Five Year Plan (7FYP) 2016–2020 reveals that the government has undertaken some programmes in line with the SDGs, especially in the areas of 'ending poverty' and 'social inequality in all its form'. Notwithstanding this plan's claim that Bangladesh has achieved remarkable success in reducing the poverty in last few decades, about a quarter of the population of 160 million still lives below the poverty line; hence setting that target is a necessary condition for attacking poverty with delineating 7.4% expected annual growth rate that was suggested by SDGs (Planning Commission 2017). Table 21.11 summarizes the relevant SDG targets as synergized with the government's 7FYP.

Reducing inequality has remained a formidable challenge for Bangladesh and ought to constitute a key national focus. The country has been maintaining stable economic growth of more than 6% yearly. The growth has contributed to the country's drive in poverty reduction. A critical look, however, reveals that improper distribution and lack of

Table 21.11 Development targets of Bangladesh in SDG (Goal 1 and Goal 10)

SDGs	Integrated issues (Development targets)
SDG 1. End poverty in all its forms everywhere	1. Reduction in the head-count poverty ratio by about 6.2 percentage points (from 24.8% to 18.6%) 2. Reduction in extreme poverty by about 4.0 percentage points (i.e. 8.9% in 2020) 3. Spending on social protection as a share of GDP to be increased to 2.3%. iv. Creation of a lagging region fund
SDG10. Reduce inequality within and among countries.	1. Spending on social protection as a share of GDP to be increased to 2.3% of GDP. 2. Reduce income inequality from 0.458 to downward.

Source: EquityBD (2017)

pro-poor development initiatives have prompted inequality on a rising trend. People may have been able to come out of extreme poverty, but this has happened due to trickle-down effect of the country's growth benefit; the resources have been concentrating among a few rich individuals: 10% rich people own one-third of asset in Bangladesh. The standard measurement indicator of inequality (Gini-Coefficient is 0.45) reflects an increasing trend—indicating a growing income gap between the rich and the poor in the country. The country has also witnessed the government's reduction of budgetary allocation in education, health and other common pro-poor development activities during the Milleneum Development Golas (MDG) period. Instead, the national revenue is going more to the sector of export promotion, re-financing the government banks, where there have been considerable cases of monetary embezzlement by the unscrupulous political and commercial quarters.

To consider the overall situation, the government policy option towards poverty alleviation and reducing social inequality will be more concentrated towards the distribution of the gains from growth; a regional, context-specific approach needs to be considered that would address and provide for, inter alia, the coastal vulnerability, income-generation opportunity for the extreme poor, special services to the most vulnerable groups, notably, people displaced by climate change and natural disasters, destitute women and disabled people, vulnerable elderly, indigenous people, and households in the hard-to-reach areas. Some of the pro-poor development strategies with relatively effective outcomes—such as social safety net programmes, allocation to community public health, vocational education and job-oriented training need to continue and further expand in order to improve the current situation of inequality. The government should forge a partnership plan and facilitative strategy so that the community-based and non-government organizations can play a more active and supplementary role, especially in providing basic social and livelihood services at the community level. In the long run, addressing inequality would require reforms at a more fundamental and structural level (for example, land distribution in favour of the poorest, targeted provisioning of income for areas worst hit by regional disparity); and this calls for systematic efforts and sustained commitment from the top political leadership.

Acknowledgments This chapter is based on the literature review section of the project 'Poverty and inequality in Bangladesh: Regional disparity in particular reference to the Sustainable Development Goals (SDGs)' funded by the Ministry of Education, Government of the People's Republic of Bangladesh and organized by the Bangladesh Bureau of Educational Information & Statistics (*BANBEIS*). The authors are grateful to the funding authority and organizer.

Disclosure Statement No potential conflict of interest was reported by the authors.

Funding The Ministry of Education, Government of the People's Republic of Bangladesh.

References

Abusaleh, K., & Islam, M. R. (2019). Traditional practices of Panchayat system among ex-enclave dwellers in Bangladesh. *GeoJournal*, 84(3), 657–669.

Afsar, R. (2010). Poverty, Inequality and the Challenges of Pro-Poor Governance in Bangladesh. *Journal of South Asian Development, 5*(2), 187–219.

Ahmed, S. M., Alam, B. B., Anwar, I., Begum, T., & Huque, R. (2015). Bangladesh Health System Review. *Health Systems in Transition, 5*(3), 1–186.

Al-Samarai, S. (2008). Governance and Education Inequality in Bangladesh. Retrieved from http://unesdoc.unesco.org/images/0018/001800/180086e.pdf.

Bangladesh Bureau of Statistics (BBS). (2010). *Bangladesh Household Income and expenditure Survey*. Dhaka: BBS, Bangladesh Planning Commission.

Bangladesh Bureau of Statistics (BBS). (2013). *Bangladesh Labour Force Survey*. Dhaka: BBS.

Bangladesh Bureau of Statistics (BBS). (2015). *Bangladesh Labour Force Survey*. Dhaka: BBS.

Bangladesh Bureau of Statistics (BBS). (2016). *Preliminary Report on Bangladesh Household Income and Expenditure Survey*. Dhaka: BBS, Bangladesh Planning Commission.

Bangladesh Bureau of Statistics (BBS). (2017). *Preliminary Report on Household Income and Expenditure Survey 2016*. Dhaka: Planning Commission, People's Republic of Bangladesh.

Barkat, A. (2004). *Poverty and Access to Land in South Asia: Bangladesh Country Study*. Prepared for the National Resources Institute, University of Greenwich, UK.

Bastiaensen, J., De Herdt, T., & D'exelle, B. (2005). Poverty Reduction as a Local Institutional Process. *World Development, 33*(6), 979–993.

Blackburn, R. M. (2008). What Is Social Inequality? *International Journal of Sociology and Social Policy, 28*(7/8), 250–259.

Blau, P. M. (1977). *Inequality and Heterogeneity: A Primitive Theory of Social Structure* (Vol. 7). New York: Free Press.

Burt, R. S. (1982). *Toward a Structural Theory of Action Network Models of Social Structure, Perception, and Action*. New York: Academic Press.

Carter, P. L., & Reardon, S. F. (2014). *Inequality Matters. William T. Grant Foundation Paper*. New York: William T. Grant Foundation.

Chowdhury, S. R., Wahab, H. A., & Islam, M. R. (2019). The role of faith-based NGOs in social development: Invisible empowerment. *International Social Work, 62*(3), 1055–1074.

Davies, J., Lluberas, R., & Shorrocks, A. (2014). Global Wealth Report 2014. The Year in Review, Credit Suisse Research Institute.

Encyclopedia of Sociology. (2001). *Social Inequality*. New York: Blackwell.

Ferdaush, J., & Rahman, K. M. (2011). *Gender Inequality in Bangladesh*. Dhaka: Unnayan Anneshan.

Goetz, A. M., & Jenkins, R. (2005). Reinventing Accountability. Making Democracy Work for Human Development. PrimeraEdición. *Gran Bretaña:* Palgrave Macmillan.

Green, M., & Hulme, D. (2005). From Correlates and Characteristics to Causes: Thinking about Poverty from a Chronic Poverty Perspective. *World Development, 33*(6), 867–879.

Hoffmann, R. (2008). *Socioeconomic Differences in Old Age Mortality* (Vol. 25). Springer Science & Business Media.

Hradil, S. (2001). *SozialeUngleichheit in Deutschland* (8th ed.). Opladen: Leske + Budrich.

Islam, M. R. (2016). Hunger reduction in China: An analysis of contextual factors. *Asian Social Work and Policy Review, 10*(3), 295–310.

Islam, M. R. (2017). Population shifting and risks of street children in Asia: Possible social work interventions. *Global Social Welfare, 4*(3), 149–157.

Islam, M. R., & Morgan, W. J. (2012). Non-governmental organizations in Bangladesh: their contribution to social capital development and community empowerment. *Community Development Journal, 47*(3), 369–385.

Islam, M. R., & wa Mungai, N. (2016). Forced eviction in Bangladesh: A human rights issue. *International Social Work, 59*(4), 494–507.

Joffres, C., Mills, R., Joffres, M., Khanna, T., Walia, H., & Grund, D. (2008). Sexual slavery without borders: Trafficking for commercial sexual exploitation in India. *International Journal for Equity in Health, 7,* 1–11.

Kamruzzaman, P. (2015). An Overview of Understanding Poverty from Diverse Perspectives. In *Dollarisation of Poverty: Rethinking Poverty Beyond 2015* (pp. 7–19). Palgrave Macmillan.

Kerbo, H. R. (2003). *Social Stratification and Inequality: Class Conflict in Historical and Global Perspective.* New York: McGraw Hill.

Khan, M. M. H., Krämer, A., Khandoker, A., Prüfer-Krämer, L., & Islam, A. (2011). Trends in Sociodemographic and Health-Related Indicators in Bangladesh, 1993–2007: Will Inequities Persist? *Bulletin of the World Health Organization, 89*(8), 583–593.

McCarthy, F. E., & Feldman, S. (1988). Processes of Impoverishment in Bangladesh: Reconceptualizing Poverty and Resources. *The Journal of Social Studies, 39,* 1–21.

Nguyen-Marshall, V. (2008). *In Search of Moral Authority: The Discourse on Poverty, Poor Relief, and Charity in French Colonial Vietnam.* Peter Lang.

Ossowski, S. (1963). *Class Structure in the Social Consciousness.* New York: The Free Press.

Planning Commission. (2017). *7th Five Year Plan 2016–2020.* Dhaka: Planning Commission, Bangladesh.

PPRC and UNDP. (2016). *Bangladesh 2016 Politics, Governance and Middle Income Aspirations Realities and Challenges an Empirical Study*. Dhaka: PPRC and UNDP.

Reza, M. M., Subramaniam, T., & Islam, M. R. (2019). Economic and social well-being of Asian labour migrants: A literature review. Social Indicators Research, 141(3), 1245–1264.

Ruiz, E., & Praetorius, R. T. (2016). Deciphering the lived experience of Latinos with diabetes and depression: A qualitative interpretive meta-synthesis. Social work in public health, 31(2), 70–82.

Schwartz, J. E., & Winship, C. (1980). The Welfare Approach to Measuring Inequality. *Sociological Methodology, 11*, 1–36.

Schuman, D. (2016). Veterans' experiences using complementary and alternative medicine for posttraumatic stress: A qualitative interpretive meta-synthesis. Social Work in Public Health, 31(2), 83–97.

Scopus. (2018). *Documents serach*. https://service.elsevier.com/app/home/supporthub/scopus/.

Sen, A. (1976). Welfare Inequalities and Rawlsianaxiomatics. *Theory and Decision, 7*(4), 243–262.

Sen, A. (1981). Issues in the Measurement of Poverty. In *Measurement in Public Choice* (pp. 144–166). Palgrave Macmillan.

Tilly, C. (1998). *Durable Inequality*. California: University of California Press.

UNICEF. (2018). Maternal and newborn health disparities Bangladesh. Dhaka: UNICEF.

Villar, A. (2017). Introduction. In *Lectures on Inequality, Poverty and Welfare* (pp. 1–15). Springer International Publishing.

Wade, R. H. (2014). The Piketty Phenomenon and the Future of Inequality. *Real World Economics Review, 69*(7), 2–17.

Web of Science. (2018). Search al databases. https://clarivate.libguides.com/webofscienceplatform/alldb.

World Bank. (2016). Poverty Headcount Ration. Retrieved from https://data.worldbank.org/topic/poverty?locations=BD.

Yunus, M. (2002). Toward Eliminating Poverty from the World: Grameen Bank experience. In *Making Progress: Essays in Progress and Public Policy* (pp. 371–378). Dhaka: Grameen Bank.

Part V

Contemporary South Asian Civil Society: Issues, Challenges and Concerns

Rajendra Baikady, Md Nurul Momen, and Cheng Sheng-Li

Introduction

Over the last two decades, civil society has evolved significantly and might continue to change in the ensuing years. At the global level, civil society is flourishing and contributing to the development of society in several ways. Globalization induced technological development; geopolitics and the market have created opportunities for the millions of civil societies across the globe. Civil society organizations or Non–governmental organizations essentially contribute to the development of state and society. In many countries across the globe, civil society has become an evconomic opportunity to improve the poor people's lives, and help them to uplift their socio-economic situation. The operations of civil society organizations are directed towards addressing social, human and economic issues, such as health, education, youth empowerment, good governance and women's safety. South Asian countries have experienced profound social problems, such as child labour, poverty, unemployment, illiteracy, public health issues, etc. Research has proved that South Asian countries show higher level of child labour and exploitation of children below the working age (Fatima et al. 2018). Furthermore, South Asia is home for the world's largest population with malnutrition, and stunted children under the age of five (IFPRI 2015). The number of people with diabetes, chronic heart disease and micronutrient deficiencies is on the rise (IFPRI 2015). Countries such as India, Pakistan, Bangladesh and Nepal have an alarming sex ratio at birth imbalance between girls and boys (Dyson 2012; Gupta 2006; Kaur 2016; Klasen and Wink 2002).

Over the past two decades, civil society organizations in developing as well as developed economies are receiving increasingly more attention from intellectuals, activists, donors and recipients of development assistance and political parties and leaders. Further, in many Western democracies, democratic consolidation has been achieved by the intervention of civil society. Emma Porio (2014) argued that civil society in Asia is still a central force in the process of democratic consolidation across different political regimes. For a better understanding, the growth and development of the civil society must be viewed from the perspective of the relationship between the state, the market and society (Emma Porio 2014). The process of globalization has impacted the growing numbers and influence of international and local Civil Society Organizations in many countries across the world.

Civil Society Role in Welfare: A Profile

Civil society in a broader sense is regarded as 'the area outside the family, market and the state' and surrounded by a spectrum of civil society actors with a wide range of purposes, structures, degrees of organization, membership and geographical coverage. However, today the definitions of civil society are changing, as the concept in present day includes a wide and vibrant range of both organized and unorganized groups which are far more than mere a 'sector" dominated by Non-governmental Organizations (NGOs). Along with the definition, the role played by civil society is always changing across societies the world over, and they are being shaped by the contemporary world phenomena. In the present society, civil societies are demonstrating their value as facilitators, convenors, innovators and service providers and advocates. Further, civil society organizations in most OECD countries account for 5–10 per cent of GDP (see Anheier 2014); in addition, they get more attention in the context of civic participation and people engagement. Civil society organizations in most of the countries are regarded as a source of social innovation. In many developing countries, civil society plays a considerably high role in dealing with social issues, such as poverty, child labour, women's rights and refugees. The development with civil society across the south Asia and rest of the world is taking place, irrespective of social economic and political differences among the countries. Civil society organizations in many developed countries as well as countries in economic transition have widely seen the contribution of civil society in the field of health, and social, educational and cultural services to its citizens. Civil societies are majorly working in collaborative framework and in partnership with other stakeholders such as from business, government and international organizations in dealing with the current societal issues and development. These partnerships and collaboration across public and private sectors are important in the context of sustainable development as no one sector can solve the most pressing challenges faced by a society in isolation.

Civil society in the South Asian region is playing an essential role in societal development, and in promoting good governance across the region. Characterized by high level of corruption, absence of adequate budget and financing and monopolistic approach, governmental structures in South Asian countries are unable to cope with the functions assigned to them. Hence, the role of civil society in the process of governance is receiving obvious emphasis in the region. Further, during 2007–2009, 75 per cent of the World Bank projects were designed and handled by the civil society organizations. Civil society participation in consultation on poverty reduction strategy papers (PRSPs) rose a 100 per cent, whereas 87 per cent rise was seen in country assistance strategies (CASs) (World Bank Civil Society Engagement 'Review of Fiscal Years 2007–2009).

Despite the contributions, civil society organizations across the region continue to face challenges in their functioning. In countries like Pakistan, the progress and functioning of civil society has been flawed since its independence in 1947. In Pakistan, civil society consists of a range of non-state and non-market citizen organizations and their initiatives functioning for the betterment of social, economic and cultural aspects of the country (Bhattacharya 2016). The civil society function of challenging the state and advocacy has been curtailed in Pakistan, whereas CSOs in the West majorly perform these functions as part of their role as watchdogs (Bhattacharya 2016). Pakistan, similar to other South Asian countries, being a developing economy, needs CSOs in dealing with numerous social-economic setbacks requiring government attention. The poor performance of the state in delivering the basic services, such as health care, education, and security has been taken care of by civil society to some extent (Bhattacharya 2016). While description for civil society varies across the countries, the typical civil society includes NGOs, non-profit organizations and CSOs. These are typically the group of organized and structured activities carried out by organizations dedicated to the public good. Some researchers also associate CSOs with the democratization process of the states, whereas other researchers have highlighted the impacts of globalization on development of civil society (see Dzwończyk 2016).

In India, CSOs were formed shortly after independence in order to support and intervene in the problems faced by migrants and their families. These areas expanded to family service, poverty and well-being of the people and social welfare services (Tahmina 2011). Failure of the state in delivering affordable and accessible services, such as health care, education, water and sanitation resulted in the development of CSOs (Luqman et al. 2013). Furthermore, CSOs are important aspects in the democratic governance process. Many of the CSOs are not only dedicated to enhance the democratic governance, but have also strengthened their capacity for upliftment of the poor and address the suffering of people in diverse societies.

According to the World Economic Forum (2013), CSOs perform the following roles:

Watchdog: promoting and ensuring transparency and accountability of all government institutions, and other aspects of both state and non-state institutions.

Advocacy: advocating for the changes and creating awareness among individuals, communities, and societies at large regarding societal issues.

Service provider: helping the members in society to avail of the services, such as education, health care, food and security, disaster management and rebuilding, preparedness and emergency response during and after disaster.

Expert: influencing the policy and developmental strategies through unique and rich knowledge, skills and experience and helping in identifying the problems and finding a suitable solution for issues faced by society.

Capacity builder: helping the people in the community in building their capacity, and supplementing with different trainings and learning opportunities through creating effective education and training environments.

Incubator: developing solutions that may require a long gestation or payback period.

Representative: representing the voice of marginalized groups or unheard voices and helping them to be heard in the larger decision-making process.

Apart from these functions, CSOs are also expected to provide their support in installing fundamental universal values, and encourage citizen engagement, support and protect rights of the citizens. Researchers have noted remarkable changes in the public policy process in the past 15 years (e.g. Davies et al. 2000; Oakley 2002; Davies 2004; Nutley et al. 2007). Civil society engages people in discussion, and thus encourages evidence-based policy formulation. Evidence-based policy making with peoples' participation allows people to make well-informed decisions about policies and programmes (Davies 2004, p. 3). Evidence-based policy making has emerged as an international phenomenon in the recent past (Davies 2004, p. 1). Further, evidence-based approach in policy making has become part of the political discourse in many European and OECD countries (e.g. Ozga et al. 2006; European Commission 2007; Eurydice 2007; OECD/CERI 2007).

Issues and Challenges

CSOs across the globe are working for the promotion of fair and equitable economic development, gender equality, and human rights. In many developing as well as developed countries, CSOs are working for the upliftment of marginalized people affected by poverty and other societal inadequacies. However, majority of these CSOs across the globe are facing numerous problems in executing their functions, ranging from inadequate funds to lack of internal democracy; identity problem; and other kinds of serious threats. Also, in some countries the CSO community faces ever-tightening restrictions. The governments through regulations and restricted media oversight control the freedom of the third sector and their activities by impacting the smooth and effective functioning of CSOs in many parts of the world. Thus, the space for civil society in the areas of advancing human rights and democratic principles are largely affected. Another important problem faced by the CSO community is to provide sufficient, appropriate and continuous funding for their activities. Further, accessing donors and funding their initiatives is a challenging task for many NGOs and CSOs because of limited resource mobilization capacity. In many cases, CSOs have to shift their intervention policies and strategies according to the donor priorities, which may be unethical behaviour and against the principle objectives of the CSO.

Lack of financial sustainability is the major issue faced by a large number of CSOs in developing countries. Most of these problems are because of their high dependency on foreign donors. The key aspect of sustainability of these CSOs is dependent upon how they are financed and where the fund comes from (Susan Phillips and Tessa Hebb 2010). However, charity and support from their respective governments remain crucial issues to their activities. Traditionally, these initiatives were funded by the government and through philanthropic support and later the new form of revenue-generating social enterprises. It is worthy to note that at present, the financing with regard to CSOs is undergoing significant transformation (Susan Phillips and Tessa Hebb 2010). Limiting access to national or foreign funding, erecting barriers to mobile communications, and applying onerous, arbitrary or poorly administrated registration processes are the steps to suppress or curb the freedom of CSOs and limit their functions.

International civil society leaders have identified a declining trend in funding for the civil society activities that shrinks their space in advocacy. Even though the national governments are speaking of the issues related to funding in many countries, international or foreign funding is facing serious issues in all parts of the world. Further, the 1990s' government retrenchment and fiscal restraint still has its impacts on the short-term funding horizon, greater competition and more stringent accountability requirements (Susan Phillips and Tessa Hebb 2010). Further evidences also show that the problems and challenges faced by the CSOs range from financial problems, accountability, lack of synergy and widespread corruption (Akindele and Ayoola 2017). Fund-raising income and the donations for CSOs have declined significantly across many countries in the world. Investments and endowments were hit hard and grant income declined considerably. However, scholars have noted a considerable increase in the demand for CSOs' service (Charity Commission 2009; Salamon et al. 2009). A study reported that 53 per cent of civil society revenue comes from fee, 35 per cent from government, and 12 per cent from private philanthropy (Salamon et al. 2003). In many countries, the new financial instruments, introduced by the government are bringing a considerable promise to enhance the productivity of the third sector (Young 2007). Despite the enormous effort by governments and the international civil society actors, CSOs are still facing some constraints, such as lack of funds, corruption, government interference, identity and lack of cohesion in the performance of their duty. As noted by Susan Phillips and Tessa Hebb (2010), ensuring productivity, effectiveness and sustainability of the civil society organization is an important public policy issue in contemporary society. CSOs face issues related to resource inadequacy and efficiency problems, while being contributors to society (Anheier 2014). The capacity of CSOs to perform is inexorably linked with the means of financial support received from external sources in order to sustain their activities, especially in countries like Bangladesh and Pakistan, where CSOs largely depend on external funding, and sustainability of these organizations are dependent such international funding (Bhattacharya 2016). At present, many of the foreign funding agencies are not offering funds, because they expect the government to fund the CSOs and make right-based interventions in their own societies.

The sustainable development goals—as defined to transform the world—are formulated with a target to achieve by 2030. Given the context, civil society leaders are required to understand how to shift their external context and relationship with the other actors, such as government and business. CSOs must also understand that in a constantly changing society no one can work in isolation, and collaboration with the business and governments is a must for the survival and functioning of the future advocacy programmes and right-based intervention of CSOs.

This part of the book focuses on the challenges and issues faced by the civil society organizations in the South Asian region. A contribution in these sections discusses the major thematic issues faced by the CSOs, and their solutions in the national and regional contexts. The last two chapters provide a comparative overview of CSOs in the South Asian countries, while touching on the issues and problems faced by these CSOs. Chapter 22 tries to conceptualize the theoretical underpinnings of the emergence of civil society movements, and their growth and development in South Asian countries. This includes identifying key issues and making thematic analysis of and understanding civil society movements in South Asia based on their different trajectories of regime evolution and growth. The chapter concludes with observations on future prospects for civil society movements in different countries of South Asia, making a special analytical framework on the trends of commonalities and differences. Chapter 23 explores the current state of slum women's participation in politics—at national level as well as at the local and community levels in Bangladesh. The focus of the chapter is to understand the conditions of women living in slum areas, and assessing the level of their involvement in politics. Chapter 24 highlights the promising practices with which civil society addresses governance problems in Bangladesh. The chapter also highlights the challenges faced by civil society itself in Bangladesh. Chapter 25 discusses the role of civil society in the developmental discourse of the South Asian region. The chapter makes a humble attempt to highlight the role of civil society organizations in the socio-economic development of South Asia. Chapter 26 is an attempt to provide a comparative illustration of CSOs in Pakistan, Afghanistan and Nepal. The chapter provides a conceptual outline of the civil society in social, economic and political settings, and then moves to explain the interconnection between civil society, law and media. Chapter 27 spells out the comparative analysis with respect to the possible role of civil society in maintaining political peace and stability between Bangladesh and Sri Lanka, while at the same time pointing to the difficulties faced by civil society taking on this role.

References

Anheier, H. K. (2014). *Nonprofit Organizations: Theory, Management, Policy* (2nd Revised ed.). Abingdon, Oxon: Taylor & Francis Ltd.

Bhattacharya, S. (2016). Civil Society in Pakistan: Functioning and Challenges. *South-Asian Journal of Multidisciplinary Studies, 3*(3), 24–45.

Charity Commission of England and Wales. (2009). *Charities and the Economic Downturn*. London: Charity Commission. Retrieved from http://www.charity-commission.gov.uk/Library/enhancingcharities/pdfs/downturn.pdf.

Davies, H. T. O., Nutley, S. M., & Smith, P. C. (Eds.). (2000). *What Works? Evidence-Based Policy and Practice in the Public Services*. Bristol: Policy Press.

Davies, P. (2004). Is Evidence-Based Government Possible? Paper presented at: 4th Annual Campbell Collaboration Colloquium, Washington, DC, February 23–24, 2004.

Dyson, T. (2012). Causes and Consequences of Skewed Sex Ratios. *Annual Review of Sociology, 38*, 443–461.

Dzwończyk, J. (2016). The Paradox of Global Civil Society. *Zesz. Nauk. UEK, 12*(960), 5–19. https://doi.org/10.15678/ZNUEK.2016.0960.1201

Emma Porio. (2014) Civil Society and Democratization in Asia: Prospects and Challenges in the New Millennium.

European Commission. (2007). *Towards More Knowledge-Based Policy and Practice in Education and Training*. Brussels: EC.

Eurydice. (2007). *The Relationship between Research, Policy and Practice in Education and Training*. Comparative Summary. Brussels: Eurydice.

Fatima, F., Fatima, S., & Ali, S. (2018). Prohibition of Child Labour Leading to Delinquent Behavior in the Constitution of Islamic Republic of Pakistan 1973. *Pakistan Administrative Review, 2*(4), 383–396.

Gupta, M. D. (2006). Cultural Versus Biological Factors in Explaining Asia's "Missing Women": Response to Oster. *Population and Development Review, 32*, 328–332.

International Food Policy Research Institute. (2015). *Global Nutrition Report 2015: Actions and Accountability to Advance Nutrition and Sustainable Development*. Washington, DC: IFPRI.

Akindele, I. T., & Ayoola, D. O. (2017). Challenges of Civil Society Organisations (Csos) in Influencing Democratic Governance in Nigeria. *African Journal of Management, 2*(2), 51–65.

Kaur, R. (2016). *Too Many Men, Too Few Women: Social Consequences of Gender Imbalance in India and China*. Hyderabad: Orient BlackSwan.

Klasen, S., & Wink, C. (2002). A Turning Point in Gender Bias in Mortality? An Update on the Number of Missing Women. *Population and Development Review, 28*, 285–312.

Salamon, L. M., Anheier, H. K., List, R., Toepler, S., Sokolowski, S. W., & Associates. (2003). *Global Civil Society: An Overview* (p. 28). Baltimore: Johns Hopkins Center for Civil Society Studies.

Luqman, M., Shahbaz, B., & Ali, T. (2013). Civil Society Organizations: Nature and Their Role in Rural Development (A Case Study of Hazara Region of Pakistan's Khyber Pakhtunkhwa Province). *Life Science Journal, 10*, 91s.

Nutley, S., Walter, I., & Davies, H. T. O. (2007). *Using Evidence: How Research Can Inform Public Services*. Bristol: Policy Press.

Oakley, A. (2002). Social Science and Evidence-Based Everything: The Case of Education. *Educational Review, 54*(3), 277–286.

OECD/CERI. (2007). *Evidence in Education*. Linking Research and Policy. Paris: OECD/CERI.

Ozga, J., Seddon, T., & Popkewitz, T. (Eds.). (2006). *Education Research and Policy*. London: Routledge.

Salamon, L. M., Geller, S. L., & Spence, K. L. (2009). *Impact of the 2007–09 Economic Recession on Nonprofit Organizations*. Communiqué No. 14, Listening Post Project. Center for Civil Society Studies, Johns Hopkins University. Retrieved from http://www.ccss.jhu.edu/pdfs/LP_Communiques/LP_Communique_14.pdf.

Phillips, S., & Hebb, T. (2010). Financing the Third Sector: Introduction. *Policy and Society, 29*(3), 181–187. https://doi.org/10.1016/j.polsoc.2010.07.001.

Tahmina, R. (2011). Radicalisation of Civil Society: A Case Study of Pakistan. In S. Pattanaik (Ed.), *South Asia: Envisioning a Regional Future* (p. 155). New Delhi: Pentagon Press.

World Economic Forum. (2013). *The Future Role of Civil Society*. World Scenario Series. Geneva, Switzerland: . Retrieved from http://www3.weforum.org/docs/WEF_FutureRoleCivilSociety_Report_2013.pdf.

Young, D. R. (2007). Toward a Normative Theory of Nonprofit Finance. In D. R. Young (Ed.), *Financing Nonprofits: Putting Theory into Practice* (pp. 339–372). Lantham, NY: Altramira Press.

22

Civil Society and Governance in South Asia: Issues and Challenges

Abhishek Pratap Singh

The concept of civil society has its origin with the development of modern-state. It is mostly understood as a community of citizens linked together by common interests and purposes. Originally identified as 'social mutuality', later on it was understood in terms of 'social space' that existed beyond the realm of the state. Resurrected in the 1970s, it has now become a dominant theme in the contemporary political discourse on the relevant issues like social and public good, voluntarism and participation, individual rights, the pursuit for modernity, development and inclusive growth. It denotes a 'community of citizens' with common cause and unity. Its formation may be spontaneous or evolutionary. It also forms a source of moral authority and value. It draws from the belief and practice of collective consciousness among the community beyond the limits of the state towards any larger objective of public interest or welfare. It consists of a diverse set of institutions covering a range of

A. P. Singh (✉)
Deshbandhu College, University of Delhi, New Delhi, India

© The Author(s) 2020
Md. N. Momen et al. (eds.), *Building Sustainable Communities*,
https://doi.org/10.1007/978-981-15-2393-9_22

activities, which are not taken up by the state and its institutions. In a democratic set-up, the presence of vibrant civil society is a positive constituent factor.

Over the years, the role of civil society has been a critical factor towards democratic consolidation in developing countries, including South Asian countries. The participation of citizens at the different levels of the political system and its decision-making process is regarded as one of the fundamental necessities for the functioning of democracies, and towards this, the role of civil society remains critical. Hence, the presence of civil society is also attributed as a basic pillar of democracy. In the case of South Asia, the growth and development of civil society have been fragmented and challenging, being shaped by a variety of factors like history and culture, levels of political democracy, economic development, demographic and social composition, mobility patterns, value system and effect of globalization.

More or less, civil society movements constitute a form of interest representation allowing different levels of engagements beyond the nation-state framework. The idea has seen growing importance with the development of the third wave of democracy in newly independent states. Besides, the changing nature of the state in the age of globalization and neoliberal world order has redefined the meaning and understanding of state–civil society relations. This has also been a phenomenon in South Asian countries.

State and Civil Society

Before looking into the analysis of civil society and governance in South Asia, it is necessary to understand the conceptual relation between state and civil society. In the present times, 'Civil society' has become one of the widely used terms, and which is essentially identified as a very common and 'big idea' (Edwards 2004: 2).[1] Broadly identified as the 'area outside the family, market and state', civil society encompasses a wide range of organization, purposes, membership and geographical coverage.

[1] Edwards, M. 2004, *Civil Society*, Cambridge Press: Polity.

The historical evolution and development of civil society have been largely associated with the development of the capitalist state. Certainly, the nature of the state has its level of influence in the emergence of civil society. Over the years, state–civil society relation has seen a complimentary level of interactions. The style of governance also has its source in the character of the state and primarily functions within the varying levels of state–civil society interactions in any political set-up. These are some very basic and critical realms of any socio-political activity. State–civil society relation varies between different regions and countries across the world, also being influenced by immediate sociopolitical and economic exigencies. It also carries a historical impression and cultural legacy. Complexities abound concerning the conceptualization and interpretations of the relation between the state and civil society (Jessop 1990; Kaviraj and Khilnani 2001).

Much of the contemporary work on the subject tends to identify how the state and its agencies face critical challenges under neoliberal conditions in the globalized world. This also affects the nature of state–civil society relations in different countries, including South Asian countries. With the development of 'retreat of state' and limitations on welfare politics by the public agencies, the new social space has been created to allow the emergence of civil society activism and movements in countries (Fukuyama 2001; UNDP 2002). Over the years, the concept of civil society itself has undergone various interpretations and achieved different meanings. In the current age of Information and Communication Technology (ICT) Revolution, civil societies are becoming more diverse and expansive comprising a different set of individuals, communities and organizations.

Neoliberal state order has allowed 'new spaces' of activity and engagement for civil society activism in South Asian countries, enabling them to build networks across geographical, social and physical divides. It also reflects vivid manifestations of changing social values and adjustments spurred up by the impact of modern information systems. The convergence of the nascent level of legislative reforms, the role of market economy and civil society also underline the pertaining confusion about their roles and responsibilities towards social empowerment and nation-building in South Asia. The insecurity complex has caused some levels of appropriation of civil society by neoliberal institutions as an alternative to the state, and also the resurgence of revivalist tendencies into the public sphere. Given the cultural and ethnic diversity in South Asian countries, this trend becomes challenging for democratic processes.

Civil society is a rather diverse concept, which has drawn different meanings and interpretations based on the subject of discussion. A survey of available literature on civil society suggests that this subject contains elements that are diverse, complex and contested. Perhaps, due to this, there is no commonly agreed-upon definition of civil society, and much a general understanding about its characteristics as a domain of voluntary and non-coercive collective action based upon shared interests and values (Merkel and Lauth 1998: 10).[2]

Over the years, the concept of civil society itself has undergone different interpretations and meanings. In the liberal school of thought, it has been understood as a domain of rights protection of individuals against any kind of arbitrary exercise of state power. In this way, it has found a new source and strength against neoliberal state policies. According to Marxist school, state–civil society relation has to be understood in the context of capital accumulation. Marx and Engels both described the modern state as 'a committee for managing the common affairs of the whole bourgeoisie', and endorsed for the vital role of civil society cutting across the state and nation boundaries (Marx and Engels 1848).[3] However, during the past three decades, the understanding of civil society and its role has been made to detach from the functioning of the State. There are also efforts to transpose the state in the Third World from its developmental practices (Blaney and Pasha 1993; Wickramasinghe 2005). In the name of promoting 'good governance' and 'developmental democracy', a large number of networks formed of non-governmental organizations (NGO) have found new roles and importance in the name of civil society activism (Freedom House 1999). In the context of South Asia, at times they are at loggerheads with the state and government, given their objectives, financing and work practices.

[2] Merkel, W., and H. Lauth. 1998. Systemwechsel und Zivilgesellschaft. Welche Zivilgesellschaft braucht die Demokratie? *Aus Politik und Zeitgeschichte*, 6 (7), pp. 3–12.

[3] Marx and Engels, (1848) *The Communist Manifesto*, revised edition 2004, Penguin Classics.

In recent years, the role and functioning of civil society are being largely understood beyond the domain of political state. In the eighteenth century, the western philosophers like Aristotle, Rousseau and Kant had identified the notion of civil society within the framework of the political society or state (Keane 1988: 36). It tended to be an expression against any kind of state arbitrariness and towards limited government. A major shift in the understanding of civil society took in the mid-half of the eighteenth century, when thinkers like Adam Ferguson and Thomas Paine conceptualized civil society much discerned from the state, and endowed with its forms and values. In later years, civil society has been conceptualized with the liberal state order but also counteracting it towards rights protection as well.

Beginning from Locke's idea of 'civil society being separate from state'; Montesquieu notion of 'separation of power' making a distinction between civil and political society; Hegel's idea of 'civil society being the historical product of economic modernization' (Keane 1988: 53); Marxist notion of civil society much like the state being an instrument of class domination under capitalist model; Gramsci's analysis of civil society being part of superstructure forming a source of ruling-class hegemony to Jurgen Habermas' (1992) idea of civil society playing a key role for the communication in the public sphere.[4] From the perspective of critical theorists of Frankfurt school, the notion of 'public sphere' was to be located within civil society forming source for mutual concern, critique, accessibility and expression of interests. Neo-Marxists scholars emphasized on new insights on the subject that 'transcended the conventional paradigm of state and civil society' (Seethi 2009).[5] In any case, the concept of civil society has drawn different meanings and relative interpretation concerning the state.

Over the years, with the formation of 'modern nation-state' in the sixteenth century towards its transformation into 'welfare state', and now with the emergence of 'neo-liberal state', the concept of civil society evolved and had been shaped by the changing nature and character of the

[4] Jurgen Habermass (1992) *The Structural Transformation of the Public Sphere*, MIT Press.

[5] Seethi, K. M. (2009). Reinventing Public Sphere. *Indian Journal of Politics and International Relations*, Vol. 2, No. 1: 174–91.

state. Civil society has found different levels of expression, depending on the nature of the state from being liberal, socialist, communist or minimal. The state does not remain a fixed ideological entity. Rather it embodies an ongoing dynamic process, with a change in objectives as it engages other social groups (Migdal 1994: 2001).[6]

In the context of South Asia, it is interesting to understand that the role of civil society was minimal in the early years of nation-building and development post-independence. The state was identified as a primary actor and driver for growth, making efforts for better allocation of goods and services within the society. In later years, with the increased level of democratization, the role of civil society found expression in the form of different webs of autonomous associations independent of the state coming in place and also having an effect on public policy (Taylor 1997).[7] This not only caused public policy adjustment in response to civil society demands in South Asian countries, rather formed an important instrument for the democratization of state–civil society relations. However, this is not true of all South Asian countries, given civil society formation on the revivalist's line causing 'value conflict' in countries like Pakistan and Bangladesh. Also, the lack of democratic space allowed certain levels of contestation between state and civil society. In this case, South Asian governments have often seen civil society groups as undermining state power and legitimacy and have attempted to bring them under control. While the presence of civil society allowed 'the possibility of democratization' in South Asian countries, often it also led to 'violent responses' from the state (Chandoke 1995: 51).[8]

Nevertheless, given the debate on state-civil society relations, a certain level of common ground can be found to understand the structure and level of interaction between both. To say being domain in itself, civil society comprises a large number of voluntary and autonomous

---

[6] Migdal, Joel S, and Kholi, Atul eds (1994) *State Power and Social Forces: Domination and Transformation in the Third World*, Cambridge University Press.

[7] Taylor, Annie (1997) NGOs and International Institutions: the Trade and Environment Debate. Paper presented at the International Studies Association 38th Annual Convention, Toronto, 18–22 March, 1997.

[8] Chandoke, Neera (1995), *State and Political Theory: Explorations in Political Theory*, New Delhi: Sage Press.

associations in the public sphere with competing and often conflicting interests and ideologies. While not purely driven by private or economic interests, it is largely directed towards some common social interest. It is both different and independent from the state and its political sphere, but given the fact that it influences the formation of public policy and state actions, it is oriented towards and interacts closely with the state. In this sense, it forms the intermediate domain of interaction between state and society at large. And towards this process, it remains a source for political democratization and transition, social empowerment, development and economic growth.

Civil Society Movements in South Asia

The development of civil society in South Asia is largely linked to the process of the third wave of democratization in post-war era when most of these countries became free and independent. The development process in the post-independent era led to the emergence of contingent forces forming the source for civil society movements in South Asia. In the later years, civil society organizations came to be considered as a key constituent for the survival of democracy. Moreover, they were identified as 'necessary links' in South Asian countries to build and attain a Western form of liberal political democracy. It was also found necessary to broaden the horizon of social welfare in the process of political modernization.

In the case of South Asia, there have been several debates if the concept of civil society can be made applicable, given the case of it being a non-homogenous political entity (Brunn and Jacobson 2000: 34).[9] In another major work on the subject, Partha Chatterjee (2008: 227) makes a distinction between 'civil societies' consisting of citizens, who are mostly middle class and a 'political society' consisting of people from ruling establishments.[10] In the context of South Asia, the character of regime differs, given the presence of formal democracies, military dictatorships

[9] Bruun, O and Jacobsen, M, eds, 2000. *Human Rights and Asian Values: Contesting National Identities and Cultural Representations in Asia*, London: Curzon Press.

[10] Chatterjee, P, 2008. Democracy and Economic Transformation in India, *Economic and Political Weekly*, April 19, pp. 227–282.

and monarchical systems. The level of political freedom and guarantee of fundamental rights that define a democratic form of government also vary widely within the South Asian continent. This is not only shaped by the nature of the political system rather is also defined by the varying levels of socioeconomic development. Many of the states have been victims of colonial rule, though how this has impacted their polity and civil society varies widely. Interestingly, the political institutions, which make the civil society, are equally present in South Asia. However, what varies from state to state is how they interact with each other and within the given political system.

One important factor to understand the politics of civil society movement in South Asia is the range of non-governmental actors, which exist at different levels and struggle for the 'democratic change' and space in the political system. In case of a more restricted and non-democratic nature of the political regime, much like in Pakistan and Nepal or Sri Lanka, this kind of civil society activism is in place. In the case of India, this kind of activism tends to seek more representation within the political and legislative framework. In this case, it is interesting to see the different factors shaping the politics of civil society and their relationship with the state. The role of the state is not of a 'neutral actor', standing apart from these political developments. Rather, the state is much involved and is often interacting with these agencies forming a role in civil society movements.

Another factor which relates to civil society movements in South Asia is that of whether any people's movements in the form of being an agent of social progress or change is primarily civil or not (Alexander 2008: 185).[11] For instance, the Liberation Tigers of Tamil Eelam (LTTE) movement while standing for the legitimate rights of the Tamil people against the indifference of the Sri Lankan state was largely guided by the use of force and violence. The same kind of logic applies to Maoist guerrillas groups in India, which fail to acknowledge the importance of democratic change and attempt to make their way in confrontation with the administrative apparatus of the state causing huge loss of lives and threat public

---

[11] Alexander, Jeffrey C, 2008. Civil sphere, State, and Citizenship: replying to Turner and the fear of enclavement, *Citizenship Studies*, vol 12, no 2, pp. 185–194.

order. In this situation, irrespective of the limited progressive nature of these movements, they fail to pass the test of being civil given the nature of its organization, functioning and operations. While adherence to public law cannot be the only hallmark for civil society activities, the protection of public order forms a key component in the regulation of state–civil society relations.

In the post-colonial era, the idea of the state as a most paramount agency for social development dominated the public discourse in South Asia. Certainly, the limitations of public policy and the inability of the political system to respond have caused civil society activism in the region. In South Asia, the formation of state processes and particularly the legal and political regimes under which activists operate makes a significant difference for civil society movements. At the same time, there are inherent cultural and historical commonalities in these countries. For example, in the case of India and Nepal, the relationship between the national and ethnic or tribal is a key factor in the process. In countries like Pakistan and Sri Lanka, difficulties of regional discrimination and representation remain more challenging.

Besides, a many South Asian countries are grappling with the gradual institutionalization of 'modernity'.[12] The process of modernization is still unfolding at different levels in South Asia, although they have attempted to a certain degree of institutionalization in terms of separation between political, economic and other associations. The civil society has to face a contest between 'pre-modern ascriptive identities' and tendencies of modernism. Civil society groups often tend to form a link between formal state framework and informal social expressions. Given this context, civil society is largely understood in South Asia as a fragmented and politically contested space forming challenge to understand changing local, regional, national and transnational processes.

In South Asia, civil society formation also faces some issues and challenges. To say, with the development of the neoliberal framework in these states, there is an 'essential contest' between three related discourses on policy, structures and value system. The affluent elite with transnational character is more interested in policy formulations to serve the interest of

[12] Elliott, C.M. (2003) eds. *Civil Society and Democracy: A Reader*, Oxford University Press. p. 51.

new economic forces in the country. However, in that process, it comes often in conflict with some established state structures and value system. This contest finds expression in civil society movements in South Asia. In the name of promoting economic growth, mostly understood in numbers, the loss to the environment, violation of rights and ethics and craving consumerism causes scope for civil society activism. If found disturbing for social stability, states often resort to response mechanisms, causing frictions in state–society relations.

Much of the civil society activism in South Asia has its root in the failure of the welfare state model to provide for basic access to education, health and food. Reasonably, civil society movements often take the form of being an important instrument to acquire political power in a democratic setup. It is critical to recognize that civil society, however we conceive of it, is not free from power struggles.

In South Asia, another issue before the civil society is to draw a balance with the state goals. Many of these countries are tilted towards the practice of crony capitalism, authoritarian ruling structures and business-politics link up. In the formulation of the development process, the desire to capture a larger share of state resources often leads to regional diffusions and economic inequality. The process of civil society activism in South Asia also has to do with the limitations of nation-building process. Whether to adopt a liberal plural model based on recognition of the multi-cultural and ethnic character of these states or to foster a more homogeneous nation-building process also forms a key role behind civil society activism. As the state failed to build 'good governance' and equitable development model, state–civil society relations become more contested allowing scope for activities in the informal sector.

Despite these issues, over the years civil society in South Asia has emerged as a powerful sector outside the established formal framework of politics (state) and economy (market). It is also playing an active role in promoting development and democracy, in the form of having various sets of activities within the public sphere. In South Asia, it must be underlined that given the diverse composition of these states, the civil society processes must remain largely pluralistic. Nonetheless, its formation can include both civil and uncivil elements within its domain, towards economic development, democracy and political change.

Governance in South Asia

It is not only in terms of civil society activism rather in the areas of governance that South Asian countries show certain levels of commonalities and patterns. To say, the notion of governance had a rather peripheral role in the social science discourse until the 1960s. Rather, it was only in the early 1990s that the meaning and understanding of governance has become more pervasive and critical in a different context (Rhodes 1997: 5). The concept of governance forms a key place in the contemporary discourse on developmental politics and change. The World Bank introduced the more normative concept of 'good governance' in its report in 1989.

Despite sharing a common history of colonial rule and suppression, South Asian countries show different levels of governance trends among themselves. While India showed development of mature democracy, Bangladesh and Pakistan are affected by military coups, Nepal is struggling with the formation of a democratic government since 2008 and the Maldives is having its first multi-party election in 2008. Bhutan has been able to maintain political stability, while Sri Lanka has to face consistent violent struggle from the Tamil separatists. However, despite these developments, some common trends also form the process of change in terms of governance in South Asian countries. Different sets of reforms, institutional changes and policy formulation have been tried to streamline public administration and governance mechanisms both at the central and local levels. In any case, the larger challenge relates to political corruption, poor delivery of services, unemployment, food security, access to basic amenities and ineffective public policymaking. Much of this stems from the fact that in the South Asian region strong loyalties based on parochial interests play a key role, unlike West where rationality, as it is understood, is based on principles of neutrality, universalization and formalism. Despite the many efforts by the leadership and government itself, in South Asia, challenges remains to establish a good governance model.

There are two specific dimensions under which the development of governance in most of the South Asian countries can be understood. The first dimension can be identified with the informal and formal authority system in the governance, depending on the means and ends. To say, how the organizational structures and authority are set according to the established patterns. The second dimension relates to the top-down manner of authority and how the stability has to be maintained in the system.

In the case of South Asia, paternalism has found a key role and place in the governance model of different countries. It refers to an authority pattern where obedience and loyalty to an individual or family are more important over any impersonal office or position. It is something that Max Weber has been referring to as the traditional authority system. Under the system, certain levels of social privileges have to be maintained based on parochial identities. The nature of leadership is personalized and the authority pattern is hierarchic and top-down. This kind of leadership patterns has led to dynastic rule as evident in Bangladesh, India, Pakistan, Sri Lanka and even in Nepal. Governance in such kind of system is based on informal and hierarchic model based on family and kinship lineage. This hierarchic relationship is often characterized as a patron–client relationship (Jamil 2007: 8).[13] For example, the Nehru–Gandhi family in India, Mujib and Zia families in Bangladesh, the Bhutto family in Pakistan, the Bandaranaike family in Sri Lanka, and the Koirala family in Nepal.

The practice of paternalism affects the functioning of governance in South Asian countries in different ways. *Firstly*, accountability in the governance requires adherence to formal rules, which often becomes a subject of personal obedience. *Secondly*, due to the practice of paternalism distinction between impersonal and private rules often becomes blurred and ignored. *Thirdly*, the question of merit often is subjugated to favour someone loyal and subservient. In all three cases, the larger damage is done to the cause of institutional legitimacy and crises of governability.

The second tendency, which has found its importance in the governance processes in South Asia, is the notion of an administrative or bureaucratic state. It comes close to the Weberian notion of legal-rational authority. Given the long period of colonial rule in the Indian

[13] Jamil I (2007) *Administrative culture in Bangladesh*. A H Development Publishing House, Center for Development Governance, Dhaka.

subcontinent, a strong system of bureaucratic norms and order has been built into the political system and governance of the South Asian countries. This system of governance is identified with formalization of organizational rules, rational human action, common and more predictable administrative order. In the case of India and Pakistan, a strong presence of bureaucratic order has been both positive and negative for the governance. While in one way it contributed to building a rule-based framework for governance, on the other hand, it has come in question given paternalistic system in South Asia, restricted entry and exclusiveness and devoid of informal interference.

In recent years, both these dominant tendencies in the governance of South Asian countries have to face some retreat and reform in their practice. The forces of the neoliberal market economy are more organized and influential to shape the forces of governance in South Asia. The alliance between politics and business has become a new trend in the politico-economy of these countries. The business community has seen an increase in political representation drawing a new role in public policy. While most of the South Asian countries still face the process of incomplete political modernization and lack of economic transition, it becomes challenging to see the shrinking space for a welfare state.

However, in the context of the new liberalism, the market economy and globalization, South Asian countries are also adapting, though slowly, to these new modes of governance resulting in less government and more governance. The focus now remains to build an effective, citizen-centric and development model of the governance system to improve the welfare and quality of life of citizens. The New Public Management school has inspired this thought process and is increasingly being embraced by the South Asian nations (Haque 2003: 942). As the change unfolds, it is interesting to see how the new force of change and transformation will affect the governance trends in South Asia. And in this process, the role of civil society cannot be overlooked. In line with liberalization policy, non-governmental actors and civil society have become increasingly involved in governance in South Asia. Rather than just relying on the donor discourse, it is necessary that more homegrown ideas must be given a place in the process of governance in South Asia towards development.

Conclusion

In the context of South Asia, its political and sociocultural values pose certain challenges to the formation of vibrant civil society and changes in governance. In Nepal and Sri Lanka, post-conflict state processes are finding it tough to strike a workable balance between different contending groups in the governance. Similarly, civil society has also failed to take off given the ethnic and linguistic differences. While it has found some space in the urban sphere, the larger role for civil society is yet to be evolved in practice. In terms of promoting development politics, civil society must take the moral high ground by involving direct citizen action to question the causes of poverty, misery and absence of social security for the masses. Technology, geopolitics and the markets have created opportunities and pressures both for civil society in South Asia, and it needs to respond accordingly considering the larger social benefits. The new digital age and social media provided ample scope for civil society groups to organize and fight for people's interests. Similarly, for the reforms in the governance, efforts must be directed to build functional democracy based on well-protected rights-based regime in South Asian countries.

References

Blaney, D. L., & Pasha, M. K. (1993). Civil Society and Democracy in the Third World: Ambiguities and Historical Possibilities. *Studies in Comparative International Development, 28*, 3–24.

Freedom House. (1999). Democracy's Century: A Survey of Global Political Change in the 20th Century. Retrieved from http://www.freedomhouse.org/reports/century.pdf.

Fukuyama, F. (2001). *End of History and Last Man*. Free Press United States.

Haque, M. S. (2003). Reinventing Governance for Performance in South Asia: Impacts on Citizenship Rights. *International Journal of Public Administration, 26*(8), 941–964.

Jessop, B. (1990). *State Theory-Putting the Capitalist State in Its Place*. Penn State: Press.

Kaviraj, S., & Khilnani, S. (2001). *Civil Society-History and Possibilities*. Cambridge: Cambridge University Press.
Keane, J. (1988). *Civil Society and the State*. London: Verso.
Rhodes, R. A. (1997). *Understanding Governance: Policy Networks, Governance, Reflexivity and Accountability*. Buckingham: Open University Press.
United Nations Development Program (UNDP). (2002). Human Development Report: Deepening Democracy in a Fragmented World, New York.
Wickramasinghe, N. (2005). The Idea of Civil Society in the South: Imaginings, Transplants, Designs. *Science and Society, 69*(3), 458–486.

23

Role of Civil Society in Bangladesh: Challenges and Prospects

Md. Awal Hossain Mollah

Introduction

There are verities of opinions on the origin and development of the concept of civil society. However, the thinking goes back many centuries in Western thinking with its roots of Ancient Greece Roy (2015: 32). The idea of civil society is considered as a modern concept since the eighteenth century (at the time of political thinkers Thomas Paine to George Hegel) and the notion of civil society is used as a separate domain parallel to the states (Carothers 1999). It is one of the vital instruments of social life that is voluntary, self-generating, self-supporting, autonomous from the state and bound by a legal or shared set of rules. This chapter highlights the role of civil society in various aspects of the governance of Bangladesh. It is found that there are lots of challenges that have been facing civil society in Bangladesh since its independence but their contribution to development and governance is enormous. In recent days, the respect and concentration of government and political parties to the civil

Md. A. H. Mollah (✉)
Department of Public Administration, University of Rajshahi,
Rajshahi, Bangladesh

© The Author(s) 2020

Md. N. Momen et al. (eds.), *Building Sustainable Communities*,
https://doi.org/10.1007/978-981-15-2393-9_23

society organizations (CSOs) are increasing. This is a light of hope for the democratic values of Bangladesh.

Civil Society: A Conceptual Voyage

The term *civil society* is a much talked and articulated word, and originates from Aristotle's expression '*koinōnía politikḗ*', which means a network described by a mutual understanding and shared values based on civil law (Nezhina 2017). The idea of civil society is usually defined as space and buffer between the state and market where individuals publicly debate with political and national issues related to the interest of common people (Walzer 1995). There are numerous varieties in the definition of civil society and there is no universally accepted definition. According to Veneklasen, "Civil society is a sphere of social interaction between the household (family) and the state which is manifested in the norms of community cooperative, structures of voluntary association and networks of public communication norms are values of trust, reciprocity, tolerance, and inclusion, which are critical to cooperation and community problem solving, structure of association refers to the full range of informal and formal organization through which citizens pursue common interests" (Veneklasen 2003). Civil society, according to Harry Blair (2000), "comprises the collective of those social organizations that enjoy autonomy from the state, and have as one important goal, among others, to influence the state, one behalf of their members". Connor has also expressed a similar view, "civil society is composed of autonomous associations which develop a dense, diverse and pluralistic network. As it develops, civil society will consist of a range of local groups, specialized organizations and linkages between them to amplify the corrective voices of civil society as a partner in governance and the market" (Connor 1999). The World Bank defined from a different viewpoint, according to the bank, "Civil Society Organizations (hereafter CSOs) refers to a wide of arrays of organizations: community groups, non-governmental organizations (NGOs), labor unions, indigenous groups, charitable organizations, faith-based organizations, professional associations, and foundations" (The World Bank 2010). Some cardinal features of civil societies have been identified by reviewing various definitions:

1. Separated from but buffer between the state and the market;
2. Formed by the people who have common needs, interests and values like tolerance, inclusion, cooperation and equality;
3. Non-political and non-bureaucratic;
4. Non-governmental and non-profit organization but work for the wellbeing of common people; and finally,
5. Develop by an autonomous process that cannot easily be controlled from outside.

Thus, we can say, civil society encompasses all non-political, non-bureaucratic, non-military, non-governmental and non-profit organizations and institutions that manifest interest and will of citizens; individuals and organizations in a society which are independent of the control of the government. It criticizes constructively against any public interest activities of the government. Strong civil society is pivotal for accountability and transparency, democracy, sound law and order situations, welfare or sustainable development, democratic leadership and the absence of arbitrariness, which are the core elements of good governance.

For better understanding, Hye (2000) mentioned a profile of civil society, which indicates various types of civil society's organizations and their activities which are worth mention (Table 23.1).

From the aforementioned profile, it is clear that members of civil society are active in diverse and wide-ranging areas and ultimately contribute to better governance. It is significant that in almost all countries, the number of CSOs is increasing by the scope of activities. To the extent of civil society expands, the government can withdraw into core areas of competence and where lumpy investment is required (major public infrastructure projects). Not only as a substitute for government in selected areas but also for articulating demand by their members for services provided by government agencies, civil society can assume a crucial role. On the other hand, in cases where some of the institutions in civil society provide limited public goods (primary health care, primary education, etc.) and consumer goods, they set a standard of quality and price-fixing

Table 23.1 Civil society profile

Organizations of civil society	Activities	Out comes
1. Media (Press, Radio, TV) 2. Trade Union 3. Professional Bodies 4. Private Voluntary Organization 5. Sports and cultural bodies 6. Social welfare organization	1. Public opinion forming, information dissemination 2. Collective bargaining and awareness creation among members about their rights and public policies relating to them. 3. Maintaining high standard of the profession, protecting the rights of the members, promotion of interests, etc. 4. Advocacy of social change in particular areas, provision of limited public goods to their clientele (credit, primary health, adult literacy, etc.) manufacture and distribution of consumer goods to member of the public (milk processing, fisheries, textile etc.) protection of the rights of disadvantaged groups (e.g. minorities, tribal people, destitute women, street children, etc.). 5. Promotion of sports and culture through the activities of their members. 6. Promotion of social welfare of members or limited clientele in particular areas through voluntary services of various types.	Better Governance

Source: Hye, Hasnat Abdul (2000: 27)

against which performance of public sector agencies and private companies can be compared. Because of this, actual and potential role expansion of civil society is an effective mechanism for the promotion of good governance. In recent years, the role of civil society has been seen as contributing to the accumulation of 'social capital', which is the community's capacity to participate in and benefit from development activities. Related to the concepts of social capital is a useful idea that it functions through and with government and markets, not in place of them. Thus, the apparent competitive or substitute relation between civil society on the one hand, and government and private sector on the other, dissolves into an interdependent network of synergistic development potentials. Good governance thus can be promoted not only through the autonomous emergence of civil society as a dynamic sector of development and change but also based on active promotion by the government through public policies and healthy competition with the private sector. Civil society as a participant in the business of governance has to demonstrate many of the same attributes applicable to the government and private sector. Among these attributes, accountability (to the clientele/community), transparency (to the greater public and the clientele), participation, empowerment, efficiency, the fairness of price and fees charged (e.g. interest rate for microcredit) may be mentioned (Hye 2000). On the other hand, the role of mass media as civil society organizations has a great impact on governance. It is now shaping the public opinion on every single issue of our existence-national as well as international.

Finally, the expansion of civil society is a key component of political liberalization, which promotes such civil rights as freedom of opinion, speech, faith and assembly; and the right of life, security and the ownership of property, which ultimately contributes to good governance.

Role of Civil Society in Bangladesh

The role of civil society is increasing through globalization; use of information and communication technology (ICT); right to information; political; and many other public interest activities of Asia, Africa and Latin America (Fisher 1993; Brown and Korten 1989). From poverty alleviation to good governance and human rights ensuring programmes have been initiated by CSOs. Thus, civil society has been widely recognized as the 'third' sector of a state. In Bangladesh, NGOs are contributing significantly to alleviate poverty through microcredit programmes. Besides, many other CSOs are working for women empowerment, human rights, better governance and democracy. The contributions of civil society in some aspects of state management in the context of Bangladesh have been analysed further.

Democratic Governance

In Bangladesh, many CSOs are working for democratic governance. For example, Shushashoner Jonno Nagorik (Citizens for Good Governance), known as SHUJAN, Janipap, Broti and Election Working Group (EWG) are working for democratic governance. They work as election observers to watch and evaluate the fairness and credibility of elections. The EWG mentioned that in the recent election in Gazipur City Corporation held on 26 June 2018, around half of (46%) poling centres found irregularities (EWG 2018). The CSOs also works for level playing field before the election. They campaign for awaking people so that every person can apply their voting right. The SHUJAN also arranged dialogue programmes among various political parties so that election can hold freely and fairly based on consensus. Thus, CSOs are working as an important agent for promoting good governance like transparency and accountability of the government. However, the pathetic side is very few ruling party or government never care and respect to the suggestions of CSOs. As a result, the tenth parliamentary election held in 2014 was neither free nor competitive or participatory. Only 39% reported by Election commission but 25% of votes were cast reported by election observers CSOs (Mollah and Jahan 2018). Khan (2015) mentioned that after the disputed elections on 5 January 2014, many governments, inclusive of the United States, Canada, Germany and others, called for credible elections to remedy the ongoing crisis.

Apart from these, CSOs always criticize government activities and policy initiatives that are against the public interest. Some CSOs like Transparency International Bangladesh (TIB), Center for Policy Dialogue (CPD), periodically disclose their reports publicly about the corruption of public sectors and economic situation. Thus, CSOs are working on accountability, transparency and various aspects of governance in Bangladesh.

Socio-Economic Development Through Microcredit

Thousands of NGOs are working in Bangladesh for socio-economic development and women empowerment since independence. Here all those NGOs are considered as part of CSOs that are working for not profit and wellbeing of society. By this time, Dr Yunis, the founder of Grameen Bank, has been awarded Nobel Prize in peace and Dr Fazle Hossain Abed, founder and chairman of BRAC, received several world-renowned awards for contribution in poverty alleviation. These two persons have been recognized worldwide for microcredit programmes to alleviate poverty from rural areas of Bangladesh. This is now used as a model throughout the world. Like these, many other NGOs (ASHA, TMSS, MOUSUMI, NEJRA KORI, etc.) are working for socio-economic development in Bangladesh. Though there are a lot of criticisms against the microcredit programme, it has a great positive impact which I have found in many known families who are living in villages of Narail district in Bangladesh. In a study, Roy et al. (2017) mentioned that nearly four million villagers borrowed microcredit from BRAC and the number of Grameen Bank microcredit borrower was 8.64 million by the year 2014. These microcredit borrowers have changed their fate by using this money. Similarly, TMSS and other NGOs are also working on poverty alleviation and economic development. Apart from microcredit, these NGOs are contributing to the development of the social, education and health sectors. For instance, private medical college and nursing institutes of TMSS, BRAC University, BRAC schools for poor and vulnerable children have established institutes of NGOs in Bangladesh.

Rule of Law and Human Rights

Rule of law and human rights are the two basic ingredients of democracy and good governance. In Bangladesh, some Human Rights defenders organizations are working as part of CSOs for ensuring the rule of law and human rights in Bangladesh. These organizations are very active against any kind of rights-violating activities of government or any other private parties or individuals. For instance, Bangladesh Mahila Parishad (BMP), Bangladesh Legal Aid and Services Trust (BLAST), Ain o Salish Kendra (ASK), BRAC and Nijera Kori are working for promoting the rule of law and human rights in Bangladesh. Besides, they are working for combating corruption and misuse of public power like extrajudicial killing by Rapid Action Battalion (RAB) and other law enforcement agencies in the name of crossfire or encounter in Bangladesh. In a report,

ASK, one of the CSOs, mentioned that a total of 275 people have been killed by law enforcers through open crossfire, shootout and gunfight during 1 January to 30 June in 2018 (Daily Star 2018). The CSOs also challenged the authorities' failure to address extrajudicial punishments imposed by *shalishes* in the name of fatwas, opinions that are supposed to be issued by Islamic scholars (Mollah 2014). These private punishments are significantly harmful to women's and girls' lives and health.

Public Interest and Protection of Environment

Public interest including sound environment is one of the requirements of sound people, development and good governance. Some CSOs in Bangladesh like BELA and BAPA are working to protect the environment, generate awareness, educate the actors and create values in the society. The CSOs have filed around hundreds of public interest environmental litigation (PIEL) cases and the judiciary delivered verdicts related to the protection of environmental rights of citizens (Mollah 2014). In the case of *Dr. Mohiuddin Farooque* v. *Bangladesh and Others*, the court declared, "right to life includes right to fresh air and water in which one can expect normal longevity of life". Similarly, the High Court also directed the concerned authority of the government to remove all the ternary farms and industries from Hazari Berg (Dhaka, capital of Bangladesh) to Savar. Some CSOs, for instance, the *National Committee to Protect Oil, Gas, Mineral Resources, Power and Port (Lead by Anu Muhammad)*, are fighting against the Rampal coal power plant project to save the Sundarban, the largest mangrove forest in the world. According to them, the project will destroy the ecosystem and adversely affect more than two million people who depend on the Sundarban region, such as woodcutters, fishermen and honey gatherers.

Policy Dialogue

Policy dialogue is one of the very crucial instruments of CSOs to make a government transparent, accountable and participatory to common people. In Bangladesh, some CSOs like TIB, CPD, and most of the TV channels have been playing this role for the past ten years or more. They arrange programmes on contemporary issues of politics, economics and governance in Bangladesh. Talk show on TV channels is very active in Bangladesh. Every day most of the TV channels conduct talk show at midnight and re-telecast the next day. For example, the Rohingya issue is now burning and most prioritize talk shows in Bangladesh. In a talk show, at least two experts may be politicians, retired bureaucrats, practitioners, MPS, Ministers, journalists university teachers or members of any other CSOs attened and present in a TV studio on a given topic and critically analyse various issues. Therefore, it is a very important way of disseminating information to the people and makes a solution to any problem of state affairs. Media as a part of CSOs is playing a very significant role in policy dialogue.

Apart from this, media disclose every incidence positive or negative to the public by their reports print or electronic ways. Through media reports, people can know every step good or bad of government. Thus, CSOs have been ensuring transparency and accountability of government.

How far the rule of law, independence of the judiciary and human rights are ensuring or violating is known by media. Therefore, media is the nerve centre of communication and administration.

However, a pertinent question is augment—how far the media is independent in Bangladesh? Since its independence, the government media TV and radio channels are completely controlled by the government but most of the private channels and newspapers are almost autonomous to broadcast their news and reports. However, critics and owners of private channels are still claiming and blaming that they are not completely free from the intervention of government. Besides, it is noticed that most of the think tank civil society members, daily newspapers and satellite TV channels are polarized by the ideology of politics. They are influenced by two mainstreamed politics—lefties (lead by AL allies) and rightists (lead by BNP allies). Therefore, the civil society of Bangladesh is criticized by the common people as they lost their fairness and neutrality. If we want to evaluate the effectiveness of CSOs' role, how far the opinions or suggestions are accepted by the government or other political parties? For the past 40 years, it has been witnessed that governments have never cared (or little care) to follow their suggestions.

Most of the talkers of TV talk show have been used as a paid think tank of political parties.

Movement for Changes

Some times CSOs are playing a key role to change society or state. In Bangladesh, we have some realities like this. Before the independence of the country, the intellectual, press and media, writers, journalists, poets and novelists even students also play their role as civil society, for example, at the time of language movement in 1948, mass movement 1969 for self-autonomy and liberation war in 1971 against Pakistani ruler. Recently, internet-based social media like Facebook is playing a revolutionary role in changing the social and political environment in Bangladesh. Facebook is using as a platform of all sorts of people, including students, school teachers to university teachers, politicians, bureaucrats, change-makers, environmentalists, players, poets, writers, journalists, columnists, scientists and like these who are very active in social media. We can mention an instance; on 29 July 2018, two college students (Dia Khanam Meem and Abdul Karim Rajib) of Shaheed Ramiz Uddin Cantonment School and College were killed in an accident by a reckless bus driving on Dhaka Airport road (*Dhaka Tribune*, 29 July 2018). Millions of students were gathered instantly with placard, banner, festoon and slogan through a Facebook message to protest road accidents. Many university teachers, political leaders, some ministers and human right activist and common people also joined with them. Their gathering turned as a mass movement against road accidents, for optimum punishment and a new law for road safety. The government compelled to accept their demand within a week and took initiatives for road safety, and approved a draft of Road Transport Act 2018 in cabinet meeting with a maximum penalty of five years jail and a fine of Bangladeshi Taka (BDT) 500,000 ($5000) for driver involved in reckless driving (*The Daily Star*, 7 August 2018). The government also compensates two million Bangladeshi Taka per victim. Along with these, five buses have been donated for the students of the victims' school and college mentioned earlier. Similarly, Gonojagoran Mancho, a Facebook-based organization, is working pro-actively against any anti-public interest activities of public or private individuals like violence against women (rape), extrajudicial torture, punishment, killing, leakage of questions before the public exam and like these. Thus, CSOs are working for social, political and policy changes.

Challenges and Prospect of Civil Society in Bangladesh

Though CSOs (especially right-based) in South Asia including Bangladesh are working very actively, they are facing a lot of challenges. A few can be mentioned here.

Firstly, CSOs are not institutionalized for even years. Since independence, Bangladesh started its' journey as a democratic state but still suffering from a lack of democratic practice. The scope of people's participation and participation of civil society in governance is limited. Ill and chaotic politics dominated by political leadership damaged and dwarfed these institutions. The fragility of democracy is rooted in the politics of division and confrontation in Bangladesh, which is a major impediment of development. CSOs can play a significant role in reshaping these realities. They can contribute to improving the malpractice of politics and democratic culture in the greater interest of society as they more or less linked with a political ideology and philosophy (Shelley 2017). However, most of the CSOs are divided by political ideology and expressed their views and opinions as a tail of political parties. The members of CSOs are talking in various forums, meetings, sittings, seminars and symposiums basically for gaining their interest, not for the public interest. For instance, to get favour from ruling party or when the party will come in power, they will get a position like a member of Election Commission, Chairman of Human Rights Commission, Governor of Bangladesh Bank or other.

Secondly, grass root people are still under the light of the spirit of CSOs and they have little knowledge about CSOs and they do not have adequate knowledge about the benefits that can be received from these organizations.

Thirdly, the majority of the members of civil society are emerged and playing a role as retired bureaucrats or military bureaucrats in Bangladesh. As a result, their mission and vision are more or less for the interest of the Elite class instead of common people. Therefore, the impact of the policy outcomes of CSOs is reflected in Bangladeshi governance to ensure the dominating class of society.

Finally, the CSOs who are solely working for the common people or vulnerable sections of society like human rights defender organizations and NGOs are facing numerous challenges to work independently. Sometimes government arrested the activist like Adilur Rahman Khan, secretary of Bangladeshi human rights organization Odhikar, and Mahmudur Rahman, owner and editor of *Amar Desh* daily newspaper on charges of 'fabricating information' about atrocities committed by state security forces though those organizations rejected this claim (The Guardian 2013). Also, it has been found that the organizations that are working for violence against women, children and extrajudicial killing are also pending for years to implement the verdict of the Supreme Court. Thus, the CSOs are facing challenges in Bangladesh.

Apart from this, few organizations are very actively working for public interest and promoting law and order situation in Bangladesh instead of misuse and colourable exercise of government power. CSOs took a lot of initiatives to protect human rights, protection of the environment, arbitrary arrest and detention, combating violence against women and children, and extra-judicial killing. Each year they publish their annual reports disclosing total incidences of female trafficking and acid attack on women, extra-judicial punishment, killing, disappearance, torture in custody and other forms of violations of human rights. They also help victims with legal assistance through public interest litigation and awareness-building programmes. A lot of NGOs are working as non-profit organizations in Bangladesh and there is no scope to deny their contribution to alleviating poverty and empowering women in Bangladesh. Day by day, common people, media and even government are awaking of the importance of CSOs. The government took several initiatives for developing CSOs like NGO Affairs Bureau and created essential laws and rules to institutionalize this sector.

Apart from these, as part of CSOs, the intellectuals, poets, student organizations and social media like Facebook activists are very active against any kind of inhuman and violence against human rights. For instance, the role of students was very crucial in the 1990s to fall of autocratic government and restored democracy. In recent years, Gonojagoran Mancho, a Facebook-based organization is working proactively against any anti-public interest activities of public or private individuals like violence against women (rape), extrajudicial torture, punishment, killing and like these. Thus, the prospects of CSOs are very bright in Bangladesh.

Conclusion

From the review of this chapter, it has been revealed that the CSO is one of the vital instruments of development, poverty alleviation, advocating policy change and contributing to democratic governance. The CSOs in this study have been found as a form of combined and cooperative roles separate from the state, market and family. Besides, political actors and ideology have influenced the civil society in Bangladesh. It has been politicized and polarized by the political parties. Eventually, the capacity of CSOs has been abridged to generate public trust, democratic norms and articulate public interest and monitor government actions. Therefore, civil society is yet to emerge as a vigilante force to uphold good governance and consolidate democracy. The CSOs are facing numerous problems and challenges in Bangladesh but there are ample prospects too. For better governance, democracy and sustainable economic development in Bangladesh there is no alternative to free and fair civil society. Civil society and its prospects to contribute to democratic governance are determined by a blend of diverse factors like history, politics, social, culture, external influence and regulatory framework. Therefore, all the CSOs should act as all kinds of influence free independent non-profit organization. The strong voice of the vibrant civil society is inseparable for ensuring democracy and good governance. The role of civil society has a great impact on the governance of Bangladesh. And it is expected to the political government that they should have respect to the civil society and make sure their participation as far as possible in public welfare-related policy issues for ensuring good governance which mass people expect.

References

Blair, H. (2000). Civil Society, Democratic Development and International Donor. In R. Jahan (Ed.), *Bangladesh Promise and Performance*. Dhaka: UPL.

Brown, D. L., & Korten, D. C. (1989). *The Role of Voluntary Organizations in Development*. Boston, MA: Institute for Development Research.

Carothers, T. (1999). *Think Again: Civil Society*. Washington, DC: Carnegie Endowment.

Connor, D. (1999). Public Participation. *Constructive Citizen Participation, 27*(3) 6–10.

EWG. (2018). GCC Polls: EWG Finds Faults at 46% Centres. Published by the *Daily Star*, June 28. Online. Retrieved from https://www.thedailystar.net/politics/gcc-election2018-ewg-finds-faults-46-percent-centres-1596700.

Fisher, W. F. (1993). *The Road from Rio: Sustainable Development and the Nongovernmental Movement in the Third World*. Westport: Praeger.

Hye, H. A. (Ed.). (2000). *Governance: South Asian Perspective*. Dhaka: UPL.

Khan, A. A. (2015). The Politics of Constitutional Amendments in Bangladesh: The Case of the Non-political Caretaker Government. *International Review of Law, 9*, 2–16. Retrieved March 13, 2017, from https://doi.org/10.5339/irl.2015.9.

Mollah Md., A. H. (2014). Judicial Activism and Human Rights in Bangladesh: A Critique. *International Journal of Law and Management, 56*(6), 475–494.

Mollah Md., A. H., & Jahan, R. (2018). Parliamentary Election and Electoral Violence in Bangladesh: The Way Forward. *International Journal of Law and Management, 60*(2), 741–756.

Nezhina, T. G. (2017). Civil Society in Post-Soviet Countries. In A. Farazmand (Ed.), *Global Encyclopedia of Public Administration, Public Policy, and Governance*. New York: Springer International Publishing.

Roy, A. (2015). Civil Society-Its Role in Good Governance: Participation, Accountability and Transparency. *Scholarly Research Journal for Interdisciplinary Studies, 4*(27), 32–48.

Roy, I., Al Raquib, T., & Sarker, A. K. (2017). Contribution of NGOs for Socio-Economic Development in Bangladesh. *Science Journal of Business and Management, 5*(1), 1–8.

Shelley, M. R. (2017). Democratic Transformation of Bangladesh: Challenges of Civil Society Intellectuals. *The Daily New Age*, October 27. Online. Retrieved from http://www.newagebd.net/article/27003/index.php.

The Daily Star. (2018). 275 Victims of Extra-Judicial Killings and Custodial Deaths, 427 Women Raped in 6 Months. *The Daily Star* Online Report, June 30. Online. Retrieved from https://www.thedailystar.net/country/extra-judicial-killings-in-bangladesh-human-rights-situation-report-1597750.

The Daily Star, 7 August 2018.

The Guardian. (2013). Bangladesh Police Arrest Activist Over 'Fabricating Information' on Atrocities, 11 August, 2013. Online. Retrieved from https://www.theguardian.com/world/2013/aug/11/bangladesh-arrest-activist-fabricating-information-atrocities.

The World Bank. (2010). Defining Civil Society. Retrieved from http://go.worldbank.org/4CE7W046K0.

Veneklasen. (2003). West African Civil Society Position on the Millennium Development Goals and Their Implementation, Paper Presented in West African Civil Society Forum on Millennium Development Goals, Dakar.

Walzer, M. (1995). The Concept of Civil Society. In M. Walzer (Ed.), *Toward a Global Civil Society*. Providence: Berghahn Books.

24

Proactive Functions of CSOs in Post War Situations: Importance of Having Branded CSOs (Learnings from Sri Lanka)

Jeganathan Thatparan

Background

State sovereignty is the concept that states are in complete and exclusive control of all the people and property within their territory. In Sri Lanka, both natural and man-made disasters (tsunami and armed conflict) created a pathway to question sovereignty, and weakened existing systems by creating dependency on welfare supports/response. And also, this situation has significantly damaged existing social protection agencies which have been already seen as politicized. Further, the overlap strategies of state agencies, line ministries, national authorities, police, law and enforcement unit centres have confused the general public. Though there is an umbrella organization called NGOs Secretariat established by the Government of Sri Lanka, many INGOs, NGOs and CBOs remain as

J. Thatparan (✉)
Attorney-at-Law, Notary Public, Colombo, Sri Lanka

© The Author(s) 2020
Md. N. Momen et al. (eds.), *Building Sustainable Communities*,
https://doi.org/10.1007/978-981-15-2393-9_24

unregistered but engaged in implementing several humanitarian projects for more than 25 years.

In 2016, the state re-established the NGOs Secretariat for coordinating all NGOs' work in Sri Lanka. But NGOs in Sri Lanka highly rely on foreign funding for their plans and activities. Thus, the rules and controls were slightly changed to accommodate more investment and income through humanitarian aid while ensuring non-interference into the internal affairs (state sovereignty).

However, little consideration for changing the paradigm of the civil society organizations was considered in the past 30 years other than centralization. This trend of state personality created because of two reasons:

1. The Government of Sri Lanka (GoSL) tightened the acts of NGOs in November 2006 when the GoSL and Liberation Tigers of Tamil Eelam (LTTE) has announced their withdrawal from peace negotiations and commenced the war.
2. Though certain activities of the NGOs are connected to the subject matters which have been devolved[1] to the sub-national level state authorities in Sri Lanka, almost all the NGOs and their acts were fully controlled, well-coordinated and monitored by a central authority.[2]

By citing these, two different case studies describe the past and deal and spell out the possible role of CSOs in the Transitional Justice and Peace Building process. Further, those two case studies elaborate on the dynamics of civil society in situations of armed conflict or natural disasters and look at how the international community and donors can ensure their policies adequately empower citizens and civil society in all stages of peacebuilding and development.

[1] Accordance to the 13th amendment of the Constitution of Sri Lanka.

[2] NGOs Secretariat and Presidential Task Force for North & Eastern Provinces specially to monitor the development projects implemented in Northern and Eastern provinces.

Case Study One: State Recognition of the Civil Society Organizations in Sri Lanka

Sri Lanka[3] is a country that is recognized as a free, sovereign, independent, and the democratic socialist Republic; further, the constitution assures that sovereignty is in the People and it is inalienable which can be exercised and enjoyed in a certain manner specified in the constitution. This manner should be exercised or enforced through the supreme body such as a parliament. Sovereignty includes the powers of government, fundamental rights and the franchise.

In the past several decades, it should be noted that many laws introduced by the government concerning NGOs work; they mainly introduced to govern NGOs or the community-based local civil society organizations (CSOs/NGOs).

Registration of such organizations comes under the purview of the government agent (district or divisional level), or Ministry of Social Services at the provincial level. However, it is expected that all the NGOs must be registered under the NGOs Secretariat. The NGOs Secretariat was established in 1996 under the purview of the Ministry of Health, Highways & Social Services; in November 2020, the National Secretariat for Non-Governmental Organization function under the purview of the Ministry of Defense; shifted back to the Ministry of Defense from Ministry of National Co-existence, Dialogue, and Official Languages.

This is to ensure a conducive environment for national and international communities to contribute to the relief and development activities and also for better coordination among NGOs, government sector institutions and with the people who require the services of the NGOs. All those changes were made by the president through his executive power.

Although, there was a reason for such a move, the acts of NGOs were carefully monitored and screened. As NGOs provide analysis and expertise, serve as watchdogs and help agencies to monitor and implement international agreements, the interlinkage between NGOs Secretariat

[3] Known as the Democratic Socialist Republic of Sri Lanka Article 1 of Sri Lankan Constitution.

and Ministry of Defense is considered a challenging environment. NGOs' relationships with offices and agencies were watched carefully by the state with the name of national interests and security. Between 2009 and 2012, the situation worsened, as the NGOs projects had no interconnections with international law, human rights, and defending the equal rights philosophy of the United Nations system.

The aforementioned move and changes question in reality the context of sovereignty of the parliament. As described in Diceyan view, parliament is supreme, and supreme body[4] which makes laws for people. This can be further illustrated in two ways. One is the principle nature of the Diceyan view of "parliament as a law-making body" and second is the check-in balance plus functional elements of the state organs. Though, the PTF (as a mechanism to control NGOs) is considered as a parallel mechanism introduced by the president, while the parliament's legal unit (NGOs Secretariat) paralyzed due to the interference of other state organ. Meanwhile, the NGOs Secretariat also has taken measures to interfere with certain subject matters by establishing district level coordination bodies with a state officer appointed as the coordinator paid by the government. There was not a system to prevent one branch of government from having more power over another; instead, the government would be controlled by one group of people. It would not be fair to the people of Sri Lanka. No, any NGOs raised their strongest voice collectively to discipline state organs in this regard.

Case Study Two: State Personality and Strategic Plan

Globally, the corporate sector's partnership, NGOs' involvement and public–private partnerships are considered as the main pillars for sustainable development. Such groups, agencies, committees and institutions should be allowed to work independently. To ensure better-coordinated service delivery, coordination with the line ministries expected, however,

[4] Firstly formulated by John Austin in his book, titled as *The Province of Jurisprudence Determined*, published firstly in 1832 and then detailed by Dicey.

non-inference into each other's discretion must be assured. This situation was changed in 2010, when PTF gave instructions to NGOs and UN agencies, and facilitated and guided for the development of joint action plan.

The guidelines[5] for developing Joint Action Plan by NGOs and UN for 2011, restricted by a circular issued by PTF, letter dated 26 November 2010, sent by PTF. All the NGOs and the UN agencies were requested to adopt the guidelines, and short deadlines were given to submit their joint action plan for 2011 in a template, under those subtitles.

Evidence has been proven as described earlier for the NGOs and CSOs were supervised and monitored by PTF or by the GoSL. Especially, it should be noted that a major strategic review of NGOs' work during the Tsunami period laid down several challenges, including listening to the community more, helping them promote themselves and become independent, and helping CSOs hold donors and GoSL to accountable. Almost ten years on these empowering ideals still present several different challenges for GoSL and raise questions about how it collaborates and builds relationships with others. Sri Lanka joint with the Open Global Partnership (OGP)[6] in 2016 and actively get engaged through the Ministry of External Affairs and other cabinet ministries. OGP was launched in 2011 to provide an international platform for domestic reformers committed to making their governments more open, accountable and responsive to citizens.

In addition, the Consortium of Humanitarian Agencies (CHA) and National NGO Action Front are other two national platforms established collectively by NGOs to work collaboratively. Though, they have raised voice (muted or well-tuned) for the benefits of the civil society organizations in general, not on those challenges listed. In summary, the national forum or any other CSOS association couldn't imagine or support to act proactively. Instead, it started to function as implementing agencies of institutional donors or funding agencies. The frequent changes at the ministry level—the inevitable breakdown of systems that occurs in conflict situations—deeply damaged the functions of civil society. The

[5] "Guidelines to Plan Action Programs for the year of 2011".

[6] https://www.opengovpartnership.org/participants.

dilemma of political impartiality is an ongoing challenge for all development actors in situations of conflict, with CSOs speaking about it very differently, depending on the context. This situation has been considered as so weaken in those locations affected by armed conflict as the people who are associated with the CSOs have also been affected by the conflict and were not in a position to overcome and then to work for the betterment of the people.

Due to those issues listed earlier, fear prevents people from speaking out and, in turn, also reluctant citizens or civil societies' participation in decision-making processes. CSOs in 2009 and 2010 are requested to report to the district coordinator at the district secretariat. But the approval process were centrally managed. In the meanwhile, the district level council of NGOs in many parts of the country had their own forum. All three levels have experienced with weak political systems, lesser supervised projects, and high levels of bribe/corruption have been most significant for getting approval, leaving negative long-term effect on civil society during the time of armed conflict/post armed conflict.

Conclusion

Therefore, the two case studies highlight the importance of promoting country-level strategic partnerships among state actors and civil society groups to downsize the negative consequences on the lives of the people especially the ones who have gone through severe traumatization in the past. Further, this partnership should be built for the realization of social harmony (which may give a pathway for peacebuilding). And also, the CSOs and NGOs must be capacitated immediately not only to prepare themselves in line with the gradual move towards developing nations but to proactively function towards the betterment of the general public and rule of law, and to fight for the better realization of human rights.

25

Scaling Up NGO Impact on Development in South Asia

Firdous Ahmad Dar

Introduction

The idea of NGOs is easy to comprehend but very complex to delineate in precise terms. Scholars across the spectrum have made some genuine efforts to conceptualize the idea of NGOs and have proposed wide-ranging typologies of NGOs. According to some definitions put forward by the famous sociologists like David Sills and Lord Beverage "NGO is an organization in which whether its workers are paid or unpaid is initiated and governed by its own members without external control". Adding it further, David Sills also explains it as "a group of persons organizing on the basis of voluntary membership without state control for the furtherance of some common interests of its members". The World Bank defines NGOs as "Private organizations that pursue activities to relieve suffering, promote the interests of the poor, protect the environment, provide basic social services, or undertake community development".

F. A. Dar (✉)
Department of Politics and Governance, Government Degree College Kulgam, Kulgam, India

© The Author(s) 2020

Md. N. Momen et al. (eds.), *Building Sustainable Communities*,
https://doi.org/10.1007/978-981-15-2393-9_25

Formally the term "NGOs" came into existence with the establishment of the United Nations in 1945 (Fonjong 2007). With the demise of the USSR, NGO sector has bloomed across the world gaining popularity in the public domain. Moreover, the process of globalization also propelled the growth and dynamism of NGOs. This paradigm shift has cultivated a new sense being seen as the triumphant of western liberal democracy. Consequently, it helped in proliferation of democratic forces across the whole Eastern Europe and Central Asia. The dominant discourse that stands and witnesses this phenomenon was observed in the scholarships of Denial Bell and Francis Fukuyama. Francis Fukuyama marked this shift as the end of ideology and correlated it with third wave of democracy swept across more than two dozen countries in the Eurasian continent which finally paved the way for more or less democratic states (Huntington 1993). And for others, it reflected the end of ideological confrontation gave extra momentum to the emergence of civil society at global level (Kaldor 2003). People across the world have witnessed not just the end of cold war; rather it was epitomized as universalization of western values, liberal democracy and capitalism (Howell and Pearce 2001). Civil society forms the epicentre of this change, questioning and revisiting the old developmental works in the wake of numerous transitions all over the world. Under such circumstances, NGOs have gained the currency with multiple functions stretching from socio-economic to political domains. Since then, NGOs are undisputedly important actors from being charity organizations that raise funds to the agents of advocacy for marginalized, disadvantaged sections of society. Over the past years, undoubtedly NGOs have reached out to the peripheral regions performing multiple functions of service delivery, environmental protection, social justice and economic development. It is widely believed that NGOs are now addressing and working on many issues like charity, welfare services, relief and rehabilitation (Edwards and Fowler 2002, p. 187). The fact of the matter is that NGOs exist and operate in every kind of society aiming at performing tasks of humanitarian work, providing basic services to the people, build and overcome obstacles in the way of democratic societies. However, in some countries with poor democratic record, these organizations are facing tremendous challenges.

While looking towards South Asia, in addition to country-specific culturally embedded reasons, the rise of NGOs can be attributed to the number of distinctive features of the region. Arguably saying if one digs deep, the reasons for the rise of NGOs may vary across the region but the common thread that motivated their eruption is the uniqueness of problem. These public-spirited organizations emerged to meet the community needs, defend interests of weaker section and promote new policies (Ben Néfissa 2005, pp. 29–41). Although concept of civil society and NGOs are of recent origin in the development discourse, but voluntary efforts as such are deep rooted in Asian culture. Indeed, the change has brought a new kind of thinking and in the light of such changing perception, conscious citizenry not only realized the importance of NGOs but learnt to form associations to be accredited by the state and same has been formalized through laws implemented on ground. Moreover, in the hope of promoting democratic culture and transparent government, these civil society groups have received huge amount of support from the society. And in many parts, it led to the strengthening of indigenous democratic culture based on *panchayat* raj system (Parnini 2014, p. 109). But the major reasons prevalent across the region are the presence of weak and economically fragile South Asian states. Secondly, weak governance, inadequate public funding and donor community priorities also contributed to the process of NGO formation in the region. Last but not least, owing to the weak and fragile states, the conditionality of foreign aid led to the diminishing role of state. This also contributed to the emergence of NGOs. Under such an environment, NGO community has got ample opportunities to enhance their role. By the late 1990s, NGOs have witnessed tremendous growth both in scope and role. Resultantly their contribution in the overall socio-economic development of South Asia have increased significantly, thereby, pursuing goals of woman's rights, highlighting alcohol abuse, focusing on poverty issues, building democratic alliances and many more (Lewis and Kanji 2009, p. 34). Since the late 1990s, developmental NGOs have begun to take-up new developmental micro projects in both rural and urban areas implementing several welfare-related works (Pearce 2000). These NGOs are now not only engaged in social sector addressing issues like poverty, health and education but also highlighting politically sensitive subjects like human rights, democratic deficit and poor governance. There is a growing acceptance towards their role and the region has witnessed a new trend of consensus building around certain issues between NGOs and government agencies. In this process of consensus building, governmental agencies are showing the attitude of willingness coming forward with the proposals to coordinate the developmental work in collaboration with certain NGOs in order to take forward the developmental works. In Pakistan, NGOs like Aga Khan Rural Support (AKRSP) and Action Aid are taking some necessary measures in order to improve the quality and accessibility of life in rural areas especially Northern areas. Many see NGOs as essential pillars of delivering not only development aid, but development itself. In order to be more effective and acceptable, apart from humanitarian work, these civil society groups need to highlight rights of voiceless people and monitor government policies. It is only after such change, the prevailing culture of scepticism within the larger sections of people may change. At the same time, it is unfair to expect that NGOs would fix all the problems or brush them all with one colour or putting NGOs under one category and dub them as biased with malign interests.

From late 1990s until today, NGO community have secured a notable position and are active on wide range of issues such as emergency response, humanitarian crisis, building democratic societies, building trust and managing conflict, cultural preservation, environmental activism and so on. In some countries like India, Bangladesh and Sri Lanka, governments are gearing up to share responsibility with NGO community and in so doing delegating effective powers of decision making and policy implementation with them. In this way, NGOs are seen as effective tools of policy execution both at local and central level. Here NGOs are enthusiastically engaged in the formulations and execution of social welfare programmes, and coordinate developmental activities in the arenas of health, education, disaster management, women empowerment, bringing environmental and human rights issues in the public domain. NGOs are organizing people at grass root level, emphasizing on collective actions redefining their role in society. As per the reports of World Bank, the Aga Khan Rural Support Programme an NGO operates in Pakistan's rural areas runs many kinds of community development programmes in many parts of the country. Apart from community development programme, it also addresses issues of women empowerment development of infrastructure, resource management and enterprise trainings. In the same way, in Bangladesh, a significant number of NGOs have done impressive works engaging considerably in rural areas. These NGOs are notable for their innovative methods for the rural development. At this point of time, both indigenous and international NGOs are working in close coordination while addressing issues of downtrodden and economically poor sections of the society (Eade and Leather 2005). In fact owing to the diminishing role of state in the region, NGOs like Bangladesh Rural Advancement Committee (BRAC), Association for Social Advancement (ASA), SAWA, Self Employed Women's Association, Smile Foundation in India, *Sarvodaya Shramadana* movement, SANASA Development Bank in Sri Lanka, Aga Khan Rural Program and Orangi Pilot project in Pakistan are carrying out several developmental works both at state and society level. In many cases, NGOs are building coalitions and coordinating with state in order to implement social welfare programmes in public healthcare, education, improving governance, building alliances for politically sensitive issues and so forth. As a result of this, there is some evidence that the traditionally marginalized people become more participative, become more involved in planning and budgeting process. Such is the change the region as a whole witnessed as a result of work undertaken by NGOs. In Bangladesh, there are NGOs funded by the Western countries engaging in educating people on elections and democratic values. Moreover, sometimes these NGOs intervened to break the political deadlock. However, this does not testify that NGOs are operating in a peaceful environment, but at many instances, the government of the day has often arrested some prominent NGO activists especially those who work in the field of human rights for being critical of its working. The fear of reprisal is always present in the mind of the NGOs working in the politically sensitive arenas. It could also be argued that most of the NGOs proved countervailing force to the overwhelming force of state. People started to question the authority and power of the state through collective actions claiming their rights. During the humanitarian crisis in Sri Lanka, NGOs have handled the situation carefully by starting many small projects for refugee and displaced people focused on citizen involvement and reconciliation process.

Analysing NGO: Implementers, Catalysts and Partners

Given the varied experiences exhibiting the region on account of diverse socio-economic and political conditions, one can argue that whole South Asia cannot be narrowed down to strait jacket. In some countries centralized regulatory frameworks tend to dominate, while in some others, civil society/NGO participation is quite smooth. Similarly while in others, civil society organizations were required to operate within approved regulatory frameworks and the space available to them was constrained by legal, political and socio-cultural pressures. Of course, the common feature that pervades the entire region is the nature of poverty stands next to sub-Saharan countries. This could be cited as a major motivational ground which invited the attention of diverse NGOs and civil society groups towards the region. Undoubtedly, the region witnessed the largest number of NGOs in the developing world who are engaging in whole spectrum of activities ranging from rural development, healthcare system, education, environment and much more. Initially, these NGOs began to work on some developmental activities ranging from delivering development through good governance like health care, education and infrastructural development to the running of awareness programmes, institutionalization of democratic norms and practices (Muni 2009). It is also believed that these developmental NGOs have spent more official development assistance than the entire UN system (excluding the World Bank and International Monetary Fund) (Jarvik 2007). Another source cites that there are thousands of NGOs engaged in developmental works in South Asia alone. For Instance, Bangladesh being the economically fragile state witnessed huge number of NGOs and some of them invited worldwide attention for their admirable contribution to the overall development of the society. Similarly, in India, for example, as per the reports of the Central Statistical Institute around 3.3 million NGOs are registered and literally it means one NGO for every 400 Indian citizens recorded in 2009 (Civic Freedom Monitor 2017). Among them a good percentage operates only in urban areas with least interest to spread their networks in the rural societies. And also within them very small percentage is involved in some development-related activities. Although millions of NGOs exist in South Asia but very small percentage of them have learnt to collaborate with community-based organizations partnering with them in implementing small micro projects. But at the same time, it must be accepted that NGOs are bringing some visible changes expanding and redefining their roles. Now NGOs are reaching out to those marginalized sections of the society which are yet to be connected remained unheard and voiceless within the governmental circles. In many packets this has led to the development of indigenous character of civil society sector based on community involvement and activeness of youth in the formation of NGOs, increased women's participation and indigenous people's groups like that of farmers and student movements. Given the peoples involvement and the space they are occupying in the public affairs, one can see this positive trend for social-economic change (Lewis 1998). For example, the *Kerala Sastra Sahitya Parishad* (KSSP) has done a remarkable work in the state of Kerala and its work has gained worldwide admiration. Its role in mobilizing public opinion was also reported by the United Nations in its successive reports. Its task of public education is appreciative and being considered as one of the best-informed and best-organized grassroots movement in India, with over 20,000 members bringing about societal transformation within a democratic framework of India. There is now growing and visible influence of NGOs in many respects like formation of public policy, mobilization of public opinion and execution of public welfare programmes. NGOs not only deliver developmental goals but also generate alternative space for debate and discussion and in this way problem of marginalized people. So this is an alternative way of looking issues thereby emphasizing on peoples-centric development with grassroots participation. Similarly, at interstate level, apart from state perspective, NGOs lay emphasize on evolving people to people initiatives with a rationale to promote alternative perceptions on issues so far dominated by state discourse. Although these are small steps but surely will have a strong impact on overall development of South Asian region. If followed with sincerity, perhaps these small steps definitely will contribute some valuable service in bringing peace, may increase business to business and people to people contacts. In some cases, like environment and gender issues, NGOs are working to ensure better and effective legislations. In others, they do monitor state policies with regard to women's rights seeing country's environmental compatibility with the international standards and so on. For example, the petition to the international human rights organizations is gaining strength bringing complaints before the commission on rights of minorities and women. And at many instances, this has created a sort of rift between civil society groups and state led agencies. Even some NGOs like Asian Human Rights Commission in partnership with International Commission of Jurists are conducting fact finding teams investigated issues of human rights violations (Dias 1993).

At the same time, one should not assume that everything is well with these public organizations. There is other perspective as well having different point of view with regard to the role of NGOs. At the societal level, there are also some fundamental questions in the minds of common masses regarding the role of NGOs. The common perception that pervades the entire region is about the selective approach of NGOs. In fact, it is not too easy to conclude but people are doubtful posing questions like, whether these public bodies are genuinely becoming vital in matters of administration, deliverance of services, voicing the demands of the marginalized? Or is it just an emerging trend addressing issues of few, high-status groups, under the guise public spirit organization? Justifying the same on ground, some critics accuse them of representing the interests and agenda of neo-liberal world of North. As this neo-liberal project expressed through structural adjustment is being carried out by NGOs, curbing the role of state restricts spending on welfare functions and a rolling back of social sector investment. Critics also argue that NGOs are often donor driven promoting free liberal values having no real purpose to uplift the poor and downtrodden people therefore not representing the genuine needs of the people. They also argue that in this process, local elites have got an upper edge collaborate with these donor-driven NGOs for their own selfish ends.

The fact that sometimes NGOs do operate in a very hostile environments which influence their work and role. The hostile environment is either due to bullying or excessive intervention of state and at times severe laws in place. Apart from governments' hostile attitude, NGOs do confront some serious issues like the heavy dependence of NGOs on western donors, lack of accountability within and a doubtful mandate often small scale and difficult to sustain. Even sometimes collision of civil society groups with the local elite complicates the problem further in many cases within and outside government circles, there is an apprehension that NGOs are representing a donor community and are money grabbers. Another potential tension is the belief that NGOs are propagating a particular faith as in the case of Pakistan, some civil society groups that receive funds from Middle East are believed to be indulging in such kind activities.

Some other critics are also doubtful with regard to their legitimacy within the public domain perceiving NGOs of lacking credibility and financial transparency. Whom do they represent, donors or beneficiaries; do they have hidden agenda under the garb of public civil service? Of course, these are some genuine queries that need to be substantiated with facts. Overall the track record of NGOs seems to be quite satisfactory and still then NGOs should do more to increase their efficiency and transparency. Nowadays, there is a huge pressure on working of NGO's and sometimes this has significantly dented their credibility within the public domain and government circles. As the present Indian government is heavily criticizing certain NGOs questioning and suspecting their work and their existence on Indian soil. The government often accuses NGOs of non-compliance working against the core national interests of India. Whether NGOs are complying or not complying depends on the type of NGO. Even some times government does possess dubious interests in curtailing the smooth work of NGOs for their own political ends. Moreover, NGOs defend themselves as victims of compliance creep; the *Modi* government has culled nearly two-thirds of the 33,000 NGOs it inherited from its books. The *Modi-led* establishment cancelled Green peace's licence and blocked its bank accounts under the Foreign Contributions (Regulation) Act (FCRA), asserting that these environmental-related NGOs were working against "the country's interest" and accusing them of violated laws and diverted foreign donations without informing the concerned authorities. Foreign observers struggle to interpret the situation. But in most of the cases only those civil society groups are being targeted who oppose government plans and these NGOs are unlikely to get any political support. Civil society actors also blame the present dispensation that the FCRA was promulgated to control the flow of foreign funds to civil society groups during the emergency period. But the irony is that the successive governments invoke it to curb and keep surveillance on the activities of NGOs.

On the other hand, there is a growing profile of NGOs complimenting the work of government agencies, initiating dialogue on issues confronting people in rural areas like farmer crisis and women security. In an attempt to bring change at grass roots level, some local NGOs have initiated the process of mobilizing the public opinion, advocate for reformation and participatory governance. Due to localization of NGOs, the whole region witnessed a lot of improvement in the delivery of humanitarian services especially in health, education and poverty eradication. Again one can easily infer from the sources reflected in the annual reports of FCRA of 2011–2012, rural development received the highest fund from donor community through the NGOs followed by the welfare of children, construction and maintenance of schools and colleges and research. As per the reports, the CBI found 30,81,873 NGOs operating in India, mostly in rural areas. Despite this huge number, the report alleges that a very small percentage exists in government official papers and among them a meagre percentage files their annual reports in the concerned offices. Out of 41,844 NGOs registered under FCRA Act, 16,756 have not filed their annual returns. The fresh wave of civil society groups in mobilizing public support for various causes like farmer suicide, drought mitigation, environmental protection and women security constitutes the power and resilience of NGOs. Bangladesh was flooded with NGOs and it is argued that 20–35 per cent of population is beneficiaries of NGOs' work and aid. This vibrant voluntary sector is largely based on and dominated by indigenous NGOs receiving aid from foreign donors working in the field of health, education and credit system (Shaffer 2013). In doing so, NGO community is showing an ability to negotiate conflicts within the society through peaceful methods.

The tension over the allegiance of NGOs towards donor community is sometimes hyped out of proportionate. The public hoax is being created to defame the NGOs otherwise NGOs have done some amazing work filled up the gap of irresponsive state. For instance, MYRADA a multi-project NGO working on numerous integrated rural development projects simultaneously operating in several states of India. As of now it accepts funds from various donor organizations. It is important to mention here that this multipurpose NGO is adopting numerous developmental programmes, experimenting new techniques and ideas in the rural areas.

Similarly, NGOs are also adopting different kinds of strategies in order to neutralize the undesirable effects of structural adjustment programme. For instance, experimenting social safety-net projects and social investment funds are of prime importance. In order to be more sustainable and well placed, both the government and NGOs need to cooperate and build partnerships. NGOs being prompt in action may prove suitable roadmaps in addressing the issues of those who are risk and lagging behind—or for whom the net was simply never designed to protect (Lewis 1998).

Over the years the positive trend is the growing acceptance and realization about the role of NGOs. Moreover, in majority of cases, governments across the region are now acknowledging the role of NGO's as collaborators, facilitators reducing human suffering and much more. Even the perception of people is changing and they see them as agents of change and having the potential to play a vital role if occasionally backed by political support. Many argue that NGOs have the potential to hold the governments accountable and have been instrumental in pushing the Government in India towards legislation of RTI Act 2005, bringing Lokpal Legislation and more importantly NGO's have been stellar performers when it comes to the issues of environment. The new approach is to help planners draw upon their views and "represent concerns of the citizens in a way that would be better understood by the policymakers". Many NGOs are serving public administration in designing schemes better and more efficiently. "NGOs have a responsibility to ensure that both government and corporate sector should act with fairness, accountability, transparency", says N.R. *Narayana Murthy*, chief mentor at Infosys. Yes, there are political and bureaucratic pressures, despite this, the credibility of NGOs is improving in the public domain.

India's flagships programmes like Rural Employment Guarantee Act, Food Security Act, National Rural Health Mission and the Right to Education have witnessed not only active participation of NGOs but proved instrumental in their implementation as well. These grassroots-level NGOs are engaged in providing services to the underprivileged, poverty lessening, rural development, public health care and environmental protection. It has also been argued that poor quality of life in the region is an obstacle in the way of building democratic societies so donor community started poverty mitigation progammes. In this way, considerable amount of the EU's aid package to South Asia channelled and utilized through NGOs especially those that work in the field of poverty alleviation and economic backwardness. So there seems to be sort of correlation between poverty and in order to build democratic societies, NGOs must lay emphasis on poverty mitigation programmes. Though human rights constitute an indispensable component of democratic societies but strengthen democracy by protecting human rights on an individual basis does not serve us in the long run. Therefore, emphasis must be on creation and sustaining human rights institutions, like the establishment of National Human Rights Commissions, robust rule of law and a dynamic media.

Unlike Pakistan, NGO sector in Bangladesh is unique and distinct. The country's NGOs sector has evolved both in the strength and role (Chowdhury and Mahmud 2008). There are some of the famous NGOs engaged at the grassroots level addressing issues confronting people like poverty, micro finance, women empowerment, education and health. The BRAC (1972), the *Grameen* Bank (1976) and *Proshika* (1976) are some of the famous NGOs operating at local level with a huge network of activities contributing in the overall social-economic development of the country. For instance, both BRAC and ASA have done some amazing work and established a strong network of microcredit schemes. These NGOs are focusing on agriculture, fisheries and social forestry. In fact, Bangladesh is the home for 19,000 action-based think tanks working in the field poverty mitigation schemes, healthcare system, affordability of education for poor and downtrodden and others. Among many reasons, the prevalence of weak state and the preference of donor agencies to provide funds through think tanks. It was also in response to inefficient and stagnant delivery state system that the donor society gradually shifted and activated these action-based think tanks running small developmental programmes in the areas of healthcare, education, poverty alleviation and microcredit (Nobusue 2002). They have developed many innovative methods and some have been emulating in other parts of the world. In addition to Bangladesh, BRAC has introduced micro schemes in various parts of South Asia to mitigate the problems of education, poor health facilities and rural development, this kind of third world. Many believe that this model in developing countries due to its success should be replicated to developed countries also. Over the years this participatory method of NGOs brought some visible changes at the grass roots level. Now these voluntary organizations are using mobilization as a process of empowerment of local communities, strengthening democratic base and building trust of people in the government. As a result of this, local community development groups are emerging around certain issues confronting the people. And at many places, these local civil society groups try to hold government accountable and transparent. It not only led to the improvement of governance at local level with a special focus on women but also preparing people for change (Panday 2016). NGOs taking environmental issues are gaining currency in the region thereby asserting their role and their number in the region is growing steadily during the recent past supported mainly by donor funds. Bangladesh alone witnessed more than 10,000 environment-related NGOs, of which about 250 receive funds from foreign sources (Tolba and El-Kholy 1992). However, critics also pointed out in majority of the cases these environmental NGOs do succumb before the authorities.

As per the International Crises Group (ICG) Report of 2008 in Bangladesh, although a plethora of international NGOs do exist in the country believed to be working for promotion of democracy and human rights but no one actually opposed the caretaker government and its efforts to change the contours of the party structure and political dynamics. It is also believed that these NGOs are not coordinating properly and are intolerant towards the other similar kinds of groups. But despite these lacunas, NGOs are performing many roles simultaneously, as service providers, advocates and innovators and finally *monitors* (trying to ensure that government and business do what they are supposed to be doing). Despite this progress, their liberty is regulated by the authoritarian power of the state as sometimes all these non-governmental bodies are subjected to state regulation. Both Pakistan and Bangladesh have witnessed long periods of military regimes under which operational aspect of sector got a big blow. The democratic interval under the Benazir Bhutto and Nawaz Sharif regimes witnessed the birth of a number of civil society groups addressing a wide range of issues, from development to environment, consumer rights and education.

Considering the role of NGOs as developmental partners, at many levels government partnered with them to deliver services based on NGO models of group-based credit. NGOs have been successful in stimulating local initiatives for meeting the needs of local people. There is increasing acceptance of NGOs to be treated as agents of change effecting community development goals, facilitating people's participation, addressing gender, poverty and environmental issues. Noted NGOs like PLAN and other international NGOs are working with Village Development Committees to improve the lot of people in rural areas in Nepal. Although the environment for humanitarian NGOs remained conducive to a large extent, but NGOs have largely suffered due to lack of resources, donor dependence and less participation of women.

It would be naïve to downplay the role of the international community in Sri Lanka. Here, NGO community remained very effective in the post-war peace building process in. Apart from this, NGOs have remained engaged in a wide range of activities such as emergency response, democracy building, human rights, environmental protection and conflict resolution. Initially, Sri Lankan government's attitude towards the work of civil society groups was quite precarious, alleging them of having LTTE connections and anti-state agendas. It was against this understanding that Sri Lankan government restricted the role of NGOs especially in the conflict-ridden Jaffna area by issuing a new regulation concerning NGOs along with Voluntary Social Service Organizations Act. Compared to Bangladesh and Pakistan, due to democratic environment and literacy, civil societies flourished and multiplied effectively.

In Afghanistan, Pakistan and to a lesser extent Nepal, given the extent of the international aid bureaucracy, the prevalence of violent peace or the return of war despite extensive international funds and expertise engaged raises fundamental questions about the relevance and impact of international aid on conflict transformation. The analysis in this chapter suggests that both the international-aid paradigm and delivery process in several South Asian contexts require fundamental reform for sustainable peace building (Pattanaik 2011).

Over the years, the process of localization of NGOs and partnerships with governments has evolved. Now NGOs have a role to play and they are now engaged in environmental protection, development works and delivering basic services. These services were taken by the NGOs due to the inability of governments on account economic backwardness. As compared to Central Asia, women in whole South Asia underrepresented and even at many instances, their absence led to the male-centred development process. South Asian governments should look for ways to improve the environment for women so that growth and development of NGOs and of the whole region should be inclusive. There is need for training programmes for local-level community groups, development of their institutional capacity, participation and mobilization power (World Bank 1994).

Critics pointing out that most of the NGOs do focus more on their procedural efficiency but lack answerability on ground. The region has witnessed the excessive use of NGOs by politically motivated groups for propagating and manufacturing false consciousness. In most of the cases, they are operated and owned by donor community thereby losing social relevance. This kind of thinking has tarnished their integrity among people and this trend is gaining the ground in whole South Asia. And it weakened the influence of the pro-NGO social-development advocates within those institutions.

Rather than seeing NGOs as collaborators of development, sometimes the rupture of state and civil relation creates negative impression, which distracts them of their genuine job. State has an important role to give effective and democratic space to NGOs so that every organ of the state functions normally and hence the rights of citizens can be held high.

The general perception prevails across the region concerning the role of NGOs is that the majority of NGOs lack citizen engagement, having donor and elite connections and even some argue that the NGOs are not responsive to the genuine needs of people. Some other critics believe that donor dollars are being diverted to other anti-social and anti-state activities, money coming from Middle Eastern countries, especially Saudi Arabia, has never been used for some missionary activities. But the other side of the fact is that the NGOs do not have quick fix solutions, and given the regions backwardness, it is immature to deduce conclusion. And even to improve the overall situation needs political and economic stability. NGOs must understand that they operate with the blanket concept of state and any kind of malfeasance from NGOs should be avoided. The basic purpose should be to strengthen the state–society relationship rather than trying to trample the domain of any other organ of the state.

Conclusion

With the larger context of South Asian region, the emergence and role of NGOs is quite distinct and unique. This is reflected in the manner in which NGOs do work and create constituencies of development and influence. In case of India and Sri Lanka and to a larger extent Bangladesh as well, one can project larger role of NGOs. In SAARC region, NGOs remain sine qua non for development. We must import best practices from other regions and try to adapt them to suit the local conditions. For instance Bangladesh's Micro-Credit experiment under the aegis of Muhammad Yunus has brought laurels to the rural population and what stops other neighbouring countries from emulating the project. In case of India, the Rights to Information Act, the MREGA and Food Security Act are obvious evidences of some of the victorious attempts by NGOs to advocate human rights and improve living standards.

This includes significant expansion of human rights advocacy groups, women empowerment, ecological balance, highlighting issues of public concern, good governance and promotion of democracy, peace, national reconciliation and more dynamic and nuanced understanding of the civil society sector. On the other, the experience of NGOs in Pakistan, Myanmar and Nepal despite the shift in political landscape, their roles are limited to social humanitarian goals rather than bringing political changes. However, the overall impact of NGOs can be denied as number is growing exponentially gearing towards social and political development. The NGO sector still lacks capacity and human resources necessary for a more coordinated and collaborative force to generate effective and vibrant impact in the three areas like promotion of democracy, good governance, peace and national reconciliation.

References

Ben Néfissa, S. (2005). *NGOs and Governance in the Arab World*. Cairo: American University in Cairo Press.

Chowdhury, A., & Mahmud, W. (Eds.). (2008). *Handbook on the South Asian Economies*. Edward Elgar Publishing.

Cyrill, M., & Pitnum, A. (2017, October). Retrieved March 15, 2019, from www.icnl.org/reserach/monitor/india.html.

Cyrill Melissa and Adam Pitnum, (2017, September), https://www.india-briefing.com/news/fcra-compliance-india-how-24000-ngos-lost-their-license-15287.html/

Dias, C. J. (1993). The Role of NGOs in the Protection and Promotion of Human Rights in Asia. *Asian and Pacific Migration Journal, 2*(2), 199–222.

Eade, D., & Leather, A. (Eds.). (2005). *Development NGOs and Labor Unions: Terms of Engagement*. Bloomfield, CT: Kumarian Press.

Edwards, M., & Fowler, A. (Eds.). (2002). *The Earthscan Reader on NGO Management*. Earthscan.

Fonjong, L. N. (2007). *The Challenges of Nongovernmental Organisations in Anglophone Cameroon* (Vol. 2). Nova Publishers.

Howell, J., & Pearce, J. V. (2001). *Civil Society and Development: A Critical Exploration*. Colorado: Lynne Rienner.

Huntington, S. P. (1993). *The Third Wave: Democratization in the Late Twentieth Century* (Vol. 4). University of Oklahoma Press.

Jarvik, L. (2007). NGOs: A 'New Class' in International Relations. *Orbis, 51*(2), 217–238.

Kaldor, M. (2003). *Global Civil Society: An Answer to War Polity*. Cambridge.

Lewis, D. (1998). Development Policy and Development NGOs: The Changing Relationship. *Social Policy and Administration, 32*(5), 152–174.

Lewis, D., & Kanji, N. (2009). *Non-governmental Organizations and Development*. Routledge.

Muni, S. D. (2009). *The New Democratic Wave and Regional Cooperation in South Asia*. Institute of South Asian Studies.

Nobusue, K. (2002). Bangladesh: A Large NGO Sector Supported by Foreign Donors. In *The State and NGOs: Perspective from Asia* (pp. 34–52). Singapore: Institute of Southeast Asian Studies.

Panday, P. (2016). *Women's Empowerment in South Asia: NGO Interventions and Agency Building in Bangladesh*. Routledge.

Parnini, S. N. (2014). *Poverty and Governance in South Asia*. Routledge.

Pattanaik, S. S. (2011). Moving Beyond the Rhetoric of Cooperation. *South Asia: Envisioning a Regional Future*. New Delhi: Institute for Defence Studies & Analyses.

Pearce, J. (2000). *Development, NGOs and Civil Society*. Oxfam GB.

Shaffer, R. (2013). Deepa M. Ollapally. *The Politics of Extremism in South Asia*. New York: Cambridge University Press, 2008. xi + 239 pp., $95.00 Hardcover. ISBN: 978-0-521-87584-4. *Terrorism and Political Violence, 25*(5), 864–866.

The times of India, September 19, 2015, https://timesofindia.indiatimes.com/2015/9/19/archivelist/year-2015,month-9,starttime-42266.cms

Tolba, M. K., & El-Kholy, O. A. (1992). National Responses. In *The World Environment 1972–1992* (pp. 693–736). Dordrecht: Springer.

World Bank. (1994). Poverty Reduction in South Asia, promoting participation by the poor.

26

Making Sense of Civil Society: The Experience of Pakistan, Afghanistan and Nepal

Debasish Nandy

Introduction

Civil society and democracy are interrelated. The notion of 'civil society' is associated with the emergence of capitalist economy. Until the eighteenth century there was hardly any difference between state and civil society. John Locke was the first political philosopher who had clearly differentiated between state and civil society and also introduced the concept of 'political' and 'civil' society. In between 1750 and 1850 the idea of civil society had emerged in English, French and American political history. Hegel had modified the concept of civil society in a sophisticated manner. Karl Marx had interpreted civil society as nothing but a bourgeoisie society. After the end of the Second World War a number of Afro-Asian states had emerged in global politics with Western-model of democracy. Like other post-colonial states, South Asian States have also introduced Western democratic system without ensuring the separate

D. Nandy (✉)
Department of Political Science, Kazi Nazrul University,
Asansol, West Bengal, India

© The Author(s) 2020

Md. N. Momen et al. (eds.), *Building Sustainable Communities*,
https://doi.org/10.1007/978-981-15-2393-9_26

space for the civil society. Pakistan has witnessed the ethno-political and religious violence over the decades. It has gone through the military regimes or mock elections in regular intervals. Voice of civil society is not satisfactory in Pakistan. In case of Afghanistan, the role of civil society is more vulnerable and alarming. In the Nepali case, it is not only complicated and unclear but unorganized too. The separate apolitical and responding identity of civil society is a pre-requisite for democratic vibrancy. This chapter intends to critically investigate the role of civil society. It also aims to identify the challenges of civil society in Pakistan, Afghanistan and Nepal respectively.

A Conceptual Outline of Civil Society

The notion of 'civil society' has been an integral part of a democratic society. In their research paper R Islam-Fakhr-ul and Farmanullah have argued that "[c]ivil society does not include the state and public agencies even though, through its judiciary and regulatory function, the state upholds the rule of law, social order and other essential components of society and civility. The legal system and specific laws dealing with civil society institutions and organizations can be included. And finally, it does not include the family. *Cross-cultural family* forms vary significantly and tend to imply different demarcation lines between the private sphere of the family however defined and the public sphere of the wider society" (Islam-Fakhr-ul and Farmanullah 2015: 251). Civil society is a non-governmental forum where a wide range of apolitical citizens united for greater interests of the society. In a broad spectrum of social, economic and cultural fields, the conscious and responsive citizens form a civil society for various interests. According to Nikhat Sattar, "civil society is a combination of formal institutions, non-governmental organisations, trade unions, professional associations, philanthropies, academia, independent pressure groups, think tanks, and traditional informal formations, such as faith-based organisations, shrines, seminaries, and neighbourhood associations" (Sattar 2011: 5). Civil society can be classified into two categories—formal and voluntary.

The intellectual discourse of civil society had started in eighteenth-century Europe. Civil society is a specific forum within the greater society in which citizens of a country seek relief against the authority on contemporary issues which are very much relevant for the society. In earlier times, there was no space between the society and state. Basically, these two were synonymous. The voice of citizens was not allowed to reach the state authority as well as the bigger society. Noted Economist Adam Smith wanted to relate civil society with economy. According to Adam Smith, "trade and commerce among private citizens created not only wealth but also invisible connections among people—the bonds of trust and social capital in today's terminology." Political philosophers like John Locke and Alexis de Tocqueville established in their writings that civil society is much more related to the society rather than the economy. Friedrich Hegel is considered as one of the key exponents of civil society. He had given a clear-cut outline of the civil society, where it is described as a 'self-organizing' and 'self-regulating' unit of the society. Hegel also advocated the requisite of the state to regulate the society. He never wanted to delegate autonomy to the civil society which can freely exercise its role. He had given the interdependent relational model of society and state. The relation between state and society is complicated and subject of tension. He had suggested to make a balance between the two, but it is quite difficult. In the writing of Montesquieu and Von Stein, the state was given more emphasis than the civil society. Both of them had given stress on 'rule of law'. According to them the regulating authority of the rule of law is the state. So civil society was not given any particular space to raise voice against the state. But they had mentioned the importance of the civil society.

In the twentieth century, the meaning of civil society had implied two things: (1) popular participation and (2) civic mindedness. C. D. Verba, Jürgen Habermas, George Putnam, Antonio Gramsci and Etzioni had elaborately researched on civil society through different dimensions and manners. They had viewed the civil society from different angles. They had shed light on social capital, the public sphere, social capital, culture and community respectively. There are various aspects and concepts and elements of civil society. The major aspects of civil society are values, norms, tolerance and so on. The role of the mass-media and intellectuals are vital to run a civil society very well. The vibrancy of a civil society depends upon peoples' participation and civic engagement.

Apart from liberation struggles and movements for restoration of democracy, civilian movements on various socio-political and economic issues in South Asian countries have manifested itself in various forms. The nature of civil society and mode of activities are quite different from each other. The formation of civil society in this region is very complicated because of differences of nature of the states. If democracy is understood as evolving through social contestations and the battle of claims and counterclaims, readymade recipes of democratization are likely to falter at the threshold of authenticity. The civil society and non-governmental organizations (NGOs) are playing very significant role in few countries. In this chapter I will discuss about civil society of Pakistan, Afghanistan and Nepal. I have chosen these three countries to understand the different nature of civil society.

Pakistan

The existence and role of civil society in South Asia are mixed in nature. Civil society movement in Pakistan is vibrant and active, but for few countries its role is questionable. Some South Asian Scholars rather interested to interpret the role of civil society for some particular countries as infant stage. Pakistan is a warrior state as described by T. V. Paul. He had observed that since its independence the basic tendency of Pakistan state is to spend much more money in military sector. The citizens of Pakistan have been ruled by the military for more than 30 years. Inter-religious riots, ethnic conflicts and continuous state-sponsored terrorism have collectively jeopardized the society. In the absence of democracy, human rights, accommodative policy and human security, the formation of civil society in Pakistan is a natural phenomenon. Nikhat Sattar, a Pakistani scholar, has rightly observed that in his working paper in 1980s, during Zia-ul-Haque's regime, some significant movements took place in different parts of Pakistan. The NGOs had played the key role in those movements. The basic demands of those movements were to ensure human rights. The 1990s had witnessed the impressive growth of NGOs. During that time, most of the NGOs worked on the issue of development and community based-organizations. Due to the organization of massive civil movements on different issues, a number of civil activists, writers and intellectuals, journalists had been sentenced to different prisons. Sanction was imposed on free thinkers and media. The federal administration had taken several brutal policies with the help of corrupted police force (Sattar 2011: 7).

Pervez Musharraf, the former military President of Pakistan has stated in his memoir that "I ardently believe that no country can progress without democracy, but democracy has to be tailored in accordance with each nation's peculiar environment. Only then it can be a functioning democracy that truly empowers the people and produces governments to address their needs. If it does not function, then it merely creates a façade without sprit or substance" (Musharraf 2008: 154). It is very positive statement and at the same time it is also puzzling to listen such type of statement from a military president. It is very difficult to play the active role by the civil society. The governmental non-democratic and perspective attitudes often make challenges to the civil society members. Being a disrupted democracy Pakistan faces lot of questions from world community. The economic vulnerability has been more acute for uneven and biased developmental policy. People of Pakistan are living with discontents.

In Pakistan, the elitist section of the civil society is very strong, but they rarely show their performance for greater interests of the society. Ayesha Siddiqa has explained the relationship between the military and civil society. She seemed that presence of active civil society and civilian institution is possible in stable democracies. It is expected the political environment should be controlled by the rational civil society. Political environment should not be controlled by army. She further said, "The military's sense of professionalism and restriction to an external security role can be attributed to the strong civil society and democratic institutions such as the media, judiciary, human rights organizations, election commissions, political parties and government audit institutions. The media in particular are quite strong, which makes it imperative for the armed forces to operate in their well-defined area of operations" (Siddiqa 2007: 36–37).

Civil Society, Law and Media

Asim Sajjad Akhtar has described the Pakistani civil society from a new outlook in his article published in Pakistan's premier newspaper *Dawn*. He wrote that "notwithstanding the simplistic notion of 'civil society' propagated throughout the 1990s. Western social theorists from Hegel to Marx to Gramsci had developed complex philosophical treatises on the subject over the previous century and a half". He further explained that in Pakistan, 'civil society' had reached in a very good position layers' movement. The lawyers had raised their voice against dictatorial regime and anti-democratic stance of the Pak Federal government. Same thing had happened in Pakistan in August 2008 against General Parvez Musharraf regime (https://www.dawn.com/news/1353803). The lawyers of Lahore had shown agitation against Lahore police as a part of civil disobedience. The 'Lawyers' Movement' greatly popularized among the common masses as it was considered as the most significant 'civil society' movement in Pakistan. Until the early 2000s, the term 'civil society' was conceptualized and propagated among the urban, secular donors of the NGOs. Before the commencement of Lawyers' Movement, the role of civil society was confined within a small span. President Zardari, however, did not want to recognize the civil movement of lawyers. He had mentioned the Lawyers' Movement as the movement of political jokers. It is true that in the twentieth century, the flourishing of the concept of civil society was considered as the important measure of the growth of democracy. The social scientist could not agree to define 'civil society' in a single dimension (https://nation.com.pk/10-Feb-2010/civil-society-in-pakistan). In 2007, Pakistan had witnessed the Lawyers' Movement against the dictatorship of the government. In case of Pakistan, it was very positive reaction emanated by intellectual class. Undoubtedly, in Pakistan, corruption is a significant obstacle for good governance, supremacy of law and rational use of authority to run the affairs of state (Nandy 2017). The political leaders and head of the states often charged with corruptions and big scandals. The voice of media and lawyers are very essential to make a greater resistance against these illegal means.

Student Movements as a Part of Civil Society

Student movements as a part of civil society's struggle to achieve freedom and democratic ideals started before the creation of Pakistan. Muslim Students Federation played a very important role in mass mobilization for the creation of Pakistan. Quaid e Azam envisioned student's engagement in politics to be temporary and for achieving limited objectives. He thought that as a part of struggle for freedom students should participate in supporting the national ideal of freedom. Once this objective is achieved, they should continue their support for nation building by paying full attention to their studies as the young country badly needed educated minds to lead it through the years of adversity and grave challenges.

One of the strong civil society movements in Pakistan is students' movement. In South Asian perspective, student's movement is closely attached with a particular political party. In Pakistan, students' wings are backed by the political parties. A particular students' wing believes on a certain ideology. All political parties worked closely with their student wings. It can be stated that like other civil society fronts students' wing is not always apolitical. For Pakistan, there was an exception. There was a short-lived wing namely Student Own Power (SOP) in Punjab University. That wing was not attached with any political party. As a part of Muhajir Qaumi Movement (MQM), Muhajir Student Organization (MSO) had started to move against the Sindh Provincial government as well as the central government. This students' movement was basically moved for their community interest rather than the student interest. After the separation of East Pakistan, many youth organizations carved space for expression of cultural and social identity through works of art, literature, theatre, poetry and music. In fact, student movements like mainstream political parties have been continuously struggling to grapple with the question of identity. After the separation of East Pakistan, many youth organizations carved space for expression of cultural and social identity through works of art, literature, theatre poetry and music.

Social Structure and Civil Society

It is only in Pakistan where elections have periodically been conducted under military rule, that public opinion matches the external impression: a majority of the people either keep silent about or regard elections as unfair. People's dissatisfaction with the working of democracy may be related to a common abhorrence for power mongering, the amoral approach of the political class, scandals and corruption, and other such negative factors. In Pakistan large section is dissatisfied because of lack of basic necessities. In Pakistan, the absence of equal rights and the absence of freedom of speech are more important factors that add to the dissatisfaction (State of Democracy in South Asia 2008: 138). Against this backdrop, a question naturally raises to the civil society of Pakistan that what it is doing for the protection of fundamental rights? There is no existence of a unified civil society in Pakistan. The ethnic conflicts, intra-religious sectarian cleavages and orthodox religious rule over the society have jointly made an obstacle in making of a vibrant civil society in Pakistan. The direct or indirect influence of army on state and society creates a clear-cut restriction in free functioning of the civil society. It is very stringing to imagine that in the era of globalization Pakistan is continuing its feudal system. Most of the ruling elites are from feudal background with a strong connection with military personnel and religious masters. They are collectively controlling the society without giving any free space to society. The governmental perspective rules and regulations have tactfully intended to restrict the free activism of the civil society.

Judiciary and Civil Society

The judiciary of a country is the protector of fundamental rights. It was expected that with proper implementation of 'separation of powers' Pakistan would ensure the people's rights, but unfortunately, it did not happen. Since the inception, the judiciary of Pakistan could not function freely. It was dominated and controlled by the executives. From Ayub Khan to Pervez Musharraf, the same things happened. On many occasions, military coups exiled democratically elected governments. During these situations the judicial system was not only unsuccessful to crisscross unconstitutional regime modification, but also authorized and supported the amalgamation of illegitimately gained power. In November 2007, Chief Justice Iftikhar Muhammad Chaudhry and several other Supreme Court judges had declined to sign Musharraf's decision to hang the constitution and rule by pronouncement. In most of the cases, the entire administration keeps silent. The members of the civil society usually play very nominal role in Pakistan. The voice of civil society is to be lauded in order to prevent governmental non-democratic decisions and executions. When voice of the civil society is mute, the judiciary can show activism for the sake of the common interest. The chief justice of Pakistan's Apex Court had rightly demanded far larger answerability from higher-level government personnel, police administration, detective agencies and other governmental agencies. It was quite unthinkable step in Pakistan before him. The chief justice had received support of Pakistan's lawyers. The lawyers have a good network and support-base across the country. When judiciary is the voice of the society, it is duty of the civil society to respond on the matter and make it stronger. The activism of the judiciary is not ever-active due to some compulsions of state controlled machineries. The civil society members feel insecure to move against the state authority. In rural Pakistan, orthodox and religious masters-led society is responsible for the creation of inadequate and invisible portrait of 'rural civil society'. On the other hand, the 'self-centric salaried urban civil society' has a common phobia about the state authority. Due to the differences between the two civil societies they could not be united on any greater issue related to the country's democratic interest.

Radical Islam and Army

Before the secession of East Pakistan from West Pakistan, the role of Dhaka-based civil society was clearly manifested as a protest movement by Bengali intellectuals against the forcible imposition of Urdu on East Pakistani Bengali ethnic group. Two major incidents had restricted the famishment of civil society—one is radical Islamic force and another is

Islamization of army. The process of Islamization in Pakistan was basically state-led. During that time that process was appreciated by a section of Westernized elite-class people. Farzana Shaikh has rightly made a comment in this connection, "the latter replied on a statist interpretation of Islam to oppose the populist policies favoured by Ziulfikar Ali Bhutto and, in time, increasingly espoused a culture that was distinctly Islamic tone" (Shaikh 2009: 107). Towards the end of the 1960s religion-based political parties propagated the concept of the 'Ideology of Pakistan' when radical movements emerged to threaten the social order in West and East Pakistan. This period witnessed the rise of the Jamat-i-Islami's ideology among the intelligentsia. General Zia-ul-Haq had come into power through a coup in 1977. He stated that sovereignty lay with *Allah*, and as his representative, he was entitled to wield this power (Chengappa 2004: 24–25). He used Islam to legitimize his rule. Through the Islamization of army, he encouraged the jihadist to play an active role. In the absence of free space in society, civil society cannot not grow.

Civil Society and Media

The relation between civil society and mass-media is interrelated. The mass-media can take a significant role in active participation of civil society. Through the positive circulation of civil society's activities, it can be popularized and justified in the society. The role of media in fostering civil society was not initially very low. Later on the role of media in Pakistan has been a bit different. Actually, Pakistan is not a safe place for free media. In Pakistan, anti-government, anti-army and anti-radical broadcasting is a matter of jeopardy for media persons. Beena Sarwar, editor of Aman Ki Asha of the Jang Group of newspapers, had remarked, "Journalists in Pakistan walk a tightrope between the military and the militants…" (Sarwar 2011). Writers and journalists are integral parts of the civil society. During General Zia-ul-Haq's military rule, journalists were imprisoned and tortured and newspapers were subjected to censorship. Despite this daily repression, journalists fought a valiant battle against authoritarian controls. Beside protest meetings, demonstrations and processions, journalists had tried to make a novel way of protest by leaving 'blank space' in newspapers (Mukhopadhyay 2016: 389). During the Nawaz Sharif's regime, the relations between Jang Group and government was hostile. Due to anti-governmental news broadcasting, the Government of Pakistan had stopped the supply of news print to the newspaper and the bank account of the Jang Group was frozen by the Sharif's government. During the regime of Pervez Musharraf, it was very difficult task for civil society to make any protest movement against the government. The harsh policies of the government were continuously imposed on civilians. The democratic and rational fronts of the society had to face lot of persecution during Musharraf's rule. Media autonomy was immensely threatened in November 2007 with the imposition of emergency. So many restrictions were imposed on democratic fronts, NGOs and social activist groups.

Issue-Based Civil Society Movements in Pakistan

Despite of shortcomings and challenges, Pakistan has witnessed significant civil society movements. One of the major movements was initiated by the NGOs. The *NGO movements* happened on the basis of various issues. The basic mottos of the NGO-led movements were to enable the people to raise their voice on social, economic and political issues and to develop a healthy-workable environment in the society. A number of NGOs in Pakistan have fabricated slack associations and linkages on several matters such as child rights, females' rights, the environment and more. NGOs are very active in underdeveloped areas and tribal areas, especially, terrorist-porn areas. They are very active child rights, women's rights and environmental issues. NGOs have been the voice of civil society due to the lack of activism of so-called structured urban civil society.

Women's movement is another significant civil society movement of Pakistan. Women in Pakistan are in a vulnerable position, and on an average, social security has been challenged by non-governmental apparatus. Several associations of womenfolk and related human rights organizations had formed in Pakistan since 1980s for the protection and empowerment of women. *Environment movement* has been a more important movement demonstrated by the civil society. On the issue of deforestation, ecological destruction, and air and water pollution, the eco-friendly section of the civil society members appears in movements. Sustainable development is very emerging issue in Pakistan. The unplanned urbanization process and mass-destructive developmental projects are jointly hampering the eco-system. This movement is less political and free from religious control. This is very positive thing that the Nation-wide Preservation Policy procedure played a major role in associating and carrying together a number of dotted environmental and sustainable development issues under one umbrella.

Muhajir Qaumi Movement as a Part of Civil Society

Muhajir Qaumi Movement is one of the burning ethno-linguistic movements of Pakistan which manifested in a manner of civil society movement. Muhajir is a sect of Muslim refugee, during partition who were shifted from India and settled into Pakistan, especially, in urban Sindh. Ethnicity plays a vital role in politics of Pakistan. Korejo and Syed have opined that "the leading Sindhi nationalist leader, also fails to go beyond the rhetoric where Muhajir identity is concerned and brings no fresh evidence on the role of the Sindhi language in ethnic identity formation or assertion" (Korejo 2000). In March 1984, a new force emerged on Pakistan's political stage; the Muhajir Qaumi Movement, championing the interests of the Urdu-speaking Muhajirs (migrated from India) settled in Sindh province. The MQM became a political force of a considerable magnitude in the post 1984 phase. In 1986 a violent ethno-communal riot took place at a time in Karachi and Hyderabad between MQM supporters and Pashtuns. The Pashtuns settlers in Karachi suddenly attacked MQM supporters. MQM supporters were also engaged in several communal riots with Sindhis. In 1988 general election, MQM became third majority block in National Assembly. It also got one-third majority in Sindh Assembly. In that election Benazir Bhutto came to power. MQM leader Altaf Husain supported this government for fulfilling their demands. But not a single demand by MQM was fulfilled. After the general election of 1990 MQM came close to Nawaz Sharif's government but after a short while, MQM's honeymoon with the Sharif's government ended. The MQM party had gained momentum by the leadership of Altaf Husain. He had claimed to declare Muhajir as a different ethno-national group which has its own language (Urdu). On the eve of this turmoil, Sindh community also started a movement (*Jiye Sindh Movement*) against the central government of Pakistan on the issue of discrimination. The Sindhis claimed that they have been deprived and discriminated in all respects.

Afghanistan

Afghanistan is a lucrative place for the scholars and experts to endeavour to enigmatic socio-political and security-related issues. Afghanistan is a land of multi-ethnic groups and orthodoxies, which is struggling for a modern framework of statehood and values of modernity, the archaisms of tradition and the corruption of the state system. On a world map, Afghanistan appears as a somewhat marginal place that may alternatively be considered as a part of South Asia. Afghanistan has had a different experience of colonial domination than other South Asian regions. It was limited to the 40 years during which it maintained internal sovereignty but its foreign relations were under British rule. Throughout history, the sovereignty of the Afghanistan and its capacity to deliver services to the population have been shaped by external intervention that may limit state sovereignty but also enhance it (Monsutti 2014: 97–112). Unlike India and Pakistan, Afghanistan had no opportunity to inherit British culture to modernize its society through Western outlook. The societal outlook of Afghan society was immensely rigid and orthodox. The concept of civil society has not grown-up under the British rule. Later on, due to the lack of education and democratic deficiency, civil society could not take any proper shape. Afghanistan has witnessed a number of ethnic conflicts over the centuries. The radical Islam and incapable civil government is collectively responsible for non-formation of an iconic civil society. It is also true that Afghanistan was socially and politically isolated in the late nineteenth century and early twentieth century that is why no external idea of civil society was inherited by Afghan people.

In Search of Historical Roots

Afghanistan, with its complex ethnography of Afghans, Tajiks, Hindukush Indo-Aryans, Turko-Mongolians and strong tribal affiliations, still locked in, what Ibn-Khaldun called, "the first or second stage of civilization", is still in the throes of transformation towards a modern nation-state. Democracy and nationalism have never penetrated roots in this country (Gull 2008: 76). The Afghan state was established in the eighteenth century, when a tribal confederation developed into a dynastic state. In 1747, Ahmad Shah Abdali led an Abdali confederation, now known as the Durrani. This dynasty was followed in 1818 by another Durani Dynasty, the Muhammadzai clan of the Barakzai tribe, which held power until 1978. Through the Anglo-Afghan Treaty of 1919, Afghanistan had gained its independence. Muhammad Nadir Shah had ruled in Afghanistan from 1929 to 1933. He wanted to make his own imperial regime by legitimate method. He had formed a Grand Assembly of the ethnic leaders, religious masters and native elites to endorse a fresh constitution. Islam got importance in formulating the new constitution.

Modern and secular outlook was not reflected due to absence of modern education and enlightenment. This constitution was a manifestation of vested interest of Pashtun community. Non-Pashtun people were excluded from educational, political and developmental facilities. Later on the Musahiban rulers (1929–1978) pursued ambivalent policies towards Islam, especially, in expanding modern educational system, which has a secular agenda and a Pashtun based nationalism (Gull 2008: 78–79). The modernization programmes were introduced in the form of new democracy. In practice, democratic institutions and cultures were not grown-up due to autocratic nature of government and radical religious societal structure. A number of movements happened in Afghanistan in ideological or religious manners, but all of those movements finally turned into ethnic or sectarian lines. Most of those movements held between Tajiks, Uzbeks and Hazaras.

Modern Afghanistan and Civil Society

In a true sense, a modern state can emerge by introducing modern education system, free press, constitutionalism, liberal secularism, nationalism and ensuring the citizens' rights with an outlook of ethnic and communal harmony. Afghanistan could do this. It had failed to implement the Islamic modernization, abolition of feudalism and abolition of kingship. In absence of these elements, Afghanistan has witnessed ethnic violence and revolutionary activities of the communist parties. Daud was the first president of Afghan republic. He remained in power from 1973–1978. He tried to propagate the notion of Pashtun state but did not succeed. He also prohibited the use of ethnic surnames. His dream was to establish a common cultural and linguistic identity on which Afghan people can be united by the sentiment of nationalism. The heterogenic nature of the Afghan society has been repeatedly puzzled by external powers also. After the Soviet invasion to Afghanistan in 1978, the growth of communism on the one hand and the America's intention to make Mujahedeen (a group of Afghan fighters) to fight against the Soviet forces on the other have jointly made the country a land of instability and uncertainty.

The traditional structure of the society has become gradually darker after the assassination of Najibullah. In the post-Najibullah era, the squabbling between power contenders resulted in widespread violence. Since the mid of 1990s under the regime of Taliban, the entire country had gone through a very lawless situation. In the name of *jihad*, Taliban had banned all of the democratic institutions along with mass-media. Public domain was irrationally closed-down. Women's were not allowed to go outside. The entire education system was run by *shariya law*. The right to protest movement was totally stopped. Actually, they wanted to propagate that the voice of Allah is the voice of the country; but they distorted the message of Islam. For Afghanistan, Islam could be used as a very popular and useful means of establishing a peaceful society, it was not properly used. Afghan rulers and religious masters of the country could not able to understand the actual message of Islam. It is a message of harmony and peace with the philosophy of peaceful existence. Due to non-understanding of Islamic philosophy properly, Afghanistan has been witnessing religious intolerance and social instability. Afghanistan might have an active civil society for proper functioning of the democratic system.

The Responsibility of the USSR and the USA in Destroying of the Civil Society

After Soviet invasion of Afghanistan, its forces were very active in Afghan politics and society. Apart from political controls, persecution on civilians had crossed the limit. Moscow's tactful propaganda on communism across the country had led the country to go forward to communism. The civil society was confused during 1979–1990 due to the Soviet forces. The autonomy of Afghan civil society has been immensely hampered due to the USA's unexpected intervention. Afghanistan could not be able to show their indignity due to the external influences. At the end of the 2001, when the American forces had started to attack on Afghanistan to finish the Al Qaida and Taliban militants, then entire country once again had been started to control by the external power. State was not able to provide any space for the civil society. Due to the continuous aerial bombardments by the US Air Force, thousands of innocent peoples along with militants were killed and many were wounded. The civil society of Afghanistan in the post 9/11 era has rarely raised its voice against the American dominance. As the USA propagated that it is doing a lot for reconstruction in Afghanistan and trying to ensure the democratic environment for Afghans, that is why; civil society of Afghanistan is in dilemma about their view about the USA. A good number of civil society members were given special privileges by the US administration. So it is quite natural that those civil society members will not be interested to think about the greater society in lieu of their personal gains. This is a common method applied by the big powers during their direct or indirect intervention to any country. In this way big powers often like to control the civil society of the invaded country. The destructive means were used by the American and British forces since 2001. The unfinished Afghan civil society was totally puzzled about their duties. In the name of combatting international terrorism and restoration of democracy, the US forces had started to intervene the entire state system. Today, Afghan civil society stands in between 'making and unmaking'. Naturally, a question has been raised by the South Asian scholars whether there is any existence of civil society in Afghanistan or not. Without arranging modern, rational and secular education system with restructured societal pattern it is very difficult to make a visible and responsible civil society.

Taliban and Civil Society

The Taliban leaders have made a deep burial land for open-minded civilians. The free space was totally abolished by the radical groups. The ironic rule and barbaric customs were jointly responsible for closing the doors of civil society. Although there were no remarkable pro-democratic civil society movements happened across the country. The scattered society has had experience of scarcity. Many innocent peoples had been arrested by the US forces and also had to face extra judicial tortures. The activism of civil society in stopping of attacks on common masses has not been shown yet. Civil society has a full range of issues in the social, cultural, religious, humanitarian and political life. Afghan leadership, civil society, particularly local and national non-governmental organizations (NGOs) are not actively projecting their roadmap regarding the responding role in society. Afghanistan has no sizable educated middle class.

In case of Afghanistan, it is very difficult to differentiate between the rational civil society and traditional civil society. Over the decades, Afghanistan is continuing the traditional pattern of civil society which is a combination of local councils (Shura/Jirga), religious institutions (mosques, madrasas etc.), religious masters and other. Like Western countries or even emerging post-colonial countries, Afghanistan has no secular and democratic civil society. These orthodox non-democratic medieval institutions are the pillars of civil society. These are very influential. Modern and rational intellectuals are unimaginable in Afghanistan. The so-called civil society is very active during the civil wars and sectarian conflicts. Regarding the human rights and democratic movements they are less interested. This is very interesting thing that religion based civil society is not united. There is no homogeneous outlook among the civil society members. Due to the divergence of interests there are differences among the radical Islamist, pro-government moderates and reformists. Throughout the history, Afghanistan has witnessed the close relationship between the government and the clergy. It is true that entire society is covered with religious dogma. There is an undercurrent between 'the modern' and 'the traditional' and often they get in conflicts. Now, a diversity of various powers encounter each other, generating strains and clashes that frequently vent into vehement hostilities among the government, non-governmental organizations and spiritual players. The inheritances of the Afghan jihad—or divine war contrary to infidel socialists—and Taliban decree have had weighty things on the role of religion in culture and politics after 2001. Now, religion is extremely lobbied, and the connection between religion and the state is very challenged, as is the role of religion in the public domain. The majority of the religious leaders were highly critical of the foreign military presence in Afghanistan. *Mullahs* and *ulema* are supportive of neither the current Afghan government nor the external powers. Afghan spiritual players and organizations presently implement civic society roles that are backing the existing harmony, growth and state formation processes, but also work alongside these procedures (Religious Actors and Civil Society in Post-2001, November 2007: 0.8).

Women as a Part of Civil Society

One of the basic pre-requisites of formation of a civil society is to spread the education throughout the country. Without modern education system with a sizeable educated people, civil society can't take any proper shape. Women education is very essential for social mobility. Perter Marden has said, "The question of education for girls has inevitably raised questions over the potential role of women outside the home. It has provoked defensive reactions from men, concerned to protect the honour of women within their families and also ensure that the traditional roles of women within the domestic sphere continues to be performed" (Marden 2010: 208). In Afghanistan, women are still neglected. No adequate policy for emancipation of the women has been taken by the government. During Taliban regime women were not allowed to go outside and take education and take entry in jobs. Due to the social pressure women's' movement could not be a part of civil society movement in Afghanistan.

Youth Civil Society Members and Protest Movements

Afghanistan is a young country: about 55 per cent are under the age of 20, and the vast majority are under the age of 40. It is also one that defines 'youth' in a slightly different way than most in Europe or the USA would expect: people as old as 35 still identify themselves as a part of Afghanistan's 'youth', thinking that youthfulness is reflected in one's outlook and lifestyle rather than a number. First, for many engaged and young leaders, this was their first chance to vote or take part in a parliamentary election, and second, because it gave them a rare chance to alter Afghanistan's politics, which is dominated by a powerful executive and a set of immovable tribal and ethnic leaders (https://www.albawaba.com/news/are-afghanistan). I would like to mention a barbaric incident which reflected the democratic urge of Afghan youth and at the same time the positive spark of the civil society. Despite receiving threats from the Taliban, 27-year-old Hafizullah went to his local polling station in Helmand in southern Afghanistan, to vote in his country's parliamentary elections. When he left the station, he was found by militants who then cut off one of his fingers that had been dipped in blue dye to show he had voted: a punishment for taking part in Afghanistan's fragile democratic process. There was a huge aspiration and excitation among the youth over the first parliamentary elections held in late November 2010. Afghan youth movement was shown just after the barbaric incident in a spontaneous manner. Afghanistan's youth movements are a chaotic assemblage of parts, glued together by a cohesive desire for reconciliation and peace and foreign funding, while simultaneously being ripped apart by local power brokers and warlords, who promise access to power and resources. They nonetheless represent the country's best hopes to reconcile itself and form a lasting peace.

Peace is a utopian concept in Afghanistan. Afghanistan has rarely witnessed societal harmony. Being a war-torn country it has experienced hunger, poverty and endless bloody war-fares. Recently, a group of youngsters has brought a ray of hope in peace-building as an integral part of the civil society. Sifatullah Zahidi has mentioned in his article, published in *Salaam Times* newspaper on 9 April 2018, that a non-violent demonstration initiated by a cluster of youngsters in Lashkargah, Helmand Province, is rapidly scattering cross-wise Afghanistan in an unparalleled movement that contains all sections of society. A large number of citizens of Dozof Helmand Province had formed a shelter in Lashkargah on March 24, a day after a car packed with explosives blew up outside of a stadium in the city, killing at least 17 civilians and wounding 55 (http://afghanistan.asia-news.com). Over the past few decades, thousands of Afghan citizens had lost their lives, thousands had been wounded and millions had lost their homes. Afghanistan is also witnessing contentious out-war migration. Being a vulnerable and war-porn country citizens feel insecure. For saving their lives and a better future, potential youth are frequently leaving the country. This is not good for the future of Afghanistan. If the educated youngsters can realize the actual necessities of the country and able to understand the deep-rooted reasons behind the national crises, then they can collectively fight against the all-oppressive forces.

However, demonstrators of the aforesaid movement called themselves the 'People's Peace Movement'. The positive side of this movement is incorporation of a large section of Afghan people. They have included tribal elders, civil society activists and relatives of victims killed in violence. This peace front got support from Afghan High Peace Council, and Afghan *Ulema* Council and local political leaders. This movement aimed to spread across the country through peaceful demonstrations. They used Twitter to propagate their visions and missions. Due to adequate communication system and restrictions on mass-media and social media, the wave of social movement could not reach across the country. Former Afghan President Hamid Karzai had tried to ensure some space for the civil society members. He had to take lot of challenges to establish minimum democratic atmosphere for the Afghan citizens. The present President of Afghanistan, Ashraf Ghani has been supported by the two major democracies of the world—the USA and India. The positive stances have been taken by the both countries in order to fostering democratic culture in Afghanistan. Pakistan played a very negative role in making of civil society in Afghanistan. The illegitimate Taliban government was blindly supported by Pakistan.

Nepal

The civil society in Nepal has evolved very slowly. The concept of civil society in Nepali discourse is contested and to some extent self-contradictory and confusing. A noted Nepali politics expert Ganga Thapa has stated that, "experiments with democracy in Nepal have often gone through difficult times largely due to political opportunism. Besides, due to overwhelming poverty or whatever, not all citizens are interested in democratic values" (Thapa 2016: 68). Nepal has gone through a number of social unrests and political crises. The Nepali people had to spend a long time under the Rana rule and oppressive Monarchical system. History of Nepali democracy is remarkable due to the long struggle of the people. In early 1990s, Nepal had gained the experience of parliamentary democracy. This experience was not so smooth. The thrust of democracy could be mitigated by the humanitarian and respective approaches of the King, but it was not happened. It was basically, Communist, in fortification of their argument of wider civil society periphery with those of outmoded civic association, proclaimed that convention of civility is not an unfamiliar notion in Nepal (Dahal 2001); and the history of indigenous volunteer organizations predates induced NGOs. In contrast, some tried to measure civil society of Nepal from the window of Western concept and made sceptic conclusions such as "absence of civil society" in Nepal (Shrestha 1998); and "genuine civil society is yet to evolve in Nepal" (Dahal 2001: 42). Minimalists viewed that traditional community associations are different from what we understand civil society and NGOs today (Gyawali 2001). NGOs are new innovation and outcome of the expansion of the global multilateral assistance programmes and the expansion of the Northern INGOs into the South (Mishra 2001; Dilli Ram Dahal 2001).

Apart from democratic movements, there are a large number of issues on which Nepali people move against the state authority. Political violence, social insecurity and justice, pollution, human rights violation and women abuse are very common issues for civil society movement. Nepal is a multi-cultural, multi-lingual, multi-ethnic and multi-religious country without equal representation in democratic system and without having democratic rights. To Ganga Thapa, "there are numerous popular movements today which are striving for long-term political and social changes". Before the creation of parliamentary democratic system, Nepal had an unorganized civil society which fought for democracy. In Nepal, through the spring movement in 1990s civil society gave an effort for introducing democracy in the country. During that time, it was very difficult task to identify the non-political civil society members from political persons. That long-awaited democracy could not be introduced without generating momentum by Nepali Congress (NC) and communist parties. The members of the civil society were attached with a particular political party. The complex relationship between the political party and civil society often creates identity crisis of civil society as a non-political platform. As Nepal is a very backward state without experience of democratic practices that is why, overlapping nature of civil society is not unnatural. The civil society of Nepal has spent a long time for getting footing in the society and political arena. To create a stronger movement and the fight against the oppressive despotic authority, civil society has had to keep connections with political parties. Here a contradiction can be raised in order to compare with Western notions of civil society.

The Hindu monarchy in 2008 endorsed new socio-political and liberal construction to take shape and new forms of local governance defined by non-dominant ethnic communities to foster (Hangen 2010). However, ethno-based civil society movements took place in a number of times in Nepal. Nepali people expected that after the abolition of monarchical system in 2008, they will get a pure democratic atmosphere. It was not happened in practice. Its true Maoists came to power through huge mandate. Within ten years a number of governments have been changed, but not a single acceptable draft of the constitution is prepared by the government yet. Some ethnic groups have been intentionally exempted from the constitutional rights. There is no provision for the refugees and *Mahdeshi* people also. Regarding the policy of exclusion, some civil society movements are going on across the country. Asis Mistry in his recent research work has rightly mentioned that ethno politics is often viewed as incompatible with democracy. In the age of globalization and technological advancement, especially in the sectors of e-communication, social and electronic media, the ethnic communities in various parts of the globe have been gradually but significantly becoming mobilized for the attainment of their political rights of self-rule or secession (Mistry 2018: 34). In fact, modernization can be extremely important for the development of nationalism, both civic nationalism and ethno-nationalism, by producing greater socio-political mobilization and increasing assimilation of those mobilized (Deutsch 1953). Against this backdrop, it can be said that in the Nepali case, it is an easier task for Nepali suppressed ethnic groups to convey their demands through civil society platforms. Thus the effectiveness of civil society has been reflected in Nepal today through various dissatisfied ethno-nationalist groups.

Saubhagya Shah had sharply analysed about the civil society movement of Nepal from 2006 to 2008. That period was so vital for Nepali society. Pro-democratic Nepali citizens had to fight against the military and oppressive rule of the King Gynendra. He said, "the role of civil society in the *janaandolan* (mass-movement) in April 2006 in Nepal, the popular movement that abolished Nepal's Hindu monarchy" (Shah 2010: 75). He has mentioned that a major reason for the visibility and effectiveness of civil society during this political movement was the material and moral support from external agencies, mainly development partners. This is partly true. Civil institutions developed in response to local demands, and they operated largely at community levels. In Nepal, civil society movement got momentum due to the citizens' huge aspirations for democracy. In case of Nepal, questions have been raised by Saubhagya Shah: is Nepal a civil or uncivil place? Is there any possibility to form and function of civil society in a pre-mature democratic country? In Nepal it was proven that civil society can be an important factor in political change. The unimpeded force of civil society was succeeded to overthrown the autocratic and inhuman kingship of Nepal. There is a narrative about the flourishing of civil society in Nepal. According to this narrative civil society in Nepal was flourished after 1990 because donors believed that civil society was needed for development and thus provided ample funding. It's true that in introducing democracy and open market, external forces were so active. Since mid of 1990s a good number of NGOs were formed in Nepal as a part of civil society. Now question is that were the Nepali people mature enough to formulate a civil society like that in West? If answer is not, then it can be said that in a prematured society civil society can face lot of problems in making of decisions. After the abolition of Monarchical system in Nepal, Maoist influence has been spread across the country. Maoist President Prachanda, Babu Ram Bhatta Rai and Khadga Prasad Oli—none of those Nepali prime ministers—could able to establish the social and political stability in Nepal. The weak judicial system is also incapable to ensure social justice for the citizens through judicial activism.

Conclusion

In my write-up, I have given an outline of the civil society of Pakistan, Afghanistan and Nepal. These three countries are different in nature. In Pakistan, a journey of democracy was started with euphoria by Jinnah. Later on, State itself created numerous problems with various targeted ethnic groups, religious sects and women. As an economically challenged state it has started to enhance its military budget in lieu of social welfare. It is quite painful to think that Pakistan has a good number of bonded labours with existing system of feudalism. The citizens of Pakistan have witnessed the oppressive rule of military and radical Islam. However, civil society of Pakistan has been formed properly, but it is unable to free functioning. South Asian civil society has a wide range of social actors. It includes lawyers, architects, journalists and students; NGOs; transnational civil society; media; and political parties. Being a war-torn state Afghanistan is passing through immense political, economic and social crises. The process of civil society making has not been completed yet.

Although some civil society movements cropped up across the country, they were largely unorganized. The democratic culture is yet to be developed properly in Afghanistan. The orthodox and medieval outlook of the Afghan society is to be removed immediately. There is a ray of hope in making of civil society that the country is moving gradually towards parliamentary democracy without caring radical forces. Nepali civil society is struggling with a number of problems. The democratization of Nepal with a policy of accommodation is the core objectives of pro-democratic civil society in Nepal.

References

Chengappa, B. M. (2004). *Pakistan: Islamization, Army and Foreign Policy*. New Delhi: A.P.H. Publishing Corporation.

Dahal, D. R. (2001). *Civil Society in Nepal: Opening the Ground for Questions*. Kathmandu: Centre for Development and Governance.

Deutsch, K. (1953). *Nationalism and Social Communication*. London: MIT Press.

Gull, S. (2008). *Afghanistan Through Passages of History*. New Delhi: Kanishka Publishers.

Gyawali, D. (2001). Are NGOs in Nepal Old Wine or New Bottle? A Cultural Theory Perspective on Nepal's Contested Terrain. In K. B. Bhattachan et al. (Eds.), *NGO, Civil Society and Government in Nepal*. Central Department of Sociology and Anthropology, Tribhuvan University.

Hangen, S. I. (2010). *The Rise of Ethnic Politics in Nepal: Democracy in the Margins*. London: Rutledge.

Korejo, M. S. (2000). *G.M. Syed: An Analysis of His Political Perspectives*. Karachi: Oxford University Press.

Marden, P. (2010). *Afghanistan: Aid, Armies and Empires*. London: I.B. Tauris & Co. Ltd.

Mishra, C. (2001). New Predicaments of Humanitarian Organizations. In K. B. Bhattachan et al. (Eds.), *NGO, Civil Society and Government in Nepal*. Central Department of Sociology and Anthropology, Tribhuvan University.

Mistry, A. (2018). *Ethnic Politics in Nepal: A Theoretical Outlook*. Ahmedabad: Blue Roan Publishing.

Monsutti, A. (2014). The Transnational Political Economy of Civil War in Afghanistan. In A. Sundar & N. Sundar (Eds.), *Civil Wars in South Asia: State, Sovereignty, Development*. New Delhi: Sage Publications India Pvt. Ltd.

Mukhopadhyay, G. (2016). Media in South Asia: Emerging Trends and Issues. In D. Mitra & D. Nandy (Eds.), *South Asia and Democracy: Contextualizing Issues and Institutions*. New Delhi: Kunal Books.

Musharraf, P. (2008). *In the Line of Fire: A Memoir*. London: Pocket Books.

Nandy, D. (2017). *Understanding Pakistan*. New Delhi: Kunal Books.

R Islam-Fakhr-ul and Farmanullah. (2015, January–June). Civil Society and Democracy in Pakistan. *JRSP, 52*(1).

Religious Actors and Civil Society in Post-2001. (2007, November). *Afghanistan Kaja Borchgrevink*. International Peace Research Institute, Oslo (PRIO). For the Royal Norwegian Ministry of Foreign Affairs.

Sarwar, B. (2011, July 16–22). Pakistani Journalists: Standing Tall. *Economic and Political Weekly*.

Sattar, N. (2011). *Has Civil Society Failed in Pakistan*? SPDC Working Paper No. 6.

Shah, S. (2010). *Civil Society in Uncivil Places: Soft State and Regime Change in Nepal*. Policy Studies 48, East-West Centre, Washington, DC.

Shaikh, F. (2009). *Making Sense of Pakistan*. London: Hurst and Company.

Shrestha, A. (Ed.). (1998). *The Role of Civil Society and Democratization in Nepal*. Kathmandu: Nepal Foundation for Advanced Studies.

Siddiqa, A. (2007). *Military Inc.: Inside Pakistan's Military Economy*. London: Pluto Press.

State of Democracy in South Asia: A Report. (2008). SDSA Team, New Delhi. New Delhi: Oxford University Press. Retrieved December 29, 2018, from https://www.albawaba.com/news/are-afghanistan%E2%80%99s-frankenstein-civil-society-movements-country%E2%80%99s-best-hope-1207730; http://afghanistan.asia-news.com/en_GB/articles/cnmi_st/features/2018/04/09/feature-02; https://www.dawn.com/news/1353803; https://nation.com.pk/10-Feb-2010/civil-society-in-pakistan.

Thapa, G. (2016). State of Political Parties, Electoral Reforms and Transition to Democracy in Nepal. In D. Mitra & D. Nandy (Eds.), *South Asia and Democracy: Contextualising Issues and Institutions*. New Delhi: Kunal Books.

27

Civil Society, Political Stability and Peace-Building in Bangladesh and Sri Lanka: A Comparative Study

Z. R. M. Abdullah Kaiser

Introduction

Civil society organizations (CSOs) are promoted as an integral player of democratic development, the establishment of good governance, the promotion of human rights, (Dagher 2016; Morgenthau 1985; Lorch 2017) and sustained political stability and peace-building (Nilsson 2012). Peace accords with involvement from CSOs and political parties in combination are more likely to see peace prevail. The inclusion of civil society has a particularly profound effect on the prospects for overall peace in non-democratic societies (Nilsson 2012); conversely, most practitioners emphasize that enhanced CSO participation makes it more difficult to reach a peace agreement (Paffenholz 2014). More specifically, the centrality of civil society in enhancing and consolidating democracy, ensuring political stability and peace-building, and providing a deterrent against

Z. R. M. A. Kaiser (✉)
Department of Public Administration, University of Chittagong, Chittagong, Bangladesh
e-mail: abdullah.kaiser@cu.ac.bd

© The Author(s) 2020
Md. N. Momen et al. (eds.), *Building Sustainable Communities*,
https://doi.org/10.1007/978-981-15-2393-9_27

abuses by the state security forces (Forman 2006; Cole et al. 2008; Fukuyama 2011) has long been recognized.

Both Bangladesh and Sri Lanka are South Asian countries and former British colonies. Bangladesh won independence from the British in 1947 as a part of Pakistan. In 1971, it split far from the western part of the nation throughout one of the world's bloodiest wars of freedom. In Bangladesh, political influence has played a hegemonic role in government machineries and other democratic organizations. Sometimes it takes a hostile turn; which is accompanied by conflict, violence and other radical misdeeds (Lorch 2017a). Sri Lanka gained independence from the British in 1948. The Sinhala-Tamil ethnic crisis in Sri Lanka climaxed in 1983 into a civil war (Devotta 2004; Rahman 2007). In Sri Lanka, the electoral process continues to be dominated by ethnopolitics, and traditional political elites. Despite all these differences and similarities, however, Bangladesh and Sri Lanka display the same ambiguous relationship between civil society development, oligarchic politics, political stability and peace-building (Moniruzzaman 2009; Lorch 2017a).

However, CSOs have a long history to play the vital role to establish political stability and peace-building in Bangladesh and Sri Lanka. Local community-based organizations, civil society actors and other interventionists are active in assuring that political stability, peace-building and good governance can be strengthened in Bangladesh and Sri Lanka. Nevertheless, the outcomes are different in the two countries; CSOs have played the potential role on (1) preventing violent conflict, (2) doing advocacy work on different political and social issues, (3) supporting negotiations and settlements, (4) endorsing reconstruction and reconciliation, (5) influencing the policy-making process (6) and public opinion (Orjuela 2003; Tasnim 2012).

The main objective of this chapter is to depict a comparative analysis of the roles of CSOs in political stability and peace-building in Bangladesh and Sri Lanka. The chapter will specifically also explore: firstly; interrelations between civil society, political stability and peace-building; secondly, attempt to identify the present status of civil society in Bangladesh and Sri Lanka; thirdly, exploring the role of CSOs on resolving the political instability and peace-building in Bangladesh and Sri Lanka; finally, identifying the difficulties faced by CSOs taking on these roles.

Civil Society

Civil society is not emerged as a new concept. It has been a part of the polity right from the times of Greek City-states. The recent development of civil society as a non-political organization is related to complex social and monetary power at work in the eighteenth century, as power decayed from rulers to prevalent assemblies. The evolution of the concept of CSOs in its various perspectives has tried to define its meaning and scope. As has been observed, the conditions for the formation and reproduction of civil society have not only been economic but also political and ideological. In the context of globalization, civil society needs to take off from here and find itself extensively inside the connection between the state, market and civil society (Dhameja 2003). On top of that, civil society is considered as a dominant element of the modern state, along with government and businesses. According to the realistic context, the role of civil society is idiosyncratic and independent. However, its power and network are not as strong as the state-level authority. Therefore, its hegemonic power cannot be replaced by the government of a state or territory. Its activities are limited to two main domains. One is the operation of policy advocacy and arranging campaigns for progressive change and the other is providing basic service to the people in need (Lewis and Kanji 2009). Thus, civil society should be an important element for a developing state such as Bangladesh and Sri Lanka for ensuring political stability and peace (Orjuela 2003; Parnini 2006).

Both countries have long been known as the countries with vibrant CSOs. In contemporary Bangladesh and Sri Lanka, the concept of CSOs can be applied and reinforced with a definitive object of changing and improving the connection between the state and citizens, the formulation and execution of public policy and the institutional landscape (Parnini 2006).

However, civil society organizations in both Bangladesh and Sri Lanka, including those formed on the basis of profession, geography, chambers of commerce and industries, socio-political ones, advocacy groups and development NGOs, help to ensure popular sovereignty, political equality and political liberty (Orjuela 2005; Azharul 2005; Tasnim 2012).

Thus, the term 'civil society' has become quite controversial in South Asia; some favour to call it as the 'citizens group' and some others mention it as a 'civilian society.'

Civil Society, Political Stability and Peace-Building

"Only a democratic state can create a democratic civil society, only democratic civil society can sustain democratic state" (Waltzer 1990). The activities of civil society and citizen activism in the space of civil society are also essential in the democratization process, governance, political stability and peace-building (Chowdhury 2018; Chandhoke 2011; Booth and Richard 1998) and it is the microcosm for the development of democratic norms and practices (Belloni 2018). Similarly, non-state actors have a limited but vital contribution to make in the transformation of internal conflict situations (Rupesinghe and Anderlini 1998). Civil society has been taking on significant roles in peace negotiations through different forms (Paffenholz 2014). Statistical evidence shows that the incorporation of civil society has constructive effects on the period of the peace treaty (Nilsson 2012). Strong civil society can contribute in numerous ways to control conflict situation. Civil society provides the opportunity for building trust in a way that is not available in the formal structure of the society. The concept of peace-building is a much-debated issue. Rather, the term is often used extensively to mean any action undertaken to stop, alleviate or mitigate the conflict. Furthermore, agencies have prevented such different functions as discussions of gender issues, health programmes, rape and torture counselling, and political mediation between conflicting groups under peace-building. A more concentrated definition of peace-building would incorporate functions that contribute to the organizational and infrastructure works connected to sustainable reconciliation and activities of social integration (Goodhand 1999). The role of CSOs in politics, democratic development, political stability and peace-building are far different from the CSOs literature would imply. However, it calls for a more contextualized and less value-laden approach to the realization of the political function of CSOs (Mercer 2002).

Political Instability and Ethnopolitical Problem in Bangladesh and Sri Lanka

A quick look at the political history of Bangladesh and Sri Lanka would confirm a clear role of CSOs on political stability and peace-building. Bangladesh and Sri Lanka have been displaying huge differences as well as similarities concerning their historical legacy, political system, ethnic composition, religion and level of economic growth. According to the Failed States Index, both countries are staying in weak state zones in the world (FSI 2018). Since the early days of independence, Bangladesh has witnessed the confrontation and use of pervasive violence in politics. Also, decades-old insurgency and bloody conflict between the tribal insurgents of the Chittagong Hill Tracts (CHT) and the government of Bangladesh is the most common phenomenon of Bangladesh which makes the society more volatile and unstable. The 'periphery of conflict' between the country's main political parties and the weakness of the state has long been two sides of one coin (Migdal 2001). Ideologically political parties would not go about as a 'reliable' opposition by taking recourse of boycotts, strikes and political turmoil when they are defeated and rely upon patron–client relationships to reward supporters and keep up the coalition with predominant elites (Brett 2017).

Bangladesh shifted from democratic to military rule in 1975, and back to democracy in 1990, followed by regular elections, although, confrontational politics is a common spectacle among major political parties. The violence in politics took the most appalling and gruesome shape in its scale and nature, following the Fifteenth Amendment in 2011 (Islam 2015). But, the unrest in 2013 and 2015 before and after the election of the tenth parliament respectively, surpassed all the previous records that moved the state to an oppressive and uncertain future. It bears to be noted that the country was stunned by violent protests with a spate of blockades and shutdowns; particularly, the nature of attacks was quite different from the past pre-election and post-election attacks (Kamal and Kaiser 2015). The human right situation indicates that the nature and the quality of democracy are not satisfactory and very much disregardful to the rule of law (Riaz 2015). Likewise, the waves of political atrocities not only affected the politicians, but the widespread bomb attacks and chaos also brought sufferings for secular intellectuals, cultural activists and general people (Obaidullah 2019).

On the other hand, Sri Lankan political conflicts represent a prime example of ethnic conflict, where minorities are severely suffering from economic, political and cultural deprivation and other grievances. Long-time deprivation has provoked the violent rebellions to organize themselves against the state (Orjuela 2003). Sri Lanka has an illiberal democracy that has denied equal citizenship rights to its Tamil minority groups resulting creation of a militant group which have long been engaged in an armed struggle for an independent homeland for the from 1983 to 2009 (Alagappa 2004). It is noteworthy that president Rajapaksa oversaw the end of the 27-year Sri Lankan civil war in 2009, launching military atrocities against the Tamil people in which, according to a United Nations (UN) estimate, more than forty thousand civilians had died. Surprisingly, President Mahinda Rajapaksa does not allow the UN to investigate the war crimes committed against Tamil Tiger insurgency, though Sri Lanka's parliamentary democracy has been practising a fusion of power since its inception. However, it is not well accepted by the country's non-Sinhala-Buddhist minority that a pro-active group has been pressing the government to press home their demands including to ensure equal rights for decades. Similarly, the mainstream population is not that optimistic for unified progress. Since 2015, although President Maithripala Sirisena was committed for reconciliation, but failed due to the unwillingness and non-cooperation of the majority Sinhala Buddhist (Padma 2018).

Role of CSOs in Bangladesh and Sri Lanka in Political Stability and Peace-Building

Civil society actors play a role to open new spaces, build relationships in and across society and advocate the state for establishing peace (Pearce 2011). CSOs of Bangladesh and Sri Lanka have played numerous functions for political stability and peace-building under the political and ethnopolitical problems. However, before going into comparative analysis, the chapter will discuss the role of CSOs in political stability and peace-building in Bangladesh and Sri Lanka.

The Bangladesh Context

Since independence, the CSOs have been working in the socio-economic development of the country, nurturing the newly established democracy and struggling for the restoration of the democracy while Bangladesh dismays many observers with its endless dysfunctionality and resulting "illiberal democracy" (Bertocci 2014; Islam 2017). Historically, during military rule from 1975 to 1990 and the period of military-backed caretaker government from 2007 to 2008, civil society assertions have attempted to resist authoritarian regimes and military dictatorships. During this long period of non-democratic rule, different CSOs, as part of their responsibilities, raised the demand for restoration of demcoracy, advocated a series of electoral and institutional reforms, and acted as pressure groups, mobilizing public opinion in support of their demand for reforms. In the 1990s, the CSOs in Bangladesh quite visibly proliferated and expanded their role in a wide range of activities; additionally, the tasks of good governance and civil society movements became more complex (Parnini 2006). At the same time, successive party governments have crippled independent state institutions by staffing the bureaucracy, judiciary and the security apparatus with their loyalists. Similarly, social services have often been channelled through the patron–client networks of the political parties rather than state agencies (Lorch 2014). Henceforth, various individuals and groups tried to nourish democratic ethos, motivate establishing a transparent and accountable government to prevent these irregularities.

The initial constructive engagement on the restoration of democracy among political parties began to evaporate within two years. Subsequently, the problem was to keep the doors of democratic dialogue open between them, the inability of which brought about a political deadlock that continued two years from 1994 to 1996, moving the state to the brink of economic deterioration and political chaos (Jahan 2005). Thus public life became gradually intolerable with blockades, collapsing public services, strikes, shutdown of universities and gunfights among factions of party cadres. In this situation, civil society was compelled to step in and play the delicate role of moderator and peacemaker through negotiation and pressuring the government in different ways. The problem was resolved when the incumbent government was compelled to add non-party caretaker government (NPCG) provision in the constitution amid the pressure of opposition parties and the CSOs. In 2011, the same deadlock returned back when the ruling party withdrew NPCG provision from the constitution. This crisis was following the pre- and post-election political turmoil in 2013 and 2015. As a result, the country witnessed that political parties had seriously violated human rights situation. Also, attacks on police and setting fire on public vehicles and goods-laden trucks were common examples of political violence. Similarly, CSOs like Centre for Policy Dialogue (CPD), media, university teachers' association and professionals groups have played an immense role in restoring political stability and peace in Bangladesh. Most notably, media has become pressure group actor of CSOs to influence the citizen's view on public affairs such an active watchdog against human rights violation and other excesses of the state power.

In 1997, during the tenure of the then Prime Minister Sheikh Hasina, the two decades insurgency and bloody conflicts between the 12 different ethnic minorities' insurgent groups of the CHT and the government of Bangladesh came to an end following the signing of an accord between the government of Bangladesh and the Parbattya Chattagram Jana Sanghati Samity (PCJSS). This peace accord represents a milestone not only in the history of Bangladesh but also in the world at large in as much as none of the 15 or so similar ethnic insurgency-related problems of the world could yet to be solved. Resolving the land issue and guaranteeing the political, social, cultural, pedagogical and financial privileges of the indigenous communities of the CHT were the focal clauses of the accord. Equal privileges of all citizens of Bangladesh and the assurance of socio-economic advancement in CHT were equally incorporated into the peace accord. However, it did not happen overnight; rather it almost took two decades to formulate the peace treaty. Diverse national and international groups had a remarkable role as facilitators to proceed with the negotiation for implementing the procedure of the Peace Accord to conciliate the insurgency groups in the CHT (Chowdhury 2005). Most notably, efforts at peace-building between the government and CHT minorities have continued until now after 21 years of signing the Peace Accord. Civil societies like Hill Tracts NGO Forum, Manusher Jonno Foundation, Action Aid Bangladesh and Green Hill have been working as pressure groups for the protection and implementation of all the clauses of the Peace Accord.

In Bangladesh, civil society actors are struggling for the inclusion of citizen participation in all democratic institutions for a stable democratic country. Scholars have similarly argued that citizen involvement in organizations contributes directly or indirectly in political participation, democratic values, democratization and economic growth (Booth and Richard 1998). At this standpoint, CSOs have not only created awareness and participation of citizens of various democratic institutions but also made attempts to ensure of accountability by the concerned officials. Besides, investigation and research-based news also conduct a vital role for ensuring accountability and transparency of government, bureaucrats and political leaders. Moreover, the political element of many CSOs enables to create awareness and to groom a more informed citizenry that participates in politics, make better voting choices and hold government more accountable and transparent consequently. One of the far-reaching initiatives of CSOs is the creation of mass awareness to develop a questionnaire of seeking information from candidates contesting in the election. All these efforts assisted in identifying significant electoral issues, which were then mentioned by the media, thus leading to education, building awareness and consequently making a movement for reform.

In recent years, some civil society institutions, such as Transparency International Bangladesh (TIB), *Ain-o-Shalish Kendro* (*ASK*) and the Human Rights Watch, have played a seminally important role in holding successive governments to account for their perceived anti-democratic practices; nor galvanizing support for institutionalizing the rule of law and human rights; which are vital to establish political stability in Bangladesh (BIGD2013). For a long, the ruling party has been maintaining an unscrupulous practice for recruiting and appointing civil servants using political power, position and favour instead of giving preferences of merit, performance and experiences (Zafarullah and Rahman 2008). Moreover, the predominance of quota policy in government recruitment policy is also seen a hindrance against the merit system (Kaiser 2015). In 2018, college and university students organized a forum against this recruitment policy, especially discriminatory quota policy. The government was compelled to amend the recruitment policy amid countrywide strong student demonstration. CSOs of the country supported their activities actively, especially the media advocated in favour of the protest and aired their all sort of protest programmes and activities. The role of civil society in Bangladesh is described concisely in Table 27.1.

The Sri Lankan Context

Civil society tradition in Sri Lanka is vibrant and intricately woven in the fabric of the nation (UNDP 2007). Sri Lankan ethnopolitical conflict turned into civil war from 1983 to 2009. Most notably, in this crisis, civil society has played a pivotal role in promoting peace-building and sustainable development through policy advocacy and grassroots initiatives by resettling displaced communities, restoring livelihoods, coordinating interventions, collaborating with decision makers and advocacy work with politicians and different minority groups (Akurugoda 2018). Peace education, courses, seminars, media campaigns, rallies, posters and pamphlets have served to attract people's attention for maintain peace and to alert people in terms of new facts and different ways of framing peace and conflict. CSOs involvement in creating policy networks helps in mobilizing local activists, social movements and other factors that can pressure governments to change their policies and practices for a stable democratic country. These measures may not lead to dramatic attitude changes but can serve to initiate discussions, public opinion, envision a future for peace and give strength to those who already have pro-peace attitudes (Orjuela 2008). When the civil war started to put tremendous impacts over the civilian lives in Sri Lanka, many CSOs were emerged and worked on peace work at all levels in all parts of the country.

Citizen Committee for National Harmony, Movement for Inter-Racial Justice and Equality, Jaffna Citizen's Committee, Mother's Front, the

Table 27.1 Role of civil society in political stability and peace-building in Bangladesh

Activities	Target group	(Intended) Impact
Public opinion formation, disseminating information, advocacy work	General people, politician, minority groups	Creating awareness; make government and political parties accountable and transparent; developing human rights situation
Negotiation work, bringing different groups together	Political parties, minority groups	Mitigating political instability; development of democratic institutions and boost up relationship between Bengali and other minority groups
Pressure group, watchdog	State forces, political parties, religious leaders	Prevent violence; protect human rights
Free-flowing discussions on diverse issues, investigation based news	Citizens, different stakeholders through media, seminars	Influence the citizen view on public affairs; checking the exercise of assuming power arbitrary by the state
Collective bargaining, awareness-raising	Trade unions, different political parties	Accountable to its citizens through elections and the rights of advocacy
Maintaining a high standard of professionalism, protecting the rights of members, promoting interests	Professional bodies, cultural bodies	Political equality; all enjoy the full range of professional and human rights
Promoting social welfare	The vulnerable group, minority communities	Changing the condition of livelihood, education of people for the betterment of the country and its democracy

Source: Azharul (2005) and Author

University Teachers for Human Rights—Jaffna, and National Peace Council were emerged as a strong CSOs, based on peace movements, worked against war and to establish peace in Sri Lanka (Aliff and Sarjoon 2010; Orjuela 2004). Moreover, some vociferous CSOs are inclined to mobilize Sinhalese nationalism. They perceive it as a commitment to protect the unity and the sovereignty of holy Buddhist country. They also think that if they go against nationalism, it may help the minorities and Tamil militants to be a stronger force against the state. To them, it could be a betrayal of the country, and it might result in a concrete separation between majority and minority (Orjuela 2005). Also they could not stop the continuation of war, the cost of war, declination of the economy and most importantly the declination of parliament democracy (Aliff and Sarjoon 2010). Despite all these failures, they did not give up their struggle for ensuring peace in the South Asian Island. Furthermore, peace organizations have attempted to improve the skills and ethical knowledge of journalists, and expand their understanding of conflict through initiating courses, meetings and paying visits to the vulnerable zone. These activities have not been able to alter the underlying structural problems due to lack of resources in media sector and the use of media for political party and national interests (Orjuela 2008). On top of that, the role of CSOs had increased when a ceasefire agreement signed in 2002 with Norwegian assistance and after the 2004 tsunami of the Indian Ocean claimed that over 30,000 lives in Sri Lanka alone and other half of million people left displaced (UNDP 2007). Many donor peace-building programmes involving components were implemented through Sri Lankan NGOs. The purposes of these programmes were to strengthen public support for peace-building; to forge connection between different ethnic groups; to minimize inter-community tensions by organizing seminars and workshops, trainings and media campaigns, and conducting research.

Likewise, international organizations develop a multilateral network with their counterparts and share information and ideas regarding the peace-building process. They also develop and maintain public support in their respective constituencies. UNDP's works on a business development ground that promotes peace and adopts a three-step strategy: (1) developing an environment for investment and trade; (2) promoting a capacity-building strategy for sustainable development in business-supporting institutions; and (3) enhancing business to work pro-actively for peace (UNDP 2007). CSOs to extend innovative projects to communicate and generate public dialogue on peace, to improve interpersonal relations and cross-cultural understanding, and to increase public participation in peace-building process. Civil society has played a vital role in promoting accountability and transparency in Sri Lankan local institutions. Their purpose was to strengthen local capacity for institutional development. Although the 2002 ceasefire was proved successful, but later the situation started getting worse for serious disagreement among actors. For example, disagreements over the distribution of aid to the Liberation Tigers of Tamil Eelam (LTTE) in 2004 and the election of 2005 caused severe violence all over the country (Bowden and Binns 2016). In Sri Lanka, popular mobilization efforts are typically undertaken by groups advocating a particular nationalist or ethnic cause. The popular support for the military victory over the LTTE in 2009 demonstrates the power of the pro-war movement where the facilitators and the international community had greatly underestimated (Paffenholz 2014).

In the post-civil war in Sri Lanka, the state and non-state are actively engaged in cooperative dialogues and agreements across the ethnic line. However, these initiatives are considered as the donor-driven actions and not recognized as an appreciated initiative in official discourse. Also, some advocacy-level initiatives have been undertaken by the Sinhala nationalist organizations and the Sri Lankan armed forces. Their aim was to make a unified Sri Lanka by reaching out to the Tamil war victims and other ethnic groups. They tried to build a common ground where participation of every actor was important (Höglund and Orjuela 2011). During the conflict period, citizens' lives were volatile due to the instability of the state, including economic decline and the dysfunctionality of the public services in different sectors such as education and health. The less skilled people suffer from severe austerity and unplanned government actions. Also, the unemployment rate of a conflict-driven region is alarmingly high. The government and the CSOs can work together to address these issues. Their bilateral relationship can arrange oral communication, entrepreneurship, critical thinking and leadership quality, as well as specific vocational skills for creating employment in different types of business (Bowden and Binns 2016). Many CSOs are working for rural Tamil women who acquired a disability during the civil war in Sri Lanka by advocating justice for gender based disability, social assistance for rights, resistance and reintegration, reconciliation and building cross-ethnic relationships. After their involvement with the organization, Tamil women had experienced significant improvements specially financial supports for their lives which is highly important to establish trust-building by reconciliation and cross-ethnic relationships for stability and peace (Kandasamy et al. 2016).

Sri Lanka had been divided on the context of ethnopolitical problem and faced crisis of democracy, where civil society could play an important role for a stable democracy. But civil society and its activities were very poor in institutionalizing democracy in Sri Lanka. After the 2015 presidential elections, the elected ruling party started using the same fear psychosis and physical violence like the previous regime. During that time, a wide spectrum of organizations and individuals comprising the majority of civil society took a leading role to establish political stability and peace (Bopage 2016). But, in 2018, the country's politicians have unleashed a democratic crisis and the legislators were engaged in fisticuffs in parliament, attacked police personnel and the speaker with parliamentary equipment. A parliamentarian was accused of bragging a knife and another was accused of attacking opponents with water mixed chilli powder. The world witnessed the incident which ruined Sri Lankan political reputation (DeVotta 2018). The role of civil society in Sri Lanka in the context of political stability and peace-building is depicted in short in Table 27.2.

Comparative Analysis

It is a tough task to compare the role of CSOs in political stability and peace-building between these two countries amid similarities and dissimilarities in their history, politics, ethnic-religious minority problem and their economy. CSOs in Bangladesh and Sri Lanka have played expanded roles for political stability and peace by their advocacy work, emerging as a pressure group, negotiation and of course for popular mobilization. In Sri Lanka, CSOs played a significant role in peace establishment in the civil war period especially in the ceasefire in 2002 and have also been working in the post-civil war period for reconciliation, building trust and development, particularly in the war-ravaged area. Similarly, Bangladesh, CSOs worked a lot for restoring democracy, singing the Peace Accord in CHT, and ensuring political stability among

Table 27.2 Role of civil society in political stability and peace-building in Sri Lanka

Civil-society activity	Target group	(Intended) Impact
Advocacy work	Politicians, the international community, different ethnic group	Put pressure on political leaders to work for political stability, peace-building, political reform, reconciliation and trust-building.
Research and information	The general public, politicians and international community through media, seminars	Increased knowledge about the background, cost and possible solutions to the conflict; increased awareness of the conflict and human rights abuses
Informal diplomacy	Key actors from the Sri Lankan government and other actors	Upholding of a dialogue between actors on different sides, conveying messages and exchange views of the other side, discussion for future solutions
Reconstruction, mobilizing people to satisfy their own basic needs	Local communities and vulnerable communities	Improvement livelihood; decreased risks of discontent and conflict
Awareness-raising and peace education	Local communities, teachers, community leaders and religious leaders	Increased dialogue and prevent the conflict from spreading; vote for peace at elections and abstain from supporting or mobilizing for violence; mobilization of people in manifestations
Organization of peace marches, rallies, and support for livelihood change	Grassroots, NGOs and other networks	Increased awareness for peace among the general public; put pressure on political leaders to work for de-escalation, peace negotiations and political reform; support victims of the civil war
Bringing persons from different ethnic groups together	Local communities, professional groups, religious leaders and political leaders	Decrease prejudices between ethnic groups; develop cross-ethnic relationships

Source: Orjuela (2003) and Author

political parties. Despite, in both country's initiatives, Sri Lanka achieved a 'victor's peace' where Bangladesh achieved 'negotiated peace' in their ethnic problem. Peace in CHT, still, sustaining and the government and CSOs are almost successful in reconciliation and trust-building among government, Bengali and ethnic minorities. Sri Lanka, on the other hand, has failed to sustain its peaceful solution in the post-civil war period; it has also failed to build trust among the government, the Sinhalese, the Tamils, and other religious minorities, although many CSOs are working in different ways to promote it.

Achieving a sustainable peace in Sri Lanka is a complex and challenging task. In post-war literature, the word 'peace' is debatable. There is no definite explanation of 'peace' in a war-torn country. Tamil minority and the Sinhalese majority have developed a different form of post-war sentiments which are moulded by their post-war experiences (Bowden and Binns 2016). Surprisingly, in both countries, public universities have done tremendous work in restoring democracy, peace-building and raising the voice in ensuring human rights. Most importantly, public universities of Bangladesh played vital roles in the historical ideological and cultural struggles of 1971 and against the military regime in between 1981 to 1990 and any crisis of the country as a dynamic subset of civil society (Maîtrot 2016). Sri Lankan university teachers' association and undergraduate associations have also done some significant role in establishing peace and protecting human rights in the civil war and have been working restlessly since the post-civil war (Aliff and Sarjoon 2010).

Moreover, the word political party and crisis have become regular companions in Bangladesh & Sri Lanka. Now Bangladeshi civil society is more concerned about political stability rather than any other problem because it is a common phenomenon during pre and post-election political impasse. In Bangladesh, non-political actors have failed to build trust among political parties; similarly, Sri Lankan politics also faced a severe crisis in 2018. CSOs in Bangladesh have almost ensured political freedom for all citizens of Bangladesh; conversely, the political space for ethnic minority rights remains highly constrained in Sri Lanka (Goodhand 2013; Wickramasinghe 2014; Byrne and Klem 2014). Since the end of the civil war, the Sri Lankan government has tried to build up a unified state and set up an authoritative power by reforming its political and economic strategy (Goodhand 2013). In the context of Bangladesh, the vigilant role in which civil society associations successfully influenced the state in institutionalising democracy is, however, restricted due to the inability of civil society to maintain its autonomy vis-à-vis the state and political parties (Quadir 2015). In sum, it can be argued that Bangladeshi CSOs are more successful and get vast freedom from the state in their activities which are comparatively better than Sri Lanka.

The section will explore problems faced by CSOs in Bangladesh and Sri Lanka in taking aforementioned roles. CSOs of Bangladesh and Sri Lanka are lagging behind due to the lack of autonomy. Apart from that national and international agencies are also dominating them. Moreover, loyalty for political parties is a great threat for ensuring political stability and peace-building (Orjuela 2003; Tasnim 2017). National and local CSOs are hardly neutral or independent and they tend to be aligned with political parties and further political interests that create crises of autonomy and freedom due to the unwillingness of the country and involved in party politics. Most trade unions are not effective in achieving the objectives for which they have been constituted. Ottaway (2004) draws the same findings that weak states civil society often lacks autonomy, displaying fluid boundaries with powerful social forces both inside and outside the state. Civil society in Bangladesh has faced numerous difficulties to play these roles such as encroaching on the autonomy of the civil society sphere by the government (Alam et al. 2011; Stiles 2002; Lorch 2017), and the dominating politicization and patronizing culture (Blair 2001; Devine 2006) diminishing the space for civil society (Maîtrot 2016). Furthermore, supporting or opposing the government and philosophical divisions are also acute within most professional associations (Quadir 2003). Comparatively, the high politicization of Sri Lankan society leaves a relatively small space for civil society activities. Most of the civil society function in Sri Lankan rural areas is introduced or controlled by local or international NGOs with a top-down approach (Orjuela 2003). Moreover, the ethnic polarization resulting from political and violent conflict also impinges on CSOs, which is ethnically divided. The Sri Lankan government has been slower to provide legal protection and proper regulation for CSOs than other South Asian countries. Walton (2008) finds in his research that although donors have increasingly favoured national NGOs in their peace-building interventions, these organizations have been particularly vulnerable in the context of crises of legitimacy.

Conclusion

Paffenholz identified seven functions for peace-building. They are protection, monitoring, advocacy, socialization, social cohesion, facilitation and service delivery (Pearce 2011). Despite having limitations and political unrest, CSOs in both Bangladesh and Sri Lanka are working to ensure the above functions for country's political stability and peace-building. Vast initiatives like advocacy work, negotiation with different groups, research, awareness-raising and trust-building programmes have been taken in this connection. Notably, in Bangladesh, CSOs have almost succeeded in resolving the decades-long ethnopolitical problem through peace accord and are now working for institutionalizing democratic institutions like the electoral system, people's participation and trust-building between political parties. Conversely, the CSOs of Sri Lanka had worked a lot during the time of the civil war for maintaining peace and protection for human rights although they have earned little success in this arena. In the post-civil war era, they had worked for reconciliation and trust-building among citizens of the country, especially the war-ravaged area. In this reality, future research should also be explored on the potential role of CSOs for political stability and peace-building.

References

Akurugoda, I. R. (2018). *NGO Politics in Sri Lanka: Local Government and Development*. Switzerland: Palgrave Macmillan.

Alagappa, M. (2004). *Civil Society and Political Change: An Analytical Framework*. Stanford, CA: Stanford University Press.

Alam, G. M., Rabby, T. G., Boon, T. L., Khan, I., & Hoque, K. E. (2011). National Development and Student Politics in Bangladesh. *African Journal of Business Management, 5*(15), 6044–6057.

Aliff, S. M., & Sarjoon, A. (2010). The Role of Civil Society in Democratic Governance: State-Civil Society Relations in Democratization Special Focus on Contemporary Sri Lanka. *Kalam Journal, 4*, 26–34.

Azharul, P. (2005). Civil Society and Democracy in Bangladesh. *Social Change, 35*(2), 85–100.

Belloni, R. (2018). Civil Society in War-to-Democracy Transitions. In A. K. Jarstad & T. D. Sisk (Eds.), *War to Democracy: Dilemmas of Peacebuilding* (pp. 182–210). Cambridge: Cambridge University Press.

Bertocci, P. J. (2014). [Review of the Book *Bangladesh: Politics, Economy and Civil Society*. By David Lewis]. *The Journal of Asian Studies, 73*(2), 563–565. https://doi.org/10.1017/S0021911814000308.

BIGD (BRAC Institute of Governance and Development). (2013). *The State of Governance Bangladesh 2013: Democracy, Party, Politics*. Dhaka: BRAC University.

Blair, H. W. (2001). Civil Society, Democratic Development and International Donors. In R. Jahan (Ed.), *Bangladesh: Promise and Performance* (pp. 181–218). London: Zed Books.

Booth, J. A., & Richard, P. B. (1998). Civil Society, Political Capital, and Democratization in Central America. *Journal of Politics, 60*(3), 780–800.

Bopage, L. (2016). Democracy & the Critical Role of Civil Society: The Lankan Experience. *Colombo Telegraph*. Retrieved January 7, 2019, from https://www.colombotelegraph.com/index.php/democracy-the-critical-role-of-civil-society-the-lankan-experience/.

Bowden, G., & Binns, T. (2016). Youth Employment and Post-War Development in Jaffna, Northern Sri Lanka. *Conflict, Security & Development, 16*(3), 197–218.

Brett, E. A. (2017). Representation and Exclusion in Partial Democracies: The Role of Civil Society Organizations. *The Journal of Development Studies*. https://doi.org/10.1080/00220388.2017.1344647.

Byrne, S., & Klem, B. (2014). Constructing Legitimacy in Post-War Transition: The Return of 'Normal' Politics in Nepal and Sri Lanka? *Geoforum*. https://doi.org/10.1016/j.geoforum.2014.10.002.

Chandhoke, N. (2011). Civil Society and Peace. In M. Edwards (Ed.), *The Oxford Handbook of Civil Society*. New York: Oxford University Press.

Chowdhury, N. J. (2005). Elites and Policy-Making: A Case of Chittagong Hill Tracts Accord Making in Bangladesh. *Asian Affairs, 27*(3), 55–71.

Chowdhury, M. H. (2018). *Democratization in South Asia: Lessons from American Institutions*. London and New York: Routledge.

Cole, E., Eppert, K., & Kinzelbach, K. (Eds.). (2008). *Public Oversight of the Security Sector, a Hand-Book for Civil Society Organisations*. Bratislava: DCAF & UNDP.

Dagher, R. (2016). Civil Society and Development: A Reconceptualisation. *Canadian Journal of Development Studies*. https://doi.org/10.1080/02255189.2016.1196167.

Devine, J. (2006). NGOs, Politics and Grassroots Mobilisation: Evidence from Bangladesh. *Journal of South Asian Development, 1*(1), 77–99.

Devotta, N. (2004). Sri Lanka: Ethnic Domination, Violence, and Illiberal Democracy. In M. Alagappa (Ed.), *Civil Society and Political Change: An Analytical Framework*. Stanford: Stanford University Press.

DeVotta, N. (2018). Sri Lanka's Crisis of Democracy. *East Asia Forum*. Retrieved January 7, 2019, from http://www.eastasiaforum.org/2018/12/03/sri-lankas-crisis-of-democracy/.

Dhameja, A. (2003). Emergence of Civil Society Organizations: The Globalization Context. In A. Dhameja (Ed.), *Contemporary Debates in Public Administration*. New Delhi: Prentice-Hall of India Private Limited.

Forman, J. M. (2006). Security Sector Reform: What Role for Civil Society. In M. Caparini, P. Fluri, & F. Molnar (Eds.), *Civil Society and the Security Sector: Concepts and Practices in New Democracies* (pp. 26–42). Berlin: LIT.

FSI (Failed States Index). (2018). Failed States Index 2018. Retrieved January 7, 2019, from http://fundforpeace.org/fsi/2018/04/24/fragile-states-index-2018-annual-report/.

Fukuyama, F. (2011). *The Origins of Political Order: From Prehuman Times to the French Revolution*. New York: Farrar, Straus, and Giroux.

Goodhand, J. (1999). Sri Lanka: NGOs and Peace-Building in Complex Political Emergencies. *Third World Quarterly, 20*(1), 69–87.

Goodhand, J. (2013). Sri Lanka in 2012: Securing the State, Enforcing the 'Peace'. *Asian Survey, 53*(1), 64–72.

Höglund, K., & Orjuela, C. (2011). Winning the Peace: Conflict Prevention After a Victor's Peace in Sri Lanka. *Contemporary Social Science, 6*(1), 19–37.

Islam, M. M. (2015). Electoral Violence in Bangladesh: Does a Confrontational Bipolar Political System Matter? *Commonwealth & Comparative Politics, 53*(4), 359–380.

Islam, M. R. (2017). Non-governmental Organizations and Community Development in Bangladesh. *International Social Work, 60*(2), 479–493.

Jahan, R. (2005). *Bangladesh Politics: Problems and Issues*. Dhaka: The University Press Limited.

Kaiser, Z. R. M. A. (2015). Bangladesh Civil Service (BCS) Examination: A Critical Review. *Public Affairs and Governance, 3*(2), 140–158.

Kamal, D. R., & Kaiser, Z. R. M. A. (2015). RMG in Bangladesh: A Study into the Effects of Pre-election (2014) Political Turmoil. *Foreign Trade Review, 50*(2), 135–147.

Kandasamy, N., Soldatic, K., & Samararatne, D. (2016). Peace, Justice and Disabled Women's Advocacy: Tamil Women with Disabilities in Rural Post-conflict Sri Lanka. *Medicine, Conflict and Survival.* https://doi.org/10.1080/13623699.2016.1237101.

Lewis, D., & Kanji, N. (2009). *Non-governmental Organizations and Development.* London and New York: Routledge.

Lorch, J. (2014). Elections in Bangladesh: Political Conflict and the Problem of Credi-bility. E-International Relations. Retrieved January 7, 2019, from https://www.e-ir.info/2014/02/02/elections-in-bangladesh-political-conflict-and-the-problem-of-credibility/.

Lorch, J. (2017). Civil Society Support for Military Coups: Bangladesh and the Philippines. *Journal of Civil Society, 13*(2), 184–201.

Lorch, J. (2017a). *Civil Society and Mirror Images of Weak States Bangladesh and the Philippines.* London: Palgrave Macmillan.

Maîtrot, M. R. L. (2016). Knowledge Creation in Bangladesh: Institutional Challenges and Personal Opportunities within Civil Society. *Development in Practice, 26*(6), 683–695.

Mercer, C. (2002). NGOs, Civil Society and Democratization: A Critical Review of the Literature. *Progress in Development Studies, 2*(1), 5–22.

Migdal, J. S. (2001). *State in Society.* Cambridge: Cambridge University Press.

Moniruzzaman, M. (2009). Party Politics and Political Violence in Bangladesh: Issues, Manifestation and Consequences. *South Asian Survey, 16*(1), 81–99.

Morgenthau, H. J. (1985). *Politics Among Nations: The Struggle for Power and Peace.* New York: McGraw-Hill.

Nilsson, D. (2012). Anchoring the Peace: Civil Society Actors in Peace Accords and Durable Peace. *International Interactions, 38*(2), 243–266.

Obaidullah, A. T. M. (2019). Paradox of Reforms: A Reflection on Present State of Democracy and Parliament. In A. T. M. Obaidullah (Ed.), *Institutionalization of the Parliament in Bangladesh: A Study of Donor Intervention for Reorganization and Development.* https://doi.org/10.1007/978-981-10-5317-7.

Orjuela, C. (2003). Building Peace in Sri Lanka: A Role for Civil Society? *Journal of Peace Research, 40*(2), 195–212.

Orjuela, C. (2004). *Civil Society in Civil War: Peace Work and Identity Politics in Sri Lanka*. Department of Peace and Development Research, Gothenburg: University of Gothenburg.
Orjuela, C. (2005). Civil Society in Civil War: The Case of Sri Lanka. *Civil Wars, 7*(2), 120–137.
Orjuela, C. (2008). *The Identity Politics of Peacebuilding: Civil Society in War-Torn Sri Lanka*. India: SAGE Publications India Pvt Ltd.
Ottaway, M. (2004). Civil Society. In P. J. Burnell & V. Randall (Eds.), *Politics in the Developing World* (pp. 120–135). Oxford: Oxford University Press.
Padma, S. (2018). Sri Lanka in 2017. *Asian Survey, 58*(1), 120–126.
Paffenholz, T. (2014). In Theory: Civil Society and Peace Negotiations: Beyond the Inclusion–Exclusion Dichotomy. *Negotiation Journal*. https://doi.org/10.1111/nejo.12046.
Parnini, S. N. (2006). Civil Society and Good Governance in Bangladesh. *Asian Journal of Political Science, 14*(2), 189–211.
Pearce, J. (2011). Civil Society and Peace. In M. Edwards (Ed.), *The Oxford Handbook of Civil Society*. New York: Oxford University Press.
Quadir, F. (2003). How 'Civil' Is Civil Society? Authoritarian State, Partisan Civil Society, and the Struggle for Democratic Development in Bangladesh. *Canadian Journal of Development Studies, 24*(3), 425–438.
Quadir, F. (2015). Civil Society in Bangladesh: In Search of Democracy and Development. In F. Quadir & Y. Tsujinaka (Eds.), *Civil Society in Asia: In Search of Democracy and Development in Bangladesh*. Farnham: Ashgate.
Rahman, T. (2007). Parliamentary Control and Government Accountability in Sri Lanka: The Role of Parliamentary Committees. In T. Rahman (Ed.), *Parliamentary Control and Government Accountability in South Asia: A Comparative Analysis of Bangladesh, India and Sri Lanka*. London: Routledge.
Riaz, A. (2015). *How Did We Arrive Here?* Dhaka: Prothoma Prokashan.
Rupesinghe, K., & Anderlini, S. N. (1998). *Civil Wars, Civil Peace: An Introduction to Conflict Resolution*. London: Pluto.
Stiles, K. (2002). International Support for NGOs in Bangladesh: Some Unintended Consequences. *World Development, 30*(5), 835–846.
Tasnim, F. (2012). How Vigilant Is the Vibrant Civil Society in Bangladesh? A Survey-Based Analysis. *Journal of Civil Society, 8*(2), 155–183.
Tasnim, F. (2017). Politicized Civil Society in Bangladesh: Case Study Analyses. *Cosmopolitan Civil Societies: An Interdisciplinary Journal, 9*(1), 98–123.
UNDP. (2007). *UNDP and Civil Society in Sri Lanka: Partnerships in Crisis Situations*. Sri Lanka: UNDP. Retrieved January 9, 2019, from http://www.

undp.org/content/dam/undp/documents/partners/civil_society/publications/Civil_Society_and_UNDP_in_Sri_Lanka_partnerships_in_crisis_situations.pdf.

Walton, O. (2008). Conflict, Peace Building and NGO Legitimacy: National NGOs in Sri Lanka. *Conflict, Security & Development, 8*(1), 133–167.

Waltzer, M. (1990). *The Civil Society Argument*. Stockholm: Gunnar Myrdal Lecture.

Wickramasinghe, N. (2014). Sri Lanka in 2013: Post-war Oppressive Stability. *Asian Survey, 54*(1), 199–205.

Zafarullah, H., & Rahman, R. (2008). The Impaired State: Assessing State Capacity and Governance in Bangladesh. *International Journal of Public Sector Management, 21*(7), 739–752.

Part VI

Child Welfare in South Asia

Md. Nurul Momen, Rajendra Baikady, and Cheng Sheng-Li

Introduction

The South Asian region is home to about one-fourth of the world's child population (UNICEF 2005). South Asia is home to approximately 627 million children younger than age 18 (UNICEF 2016), which is about 36 per cent of the total population of 1.82 billion (UN 2015). Most children in the region face problems related to acute malnutrition, under nourishment and hunger. Poor nutrition in early life jeopardizes physical growth and cognitive development of children across the globe and especially in South Asia. The South Asian region has the highest burden of stunting, wasting and anaemia prevalence in the world (Torlesse and Aguayo 2018). Earlier research had shown that children within the region, born in communities with low wealth, are the most likely to be stunted (Rama et al. 2015). Further, research studies at the global level have also shown that cognitive development of children during the early stage of life depends on adequate nutrition that has enduring positive consequences in school readiness, learning outcome, human capital formation, economic productivity and shared prosperity (Shekar et al. 2017; UNICEF 2018; Victora et al. 2008). According to a global estimate, South Asia is home to more than 40 (59 million) per cent of the world's stunted children and 53 (27 million) per cent of the world's wasted children (UNICEF et al. 2018). Most countries in the region are showing very slow progress in reducing child stunting, even though the overall rate of child stunting in the region is showing a decline (UNICEF et al. 2018). Slow reduction in child stunting among the countries within the region is expected to negatively affect the achievement of the estimated nutritional strategies of SDG (Development Initiatives 2017). Children from birth to 5 years of age are showing healthy leaner growth, which is almost similar among the countries across the world (Multi Centre Growth Reference Study Group 2006). However, in the recent past, according to global estimation, 23 per cent of children in the age group 0–59 months are chronically deprived of nutrition and show stunted growth (UNICEF, WHO, WBG 2017). Linear growth of children in the South Asian region is the result of poor maternal nutrition before and after pregnancy, poor diet of young children and poor sanitation practices among the people in the region (Aguayo and Menon 2016; Kim et al. 2017). However, these factors are context and time specific in nature, and need to be considered while assessing the issue (Aguayo and Menon 2016).

Situation of Children in the Region

Children in the region face an enormous number of challenges related to nutrition, health, safety and survival. The contributors in this part of the book make a modest attempt to understand child welfare initiatives in the South Asian region that provides a general understanding of children's issues with a central focus on health and abandonment;

issues related to health and nutrition—Children in the South Asian region have very poor dietary intake. Child feeding in Afghanistan, Bangladesh, India, Nepal and Pakistan is minimal; for instance, in Afghanistan (12 per cent) and Nepal (24 per cent), children in the age group of 6–24 months get very minimal diet (Afghanistan Ministry of Public Health and UNICEF 2013; NIPORT et al. 2015; Ministry of Women and Child Development (MWCD) 2015; Ministry of Health and Population (MOHP), New ERA, and ICF International 2012; NIPS and ICF International 2013). Furthermore, in many countries children within the age group of 6–23 months are not given the balanced amount and variety of food recommended by the World Health Organization (WHO 2015). In many countries, children are also deprived of breast milk (WHO 2015). Also, breast feeding has been reported to be one of the most cost-effective and life-saving interventions that reduce child stunting (Bhutta et al. 2008). Some other studies have also highlighted the contribution of complementary feeding (Stewart et al. 2013). Children in the region are affected by many factors, and studies (Stewart et al. 2013) suggest that knowledge gap, lack of availability of food, poor purchasing power, and social and cultural norms are the most likely to affect child feeding in the region. Further, studies have recommended specific programmes and strategies to address the underlying reasons (Lutter et al. 2013).

A number of research studies in recent years have reported child stunting and its health effects on children in South Asia (Corsi et al. 2015; Smith and Haddad 2014; Aguayo and Menon 2016; Aguayo et al. 2016; Black et al. 2013; de Onis and Branca 2016; Dewey and Begum 2011; Victora et al. 2008). Studies have found that diet during the first year of life has a major effect on child stunting in the region (Corsi et al. 2015; Smith and Haddad 2014; Aguayo and Menon 2016; Aguayo et al. 2016). Menon (2012) highlighted that the situation is critical because of the bad condition of complementary feeding in the region. Further studies noted that children who survive after stunting and nutrition deficiency, face enormous issues and damage related to school performance and cognitive development. The negative effect of stunting also continues in the adulthood in the form of reduced productivity and excessive weight gain, which may lead to short- and long-term health problems (Black et al. 2013; de Onis and Branca 2016; Dewey and Begum 2011; Victora et al. 2008).

Many studies have addressed the strategies and programmes for reducing stunting in children and improving health and nutritional status and child survival in the region (Dewey and Brown 2003; Lutter et al. 2011; Onyango 2013; WHO 2014; Mason et al. 2012; Bhutta et al. 2008; Ramakrishnan et al. 2012; Smith and Haddad 2015; Saldana et al. 2012). Some studies have highlighted the need for direct nutrition interventions in improving health and child survival (Dewey and Brown 2003; Lutter et al. 2011; Onyango 2013), whereas other studies have recommended increasing maternal nutrition as maternal health is responsible for more than 20 per cent (WHO 2014) of child stunting, and improving it helps to accelerate better child development outcomes (Mason et al. 2012). Alternatively, addressing the underlying socio-economic causes that are affecting the nutritional status of women in the region is also helpful in reducing child stunting (Bhutta et al. 2008; Ramakrishnan et al. 2012; Smith and Haddad 2015). Women's nutrition in South Asia is deemed as low priority similar to most of the other developing countries (Ramakrishnan et al. 2012; Saldana et al. 2012). Strategies to improve and enhance the nutrition intake of women in the region are an important step in reducing and preventing child stunting in the region (Ramakrishnan et al. 2012; Saldana et al. 2012).

Studies have also highlighted the need for maternal, newborn and child care practices among mothers with infants (Central Statistics Organisation (CSO) and UNICEF 2012; Black et al. 2013), nutrition education and counselling during pregnancy (Girard and Olude 2012), exclusive breastfeeding during the first 6 months (Bhutta et al. 2008; Black et al. 2013), and complementary feeding interventions (Bhutta et al. 2013; Dewey and Afarwuah 2008) in reducing child stunting and improving the nutritional level of children in the region. Some other studies have also highlighted educational interventions and bringing behavioural changes among the people even though it has low, modest or linear impact (Dewey and Afarwuah 2008). Shi et al. (2010) in China and Vazir et al. (2013) in India found that an educational approach helps in reducing stunting in children. The region accounts for 38 per cent of child stunting below the age of 5 years (UNICEF 2015). Women giving birth to a child in an unhygienic environment and the absence of unskilled birth attendant at delivery increases risks of neonatal death, as only less than half of the women population in the region gets support of skilled attendant (Central Statistics Organisation (CSO) and UNICEF 2012; Black et al. 2013). Poorly maintained hygiene, lack of adequate and poor drinking water and lack of sanitation facilities are the other reasons for infectious illness and low nutritional status in the region (Central Statistics Organisation (CSO) and UNICEF 2012). In Afghanistan, a wide existing inequality between social, economic and geographical groups determines the nutritional status (Afghanistan Ministry of Public Health, National Nutrition Survey 2013).

The South Asian region is well known for its cultural ties, and care and affection among family members. Families are a natural environment that supports and nurtures children's growth and development. Being cared for and loved irrespective of their social, economic, geographical and other consideration is the right of every child and caring and nurturing the children is the responsibility of every parent. Despite of these practices children and parents relations, many children in the region face abandonment and neglect of care by the parents and caregivers. As a result, many children end up living on the street and become vulnerable to illness and malpractices. There are millions of children in the region who don't have families, some of them have separated from the parents as the families are said to be dangerous for their lifelong emotional and physical development. An estimation in 2003 found that 48 million or more children in the region have lost their parents either both or one (UNICEF 2005). An estimated 8 per cent of the less than 18 years age group in the region accounts for the orphan population, with an estimated 6.5 per cent in Pakistan and Sri Lanka, 13 per cent in Afghanistan (UNICEF 2005). In addition to millions of children living without parents in the region many others are at risk of abandonment. Given these facts, there is a growing concern for the situation of children in the South Asian region. Children out of parental care in the South Asian region are the central attention for child neglect and exploitation due to a large increase in the number of children at risk. The need for care and protection of these children is well demanded in the region as the risk of these children's involvement in the high risk of violence, exploitation, abuse and crime.

The reasons for children living out of parental care or abandonment of children are well documented in the region. Poverty is understood to be the main reason for child neglect and abandonment, as an estimated 700 million people in the South Asian region live on less than US $1 per day (UNICEF 2006). Children in poor families are the most vulnerable as poverty leads to family breakdown which causes children to be abandoned. The abandonment of children is an extreme form of child neglect; however, armed conflict is the second most cause for the abandonment of the children in the region. Most of the countries in the region experience armed conflict internally or externally. During the last decade, a majority of them have experienced one or two armed conflicts. Conflicts lead to separation of families, increase in poverty, disruption of social welfare services and thus may also lead to children becoming associated with armed groups. Further a study reported that in Nepal children of 50 per cent internally displaced persons are not living with their parents (SAFHR 2005). Natural disasters in the South Asian region are quite frequent and these disasters worsen the conditions of children in the region by separating them from their parents. The 2004 tsunami in Sri Lanka resulted in 30,000 deaths and destroyed 80,000 households while displacing another 1 million people in the country (Norwegian Refugee Council 2006). According to an estimate, South Asia accounts for 7.5 million people suffering from HIV/AIDS (UNICEF 2006, p. 16), and the spread and maturation of this epidemic is pushing some more children in the region into orphanhood.

There are an estimated 8000 children placed in institutional care in Afghanistan (MOLSA 2006). And the studies have also reported the annual doubling in this number (MOLSA and UNICEF 2003). An estimated 49,000 children in Bangladesh are supported by more than 500 constructed institutions across the country (Dona 2003). In India, more than 50,000 children are supported by approximately 600 institutions for orphan care (Department of Women and Child Development 2001). Nepal had constructed 81 new orphan care institutions (Era and Macro 2005). Sri Lanka accounts for 11,000 children living in orphan care nationwide and only four provinces in the country have 16,000 children in institutionalized care (Jayathilake and Amarasuriya 2005). Dependency on institutional care in the South Asian region and a sharp growth in this number of children needing alternative care is alarming and disturbing in the region.

In sum, issues related to child nutrition, stunting and abandonment in the region need immediate attention alongside other important regional issues. As the region is committed to the sustainable development goals and strives to achieve all round development by 2030, care for children is an imperative need. Sustainable development of any nation or region depends on and is contributed to by the full and potential development, health and productive life of children. As the number of children out of parental care is very high and showing sharp growth, the emphasis is well deserved and many governmental and non-governmental interventions are addressing the issue in the region. Children and their nutrition in the region need the most attention as investment and improvement in child nutrition is one way to achieve sustainable and steady growth (Development Initiatives 2017). Furthermore, the Sustainable Development Agenda aims on 'leaving no one behind' (UNDP 2018), hence child development is an important aspect to bringing equality.

Part VI of the book is a modest attempt to give an account of the current situation with respect to children and their growth in the South Asian region. Two chapters in this part discuss child welfare in Bangladesh, whereas the other chapter focuses on child welfare in India. Overall, this part of the book is important in order to understand child-related issues in the South Asian region. The contributions in this part of the book analyse the situation of children in South Asia.

Chapter 28 specifically examines the rights of children, child laws and child welfare policy and provides some suggestions based on the empirical research conducted and the literature reviewed that will help to improve child welfare in Bangladesh. The discourses in the chapter encourage academics and practitioners for further discussions on conceptual implications of both practices and expand future implications for research on child welfare in Bangladesh. Chapter 29 attempts to understand the situation of Indian children through a situational analysis of child protection and legislation in India. While doing so, the author of the chapter intends to critically examine the present policies and programmes for child protection in India. Chapter 30 is an examination of non-government supports for non-formal education in Bangladesh and tries to find the path towards a better state of education which is accessible to the excluded portion of the inhabitants of the country.

References

Afghanistan Ministry of Public Health and UNICEF. (2013). *Afghanistan National Nutrition Survey 2013: Survey Report*. Kabul: Afghanistan.

Afghanistan Ministry of Public Health. (2013). National Nutrition Survey, 2013.

Aguayo, V. M., & Menon, P. (2016). Stop Stunting: Improving Child Feeding, Women's Nutrition and Household Sanitation in South Asia. *Maternal & Child Nutrition, 12*(Suppl 1), 3–11. https://doi.org/10.1111/mcn.12283.

Aguayo, V. M., Nair, R., Badgaiyan, N., & Krishna, V. (2016). Determinants of Stunting and Poor Linear Growth in Children Under 2 Years of Age in India. An In-depth Analysis of Maharashtra's Comprehensive Nutrition Survey. *Maternal & Child Nutrition, 12*(1), 121–140.

Bhutta, Z. A., Ahmed, T., Black, R. E., Cousens, S., Dewey, K., & Giugliani, E. (2008). What Works? Interventions for Maternal and Child Undernutrition and Survival. *Lancet, 371*, 417–440. https://doi.org/10.1016/S0140-6736(07)61693-6.

Bhutta, Z. A., Das, J. K., Rizvi, A., Gaffey, M. F., Walker, N., Horton, S., & For The Lancet Nutrition Interventions Review Group, and the Maternal and Child Nutrition Study Group. (2013). Evidence-Based Interventions for Improvement of Maternal and Child Nutrition: What Can Be Done and at What Cost? *Lancet, 382*, 452–477.

Black, R. E., Victora, C. G., Walker, S. P., et al. (2013). Maternal and Child Undernutrition and Overweight in Low-income and Middle-income Countries. *Lancet, 382*(9890), 427–451.

Central Statistics Organisation (CSO) and UNICEF. (2012). *Afghanistan Multiple Indicator Cluster Survey 2010–2011*. Kabul: Central Statistics Organisation (CSO) and UNICEF.

Corsi, D. J., Mejía-Guevara, I., & Subramanian, S. V. (2015). *Risk Factors for Chronic Undernutrition Among Children in India. Estimating Relative Importance, Population Attributable Risk and Fractions*. Social Science and Medicine. https://doi.org/10.1016/j.socscimed.2015.11.014.

de Onis, M., & Branca, F. (2016). Childhood Stunting. A Global Perspective. *Maternal & Child Nutrition, 12*(1), 12–26.

Development Initiatives. (2017). *Global Nutrition Report 2017: Nourishing the SDGs*. Bristol, UK: Development Initiatives.

Dewey, K. G., & Adu-Afarwuah, S. (2008). Systematic Review of the Efficacy and Effectiveness of Complementary Feeding Interventions in Developing Countries. *Maternal & Child Nutrition, 4*, 24–85.

Dewey, K. G., & Brown, K. H. (2003). Update on Technical Issues Concerning Complementary Feeding of Young Children in Developing Countries and Implications for Intervention Programs. *Food and Nutrition Bulletin, 24*(1), 5–28.

Dewey, K. G., & Begum, K. (2011). Long-Term Consequences of Stunting in Early Life. *Maternal & Child Nutrition, 7*, 5–18.

Dona, G. (2003). *Overview of the Conditions of Children Outside Parental Care in Institutions and Communities*. UNICEF.

Girard, A. W., & Olude, O. (2012). Nutrition Education and Counselling Provided During Pregnancy: Effects Onmaternal, Neonatal and Child Health Outcomes. *Paediatric and Perinatal Epidemiology, 26*(Suppl. 1), 191–204.

Government of India Department of Women and Child Development. (2001). *Convention on the Rights of the Child: India – First Periodic Report 2001*. Government of India.

Islamic Republic of Afghanistan Ministry of Labour and Social Affairs (MOLSA) and UNICEF. (2003). Afghanistan Country Report. Retrieved from http://www.children-strategies.org.

Jayathilake, R., & Amarasuriya, H. (2005). *Home Truths: Children's Rights in Institutional Care in Sri Lanka*. Save the Children in Sri Lanka.

Kim, R., Mejia-Guevara, I., Corsi, D. J., Aguayo, V. M., & Subramanian, S. V. (2017). Relative Importance of 13 Correlates of Child Stunting in South Asia: Insights from Nationally Representative Data from Afghanistan, Bangladesh, India, Nepal, and Pakistan. *Social Science & Medicine, 187*, 144–154. https://doi.org/10.1016/j.socscimed.2017.06.017.

Torlesse, H., & Aguayo, V. M. (2018). Aiming Higher for Maternal and Child Nutrition in South Asia. *Maternal and Child Nutrition, 14*(S4), e12739. https://doi.org/10.1111/mcn.12739.

Lutter, C. K., Daelmans, B., de Onis, M., Kothari, M., Ruel, M. T., & Arimond, M. (2011). Undernutrition, Poor Feeding Practices and Low Coverage of Key Nutrition Interventions. *Pediatrics, 128*, e1–e10.

Lutter, C. K., Iannotti, L., Creed-Kanashiro, H., Guyon, A., Daelmans, B., & Robert, R. (2013). Key Principles to Improve Programmes and Interventions in Complementary Feeding. *Maternal and Child Nutrition, 9*(Suppl 2), 101–115.

Mason, J. B., Saldana, L. S., Ramakrishnan, U., Lowe, A., Noznesky, E. A., Girard, A. W., et al. (2012). Opportunities for Improving Maternal Nutrition and Mirth Outcomes: Synthesis of Country Experiences. *Food and Nutrition Bulletin, 33*(2 Suppl. 1), S104–S137.

Menon, P. (2012). The Crisis of Poor Complementary Feeding in South Asia: Where Next? *Maternal & Child Nutrition, 8*(1), 1–4.

Ministry of Health and Population (MOHP), New ERA, and ICF International. (2012). *Nepal Demographic and Health Survey 2011*. Kathmandu: Nepal and Calverton, Maryland.

Ministry of Women and Child Development (MWCD). (2015). *Rapid Survey on Children (RSOC) 2013–2014*. New Delhi, India: MWCD: Government of India.

Multicenter Growth Reference Study Group. (2006). World Health Organization (WHO) Child Growth Standards Based on Length/ Height, Weight and Age. *ActaPaediatrica, 95*(Suppl 450), 76–85.

National Institute of Population Research and Training (NIPORT), Mitra and Associates, and ICF International. (2015). *Bangladesh Demographic and Health Survey 2014: Key Indicators*. Dhaka, Bangladesh, and Rockville, Maryland, USA: NIPORT, Mitra and Associates, and ICF International.

National Institute of Population Studies (NIPS) and ICF International. (2013). *Pakistan Demographic and Health Survey 2012–13*. Islamabad, Pakistan, and Calverton: Maryland, USA.

Era, N., & Macro, O. R. C. (2005). *Study of Children in Children's Homes in Nepal*. USAID.

Norwegian Refugee Council. (2006). Global IDP Project. Retrieved from www.idpproject.org/statistics.htm.

Onyango, A. (2013). Promoting Healthy Growth and Preventing Childhood Stunting: A Global Challenge. *Maternal and Child Nutrition, 9*(Suppl 2), 1–5.

Rama, M., Béteille, T., Li, Y., Mitra, P. K., & Newman, J. L. (2015). *Addressing Inequality in South Asia. South Asia Development Matters.* Washington, DC: World Bank. https://doi.org/10.1596/978-1-4648-0022-1.

Ramakrishnan, U., Lowe, A., Vir, S., Kumar, S., Mohanraj, R., Chaturvedi, A., et al. (2012). Public Health Interventions, Barriers, and Opportunities for Improving Maternal Nutrition in India. *Food and Nutrition Bulletin, 33*(2 Suppl), S71–S92.

Saldana, L. S., Bubback, L., White, M. J., Mulugeta, A., Mariama, S. G., Roba, A. C., et al. (2012). Policies and Program Implementation Experience to Improve Maternal Nutrition in Ethiopia. *Food and Nutrition Bulletin, 33*(2 Suppl), S27–S50.

Shekar, M., Kakietek, K., Eberwein, J. D., & Walters, D. (2017). *An Investment Framework for Nutrition: Reaching the Global Targets for Stunting, Anemia, Breastfeeding, and Wasting.* Directions in Development Series. Washington, DC: World Bank.

Shi, L., Zhang, J., Wang, Y., Caulfield, L. E., & Guyer, B. (2010). Effectiveness of an Educational Intervention on Complementary Feeding Practices and Growth in Rural China: A Cluster Randomised Controlled Trial. *Public Health Nutrition, 13*, 556–565.

Smith, L., & Haddad, L. (2014). Reducing Child Undernutrition. Past Drivers and Priorities for the Post-MDG Era. *World Development, 68*, 180–204.

Smith, L. C., & Haddad, L. (2015). Reducing child undernutrition: past drivers and priorities for post-MDG era. *World Development 68*, 180–204.

South Asia Forum for Human Rights (SAFHR), March 2005, A Pilot Survey on Internally Displaced Persons in Kathmandu and Birendranagar, available at: http://www.internaldisplacement.org/8025708F004CE90B/(httpDocuments)/E2F3AF8D0BB0399C802570B700599D CE/$file/IDP+pilot+survey+SAFHR05.pdf

Stewart, C., Iannotti, L., Dewey, K. D., Michaelsen, K. F., & Onyango, A. W. (2013). Contextualizing Complementary Feeding in a Broader Framework for Stunting Prevention. *Maternal and Child Nutrition, 9*(Suppl. 2), 27–45.

UNDP. (2018). *What Does It Mean to Leave No-One Behind? A UNDP Discussion Paper and Framework for Implementation*. New York: United Nations Development Programme (UNDP).

UNICEF. (2015). *The State of the World's Children: Reimagine the Future: Innovation for Every Child*. New York: UNICEF.

UNICEF. (2016). The State of the World's Children 2016. Retrieved from https://www.unicef.org/sowc2016/.

UNICEF. (2018). *Child Stunting, Hidden Hunger and Human Capital in South Asia: Implications for Sustainable Development Post 2015*. Kathmandu, Nepal: UNICEF.

UNICEF, WHO, & World Bank Group. (2018). Joint Child Malnutrition Estimates 2018 Edition. Retrieved from http://www.who.int/nutgrowthdb/estimates2017/en/.

UNICEF. (2005). *The State of South Asia's Children 2005*. UNICEF.

UNICEF. (2006). *Children Outside Parental Care in South Asian Countries* (p. 16). UNICEF.

United Nations Children's Fund (UNICEF), World Health Organization (WHO), World Bank Group (WBG). (2017). *Levels and Trends in Child Malnutrition. Joint Child Malnutrition Estimates*. New York: UNICEF, WHO, World Bank Group.

United Nations, Department of Economic and Social Affairs, Population Division. (2015). *World Population Prospects: The 2015 Revision*. DVD Edition.

Vazir, S., Engle, P., Balakrishna, N., Griffiths, P. L., Johnson, S. L., & Creed-Kanashiro, H. (2013). Cluster-Randomized Trial on Complementary and Responsive Feeding Education to Caregivers Found Improved Dietary Intake, Growth and Development Among Rural Indian Toddlers. *Maternal & Child Nutrition, 9*, 99–117.

Victora, C. G., Adair, L., Fall, C., Hallal, P. C., Martorell, R., Richter, L., & Sachdev, H. S. (2008). Maternal and Child Undernutrition: Consequences for Adult Health and Human Capital. *Lancet, 371*, 340–357. https://doi.org/10.1016/S0140-6736(07)61692-4.

WHO. (2014). *WHA Global Nutrition Targets 2025: Stunting Policy Brief.* Geneva: World Health Organisation.

World Health Organization. (2015). *Infant and Young Child Feeding.* Fact Sheet No. 342. Retrieved October 28, 2015, from http://www.who.int/mediacentre/factsheets/fs342/en/.

28

Child Welfare Practices in Organizational and Institutional Context of Bangladesh

Isahaque Ali, Azlinda Azman, Paramjit Singh Jamir Singh, Zulkarnain A. Hatta, Muhammud Shariful Islam, and Tahmina Akhtar

Introduction

Over 60 million children in Bangladesh face hazardous conditions and are deprived of their basic needs. Half of these children continue to live below the international poverty line despite the increasingly stable and growing economy in Bangladesh (UNICEF 2012). Each year, approximately 54,000 children are dying due to malnutrition in

I. Ali • A. Azman (✉) • P. S. Jamir Singh
Social Work Programme, School of Social Sciences, Universiti Sains Malaysia, Penang, Malaysia
e-mail: azlindaa@usm.my

Z. A. Hatta
Faculty of Social Science, Arts & Humanities, Lincoln University College, Petaling Jaya, Malaysia

© The Author(s) 2020

Md. N. Momen et al. (eds.), *Building Sustainable Communities*,
https://doi.org/10.1007/978-981-15-2393-9_28

Bangladesh (Daily Amader Somoy 2014). Furthermore, it is difficult to protect children from trafficking, child labor and child marriage as it has the lowest rates of birth registration in the world. A huge number of children are denied of their basic human rights due to the lack of governance; poverty; political instability; poor coordination among child welfare agencies and the lack of human, financial and targeted resources for childrens welfare. Programs and policies toward child welfare in Bangladesh are also lacking due to the lack of comprehensive national plans, independent monitoring and evaluation of such programs (UNICEF 2009a). Children are severely vulnerable in Bangladesh due to the absence of traditional values, absence of social and community awareness, ignorance of parents, death/disability of family breadwinner, parents' divorce or separation, the lack of educational opportunity for children, migration, social injustice, crime and natural disasters. Due to these complex issues, establishing an appropriate policy and programs formulation and vigorous implementation of the objectives set out in policy requires the highest level of coordination between government and civil society, including NGOs and other agencies involved in the planning and implementation process (UNICEF 2009a). While the government is committed to protect the child, the implementation of childrens development activities still lacks understanding and consistent planning. Moreover, it is not surprising that the laws in place against child welfare are hardly being implemented at both policy and implementation levels in Bangladesh.

M. S. Islam
Department of Social Work, University of Rajshahi, Rajshahi, Bangladesh

T. Akhtar
Institute of Social Welfare and Research, University of Dhaka, Dhaka, Bangladesh

Children in Bangladesh: Demographic and Socio-Economic Status and Problems

In Bangladesh, there are 66 million children below the age of 18 years. This makes up 45% of the aggregate population of the country. In this regard, children aged 5–17 (42.4 million) make up nearly one-third of the country's populations. In Bangladesh, 26.5 million out of 63 million children are living under the national poverty line in terms of Cost Basic Needs (CBN). Furthermore, 46% of the children consume less than the prescribed Direct Calorie Intake (DCL), and more than half of all households (51%) with children are considered as poor as they living under the international poverty line and earn below the USD1 Purchasing Power Parity (PPP) threshold in Bangladesh (UNICEF 2009b). Around, 58% of the children in Bangladesh do not have access to one basic need such as education, healthcare, pure drinking water, shelter, information and sanitation, while nearly 20% of the children severely deprived at least two basic needs. Moreover, 59% are deprived of basic information, 64% of sanitation services, 41% of adequate shelter and 57% of sufficient nutrition (underweight) (UNICEF 2009b). It is projected that one million children aged between 10 and 14 years are engaged in labor and more than 1.4 million children cannot attend schools due to severe poverty in Bangladesh. Moreover, nearly 3.8 million girl children are married off before the age of 18 years and 1.3 million are married before they reach the age of 15 years in Bangladesh. Consequently, the lack of child-centered policy, absence of proper birth registration, poor education, limited access or healthcare services and limited coordination with child welfare agencies and civil society have increased the probability of sexual exploitation and violations against children. These scenarios obviously show that the country has not been able to address the crucial issue of child rights violation and welfare in an effective manner despite Bangladesh's remarkable progress in achieving most of the Millennium Development Goals (MDGs). Hence, there is a need for a juvenile justice system that could empower children's rights in Bangladesh.

Material and Methods

Data for this study are derived from secondary information and secondary sources, specifically previously published research used for the literature review and for analyzing the findings. The systematic review is important for understanding some of the main philosophies debated and the underlying issues, which makes the study more significant. The secondary sources included various websites, newspaper articles, books, government reports, scientific articles, scholarly journals, United Nations International Children's Emergency Fund (UNICEF), United Nations Development Programme (UNDP), World Bank and International Labour Organization (ILO) published reports. They are used to contribute to the knowledge on child welfare practices in the local context of Bangladesh.

Review Results

Child Welfare in Constitutions and Legislations

The Bangladesh Constitution is the principal actor to protect the basic rights of the citizens, ensure equality of laws, protect liberty, give freedom from torture and ensure the protection of the law and exclusion from enforced labor (Sengupta and Hossain 2006). The provisions for fundamental rights of all citizens of Bangladesh including the children are included in the Constitution of the Peoples' Republic of Bangladesh. The Constitution of Bangladesh includes some fundamental principles as shown here:

- Liberty from any kind of abuse (Article 14);
- Supporting the fundamental necessities of citizens life that comprising shelter, food, clothing, health care and universal education, leisure and recreation, rights to rest and rights to social protection by social safety nets and in case of government assistance due to unemployment, ill health, or disability or hazards by an orphan (Article 15);

- To establish a unique, universal and people-oriented education system and extended obligatory and free of cost education for children defined by law (Article 17);
- Increasing the education level and developing health care (Article 18);
- No deprivation in term of race, religion, caste, sex of birth places (Article 28);
- Equal chances and assenting action in issues for children (Article 29);
- Ban any types of compulsory labor (Article 34) (Sengupta and Hossain 2006).

In Bangladesh, there are more than 35 laws established to safeguard children from abuse and exploitation, brutality and negligence to extend their rights and welfare (Sengupta and Hossain 2006).

The Children Act, 1974 and the Children Rules, 1976 are the two main legislations in Bangladesh that deal with children involved in authorized actions. These two legislations are shown here:

The Children Act, 1974: The Children Act of 1974 is the main law for children protection, special care and health care of children, either as the accused or as victims. The ultimate goals of the act are the empowerment of the children, to ensure the protection of children and the medical care of children (Sengupta and Hossain 2006). This act enforces several roles and responsibilities to the state that are implemented by a special court (Juvenile Courts), to be established by the safeguard of all children (accused and victims) (section, 3, Chapters 4 & 6). The trial system in the Juvenile Court is separated from the ordinary court (section, 7) and trials are conducted differently for children and adults (section, 6). Additionally, there is a need to maintain high confidentiality (section, 9, 10, 17), considered age of children, personality and report from probation officer's (section, 15), and not to execute imprisonment and penalty of death except for special incidents (section, 51). If required, juvenile offenders will be send to an authorized correctional home for their development (section, 52); releasing them on probation under the supervision of parents, guardians or probation officers (section, 53); and to protect children at danger under the jurisdiction of authorized institutions and supervision (section, 32.3–32.6) (Sengupta and Hossain 2006). The police is liable for taking care of children in a harmless location (section, 49), to approve bail to children (section, 48), inform parents (section, 13.2), rescue them from harmful situations (section, 32) and other role and responsibilities of Probation Officers (section, 50) (Sengupta and Hossain 2006). Meanwhile, the Probation Officer is responsible for conducting a social investigation (section, 15.50), sending the report to the court (section, 31.3.c), supervising children during probation (section, 53) and assisting for rehabilitation (section, 31.3.d). Saving children from dangerous conditions and sending them to the Juvenile Court are well-mentioned (section, 32) (Sengupta and Hossain 2006).

The Penal Code, 1860 mentions that a child less than the age of nine years is not responsible for the crime (section, 82) and children age of 9–12 years are nor accountable for offence due to not achieved ample psychological development (section, 83). This Penal Code clearly added a few elements of dealing with selling (section, 372) and buying of children (section, 373) for the aims of prostitution child only. Additionally, kidnapping, criminal bullying and offence of rape are mentioned (section, 366, 375 and 509) and anyone who encourages or forces a girl bellow 18 years of age into illegal sexual interaction will also be penalized under the code.

Women and Children Repression Prevention (Special Provision) Act, 2000 (Amended in 2003)

The revised act outlines the punishments for the crime of violence against children and women, including the death penalty or life sentence with hard detention and cash fine of not more than Bangladesh Taka (BDT) 100,000 (US$1250) due to any injury by acid burning (section, 4) (Halder 2015). Meanwhile, those guilty of child trafficking (section, 5) will be subject to lifetime detention or the death penalty or detention not less than 10 years and not above 20 years and payment of additional cash fine for (Halder 2015). Additionally, for child abduction, and sexual harassment and assault, the offender might face a death penalty or detention not less than 14 years with a payment of a cash fine, and severe detention (section, 7) (Halder 2015).

The Bangladesh Labour Act, 2006

The Bangladesh Labour Act, 2006 has substituted The Children Act, 1933, Factories Act, 1965 and The Employment of Children Act, 1938. This law banned the involvement of adolescents (14 years and not more than 18 years) and children (not beyond 14 years) in any organizations or occupation (section, 34). However, adolescents may join in an institutions or job if they could provide certificate from authorized practitioners concerning their ability to perform the particular task (section, 34); an adolescent cannot be involved in lubricating, cleaning or processing any device when machine is working (section, 39) (Halder 2015); and an adolescent cannot operate any machine without proper knowledge and precautions (section, 40). In addition, the act also stated about the exclusion of adolescents from engagement in water activities and deep soil (section, 42). In this regard, while children are not to be involved with any job or organization, however, they may be engaged in some light work at the age of 12 years that is not harmful to their health and physiological growth and does not interfere with their school education (section, 44). In these circumstances, their working hours might not interfere with their schooling hours (Halder 2015).

Child Welfare in Policy/Plan/Programs in Bangladesh

As a member state of the United Nations (UN), Bangladesh has signed the "Convention on the Rights of the Child", which is adopted by the UN General Assembly. Bangladesh's implementation of children's rights is based on its long implementation of various welfare-oriented programs for the wellbeing of children and their mothers. Even during the British regime in Bengali, an act titled "Penal Code" (Act No-XLV of 1860) was enacted for the first time for ensuring the rights and position of children. This is followed by "The Divorce Act, 1869" (Act No IV of 1869) to prevent sexual harassments and repression of children, and to curb in child marriage. Furthermore, through the Contract Act, 1872 (Act No

IX of 1872), and the Evidence Act, 1872 (Act No I of 1872), the government has paid the importance of eligibility of children to make contracts and as a witness. In 1875, in the Majority Act (Act No IX of 1875), a minimum age has been fixed to be an adult and the Guardians and Wards Act, 1890 (Act No VIII of 1890) was enacted for the protection of property rights of children below the age of 18.

The government also enacted an act that allows children to be relieved from attending the court for prosecuting purposes until maturity. The Act is "The Limitation Act, 1908" (Act No IX of 1908). The Mine Act, 1923 (Act No IV of 1923) forbids child Labor in the mining industry. To ensure child rights, child marriage had been prohibited after the enactment of "The Child Marriage Restraint Act, 1929" (Act No XIX of 1929). The government enacted "The Suppression of Immoral Traffic Act, 1933" (Act VI of 1933) and "The Children (Pledging of Labour) Act, 1933" (Act No II of 1933) for the protection of children's rights. In these acts, the detention of any female under the age of 18 for prostitution against her will, and the making of agreements to pledge the labors of children below the age of 15 years have been treated as punishable offences. According to the Payment of Wages Act, 1936, the right of imposing compensation on child below the Age 15 years has been prohibited. In the Employment of Children Act, 1938 (Act No XXVI of 1938), labor of child below the age of 12 years has been prohibited in the risk-prone areas. Moreover, The Dissolution of Muslim Marriage Act, 1939 (Act No VIII of 1939) provides that women are entitled to obtain a decree for the dissolution of her marriage arranged by her father or other guardians before she reaches the age of 15. According to the "Vagrancy Act, 1943", especial arrangement had been made for displaced children.

The Bangladesh Abandoned Children (special provision) Order, 1972 (P.O No 124 of 1972) provide provisions of guardianship for abandoned children and through "The Children Act, 1974 (Act No XXXIX of 1974)", the government has empowered to set up child courts and established the Juvenile Courts and other institutions for care and protection of destitute and neglected children. Later, in 1994, the government adopted a National Child Policy and constituted a "National Council for

Children" in 1995 for making policies, rules and regulations for the protection of rights of children. The government also adopted a "National Plan for Action 1997–2002". The government also adopted "Women and Children Oppression Prevention Act, 2000," under the observation of the "Child Decade from 2001–2010" for the welfare and protection of the rights of the children. The first National Child Policy was established in 1994. The significance of the National Child Policy should be considered in all national planning, development policies, budgeting and program implementation in Bangladesh. The National Child Policy shall be applicable to all children citizens of Bangladesh without any discrimination.

National Child Policy, 2011

The National Child Policy (NCP), 2011 has been introduced to implement the principal and provision of the Convention of the Rights of Children (CRC). The principal philosophies of the NCP have been identified as respecting childrens thoughts and confirming the participation of children, non-discrimination and transparency and accountability should be ensured by institutions, individuals for understanding of child rights. Special emphasis has given to education, birth registration and identity, children's right on health care, leisure and cultural activities and child protection. In addition, special rights of ethnic and minority children, rights of the adolescents and children with disabilities and their development are mentioned in this policy.

National Child Labour Elimination Policy, 2010

The National Child Labour Elimination Policy (NCLEP), 2010 has been implemented under the supervision of the Ministry of Labour and Employment (MoLE) to avert and reduce child labor, specifically, any dangerous forms of child labor. The objectives of the NCLEP are to retreating children from any form of dangerous work, labor and profession and engaging parents of working children in earning activities to move out of the vicious circle of poverty. In addition, offering scholarship for working children to continue study, special program for the children suffering due to natural calamities and given distinct attention children with disabilities and ethnic minority. In addition, it strengthens the institutional capacity for the enforcement of the laws in enacting pragmatic laws and planning and implementing sustainable strategies and programs to reduce various forms of child labor by 2015 in Bangladesh.

National Education Policy, 2010

The National Education Policy (NEP), 2010 was established under the leadership of the Ministry of Education (MOE) for the first time in Bangladesh to ensure quality education for all children. This policy set the target for 100% school admissions for children, and for them to continue studying at primary and high school levels. Extending the obligatory primary education up to grade eight and the introduction of the technical education is the greatest significant feature of the new NEP. The National Technical and Vocational Education Policy 2011 was introduced by the MOE to acknowledge the significance of technical teaching for human resources development in Bangladesh. Currently, to evaluate and modified the vocational education system, MOE is collaborating with ILO. The draft of Early Childhood and Care Development Policy has been completed and is waiting for approval from the Cabinet. The initial learning and development draft has been framed and shared with appropriate stakeholders. In this regard, this policy has recommended to reform number of issues—National Child Protection Policy, National HIV/AIDS Policy, Draft National Anti-Trafficking Strategies Plan Action (NATSPA) and Children's Code.

Child Welfare Plan in Bangladesh

The main theme of the Sixth Five Year Plan (SFYP) 2011–2015 is to highlight the significance rights and advancement of children. The SFYP mentioned that Bangladesh has made substantial the improvement in the field of child rights promotion. Nevertheless, the development of children is required to develop the future due to numerous Bangladeshi children are still endangered by ill health, illiteracy, exploitation, abuse, poverty, malnutrition and natural calamities. "A World Fit for Children" is the goal for children rights empowerment. These goals should be achieved by supporting medical care, assuring children's advancement and rights in the public policies and legislations, guaranteeing for food and nutrition that children need, and entrance to female education, development and training facilities. Additionally, engagement with education, cultural functions and sports should be confirmed for rural poor children and to provide safeguard commencing all kind of ferocity, exploitation and abuse. Further, to ensure healthy sanitation, environment and safe water, confirming assistance of caregivers and parents and extensive government assistance for survival and empowerment of children.

The recommended plans considered aspects like medical care for child, education, food and nourishment, entrance to water and sanitation, empowerment of child, birth registration, child abuse, child labor and protection for child. The NPA (2005–2010) was implemented to ensure the protection of children welfare in the context of National Poverty Reduction Strategy Papers (PRSP) and achieve MDGs. In addition, a National Strategy Paper for Bangladesh on Prevention of all forms Violence Against Children (VAC), 2011 has been prepared based on the guidelines of the country-level action points, by Ministry of Women and Child Affairs (MoWCA). The review procedures are presented by non-government organizations (NGOs), government organizations (GOs) and local team to map design for a way forward. Additionally, policies, laws and organizations have been reviewed to recognize policy, legal and intuitional plan. It also outlines the goals and strategies to curb violence against children in Bangladesh.

Institutional Child Welfare: Government and Non-Government Services

Government Sector

Shishu Sadan/Shishu Paribar (Children's Family)

These institutions were established to protect destitute children especially orphans. The concept was developed by government initiatives or social generous people and is now known as *Shishu Sadan/Shishu Paribar* (Children's Family). These institutions arrange necessary programs for education, training, recreation, food and so on. These institutions were later transformed into *Shishu Paribar* (Children's Family) to allow children to be brought up with the motherly environment in a homely atmosphere. Generally, children aged 5–9 years are admitted through an admission committee and they will reside in these institutions until they reach 18 years old. There were 12,000 children placed at 74 institutions in 2013. Meanwhile, there are a total of 10,141 children who were saved from employment, 9389 from child marriage and 30, 100 from different jobs from 2009 to 2013. It is to be noted here that throughout the 2012–2013 fiscal year, apart from government *Shishu Sadan* (Children's Family), there are 5457 non-government orphanages where a total of 40, 333 children reside.

Chhotomoni Nibas (Baby Homes)

The *Chhotomoni Nibas* (Baby Homes) were established by the government in 1962 for displaced children aged up to seven years and to provide them with motherly affections. Most children in these homes are abandoned and have no relatives to look after them. The first *Chhotomoni Nibas* is a 25-bed accommodation in *Azimpur* in Dhaka (Rana 1997). Presently, there are six centers at six towns providing 825 accommodations. The children of these centers will be transferred to *Shishu Sadan* and *Shishu Paribar* after reaching a certain age. A total of 2889 children benefited from these studies from 2009 to 2013.

Dibakalin Shishu Jotno Kendro (Day Care Centre)

This center was established for the children aged five to nine years within motherly environment, whose mothers engage services; the government established a *Dibakalin Shishu Jotno Kendro* (Day Care Centre) at *Azimpur* in 1962. At present, there are 50 seats at the Centre. Eight thousand children have been benefited from its inception.

Training and Rehabilitation Centre for Destitute Children

The center provides basic education, training and rehabilitation for children from the age of 6 to 14 years old. There are three centers with 750 places. A total of 7570 children were kept here until 2013.

Juvenile Development Centre

Children under 18 years old who are involved in different types of crimes due to trafficking; family disturbance; and abandonment from their parents, divorce, poverty, religious conflicts and illegal drugs and fire arms are placed in Juvenile Development Centres. In 2002, the first juvenile development center for females was set up at *Konabari* of Gazipur district. The children here are provided with the technical education and training along with general education for their rehabilitation. A total of 15,254 juveniles were placed in these institutions from 2009 to 2013.

Institution for Mentally Disabled

This institution was set up at *Raufabad* of Chittagong city in 1995 and provides care, education, technical training and rehabilitation to the mentally disabled children. A total of 5000 mentally disabled children are attending technical education and training in addition to basic education.

Employment Rehabilitation Centre for the Physically Handicapped, Tongi (ERPH)

The ERPH center was established in 1981 with a capacity of training 105 physically handicapped per year. The different types of rehabilitative and protective services provided to the disabled at the center are mechanical workshop training, training in handicraft, duck and poultry farming, nursery training, mobility training, music training, games, sports and physical training, rehabilitation grant, industrial production unit, hearing centers, Braille press, health clinic, hostel accommodation and many others (Bhuiyan and Hossain 2003). A total of 8855 disabled children benefited from this institute from 2009–2013.

School for Blind and Deaf-Dump

There are a total of five schools in Dhaka, Chittagong, Khulna, Rajshahi and Barisal that provide training and education for the visual-disabled children. The students in these schools are provided with special education and are taught Braille or sign language.

National Special Educational Activities

The government has organized different types of programs for visually impaired, hearing impaired and mentally disabled children to mold them into independent human capitals and elevate the burden of the society. There are separate schools and hostels for visually impaired, hearing impaired and mentally disabled children.

Protection of Child at Risk (PCAR) Project

The project titled "Appropriate Resources for Improving Street Children's Environment (ARISE)" stared in 1999, with the financial support of UNDP. The project aims to provide education, physical and mental health care, accommodation facilities and skill development for street children. Under this project, training programs, psychological counseling and legal help are provided to the street children through a "Drop-in Centre". A total of 40,500 street children were benefited from this program between 2012 and 2013.

Maternity and Child Welfare Centre

To ensure the health of pregnant mothers, a total of 64 centers have been set up at towns throughout the country. These facilities provide medical care for pregnant mothers until they deliver their baby, by providing them regular medical checkups and post-natal care for newborn baby and mothers.

Poverty Alleviation, Capacity Building and Sustainable Livelihood for the Socially Disadvantages Women and Their Children

This program helps socially disadvantaged women and their children to change the social stigma and include them in the mainstream. Providing basic education, pre-formal education, non-formal education, aiding admits local schools, accommodation facilities, technical training, health care and arrangement of employment are the major program of this project. About 70,970 disadvantaged women and children were benefited from 2009–2013

Bangladesh *Shishu* (Children) Academy

Academy was set up in November 1976 to encourage the growth of physical, mental, cultural and latent talents of *Shishu* (children) in Bangladesh. Different competitions are arranged nationally every year, which involved children from rural and urban areas. The academy also held activities in lieu with festivals, such as National Day, as well as cultural and training programs, including *Shishu Mela,*(children's fair), seasonal competition, international painting competition, children films presentation, books exhibitions, opening a children's library and museum, publishing books on children and arrangement of the study tour.

National Child Council

The National Child Council was introduced in Bangladesh to protect the interest and rights of children. The council is the highest authority for making any decisions and regulations in regard to the rights of children. The council is also responsible to look after the implementation of the CRC.

Non-Government Sector

Several international, national and local organizations are working closely for the care and protection of children in Bangladesh. Some of these organizations are discussed here:

United Nations International Children's Emergency Fund

UNICEF has been working to protect the rights of children and encourage the development of their latent talent. It provides funds and advices regarding children and maternal health, formal and non-formal education and responsibility of parents to their child. Besides, it provides supports to encourage immunization, breastfeeding and the intake of nutritional food. UNICEF has established maternity and child welfare centers in different parts of the country, empowering rural health workers and nutrition surveys, setting up tube-well for supplying pure drinking water, and distributing biscuits and milk, textbooks and other educational materials, such as sewing and retail training for women. It also publishes many research works on the protection of children.

United Nations Educational Scientific and Cultural Organization

United Nations Educational Scientific and Cultural Organization (UNESCO) is a sister organization of the UN and is working in the child education sector in Bangladesh. The organization provides financial supports and technical assistance to implement the "Education for All" program in Bangladesh.

World Health Organization

In Bangladesh, World Health Organziation (WHO) works particularly in the field of Expanded Program on Immunization (EPI), Polio Free Bangladesh, providing health care for mother and children and their nutrition, family planning, monitoring of standard of medicine and assistance in research in the field of health.

International Labour Organization

Established in 1919, ILO has been working to protect the rights of labors. It is working to prevent child labor cases in developing countries. ILO is working in Bangladesh to prevent child labor with the financial support from UNDP. Besides, it is working to protect the rights of women workers, which subsequently, help children indirectly.

Food and Agricultural Organization

Bangladesh became a member of Food and Agricultural Organization (FAO) in 1974. Since then, FAO Bangladesh has been working to increase nutritional intakes among children and mothers, food production, modernization of agricultural activities, preservation of food and living standard of people in Bangladesh.

Red Crescent Society

The society was established to provide relief for war-affected people who were the victims of the war in Italy and Austria in 1859. It offers health services on regular basis for 50,000 children across 21 maternity and child welfare Centers in Bangladesh.

SOS-*Shishu Palli* (Children's Family)

Since 1949, this establishment has been working for the destitute, orphaned, abandoned, stateless, undocumented and street children. They are brought up with motherly affection. The organization opened its door after the Liberation War of Bangladesh in 1971. At present, it has brought up more than 1080 children.

Save the Children

It is a leading independent organization creating lasting change in the lives of children in need in more than 110 countries of the world. In Bangladesh, it started in 1972 in Bangladesh and focuses on child education and health activities. It runs two core programs, early childhood development and primary education. The early childhood development program is conducted under the Home-Based Early Learning Opportunity Centre (HBELO) and Home-Based Pre-schools (HBPS). On the other hand, the primary education program caters to school children aged 6–15 years. The School Health and Nutrition program of Save the Children targets school-going children to improve their health and nutrition status. For this purpose, health education on sanitation, hygienic behavior, nutrition and reproductive health issues are provided to schoolchildren.

World Vision of Bangladesh

World Vision's project in Bangladesh started in 1973 to provide care and ensure the welfare of children. World vision works to improve primary education and health care and to provide micro-credit for improving the living standards of children and their mothers. It also provides informal education for school dropouts. World Vision provides financial supports to build a new school building, repair old school building and supply educational materials among the remote localities. Under this, basic education, training and rehabilitation programs are operated for street children. World Vision also organizes workshops, recreational activities, training and education, and establishes centers for dealing with juvenile's delinquencies (Islam 2005).

Underprivileged Children Educational Programs (UCEP)

The institution has been working since 1970 to distribute relief for the people who were affected by the cyclone and other natural disasters. It encourages educational activities among the destitute and disadvantaged children and aims to provide education, training and rehabilitation for destitute children.

Discussion

Children in Bangladesh are suffering due to poverty, lower societal and financial positions, higher dependency and living in slums. The country is also facing issues over the conflicting jurisdiction between the existing laws related to child welfare. According to these laws, only orphaned and abandoned children can be admitted into government institutions; however, some disadvantaged and poor parents sometimes feel reluctant to enroll their children as they do not have the required documentations. Furthermore, bribery and bureaucracy have denied some access to these facilities (Islam 2012). These children are not protected by the legal system, are unable to access their rights and have fear of being caught without identification documents. Juvenile offenders are also denied of independent legal counsel because of weak government institutions, fragmented NGO response, lack of legal protection and poor governance. The insufficient knowledge on child rights, specifically on protection and participation is proven by the absence of discussion on main policy, deficiency of child rights strategies, ignorance of senior management and the lack of staff members to safeguard childrens rights (Harper et al. 2010). Many children are denied the right to access education due to their socio-economic conditions, sex, language and disability, despite the establishment of CRC. Parents in the rural area are not able to send their children to schools because of transportation issues, while those living in poverty are not able to afford tuition costs, books and other related expenses (Islam 2012).

In this light, awareness-raising campaigns have been successful in bringing change in social thinking but still these are not satisfactory (WHO 2013). Therefore, children with lower economic opportunities and poor education are also more at risk. There is a need for an efficient plan to reduce poverty and creating employment opportunity for the family to reduce financial functions by children. The protection of children in Bangladesh should be ensured by social welfare (Khair 2005).

Bangladesh is implementing several development policies and programs to improve the condition of child welfare, but these policies and projects do not cover all children of the country. Moreover, it is very difficult for the government to collaborate with different international child and national agencies to operate child welfare programs and projects separately for ensuring child welfare (Nawaz 2011). The children rights treatise is not capable to solve the issue of power relations affecting children and the real needs of children in Bangladesh (Fernando 2001). The major barriers of Bangladeshi children include high mortality and morbidity, severe malnutrition, child labor, violation and child exploitation, abuse, lack of medical care, as well as access to education and nurture and care (Harper and Jones 2009).

Law-enforcement agencies and the corrective/prison staff are unable to work with the context of child rights or understand proper juvenile justice system due to insufficient wealth and training in Bangladesh. According to CRC, persons below at the age of 18 years are called children in Bangladesh. The National Child Policy and *Nari O Shisu Nirjaton Ain*, 2000 defined a child to be at or less than14 years, while the Children Act 1974 defines a child as aged less than 16 years. The lowest age of criminally liable for nine years mentioned by the Penal Code of 1860 is the youngest around the world. Furthermore, under the marriage laws, a girl has to be at least 18 years old and a boy 21, in order to marry in Bangladesh. In majority cases, the girl and guardians are not aware of the law and most of the times, young girls are forced to marry. This has proved that the law is rather loose at the execution level.

Bangladeshi ratified the Labour Act in 2006 that added a section on child labor. This current law banned the dangerous forms of child labor below the age of 18 years and employment of children aged less than 14 years old. Nevertheless, children at the age of 12 or above can involve in some light works, providing that it does not hamper their education, physical and mental improvement. It does not deliver a robust implementation tool for the provisions of child labor, and the constitution of Bangladesh does not also ban the child labor. These laws indicated some standard for the employment of children rather than banned the child labor. Meanwhile, the present domestic laws do not cover children who are engaged in household services like agricultural sector or domestic workers (Aktar and Abdullah 2013).

In Bangladesh, NEP, 2010 and National Child Act, 2011 have given more emphasis on child education. In addition, more importance is given in the government budget in term of education entrance of each child with school-age by "Education for All" campaign and indorsing ILO resolutions. Further, the government increases attention to the technical education to empower the risk children to eliminate child labor but street children still remain socially excluded and have less rights protection (Nawaz 2011). The Bangladesh National Development Plans appealed for high establishment services for vulnerable children but do not clearly mentioned approached for children like child workers (Khair 2005). Moreover, the laws are frequently misunderstood and tainted by law-enforcing agencies due to poor awareness and corruption in Bangladesh. In this light, the current legal systems emphasize merely polio victims, and the mentally sick children with additional infirmities are completely ignored (Ahuja and Ibrahim 2004).

Provisions for safeguarding the interests of children, subjected to torture and violence have been clearly mentioned in the Bangladesh Penal Code, other ordinances and laws and the Children Act of 1974. The offence for cruelty to children, guardians and parents are also liable. Nevertheless, a male child is not protected from beating under the Children Act of 1974 (Ahuja and Ibrahim 2004). There is a lack of trials and laws for rehabilitations and rescue of child from sexual exploitation. The child victims or children at risk of commercial sexual exploitation are not supported by law-implementing agencies even though they can do so under legislative power (CRGA 2012). The Committee on the Rights of the Child of United Nations has raised anxiety regarding the procedures of juvenile legal system in Bangladesh and given commendations to continue juvenile justice in the light of CRC. While it has been reviewed several times and modified and currently adopted by the Cabinet, it is not implemented under the Children Act, 1974 (PRI 2014). Bangladesh has taken several steps for the amendment of the legal framework regarding the Children Act and frequently submitted the report of the Committee on the Rights of Child. However, Bangladesh has not yet contracted the International Covenant on Civil and Political Rights (ICCPR), the Teenage Engineering (OPI) and Optional Protocol to the Convention against Torture (OPCAT), and there is a lack of coordination between the Bangladesh government and international human rights organizations (PRI 2014). Since its inception, the law of Constitutions of Bangladesh has focused on childrens protection similar to other countries in the world. There is no positive legislation or children justice law having facilities for moral, intellectual and physical improvement of children in Bangladesh. In addition, the law has not been changed to protect children after 42 years of liberty of Bangladesh. There is a frequent lack of cooperation regarding the national child protection systems in the region among local levels, district levels, departments and the ministry.

The people of this region consider that child security as a minor issue. Hence, a systematic approach for the national child protection system needs to be effectively implemented due to the lack of technical expertise, financial resources and trained staff. Qualified and experienced social workers are required for effective delivery of child protection services in Bangladesh. Social work courses are offered in higher education but only few institutions offer courses in child protection. Consequently, there is a lack of psychosocial counselor experts to handle the children who are victims of violence, exploitation and abuse. Child labor is a burgeoning problem in Bangladesh and the national legislation has prohibited child labor to some extent. An emphasis on special provisions for children should be given in the field of healthcare, nutrition, education, strategies, policies, social security and law by the policies, programs, plans and the Acts of Parliament in Bangladesh. However, the implementation of all these policies, strategies and programs has yet to be effectively due to the lack of monitoring and valuation (UNICEF 2009a). In Bangladesh, there are few laws regarding child labor and there is contradiction among the definition of a child. Safeguard children from child labor, abuse, neglect and other forms of vulnerabilities still remains a challenge despite the constitutional provisions and enactments including the Children Act. Moreover, children from the poorest families of Bangladesh are still being trafficked despite the existing laws to Pakistan, India and the Middle East for slave labor, camel riding, organ transfers and prostitution (Ahuja and Ibrahim 2004). A scenario of current skills in planning and implementation has been provided by an extensive review of all pertinent policies and programs in Bangladesh. Although, the policy and skills at the local levels are progressing, application abilities gap behind at all levels. Hence, the results of strategies policies do not reflect "hoped-for results" for child welfare in Bangladesh.

Policy Implication, Suggestions and Conclusion

The children laws should be executed properly by amending the guidelines of CRC. The age of criminal responsibility for all children should be increased from 9 to at least 12 years. The benefit of the doubt should be given to the child, where a child claims to be under the age of 9/16/18 years old in Bangladesh (in contradictory situations), (PRI 2014). However, there is no clear plan regarding how government will meet up the targets to ensure the child welfare due to the government's ineffective policies for child welfare. To ensure child welfare, the government should improve the anti-child labor laws, national development plan, birth registration, adoption policies and guardianship, as well as increase the efficiency of juvenile justice procedures in Bangladesh. Birth registration should be encouraged to help proper age determination procedures.

To further improve child welfare policies, Bangladesh should work toward bringing the relevant organizational and judicial modifications, advocating for the rights of the children, particularly those who are vulnerable and participating in capability development of child security amenities and piloting child safeguard systems (UNICEF 2012). A widespread child safety policy should be developed and action plans on child labor, child marriage and street children should be established to ensure preventive and protective measures for child welfare in Bangladesh (UNICEF 2009a).

It is important for the government to examine poverty-connected issue for child neglect, and other social issues and make desire improvement to the support of family income, reasonable housing and medical facility to protect child welfare in Bangladesh. The government is committed to protect the child, but it lacks depth of understanding and consistent planning. A multi-sectoral and multi-disciplinary approach requires child welfare, including an extensive range stakeholders, including multilateral agencies, communities, government, caregivers and families. The shortfall of the current government in Bangladesh has created additional anxiety for child welfare. In addition, police, executives, lawmakers, probation officers and judges remain insensitive to children's rights and fail to provide security, special care and treatment. It is recognized that legislation alone is not enough to address the complexities of child welfare in Bangladesh. The public and donor agencies should be given priority on resources mobilization, training to social workers, capacity builders, caregivers, counselors, police officers, lawmakers, teacher, judges, youth workers and healthcare professionals. Moreover, attention should be given on immigrations and humanitarian workers, detention facilities, childcare institutions and public policy makers/officials to confirm that the government has the ability to implement national plans in Bangladesh. To address issues of poverty and inequality that are often the root of the problem, it is equally essential to develop social and economic measures simultaneously. In this respect, it is crucial for policy for poverty reduction and empowerment at the community level not only to ensure fair delivery of wealth and delivery of basic needs for the poor but also to seek sufficient investment in safeguarding and increasing human capability, particularly for children and young people in Bangladesh.

References

Ahuja, A., & Ibrahim, M. (2004). *An Assessment of Inclusive Education in Bangladesh*. Dhaka: UNESCO.

Aktar, S., & Abdullah, S. M. (2013). Protecting Child Labour in Bangladesh: Domestic Laws Versus International Instruments. *Bangladesh e-Journal of Sociology, 10*(1), 153–172.

Bhuiyan, A. B. S., & Hossain, M. F. (2003). Institutional Care, Protection and Training Programs for the Disabled Children in Bangladesh. *Social Science Journal, 8*(1), 50–65.

Child Rights Governance Assembly (CRGA). (2012). *Situation of Child Rights in Bangladesh*. Dhaka: Ain o Salish Kendra (ASK).

Daily Amader Somoy.(2014). *Malnutrition and Child Poverty in Bangladesh*. Retrieved November 3, 2014, from http://www.dainikamadershomoy.com.

Fernando, J. (2001). Children's Rights: Beyond the Impasse. *The Annals of the American Academy of Political and Social Science, 575*(1), 8–24.

Halder, J. (2015). *Child Protection and Child Rights. Current Status and Challenges*. Retrieved May 3, 2019, from https://www.slideshare.net/JhumaHalder1/child-protection-and-child-rights.

Harper, C., & Jones, N. (2009). *Raising the Game: Mainstreaming Children's Rights*. London: ODI.

Harper, C., Jones, N., & Tincati, C. (2010). *Opportunities and Challenges in Promoting Policy- and Practice-Relevant Knowledge on Child Rights*. London: Overseas Development Institute.

Islam, M. N. (2005). *Children's Welfare and NGOs*. Dhaka: Tasmia Publication.

Islam, M. T. (2012). *Residential Childcare: The Experiences of Young People in Bangladesh*. Retrieved from https://www.era.lib.ed.ac.uk/handle/1842/8262.

Khair, S. (2005). *Child Labour in Bangladesh: A Forward Looking Policy Study*. Geneva: International Programme on the Elimination of Child Labour (IPEC), Labour Office.

Nawaz, S. (2011). *Development Policies for Street Children in Bangladesh*. Lund: Centre for East and Southeast Asian Studies, Lund University.

Penal Reform International (PRI). (2014). *A Review of Laws and Policies to Prevent and Remedy Violence Against Children in Police and Pre-trial Detention in Bangladesh*. Retrieved October 20, 2015, from http://www.penal-reform.org.

Rana, A. (1997). *United Nations Children Convention and Bangladesh Children*. Dhaka: Bangla Academy.

Sengupta, S., & Hossain, D. (2006). *Child Protection and Child Rights. Current Status and Challenges*. Retrieved May 3, 2019, from http://www.askbd.org/ask/child-rights/.

UNICEF. (2009a). *Child Poverty and Disparities in Bangladesh* (Women and Children Repression Prevention (Special Provisions) Act, 2000 amended in 2003). Dhaka: UNICEF.

UNICEF. (2009b). *Global Study on Child Poverty and Poverty and Disparities: Bangladesh National Report*. Dhaka: UNICEF.

UNICEF. (2012). *Progress in Child Well-Being*. Retrieved October 21, 2015, from http://www.unicef.org.

World Health Organization. (2013). *Child Abuse and Neglect by Parents and Other Caregivers*. Retrieved October 21, 2015, from http://www.who.int/.

29

Trafficking of Children: Social Work Strategies in Prevention and Reintegration

Lakshmana Govindappa

Introduction

The life of human beings is in a dynamic state. They are achieving new heights and what once was a dream is now a reality. Now going to the Moon, staying in the space and several other things have been very easy; subsequently life has become mechanical; accumulating money and achieving power are becoming important than humanity. Today, people around the world dream of becoming a millionaire overnight. For this, they are ready to do anything. Moral values, ethical considerations and religious code seem to have vanished. People can go up to the extent of committing a lethal crime, namely, sabotage, arson, flesh trade, smuggling of arms and ammunition, trafficking and also child trafficking.

Trafficking of human beings has become a universal problem and pawning almost all the nations into its claws. It is a form of slavery and violates human rights. Here human beings are exploited by fellow humans

L. Govindappa (✉)
Department of Social Work, Central University of Karnataka,
Kalaburagi, Karnataka, India

© The Author(s) 2020

Md. N. Momen et al. (eds.), *Building Sustainable Communities*,
https://doi.org/10.1007/978-981-15-2393-9_29

and consider them as commodities, and without their consent, they are misused and used illegally. Globally, this is rated as vastly growing crime. Due to the enormous profits involved, the people are trafficked in the same way as arms, money or drugs by the global network controlled by criminal gangs. Despite the elimination of slavery long ago, criminal groups have proven to be highly efficient in entering countries affected by armed conflicts, war or badly managed transitions, with the purpose of human trafficking (Vlachová 2005). This illegal trade has not even spared children. They traffic children for various purposes and exploit them.

> Trafficking in persons, especially women and children, for the purpose of forced prostitution has been recognized as a prevalent problem of concern in India in recent years. India has served as a destination and transit point of trafficking victims from Bangladesh, Nepal and Bhutan. (UNODC 2007)

Child trafficking is a significant issue which has achieved an alarming status, drastically. During the past two decades, this issue is receiving serious attention at the national and international level. Children are lured for money, better future and luxury; they are exploited up to a great extent in various ways. Rehabilitation of child trafficking victims is a challenging process encompassing many facets that include overcoming from trauma, recovery and reintegration into the mainstream.

Definition

The United Nations (UN) optional protocol for UN convention against transnational organized crime (2000) defines "Trafficking in person shall mean the recruitment, transportation, transfer, harbouring or receipt of persons, by means of the threat or use of force or other forms of coercion of abduction, of fraud, of deception, of the abuse of power or of a positive of vulnerability or of the giving or receiving of payments or benefits to achieve the consent of a person having control over another person, for the purpose of exploitation. Exploitation shall include, at a minimum, the exploitation of the prostitution of others or other forms of sexual exploitation, forced labour or services, slavery or practices similar to slavery, servitude or the removal of organs" *(Article 3, paragraph (a))*(UNODC, 2000) *(UN Protocol to Prevent, Suppress and Punish Traffi cking in Persons, especially Women and Children, Supplementing the United Nations Convention against Transnational Organized Crime).*

The Protocol yet illustrates that the consent of a person who is trafficked may be furnished as insignificant when obtained through improper means:

> *The consent of a victim of trafficking in persons to the intended exploitation set forth in subparagraph (a) of this article shall be irrelevant where any of the means set forth in subparagraph (a) have been used. (Article 3, paragraph (b))*

Child trafficking is prevalent in most of the countries and trafficked for various purposes. It is mostly moving a person from his original place to a place that is exploitative in nature. Often this movement is physical. This transfer does not involve the consent of the person.

The United Nations International Children's Fund (UNICEF) defines a child victim of trafficking is "any person under 18 years who is recruited, transported, transferred, harbored or received for the purpose of exploitation, either within or outside a country".

The aim of child trafficking is money-oriented and profit only for the trafficker; hampering the life of a person subjected to trafficking. Children are often trafficked for slavery, sex tourism, pornography, commercial sexual exploitation, bonded labour and other purposes. Trafficking takes places under the framework of supply and demand. Helpless, vulnerable, gullible, minor girls, young women from certain areas which suffer from industrial depression, agriculture degradation, economic depression etc. form the supply of potential trafficking victims (Khan 2014; Verma 2014; Goswami and Lagon 2013; Tsalikis 2011).

Estimation

It is difficult to estimate the prevalence of human trafficking in different countries. Collaborative efforts are made by the international non-governmental organizations and UN agencies to generate database on the profiles, incidence of trafficking, forced labour and forced marriage.

A report of the United Nations, on international trafficking released in 2000, claims that annual turnover of organized crime in trafficking could be around $7 billion, victimizing seven to ten lakhs children. According to US State Department, each year across international border, 800,000 people are trafficked, 50% of which constitute children. In 2003, the International Labour Organization (ILO) estimated that 8.4 million children are engaged in "unconditional worst forms of child labour", including 5.7 million bonded or forced labour and 1.2 million trafficked. India lacks reliable data about trafficking. According to US State Department, every year throughout or within India, 200,000 people are trafficked. As per a study by the Central Social Welfare Board, there is an estimation of about 25,000 to 500,000 children being trafficked for prostitution in India (National Crime Records Bureau 2017).

The International Organization for Migration (IOM)'s (2018) global database on victims of human trafficking contains over 91,416 cases (individual), with approximately 5000 new cases being added every year. A common questionnaire with a set of standard indicators was used in respective countries by UN Office on Drugs and Crime (UNODC) to survey trafficked victims, and then the data was aggregated in 2016. The report stated that in 2012 and 2014, a total of 63,251 victims were detected from 106 territories and countries. The same report stated that in 2014, 17,752 victims got identified in 85 countries, of which 71% were women and girls, while 28% of victims were children. The regions such as Sub-Saharan African, Central America and the Caribbean had 62% and 64% of victims, respectively.

The UNODC report states that the profile of detected trafficked victims has changed over the past ten years. Men now consist of the larger share of the total number of victims in comparison to the decades ago, whereas women and children remain the most of detected victims. In 2014, identified victims included 28% children and 21% men. All the data shows that the rate of trafficking is increasing with multiple purposes of trafficking, where no country is left immune (UNODC Global Report on Human Trafficking in persons 2016). The Walk Free Foundation (WFF) and the ILO in collaboration with IOM estimated forced slavery in more than 50 countries (2017). The report stated that in 2016, 40 million people were victims of modern slavery, wherein approximately 25 million people are in forced labour and 15 million in forced marriage (ILO 2017).

India

India is "considered as the source, transit and destination of trafficking of people for purposes such as commercial sexual exploitation, slavery, etc. Though the majority of trafficking occurs within the country, a large number of people trafficked from Nepal and Bangladesh also. One of the leading target groups is the children, where there is no national or regional estimation for the number of children trafficked each year. There is a growing demand for young girls in the commercial sex markets, where 40% of sex workers are children. It was also reported that the children trafficked for sports such as camel racing to Middle Eastern countries" (Childline India).

According to the National Crime Records Bureau (NCRB 2017), India, there were 8132 reported cases of human trafficking in 2016. In the same year, there were 15,379 trafficked people wherein more than half, that is, 9034 victims, consist of the age group below 18 years. Further, 23,117 people rescued from trafficking out of which 14,183 people were below 18 years of age. It states that the number of victims rescued is higher than the number of people trafficked which can also denote that the rescued may also include people from the previous year. Most of the rescued are trafficked for forced labour (10,509), sexual exploitation for prostitution (4980) and other forms of sexual exploitation (2590), respectively.

Reasons for Trafficking

There are various reasons for trafficking. These can be discussed under push and pull factors.

The Push Factors

The growing demand for unqualified, illicit, cheap labour, children separated from their families because of war, the death of caregivers, break-up of their family or destitution and sex trade represent a distinct group of victims. As a form of profitable business, illegal adoption of children is taken up in which thousands of children have been purchased and sold from under-developed and developing countries to developed countries.

The Pull Factors

The situation like lack of informed choice regarding the place and occupation makes the women and girls vulnerable to the trafficking. Limited access to education and information aggravate these situations. Since they are unaware of the hardships of migration, the traffickers' bait easily entraps them. The most common pulling factor for trafficking is the demand for commercial sex work. The nature of this industry necessitates the regular and uninterrupted fresh supply of women, which keeps trafficking most profitable. Besides, the growing demand for commercial sex work due to the increasing trend of migration and separated family also contributes equally to pulling out the women for trafficking. The continually growing market encourages the traffickers to lure women and girls into it (Department of Women and Child Development, Government of India, 2000 & 2006). The Fig. 29.1 descrubes the exploitation types among identified victimes of trafficking betwee 2005-2016.

The graph shows that forced labour was the highest-grossing type of exploitation faced among identified victims of trafficking from 2005 to 2016. Further, it shows a continuous increase since 2005, thereby making it the most exploited type. The other prominent form of exploitation was sexual exploitation, which was 60% in 2005 and has reduced over the years. These dynamics are changing over the years.

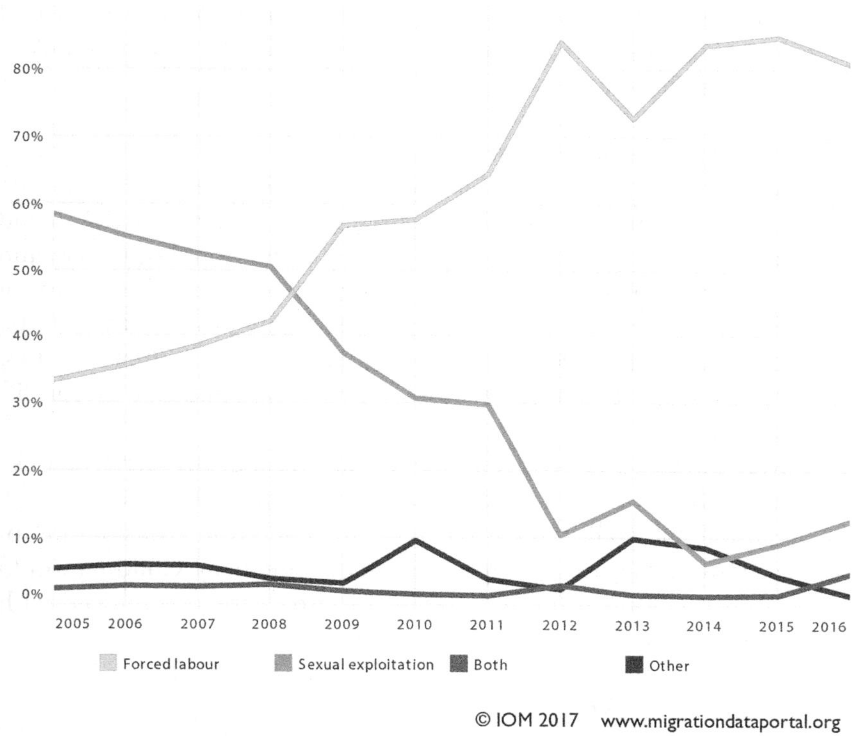

Fig. 29.1 Exploitation types, among identified victims of traficking, 2005-2016 Note: The "other" includes: forced marriage, forced military service, low-level criminal activities and trafficking for blood, organs or other body parts. (Source: IOM's Human Trafficking Global Database 2017)

Methods of Trafficking

Traffickers are very much skilled, and they use these various methods to catch the children. Most of the time traffickers are known persons, from a local area or relatives. The traffickers will have a strong network, and before attempting to traffic the child, they prepare themselves very well.

Here are some of the methods used (; Department of Women and Child Development, Government of India, 2000 & 2006 Lusk and Lukas, 2009; Verma, 2014).

1. Tempting offers of lucrative jobs in cities
   The profile of trafficked children shows that majority of the children are from low-income families. Their families are poverty-stricken and unable to feed the children. Some of the families would be in debt and looking for financial support. Traffickers study the vulnerability of these families and extend support to overcome poverty. By offering help and giving hope, parents would send their children to cities. Once these children are away from home, they would be used for different purposes.
2. Entice children to offer sexual services in exchange for lucrative jobs
   The children would be used for various purposes. Traffickers are well organized and well trained. Based on the condition and situation of the children, they use multiple methods to lure them. Children would be shown colourful life by offering lucrative jobs and enticed with sexual services. As these children are from poverty, they would be easily attracted to lucrative offers and ready to pay the price.
3. Offering costly gifts
   Traffickers would offer expensive gifts to attract these children. They keep offering gifts and establish rapport and trust. As explained, they build trust with these children and create a conducive environment. Once the children and family members trust, they would offer jobs, better lifestyle in far cities. As the trust has been built, without doubting the trafficker, children would be sent with them. If the family members are not sending children, they would be taken by these traffickers.
4. Roles in movies and so on
   Some of the studies show that young attractive girls and women are trafficked by offering colourful life once they move to the cities. They would be attracted by offering roles in movies, modelling and other ways. Once they are away from the family, they would be exploited for various purposes.

5. Arrange false exploitative marriages
   As mentioned earlier, generally traffickers catch poverty-stricken families. They would try various methods to traffic children. If the girl is of marriageable age and family is not in a condition to arrange marriage, they would offer suitable partner from the distant cities having a colourful life. They bring their own person and arrange a false marriage. Once the wedding takes place, they would take the girl to the cities and use for various purposes.
6. Dupe and trick girls and parents with false promises
   Traffickers use different methods to attract and lure the girls and seek the confidence of parents. They promise a good working environment, housing, salary and other false promises to entice them. Girls will be duped with a better lifestyle, becoming rich in a short time and others.
7. Sanction prostitution by tradition and custom
   In some of the countries, specific practices and customs allow prostitution in the name of the tradition. Once they practice this, they would be taken for commercial purposes. The young children are exploited for this purpose. For example, in India, the Devadasi system that existed earlier was a traditional exercise where children were forced to practise sexual services. Minor girls were forced into prostitution in the name of faith (marrying god) especially in the states like Karnataka, Andhra Pradesh and Maharashtra.
8. Coerce parents into selling their children due to poor economic condition or having to repay debts
   The traffickers are well trained, and they find out the parents who are need of money. They extend loans to these families. Once the loan amount is high, they demand to repay. If they are unable to repay the loan, then they coerce parents to sell off their children for settlement of the debts. If they do not agree, the traffickers use muscle power and other methods to force them to sell or send their children for forced labour and other work. More than 80% of persons were trafficked by people whom they knew at a personal level, who took advantage of their naiveté, trust and vulnerability (La Strada, Trafficking in Human Beings, 8–9; Patkar and Patikar 2003; Adenwalla 2002).

9. Dysfunctional home environment
   Most common reasons like a break-up of the family, physical abuse, sexual abuse, drug use and discrimination within family also increase the vulnerability to trafficking. An abusive family background encourages family members to leave home, thus making them vulnerable to trafficking (Department of Women and Child Development, Government of India, 2006).

Different Stages of Trafficking

There are different stages of trafficking:

Stage I: Pre-Trafficking

The first stage of the trafficking is pre-trafficking. In this stage, traffickers do the thorough groundwork before they traffic the child. In this stage, the details about the victim are collected. Here, very carefully, traffickers collect the child's background, financial conditions, child's movements and location. They confirm the status of the family and their financial situation. As discussed, they focus on vulnerable families and study their problems well. They also study the movement of the child, its condition, friends, neighbours, availability of the people in and around the child, time and other factors.

Stage II: Transit

Once the traffickers collect complete details of the child, family and other aspects, they move to the next step. In this stage, they may approach the family members with lucrative offers. Based on the family condition, they give offers. If the family is in debt, they try to help them financially; if they are poor and want to arrange a marriage for their daughter, they may offer help in the form of finding a suitable groom. If these efforts do not work, they may involve in kidnapping the child and taking under threat. Children are also taken forcefully under threatening condition. In this stage, they use their network very well. Once the child is trapped, the message will be sent to the entire chain from the starting point (child's place) to reaching the destination. It is a well-organized crime.

Stage III: Destination

Finally, the child will be taken to the destination. The destinations will be decided by the traffickers well in advance, and with the place in mind, they plan the trafficking. Trafficked persons are forced or deceived to exploit them in abusive, sometimes slavery-like conditions. These children would be used for exploitation in various forms, like commercial sex, housework, begging and slavery. Once they are trafficked, they would become commodities. These destinations are well equipped with all kinds of precautions which will not allow them to escape. They would be having goondas, inaccessibility to outsiders and other.

Stage IV: Rescue and Rehabilitation

In this stage, rescue operation would be taken up by the legal entities. The information may be leaked to the legal entities about the illegal activities. Once the information reaches them, through the police or voluntary agencies, the child will be rescued, and the rehabilitation process will be started. This is a big challenge, and with proper planning, the rescue operation has to be taken up. Once the rescue operation has been done, rehabilitation will be started. Rehabilitation is a continuous and systematic process which needs to be continued until the victim is completely rehabilitated.

Stage V: Reintegration with the Family

Finally, the rescuers may reintegrate the child with the family if they have a chance. Here, finding the family and sending the child to the family is not very easy. The issues related to psychosocial, financial, occupational, educational, emotional and other issues and concerns have to be addressed.

Social Work Strategies in Prevention and Reintegration

Social workers, play a vital role in the prevention, rescue and rehabilitation of child trafficking. Along with taking care of the victim in the protection home, social workers also help in rebuilding the children's lives and making them independent. Sensitivity about the needs, problems and background of these children is essential to work with child victims suffering from commercial sexual exploitation and trafficking. Broadly we can classify social work interventions under prevention and reintegration strategies.

Preventive Measures

These are measures that prevent trafficking of children. The following prevention measures need to be taken up by the community level, professionally qualified social workers and others who are involved in working with children.

1. Support to families and reduce level of poverty stress
   Poverty makes a person helpless, and the traffickers target low-income families. There is a need to keep a vigil on low-income families, and the movement of their children. There are various programmes and facilities to reduce poverty and support low-income families. By linking these programmes, families could be helped to cope with poverty. Mahatma Gandhi National Rural Employment Guarantee Scheme (MGNREGS) that provides 100 days of work in their native place is an example for such programmes.

2. Enhancing the housing and health facilities and proper use of available programmes for the needy
   In India, there are various programmes for the upliftment of the poor. If the programmes are appropriately utilized and made sure that they reach the needy, there would not be a situation where people fall into any traps. Health is a basic need. The Government of India and respective state governments are committed to provide the health services and facilities to the needy. The health system is equipped from the village (community level) level to the district level. The people need to make sure that they use the available services.
3. Use social security measures properly
   There are various social security measures that are meant for the ordinary and poor people. For example, old age pension, adolescent health care programmes and free schooling. There is a need to make sure that these security measures reach the common people and they utilize it.
4. Create awareness on traffickers
   There is a need to educate the widowed women, poor people, those living in slums and others about the traffickers, and their various ways of trafficking techniques. For example, checking the background of people offering jobs and marriage proposals. It will empower the community to handle and vigil about these kinds of people.
5. Train the girls to be more vigilant and aware of their rights
   In India, girls are not allowed to access equal rights like boys. There is a need to educate them on their rights and teach them to be assertive. For example, if parents are arranging marriage at an early age (child marriage, early marriage), they do not oppose them. These kinds of evils could be prevented by creating awareness of their rights and available legal options. Girls also need to be aware of the traffickers, their tactics to attract the girls and their parents.
6. Train people to educate in the community
   Periodically undertaking awareness programmes like capacity building training programmes for the youth, elders, community leaders and significant others in the community on various aspects of trafficking and preventive measures. This activity must be carried out on a

regular basis in vulnerable places. Forming anti-trafficking committees, highway mobile committees, combating cultural aspects and religious forms of prostitution with the help of community members, self-monitoring of entry and exit of strangers in the area could prevent this. There is a need to strengthen a healthy neighbourhood.

7. Healthy parenting and imparting life skills
   The organizations, community-level workers (ASHA, Anganwadi workers) need to be trained on healthy parenting and life skills. In turn, they need to organize parenting and life skills to the community people. It is not an issue since India is having a reliable system of these community workers, and it will strengthen the communities in vulnerable areas (slums, villages, tribal areas).
8. The legal system needs to be more active
   Policem can take patrolling; form specialized cells; and educate the villagers, teachers, health care workers and transport workers. This will prevent trafficking.

Reintegration Strategies

These are some measures that social workers can follow after the rescue of trafficked children (Department of Women and Child Development, Government of India, 2000 & 2006; Oram et al., 2012; Paksen et al., 2017).

Individual Level

1. Help the child to adjust to the institution, its routine and legal issues
   The child may be in shock and psychologically and emotionally in a problematic situation because of trafficking and exploitation. After

the rescue, the child would be brought to the institution for rehabilitation. In the institution, the child needs to adjust to the new environment, routine and face legal issues. These issues need to be handled sensitively without hurting the child. The social worker can help the child by explaining the real world and the purpose of the rehabilitation. During the process, he/she must use professional knowledge to help the child to readjust.

2. Maintain discipline issues
   After the trafficking, the child may not be disciplined due to emotional and psychological problems. The social worker would help the child gradually by explaining the rules and regulation, the need for being disciplined in life. Reinforcement and punishment may be used in limited forms whenever required. These children might have enjoyed more freedom while under the custody of traffickers. The same type of freedom they may expect. The social workers need to make them understand the reality and freedom with responsibility.
3. Interaction and relationship
   Due to the experience of their experience, children may not trust anybody. Building the rapport and establishing the relationship is the key to rehabilitation. By using professional knowledge, social workers can help these children to adjust well.
4. Look into general health and mental health issues
   These children may be having general health issues due to exploitation in various forms. There is a need to make sure that their general health issues are dealt with appropriately. Check for sexually transmitted diseases, physical abuse, physical wounds and other.
   Mental health issues need to be handled. There is a high chance that these children may have post-traumatic stress disorder (PTSD), anxiety disorders, adjustment disorders, truancy, attention deficit hyperkinetic disorders (ADHD) and other. Social workers have to handle these issues and help them to adjust to society.
5. Educational and vocational training
   Educational and vocational training on accessibility and options of vocational training programmes, product manufacturing, marketing and sales and services skills needs to be taught. Social workers should use strategies based on the needs of the individual child.

6. Preparing the child for the future with proper training and guidance. Future of these children seems dark for them. They need to be handled with care and build their confidence in the future by providing proper training and guidance in managing life. Social workers would do this.

Family Level

For family, it will be tough to accept that their child is exploited, abused and misused. They would be in trauma and shock. Some families may not accept the children. So as a therapist, social workers could do trauma counselling and prepare them to accept the child. There is a need to provide supportive education on various options for life and employment. Explaining the family about government programmes and policies which could help them is very important. The rescued child is always vulnerable to re-trafficking. Hence, preparing the child and family on the prevention of re-trafficking is very important. The family should be educated on the health aspects of the child, ensure regular follow-ups and deal with mental health issues of the child are essential. The family should be empowered in all these aspects.

Community Level

The community works as a protector if it is empowered. Earlier societies had close net and community feeling. Because of urbanization, individualization, globalization and other factors, communities have lost their close bonds. Now even the neighbours are not known. There is a need to empower the communities on prevention as well as promotional activities. Social workers need to prepare the community for pressurizing the rescue efforts, participate in rescues, lodge first information reports (FIRs) and public interest litigations (PILs), and take actions as informal. The community needs to support the family and the child to readjust with society.

Allied Systems

Law-Enforcement Level

Law-enforcement authorities should assist and support, avert re-trafficking and work towards prosecution. They should behave delicately while doing raids, rescue and post-rescue period. Social workers need to work with allied systems and assist them whenever the need arises.

Policy and Legal Level

Social workers should help the policy makers with their immense experience to sensitize the policy makers about the issues, the need for commitment and support from the administration, policies, structures and resources aimed at ensuring the wellbeing. Legal fraternity should design a child-friendly justice system. Social workers should work in this direction.

Research and Use of Mass Media

Social workers should periodically conduct research and collect data for the government. The social workers who are employed in this field should enrich social work practice with research. The use of mass media in an effective way, and sensitizing the media about the importance of action is essential.

Discussion

Social work is a helping profession that has many facets in it. The primary methods of social work, that is, social casework, social group work and community organization emphasize working directly with the victims. Social workers can practice the social casework method with children. With family members, peers and significant others, social group work would be more useful. The problems which they are not able to identify through casework can be elicited through social group work. Social group work can be used in institutions or in the community set-up. While working with the family, social casework and social group work can be used. Community organization shall be beneficial in mobilizing, organizing and empowering the community and in other aspects related to trafficking.

The secondary methods are beneficial. The social action method needs to be used while prevention, empowering, making sure that the victims and family members get justice. The administrative mechanisms available need to be used for empowering individuals, family and community. The social work research must be integrated to identify the research gaps, assessing the problems, evaluating the intervention programmes and other.

Conclusion

Child trafficking is a perpetual social malady with enormous magnitude. There are escalating facts about the link of child sexual exploitation and the spread of HIV/AIDS. Children forced in the sex trade are found vulnerable to HIV/AIDS, which is estimated to be around a million in a year. In this regard, social workers can contribute a lot to prevent as well as reintegrate these children into the mainstream.

References

Adenwalla, M. (2002). *Child Rights & Law - a Guide Book for Legal Intervention.* Mumbai: Jenaz.

Child Protection & Child Rights. Retrieved October 15, 2015., from http://www.childlineindia.org.in/child-trafficking-india.htm.

Department of Women and Child Development, Government of India. (2000). *Manual of Social Workers - Dealing with Child Victims of Trafficking and Commercial Sexual Exploitation*. New Delhi: Fernandes.

Department of Women and Child Development, Government of India. (2006). Manual for Counselors - Counseling Services for Child Survivors of Trafficking. New Delhi.

Goswami, S., & Lagon, M. (2013). Almost a Miracle: Encouraging Inclusive Growth in India. *World Affairs, 176*(1), 51–56.

International Labour Office. (2017). Global estimates of modern slavery: Forced labour and forced marriage. Geneva. Available from https://www.ilo.org/wcmsp5/groups/public/@dgreports/@dcomm/documents/publication/wcms_575479.pdf

International Organization for Migration. (2017). Key trends from IOM Victim of Trafficking Data. Retrieved on 18 January 2020 from file:///H:/Published/Trafficking%20final%20article%20rajendra/A4-Trafficking-External-Brief.pdf

International Organization for Migration. (2018). Global database on victims of human trafficking. Retrieved on 20 January 2020 from https://www.iom.int/counter-trafficking

Khan, S. (2014). Human Trafficking, Justice Verma Committee Report and Legal Reform: An Unaccomplished Agenda. *Journal of the Indian Law Institute, 56*(4), 567–580.

Mark Lusk, Faith Lucas, (2009). The challenge of human trafficking and contemporary slavery. *Journal of Comparative Social Welfare 25*(1), 49–57

Ministry of Women & Child Development, Government of India. Rashtriya Mahila Kosh, Annual Report 2005–2006. Retrieved from www.wcd.nic.in/sites/default/files/AR2005-06.pdf.

Ministry of Women and Child Development, Government of India. "Conference of Partner NGOs: Implementation of Policies, Schemes, Programmes for Women and Children: Challenges and Way Forward" on October 9, 2017 at the Ashok in Chanakyapuri, New Delhi.

National Crime Records Bureau. 2017. *Crime in India 2016 Statistics.* Ministry of Home Affairs, pp. 512–518. Retrieved May 18, 2018, from http://ncrb.gov.in/StatPublications/CII/CII2016/pdfs/NEWPDFs/Crime%20in%20India%20-%202016%20Complete%20PDF%20291117.pdf.

Oram, S., Stöckl, H., Busza, J., Howard, L. M., & Zimmerman, C. (2012). Prevalence of Risk and Violence and the Physical, Mental, and Sexual Health Problems Associated with Human Trafficking: Systematic Review. *PLoS Medicine, 9*(5). https://doi.org/10.1371/journal.pmed.1001224.

Patkar, P., & Patikar, P. (2003). *Child Trafficking Issues and Concerns.* Mumbai: Jenaz.

Peksen, D., Blanton, S., & Blanton, R. (2017). Neoliberal Policies and Human Trafficking for Labor: Free Markets, Unfree Workers? *Political Research Quarterly, 70*(3), 673–686.

Tsalikis, C. (2011). Human Trafficking: The Wound That Shames Our Present. *The World Today, 67*(8/9), 28–30.

UNODC. (2000). United Nations Convention against Transnational Organized Crime and the Protocols Thereto. Retrieved on 18 May, 2018 from https://www.unodc.org/unodc/en/organized-crime/intro/UNTOC.html

UNODC. (2007). *Government of India. Protocol on Inter State Rescue and Post Rescue Activities: Relating to Persons Trafficked for Commercial Sexual Exploitation*. United Nations Office on Drugs and Crime Regional Office for South Asia EP 16/17, Chandragupta Marg Chanakyapuri New Delhi.

UNODC, Global Report on Trafficking in Persons. (2016). United Nations Publication, Sales No. E.16.IV.6, Vienna.

UNODC. United Nations Convention against Transnational Organized Crime and the Protocols Thereto. Received from https://www.unodc.org/unodc/en/organized-crime/intro/UNTOC.html

Verma, D. (2014). Human Trafficking in Santal Parganas Division of Jharkhand. *Proceedings of the Indian History Congress, 75*, 1241–1246.

Vlachová, M. (2005). Trafficking in Humans: The Slavery of Our Age. *Connections, 4*(4), 1–16.

Wilson, J., & Dalton, E. (2007). Service Provider and Community Responses. In *Human Trafficking in Ohio: Markets, Responses, and Considerations* (pp. 43–50). Santa Monica, CA; Arlington, VA; Pittsburgh, PA: Rand Corporation.

30

Efforts Towards Education: Looking into Non-Government Support for Non-Formal Education in Bangladesh

Gazi Arafat Uz Zaman Markony

Introduction

> An investment in knowledge pays the best interest. (Benjamin Franklin)

As the quote indicates, investment in education brings the best output and outcome for a nation. And to gain the best output through boosting a country's development process, it is mandatory to keep this basic right within the reach of all classes of people. Considering this, education—the process of acquiring knowledge and skills—is a priority for all nations around the globe. To fulfil this objective in a sustainable manner, concerted efforts by the relevant governments are required. The outcomes of such efforts are quite tangible, as the World Bank data shows an increase from 68.9 per cent to 86.2 per cent in adult literacy rate on a global scale in recent times. In this process of transformation, non-government actors

G. A. U. Z. Markony (✉)
Department of Public Administration and Governance Studies, Jatiya Kabi Kazi Nazrul Islam University, Mymensingh, Bangladesh

© The Author(s) 2020
Md. N. Momen et al. (eds.), *Building Sustainable Communities*,
https://doi.org/10.1007/978-981-15-2393-9_30

are contributing significantly through supplementing government measures in implementing the task of offering education to all target groups.

Bangladesh achieved tremendous success in enhancing literacy rate in recent years; the available data show a remarkable jump to 72.76 per cent in 2016, which was only 46.66 per cent in 2007. During the same period, the literacy rate of young people in the age-group of 15–24 years was elevated to 92.24 per cent, which was 61.87 per cent in 2007. The rise in fiscal allocation for the sector is also noticeable. The government had a budget of USD 7.8 billion for the sector during 2017–2018 fiscal year, which was only around USD 2 billion in 2008. Formulation and implementation of policy decisions by the government as well as non-government actors' involvement through supportive programmes in the education sector have uplifted the sector to a much better state.

Despite these positive changes in terms of statistics, a significant proportion of the country's population remains outside the formal education system. Because of reasons like lack of awareness, engagement in income-generating activities from childhood, and fear of harassment of girls in rural or remote areas, a large number of students remain outside its purview. Moreover, formal education system does not cater to adult education. And in remote or 'hard to reach' areas (like wetlands and hill tracts), accessing education is harder for the local community people. For taking care of these population segments, non-formal education has been an alternative process. Both the government and non-government stakeholders are providing supports to this process through their activities, including framing policies, executing projects and programmes and building infrastructure. The NGOs are operating non-formal education projects since long by targeting groups who require this service throughout the country. Considerable contributions have already been made by them in this area till now.

This write-up incorporates a detailed study on the sector and identifies the policy and institutional frameworks for non-formal education as well as the supports provided by the NGOs. As many organizations are working on the issue and limitations in the integrated data are noticed, this study incorporates a number of international cum national organizations and their programmes as case studies. The chapter concludes by mentioning some associated challenges and possible ways to overcome those shortcomings.

Defining Non-Formal Education

Non-formal education points to the educational system that takes place outside the formal schooling arrangement. It often includes community education, adult education, lifelong education and second-chance education that may involve home-based or community-level learning facilitated by government or non-government institutions (Khasnabis et al. 2010). International Standard Classification of Education (ISCED 2011) defines non-formal education as "*an addition, alternative and/or a complement to formal education within the process of the lifelong learning of individuals*". It can also be defined as the organized education system outside formal education, which is simple and flexible and can be delivered at any place convenient to the learners.

This form of education has its major focus on creating access to education for a group of people lacking in education, disadvantaged or ignored classes of people irrespective of age or gender, and is usually run for a short or temporary duration. It can cover adult literacy, out-of-school children, as well as programmes on life-skill, work-skill and other forms of social development. And more specifically, dropped-out segments of population or people with no access to education are the beneficiaries of this coverage.

National Non-Formal Education Policy of Bangladesh provides an elaborate definition of the phenomenon. It describes non-formal education as "*a purposeful and systematically organized form of learning that generally occurs outside the formal educational institutions. It is designed to meet the learning needs of educationally disadvantaged persons of different ages and backgrounds, flexible in terms of organization, time and place, and may cover basic and continuing educational programs to impart basic literacy, including life-skills, work-skills, general culture; and facilitates lifelong learning and enhancement of earning capabilities for poverty reduction. It ensures equity in access and human resource development; it may or may not follow a 'ladder' system, and may be of varying duration*" (Ministry of Primary and Mass Education, GoB, 2006, National Non-formal Education Policy, p. 4).

Non-Formal Education in Bangladesh: Short History and Scopes

The history of non-formal education in Bangladesh goes back to the British colonial era in the Indian subcontinent. While this programme was launched first in 1918 through the opening of a night schooling system, it should be mentioned that around a thousand of night schools were functioning in this territory in 1926. A few years later in 1939, 'Each One Teach One' was a popular campaign launched by Frank Laubach. Later, efforts like establishment of 'Literacy Centre' in 1965 by HGS Beaver and the formation of the 'East Pakistan Adult Education Cooperative Society' in the eastern wing of the then Pakistan can also be recalled in the area. At around the same time, this form of education was taken up as a pilot project by the rural development academy (BARD) in Cumilla district under the patronage of directorate of public instructions. The project ran even after the liberation of Bangladesh in 1971 (Mashreque and Chowdhury 2015). Along with the continuation of such initiatives, history also indicates active non-government support to the sector. As seen in 1994, the Campaign for Popular Education (CAMPE) prepared a database of NGOs[1] involved in non-formal education programme. It was found that the number of such type of organizations were around 435 with approximately 2.5 million learners.

In 2003, the Ministry of Primary and Mass Education (MPME) of the Government of Bangladesh (GoB) formed a National Task Force for the non-formal education sector; and the task force proposed a detailed policy framework in this regard. As a part of this initiative, the Bureau of Non-Formal Education (BNFE) was established under the MPME in April 2005.

[1] Non-government Organizations.

As the National Non-Formal Education Policy states, it has major focus over the disadvantaged section of the population including children, youths, minorities, people with disabilities, marginalized groups and people living in ecologically difficult locations. Considering these facts, the scopes of the policy (Non-Formal Education Policy 2006, pp. 5–6) include

(a) Early childhood care and education (ECCE);
(b) Alternative chances for the out of formal education children;
(c) Second chance for dropped out and never enrolled adolescents and adults; and
(d) Life-long learning opportunities, access to micro-credit and skill development training programmes (like vocational, entrepreneurship development, etc.).

Policy and Institutional Framework

After the liberation of Bangladesh in 1971, the country set its focus on ensuring basic rights of the citizens and incorporated the issues of food, clothing, housing, medical support and education in the constitution. Along with other fundamental rights, the country set its target of 'education for all', and institutional mechanisms were put in place by the government to achieve this goal. And more than that, non-government supports by the civil society and other organizations have also been supplementing this endeavour to achieve the ultimate objective. With a view to ensuring education for all classes of people and establishing a mechanism for non-government stakeholders to support the sector, national and sector-based policies were prepared considering the global agenda and perspective. The relevant national policies and international agenda that have been addressing the issue are as follows:

National documents (laws, policies, regulations)	Constitution	The Constitution of the People's Republic of Bangladesh
	Policies	National Education Policy 2010
		National Non-formal Education Policy 2006
	Acts	Non-formal Education Act 2014
	Development plan	National Five Year Plans
International agenda (mandated by the United Nations)	Millennium Development Goals (MDGs), 2000–2015	Goal 2: To achieve universal primary education
	Sustainable Development Goals (SDGs), 2015–2030	Goal 4: Quality Education

The Constitution of Bangladesh (Article 15) includes the "fundamental responsibility of the state to attain, through planned economic growth, a constant increase of productive forces and a steady improvement in the material and cultural standard of living of the people, with a view to securing to its citizens:

(a) The provision of basic necessities of life, including food, clothing, shelter, education and medical care".

The constitution (Article 17) also mentions, "The State shall adopt effective measures for the purpose of:

(a) establishing a uniform, mass-oriented and universal system of education and extending free and compulsory education to all children to such stage as may be determined by law;
(b) relating education to the needs of society and producing properly trained and motivated citizens to serve those needs;
(c) removing illiteracy within such time as may be determined by law."

With a view to ensuring pro-people, well-planned and high quality education for all classes, the National Education Policy 2010 of Bangladesh set out its principles with an orientation for practicality and productivity, taking into account unequal scopes for all classes of people, standard levels in skills and competencies, education for challenged learners and special care for disadvantaged segments of population. Part-3 of the policy includes specific targets and strategies for adult and non-formal education. As the policy mentions, "*The aim of adult education is to make people literate, to offer them minimum skills in reading, writing and numeracy, to instil in them some humane values, to make them aware of health and environment, and to increase their professional skills. Efforts will persistently continue till the enrolment rate reaches 100% and all the adult people become literate*" (National Education Policy 2010, p. 11). According to the policy, students who lack regular educational programmes in schools are supposed to receive basic education or vocational studies under non-formal education system. At the same time, under this form of education, adult people are supposed to be in the programme so that they can receive basic literacy meaning minimum skill of reading, writing and numeracy along with professional skills and values. Considering all aforesaid issues, the policy devised strategies to provide

- enrolment programme for the age-group of 8–14 years;
- course materials based on the national primary education curriculum;
- basic literacy, awareness and skills for males and females between 15 and 45 years age-group;
- an integrated curriculum by including health, nutrition, family welfare, agriculture, environment;
- models for engaging educated sections of the society to be a part of the initiative;
- training for teachers to deliver adult and other types of non-formal education;
- collaborative efforts with stakeholders—government organizations and NGOs, civil societies; and
- required legal structure to deal with the issue and a national-level awakening programme for voluntary support.

The policy also focused on proper implementation through forming a national educational administration based on accountability, transparency and dynamism. It proposed a consolidated educational law and an autonomous permanent educational commission, which would be the advisory organization for the relevant institutions.

Moreover, the GoB earlier declared another policy titled National Non-formal Education Policy, 2006 with special and specific focus on the issue of non-formal education. The major objectives of the policy included

- educational programme and skill training for the disadvantaged groups;
- individual and group-based initiative towards creating empowered citizens;
- engaging non-government and private organizations/institutions for better planning, coordination and implementation of the policy; and
- institutional decentralization to create a community ownership and sustainability of the programme.

Another significant document for the sector is the Non-Formal Education Act, 2014. With the purpose of ensuring education for all as mentioned in the constitution of the country, the GoB promulgated the law in 2014. The act has its scope to work with standard education for the children dropping out from primary education, for the children who are not enrolled in primary education at all, vocational education and training for the disadvantaged or geographically vulnerable segments of children, and the aged groups. The act provides details on the issues, scopes of the law, implementation strategies and structures, establishment of relevant institutions, their tasks and so on.

The next important policy document regarding non-formal education issue is the current National Five Year Plan (2016–2020) of the country. The plan considers non-formal education as a tool of empowerment, which can contribute to human resource development and poverty alleviation. With a view to eradicating illiteracy completely based on the national education policy of Bangladesh, the plan includes the following targets and indicators:

Goals	Targets to attain	Input indicators	Output indicators	Impact indicators
Elimination of illiteracy/ providing basic literacy	32.5 million adolescent and adult illiterates.	Allocation of fund and development & printing of learning materials.	32.5 million adolescent and adult illiterates became literate.	Socio-economic condition and lifestyle changed, zeal for education arose among the people.
Establishing a community-based network of learning centres in order to create scope for ICT-based continuing and lifelong learning.	5025 (Minimum one in each union + in certain urban areas).	Financial allocation from government, community contribution.	5025 learning centres established.	Awareness of education enhanced.
Extending opportunities for effective skill training	5 million graduates of Basic Literacy Project.	Financial allocation, development of training materials, guidelines, etc.	5 million graduates of Basic Literacy Project became skilled in different grades/ occupations.	Improvement in lifestyle.
Establishment of Non-Formal Education Board	1 (One)	Financial allocation, making rules, etc.	One NFE Board established.	Scope created for attaining higher education and skill through establishment of equivalence between Formal and Non-Formal Education.

Source: 7th Five Year Plan (2016–2020), GoB p. 537

The GoB has also taken steps to implement international agenda like Sustainable Development Goals (SDGs), which include education as the fourth goal and extend the issue to different targets. As global data show, over 265 million children are currently out of school and 22 per cent of them are of primary school age; even the school-going children are lacking basic skills in reading and numeracy. Efforts are required to address such difficulties. A call for action under the 'Sustainable Development Goals' has therefore been taken up by the nations of the world to confront world's greatest problems. The fourth goal of the overall agenda points to the equitable and quality education for all children by 2030. The targets indicate that these are universally applicable to the target groups including the segments of population who are lacking basic education and/or are disadvantaged by any means. Targets mentioned in the sustainable development goals (Goal 4), which relate to non-formal and other forms of education include the following:

- Ensuring quality education and care to make children prepared for primary education and basic education (literacy and numeracy) applicable for the young and adults; equal opportunities for both men and women to access primary to university level education and skill development programmes
- Uplifting the number of both young and adult groups with skills and trainings and providing scopes for decent employment and entrepreneurship
- Special care for the vulnerable and disadvantaged groups with concentration on eradicating gender discrimination
- Providing gender sensitive and safe educational facilities with learning-friendly environment and inclusive process of education

Bangladesh also made great strides in fulfilling another implemented international agenda—'Millennium Development Goals' (MDGs). The second MDG was connected to education; under this goal, there were targets including enrolment in primary education, increase in literacy rate of children, men and women (age range 15–24) and adults (dropped out or never enrolled with 15+ age range). Bangladesh made great progress in different indicators and the education sector was one of them

(UNDP 2015). But the exclusion of physically and mentally challenged students, or adult literacy and skill development measures still remain challenging questions for the country.

To accomplish the target of providing education as a constitutional right, the GoB established the BNFE in April, 2005. BNFE has been the lead agency for non-formal education to implement Non-Formal Education Policy framework and prioritize non-formal education as a part of overall human resource development strategy. Since its inception, BNFE has been working in collaboration with actors including government organizations, NGOs, civil society and development partners. Vision of the bureau is to build a just and equitable society with productivity and responsibility and capable of facing the challenges of the twenty-first century. The mission of the bureau is to provide access and opportunities to disadvantaged people in the areas of education and skill development.

Non-Government Support for Non-Formal Education

Available literature shows that the Government of Bangladesh has taken initiatives to attain a better status in all forms of education. A number of policies and regulations have been adopted by the government as well as by development partners, with international goals set by the United Nations to take care of this phenomenon. Apart from these steps, the non-government sector has provided remarkable contributions and supports for the education sector, including in the domain of non-formal education. As has been found, both government bodies and NGOs operate seven types of non-formal education programmes in the country, namely, Early Childhood Development Education; Pre-Primary Education; Primary Education; Adolescent Education; Adult Education; Post-Literacy/Continuing Education; and Technical/Vocational Education. As a community based approach is usually followed in non-formal education, the NGOs have advantages in working on the issue at the grassroots level alongside other development programmes (like poverty alleviation, microcredit, etc.). The names of BRAC, CARITAS,

Ganoshasthya Kendra (GK), Gano Unnayan Prochesta (GUP), Swanirvar Bangladesh (SB), Rangpur Dinajpur Rural Services (RDRS), Village Education Resource Centre (VERC), Dhaka Ahsania Mission (DAM), Gano Shahajjo Sangstha (GSS), Saptagram Nari Swanirvar Parishad (SNSP), PROSHIKA, Jagorani Chakra, CMES and so on can be mentioned as the NGOs that made fundamental contributions to this sector. Available data also show that around 750 NGOs and local bodies had been implementing non-formal education programmes in the country until 2004, while around 350 organizations acted as partners to the government (Banglapedia 2012).

It has been noticed that non-formal education is not confined to a specific formula, and most of the NGOs have their projects with different priorities. In most of the cases, the relevant organizations design and implement their projects by including the following components of non-formal education programme:

- Education for target groups who never attended regular schools
- Education for target groups who dropped out from regular education
- Unemployed adult/young/adolescent groups who lack literacy
- Unemployed adult/young/adolescent groups who lack technical knowledge
- Children and adolescents in hard-to-reach areas of the country

As mentioned earlier, there are numerous organizations that work on non-formal education; among them, some international and national organizations have purposively conducted detailed studies on their contributions to non-formal education through different projects. The following international and national organizations and their contribution through relevant programmes are discussed further as successful stories of providing supports to the field:

CARITAS Bangladesh

CARITAS Bangladesh has been one of the early movers and contributors to non-formal education support programmes. The organization has numerous projects on access to education including non-formal education and quality assurance. Contribution of this organization indicates different projects with a vast area coverage of disadvantaged community, hard to reach areas, educators training, technical education and ensuring sustainability in mentioned fields. Some exemplary projects can be mentioned as a part of the organization's vital supports (Source: Project list—Education, CARITAS Bangladesh):

- With a view to supporting the *Hardest to Reach* through basic education, the organization has exclusive programme titled 'Aloghar (Lighthouse)'. This intuitive sets the prime target to ensure and increase literacy rate and other educational rights of the disadvantaged and hard-to-reach children from primary to secondary level in the target areas. Under this single project, there are a total of 993 education centres in 993 villages of 388 unions in 104 upazilas[2] under 27 districts of the country. Some notable achievements of the project have been enrolment of 86,675 students, a number of 2977 training courses on learning methods and 2971 training courses on advocacy and lobbying for participants.
- CARITAS also targets children of disadvantaged communities for providing easy access to education and strengthening of educational management for quality education and which is why runs a project titled 'Basic Education for the Disadvantaged Children'. It covers a total of 119 education centres in 156 villages under 73 unions of 21 districts in the country. Its major achievements include covering 119 schools, 13,483 students and a total of 211 annual events in the schools or centres.
- Juxtaposing the objectives to providing education supports for the students, the organizations also include 'Youth and Teachers' Program' with a motto of ensuring access to quality education, creating opportunities for continuous education, and inculcating moral values in practice. This initiative covers 702 villages in 239 unions under 104 upazilas/city corporations in 40 districts of the country.

[2] Local Government tier, Sub-district.

- With the basic target of helping ethnic students get quality education and create opportunities for continued education, the organization has exclusive programme titled 'Basic Education for Adivasi[3] Children Through Non-formal Community School' that covers an area of ten villages in nine unions under seven upazilas of Dinajpur district in the country. Major achievements of the project include enrolment of 1280 students, organizing 21 training courses and around 30 annual events in the schools/centres.
- While considering disability as well as sustainable education at the same time, there is a project named 'Sustainable Education Project for the Children with Disabilities' and it has covered the children living in tea gardens in seven remote villages of Habiganj district from 2016, and targets enrolment and retention of children with disabilities at primary and secondary level along with providing educational materials and care, as well as arrange alternative means of livelihood.
- In the hill tracts of the country, to guarantee basic and quality education for school-aged children, special programmes are required and to settle such necessities, CARITAS started 'Education Program for Ethnic Children of Chittagong Hill Tracts' which covers an area of 80 villages in 38 unions under 12 upazilas. The major accomplishments of the project include a total of 80 schools/learning centres with 26,217 students, 750 stipends for the students and 3 for teaching-learning training courses.
- Supporting the target group through creating employability skills and capacity acquisition to undertake small entrepreneurship, there are programmes like 'Regional Technical Schools' with a coverage of 22 upazilas in Dhaka, and over 100 upazilas outside Dhaka. The major achievements of the project include enrolment of 7308 students and passing out of 5999 students till 2015 in the Dhaka region.
- To touch the milestone of enhancing employment potentials for the dropped-out segments and capacity-building for entrepreneurship targeting the disadvantaged youth and women, programmes like 'Mobile Technical Schools for the Underprivileged Youths and Women' are in action through establishing 581 training units and organized 661

[3] Indigenous community.

short courses, while the number of passed out trainees was 25,592, while that of short course trainees was 9575. Even with special focus on agro-based capacity building, there is a unique schooling programme—'Mirpur Agricultural Workshop and Training School'. This endeavour includes diploma programme along with short and long courses for skill development. Until 2015, a total number of 1416 of students passed the mechanical courses, 37,850 modular courses and a total of 616 passed diploma courses.

BRAC

Another notable NGO making significant contributions in the development sector of Bangladesh is BRAC, currently the world's largest NGO. It was established in 1972 and the education sector is one of its core areas of operation. According to available data, BRAC schools have created opportunities for more than 12 million children to access educational facilities. The organization provides scholarships under its education programme in Bangladesh and Uganda. It should be especially mentioned that BRAC operates many 'Boat Schools' for hard-to-reach population in Bangladesh and the Philippines. Moreover, its youth interventions have contributed towards employable skills development and access to finance. BRAC's major contributions include students' enrolment of 3.8 million in 43,973 schools and learning centres; 396,931 children under pre-primary programme; 60,738 students under non-formal education programme; 1.4 million students in development clubs; travelling libraries and multi-purpose community learning centres; enrolment of 44,898 children with special needs; and enrolment of 40,000 students from different ethnic communities. Major exemplary contributions of the organization in non-formal education sector can be summarized as follows (Education Program, BRAC n.d.: Project documents):

- With a view to providing a second chance in basic education for the street and domestic worker children, a total of 92 schools are being operated in Dhaka city, under the project titled 'School for Street Children'. While targeting this section of population, a special supplementary material—'For a better and healthy life' has been developed for this group of students.

- Another attempt in this sector is 'Schools for Dropped-out Children'. This form of schooling includes students who have dropped out from the formal schooling system. Key features of this programme are:
  - Inclusion of children from poor and disadvantaged backgrounds, who dropped out of formal schooling between ages 8 and 12
  - Each class has 25 to 30 students and classes are held in one-room schools within walking distances
  - For getting admission, students have to face a test and the test score determines their appropriate class of study
  - Children are taught to read, write, compute simple arithmetic and acquire basic life-skills and social values; and all of these facilities are free of cost.
- The next significant initiative that can be mentioned is 'Bridging Course'. This course incorporates basic mathematics, English and science to stimulate basic knowledge and targets the group of children who have dropped out from formal schooling. It teaches students to be regular in the programme and follow certain regulations and discipline. At the same time, this form of schooling focuses on co-curricular activities and a follow-up process, so that the students of classes two to five can continue their education for a period of 7–11 months.
- While BRAC has focus over not only the academic schooling or courses, but also considering the extracurricular development, there is 'Performing Fine Arts' programme. This type of initiative was taken up with a view to developing socialization process and enhancing communication skills of children. The programme focuses on singing, rhyming, acting, story-telling, drawing, games and different practical and hands-on experiences. The basic purposes of this initiative are to: promote inner potentials of underprivileged talents on co-curricular activities; create a fun-filled learning environment; enhance creativity; and produce better-prepared future workforce.
- To create a habit of reading and boost the critical thinking capacity of students, BRAC has introduced mobile libraries for primary-level school students. Under the organization's education programme, a total of 2715 mobile libraries are being operated where both the

students and teachers get the chance to access books. Another initiative by the organization is Social and Emotional Learning (SEL). Being implemented in four regions of the country, this venture intends to build awareness and induce emotional development of students so that they are able to enhance their skills, learn morals and ethics, and develop decision-making capacity.

- An exemplary and innovative endeavour by BRAC to contribute in this sector is Boat schools, commonly known as '*Shikkha Tori*'[4] in Bengali. It was introduced to provide educational services in seasonally submerged areas of the country. Under this project, there are boats for schooling which pick up and drop registered students and teachers. As a significant portion of *Haor*[5] areas in Sylhet region is submerged for a couple of months and because of geographical isolation, children are forced to leave educational institutions. BRAC found through research that almost one-fifth of the children did not have communication facilities during the monsoon and alongside, a significant portion of this area did not even have schools, which made guardians and students reluctant to take up education. Keeping all these facts in mind, the organization introduced boat schooling with the following features:

  – One teacher operates this free of cost school and teaches different subjects for a four-year period
  – The schooling is quite flexible with no pressure of homework
  – Logistic supports (like wheelchairs, hearing kits, etc.) are provided to students with special needs
  – Government textbooks are followed for classes four and five, and BRAC's own curriculum is followed up to class three
  – BRAC arranges monthly meetings with guardians, where agenda like attendance, health and hygiene, awareness on social issues and other are covered in the discussions
  – An appointed boatman takes care of the boat including maintenance of cleanliness and hygiene.

[4] Boat of Education.

[5] Wetland ecosystem, which is physically a bowl or saucer shaped shallow depression.

Save the Children

Save the Children started working in Bangladesh in 1970 with a view to ensuring healthy life, opportunities and protection for children. Today, it delivers services to over 15 million people every year. This organization is one of the largest organizations engaged in the area of child-rights in the country as well as contributes notably to the education sector, including non-formal education. The following projects can be mentioned that function as a part of the organization's support in this area (Education, Save the Children Bangladesh n.d.: Project documents):

- As literature shows, a sponsorship programme titled 'Shishuder Jonno (For the Children)' is operated in two areas of the country to help the children who lack basic healthcare and educational facilities. The project also intervenes in the additional areas of maternal & child health and nutrition, school health & nutrition, adolescent development, child protection and so on. Major achievements of this project include development of reading and writing capacity of 81 per cent of students compared to the baseline data of 52 per cent; 94 per cent of trained teachers apply basic education programme-promoted instructional practices compared with 65 per cent baseline figure. The successes also include promotion of almost all the targeted students from grade one to grade two. A total of 152,045 children (aged 5–19 years) and adults in 849 schools are beneficiaries of this project.
- Like other organizations working in non-formal education sector, Save the Children operates 'SHIKHON (Non-Formal Basic Education for Hard to Reach Children)' which demonstrates a cost-effective model for reaching out-of-school children and children who have never been enrolled in schools. The basic objective of the project is to promote access and increase basic educational outcomes for marginalized children. It covers a geographical area of three districts (Sylhet Sadar, Moulvibazar and Habiganj) and four upazilas of Bangladesh. The project has reached 1800 out-of-school children through 60 Early Primary Education (EPE) centres; 1782 children from indigenous communities and 660 members in 60 centres have benefitted from SHIKHON.

- 'Holistic Approach Towards Promotion of Inclusive Education (HOPE)' is another attempt that targets to reduce discrimination against children with disabilities in the education system. Aiming at increased learning outcome and wellbeing of the target community, building capacity and mentoring information system, and engaging local level stakeholders in the process, the project is being implemented in three districts of the country. Under the project, implemented tasks include community survey, establishment of community-based rehabilitation centre, collaboration with partners and organizing events through collaboration.
- The next remarkable initiative by the organization to equip children in their mother language through 'Shishur Khamatayan[6]—MLE'[7] programme which is being operated in the hardest-to-reach community. Providing access to education in mother tongue along with multilingual education and ensuring supportive environment for the target children in Chittagong Hill Tracts and the Cox's Bazar district are the prime objectives of the programme. This project is being implemented in the districts of Khagrachhari, Cox's Bazar and Rangamati with language coverage for Chakma, Marma, Tripura and Rakhaine communities.
- 'Education and Protection for Refugee Children (EPRC)' is a specialized programme, as it deals with the *Rohingya*[8] community from Myanmar who fled to Bangladesh and are currently staying in the district of Cox's Bazar and adjacent areas. Through inclusion of Rohingya children, the project aims to create a protected child-friendly environment and provide them with education. It is being implemented in two registered refugee camps located in Ukhia and Teknaf upazilas of Cox's Bazar district. Major achievements of the project include enrolment of 1846 children of age three to four years in early childhood education; 6379 primary school-aged children in primary education and 641 of secondary school-aged young people in secondary education.

[6] Empowering Children.

[7] Multilingual Education.

[8] Typically Muslim community displaced from the Rakhaine state of Myanmar.

Dhaka Ahsania Mission

The Dhaka Ahsania Mission is another non-government development organization working in the development sector since 1958 with the motto of 'Divine and humanitarian services'. The organization has field-based projects and institution-based services in the education sector, including non-formal education. Relevant projects run by the organization can be summarized as follows (Major Education Programmes, Dhaka Ahsania Mission (DAM) n.d.: Project documents):

- With a wide range of area coverage, a project titled 'Unique Intervention for Quality Primary Education' is being operated in a total of 26 districts in Bangladesh with the goal to ensure educational facilities in vulnerable areas of the country for 297,467 children through collaborations with concerned formal and non-formal education providers. Achievements of the project include: assessing and including significant number of students in the process; improvement of the performance of students; mainstreaming the targeted disadvantaged students in project areas and so on.
- Another step towards development of non-enrolled or dropped out target children is 'Second Chance Education (SCE) Project' which was started with a target of reaching 100 thousand children at 37 upazilas in collaboration with other renowned organizations for giving a second chance to the children who were never enrolled or dropped out. It has already served 20,000 children by ensuring quality educational services and establishing effective linkage between formal and informal education system.
- Like other previously mentioned organizations, the organization also operates 'Joyful Journey of Out-of-School Children & Youths for Quality Learning (JOYFUL)', mainly a project designed for *Haor* areas of Kishoreganj district in the country. The project serves 5970 students participating in both formal and non-formal education for providing equitable and quality education for the target group, and arranging a safe, inclusive and effective learning mechanism at learning/resource centres.

- 'Workplace Adult Literacy and Continuing Education for Illiterate and Neo-literate Factory Workers' is another innovative idea of providing educational services to the workers at their workplaces. It was more like an experimental project, where around 50 per cent of workers in a manufacturing company of Chittagong EPZ[9] were identified who lack literacy. Dhaka Ahsania Mission developed a nine month-long course for those selected workers of the company. After completion of the course, the workers witnessed major changes in their daily life, as they did better in calculation, reading newspapers and so forth.
- The programme for 'Empowering Adolescents through Education and Vocational Skills Training to Become Agents of Social Transformation' includes beneficiaries in two regions of Dhaka city. A large number of adolescents migrating to the capital city Dhaka are disadvantaged in many aspects and lack life-skills and protection. The project catered to this segment of population to empower them through education and vocational skills training for enabling a social transformation. This initiative brought about some improvements in areas like fluent reading and writing skills of the target group, engaging some of them in the mainstream process of education and so on.
- To arrange educational facilities for the children who have dropped out from the mainstream educational process, the organizations have steps in hand titled 'Educating and Empowering Disadvantaged Children through NFE (EDUEMP)'. It aims to educate and empower them through learning centres. Major achievements of the project include a better state of literacy and health education, enhanced confidence and awareness about the rights of the target group.

UCEP Bangladesh

Another renowned NGO making significant contributions to non-formal education in Bangladesh has been the 'Underprivileged Children's Educational Program' or UCEP (Education and Skill Training. UCEP Bangladesh n.d.), which emerged as one of the earliest NGOs in the country involved in the sector. Like the aforementioned organizations,

[9] Export Processing Zone.

UCEP also has its major focus on providing developed educational facilities through 'help to learn skills to earn' process. The organization has undertaken multidimensional educational initiatives for dropped-out children and vocational training and job placement opportunities since its inception after the liberation of the country. There is a totality of 32 schools operated by this organization, where students from class five to class eight are provided educational facilities for a period of three to five years. They can also appear in primary- and secondary-level completion examinations. There is also a six-month-long short course titled 'Remedial of Literacy and Life Skills' *or ROLLS* where dropped-out children are taught basic Bangla and numeracy skills. As mentioned earlier, the organization provides vocational training in different vocations including garments, textiles, electronics, manufacturing and so on. There are scopes for job placement after completion of these six-month-long courses. Special feature of this project is to include non-UCEP participants in the training process, as well as entrepreneurship development training and support for the target groups. UCEP has enhanced its coverage through establishing UCEP Institute of Science and Technology, which offers a four-year-long diploma course in engineering. Although the students enrolled in diploma course are supposed to pay course fees, but revenue collected from the programme is cross-subsidized and spent for the disadvantaged groups who receive skill development training from UCEP. The students in UCEP programmes have a pass rate of 100 per cent in recent S.S.C. (vocational) examination, which is a landmark achievement for any organization working in this field.

Associated Challenges in Non-Formal Education Sector

Available literature evidences the dynamic role played by NGOs to make differences in the non-formal education sector of Bangladesh. Still, the organizations face challenges while implementing non-formal education projects. Based on the experiences gathered by NGOs in implementing projects, some associated challenges have been identified in case of providing support to the target group, which can be summarized as follows:

- As specific data on the issue are scarce, identifying or listing the appropriate target students is challenging in terms of required cost and time.
- NGOs have little control over communities; so, it takes time to engage the target community in the process, which delays the project timeline.
- As the programmes are non-formal, it is tough to ensure attendance of students as well as that of teachers at many places; the problem is more acute in the hard-to-reach areas because of geographical features.
- Scarcity of dedicated and quality teachers is also a challenge for non-formal education programmes.
- Most of the programmes are community-based in nature, and sometimes include numerous stakeholders; it takes time to engage all the stakeholders in the process.
- Preparing appropriate materials for the target group is another challenge; there is no certain or definite mechanism through which these materials can be integrated and distributed.
- There is scarcity of mother tongue-based study materials, which are required for the ethnic groups.
- All the projects have a certain tenure or time-period; there are usually deficiencies in their implementation and follow-up mechanisms, which may affect the sustainability of services provided.
- In urban areas, location-changes of the low-income community including children and adolescents engaged in income generating activities make the drop-out rate very high.

Concluding Remarks

Bangladesh has a track record of successive progress in education sector, and the country has its major policies and institutions in place to alleviate the curse of illiteracy. Moreover, responses from the non-government actors have been remarkable by time in this field. In spite of that, a substantial segment of the population remains outside the coverage of formal and non-formal education. As the education process includes both coverage and quality, the stakeholders must emphasize on both the issues. Besides the positive changes and impacts of both government and non-government initiatives, limitations are noticed in the implementation of policies and projects. To overcome such obstacles in order to further uplift of non-formal education in Bangladesh, proper implementation of existing policy framework is a prerequisite; and active roles of the government and non-government stakeholders are essential to meet the desired objectives. At the same time, special care should be taken and mechanisms to be put in place for providing educational facilities to hard-to-reach or geographically vulnerable areas. Focus on girls' education in rural and remote areas deserves special care. In addition, technical and scientific education should be prioritized as a part of the non-formal education process, with more concentration on the vocational courses and trainings. Students who complete technical courses or trainings are in need of job market orientation; so, relevant and responsible government organizations cum NGOs can undertake some initiatives for resolution of the challenge. In this regard, GoB–NGO collaborations might be an option to play an effective and positive role in addressing the problems faced by the sector; collaborative projects already exist though, the scopes for such collaboration need to be expanded. As the sector lacks an integrated approach as well as different organizations are working on the similar issues; an integrated approach can be developed so that all the stakeholders yield maximum benefits while implementing projects. An appropriate follow-up mechanism is another necessity for the sustainability of non-formal education programmes (like tracking the students and helping them to continue their education), an elaborate framework or database can be developed to serve this purpose. And finally, in a nutshell, organized and cooperative efforts by the government and non-government actors towards non-formal education can be the salvation from the curse of all forms of inadequacy of education and skill.

References

Banglapedia, Non-Formal Education, National Encyclopedia of Bangladesh. (2012). Retrieved from http://en.banglapedia.org/index.php?title=Non-Formal_Education.

Education and Skill Training. UCEP Bangladesh. (n.d.). Retrieved from http://www.ucepbd.org/.

Education Program, BRAC. (n.d.). Retrieved from http://www.brac.net/education-programme.

Education, Save the Children Bangladesh. (n.d.). Retrieved from https://bangladesh.savethechildren.net/what-we-do/education.

Goal 4, Sustainable Development Goals 2030, United Nations. (2015). Retrieved from https://www.un.org/sustainabledevelopment/education/.

International Standard Classification of Education ISCED (2011). UNESCO Institute for Statistics, Montreal, Quebec, Canada, 2012. Retrieved from: http://uis.unesco.org/sites/default/files/documents/international-standard-classification-of-education-isced-2011-en.pdf.

Khasnabis, C., Heinicke Motsch, K., Achu, K., et al. (2010). Community-Based Rehabilitation: CBR Guidelines, World Health Organization. Retrieved from https://www.ncbi.nlm.nih.gov/books/NBK310920/#__NBK310920_dtls.

Major Education Programmes, Dhaka Ahsania Mission (DAM). (n.d.). Retrieved from http://www.ahsaniamission.org.bd/field-works/education/.

Mashreque, Dr. Md. Shairul and Chowdhury, Dr. Md. Shakhawat Ullah. (2015, March 1). Basic Education for the Street Children, the Daily New Nation, Bangladesh. Retrieved from http://thedailynewnation.com/news/44679/basic-education-for-the-street-children.html.

Ministry of Primary and Mass Education, Government of the People's Republic of Bangladesh. (2006). National Non-formal Education Policy. Retrieved from http://www.bnfe.gov.bd/site/page/fd4b0ee8-e34e-42e9-a9db-b776233475d47/NFE-Policy

National Non-formal Education Policy. (2006). Ministry of Primary and Mass Education, Government of the People's Republic of Bangladesh. Retrieved from http://www.bnfe.gov.bd/site/page/fd4b0ee8-e34e-42e9-a9db-b776233475d47/NFE-Policy.

National Education Policy. (2010). Ministry of Education, Government of the People's Republic of Bangladesh. Retrieved from http://old.moedu.gov.bd/index.php?option=com_content&task=view&id=338&Itemid=416.

7th Five Year Plan FY 2016–FY 2020, Planning Commission, Government of the People's Republic of Bangladesh. (2016). Retrieved from http://www.plancomm.gov.bd/site/files/2eb8022b-d4fb-42e1-a67d-67bccabd8fe9.

Project List - Education, Goal 2: Project list of Quality Education, Caritas Bangladesh. (n.d.). Retrieved from https://caritasbd.org/goals/quality-education/project-list-education/.

UNDP. (2015, September 23). Bangladesh MDG Progress Report. Retrieved from http://www.bd.undp.org/content/bangladesh/en/home/library/mdg/mdg-progress-report-2015.html.

Part VII

Towards Achieving 2030 Agenda: Issues, Challenges and Concerns

Rajendra Baikady, Md Nurul Momen, and Cheng Sheng-Li

Introduction

The 2030 agenda was adopted by the United Nations (UN) on 25 September 2015. The principle aim of this agenda is centred on achieving sustainable development through a number of development goals collectively called Sustainable Development Goals (SDGs). The agenda is applicable to all countries as endorsed by a group of 193 member nations of the General Assembly, including both developed and developing countries. As rightly noted by the former UN Secretary General, Ban Ki-Moon, the 2030 agenda for sustainable development in the world is an agenda for the people and it is a promise by the leaders to end poverty in all of its form and dimensions. Accordingly, the newly adopted 17 Sustainable Development Goals (SDGs) reflected in the 2030 agenda intended to make a new impact on the development and well-being of the people. The agenda focuses on three core dimensions: economic, social and environmental development at the global level as well as the South Asian region. An important feature of the 2030 agenda is that it is built upon the gaps left by the Millennium Development Goals (MDGs) in operation since 2000–2015. South Asian region achieved remarkable progress in many areas of MDGs. The poverty reduction agenda was one central focus, and the region has shown considerable progress with regard to poverty reduction. However, development progress across the region and within the countries was uneven and incidence of poverty across the region persists. Unfortunately, South Asia represents the largest concentration of poverty and hunger in the world.

Sustainable development involves decisions based on long-term values. These decisions are based on the impact of today's actions on the future generation and making a living within the principle of environmental limits. Furthermore, it is a balanced fulfilment of social, economic, cultural and environmental needs without compromising the quality of life of future generations. It is also about building a resilient future (Hayles 2015). The major threat to the planet was experienced in the second half of the twentieth century that witnessed heavy industrialization, exploitation of natural resources and increasing number of the world population, resulting in environmental degradation all over the globe. Some studies (Baüer et al. 2006) stressed that evidences of increasing atmospheric carbon dioxide and global warming results in unpredictable climate change. Furthermore, studies have also reported the increase in air pollution and greenhouse emission (Hémous 2016; Olivier et al. 2017). Study data by NASA's Applied Sciences Program, the World Weather Watch (WWW), the Committee for Earth Observations Satellites (CEOS), and the Global Climate Observing System (GCOS) reviles the human responsibility on the planet and preserving the plant (National Aeronautics and Space Administration [NASA] 2006). Despite these documentary evidences and additionally, the 1997 Kyoto Protocol and the 2015 Paris Agreement, the progress on sustainable development and preservation of resources has shown a declining trend.

Achieving Sustainable Development: Issues, Challenges and Prospects

An overwhelming effort to pursue sustainability can be seen in many countries across the globe. Business, engineering and non-profit enterprises have stated sustainability as their universal goal. Evidently, there is an increase in the incorporation of sustainability issues in engineering and business activities in the recent past. The universal definition of 'sustainability' was given by the Brundtland Commission (formally known as the United Nations World Commission on Environment and Development (UN-WCED); the definition interpreted sustainability as the way systems can remain productive in any development type (United Nations 1987). Many scholars have documented the importance of the triple bottom line (TBL[1]) in determining sustainable development; it is argued that TBL helps in obtaining greater profit and better performance, while not compromising on social and environmental conditions (Carley and Christie 1993; Sianipar and Yudoko 2012).

In order to achieve sustainable development and ensure quality of life and well-being for all, the South Asian region needs to improve in many of its areas. The region needs to create rapid economic growth (SDG-8) through industrialization (SDG-9) for poverty alleviation (SDG-1). In order to achieve progress of these goals, the region needs to create sustained job opportunities to the growing unemployed young people. Creating sustained employment is crucial as the region's more than 80 per cent of workforce is stuck in the informal sector. Even though there is no sharp division between informal and formal sector employment and they appear as a continuum (Benjamin and Mbaye 2012a, 2012b; Steel and Snodgrass 2008; Henley et al. 2006), the unorganized sector is the largest sector providing unskilled and unsecured employment to a large number of working population in the region. This informal (unorganized) sector in all the countries of the region is showing a sharp growth. As a result, most of the jobs are being created by these informal sectors, which do not provide professional and ethical job environment for the working class.

Poverty is still rampant in South Asia and the region remains home to more than 40 per cent of the poor, living on less than US$1.25 a day (Sumner 2012). Food energy deficiency (51 per cent) and child birth with low weight (56 per cent) are also major issues in the region (Iqbal and Amjad 2010). Poverty reduction is an essential need for the region as many research studies have shown that living in poverty leads to complex mental health issues (Ackerman et al. 2004; Blair 2010; Leventh al &

[1]Triple bottom line (TBL) refers to profit, people and planet, which means an equal commitment by companies towards social and environmental concerns as equal to their concern on profit (Andrew Savitz, The Triple Bottom Line (San Francisco: Jossey-Bass, 2006)).

Brooks-Gun 2011; Raphael 2004), and increasing well-being of the population is an important aspect of any developing nation. In order to achieve sustainable development, one has to focus on poverty eradication, changing unsustainable patterns into sustainable patterns and ensuring protecting and managing natural resources. Countries across the South Asian region face the burden of ensuring food, water and energy to meet the demands of a rapidly growing population. The region needs to concentrate on ensuring sustainable use and management of water, energy, land and food along with other natural resources for achieving many of the goals set in the SDGs. The goal related to poverty reduction and zero poverty (SDG 1), ensuring food security and ending hunger (SDG 2), ensuring water security (SDG 6), ensuring access to modern energy (SDG 7), sustainable economic growth (SDG 8), ensuring sustainable production and consumption (SDG 12), sustainable use of ecology and the marine resources (SDGs 14 and 15) are interlinked and realization of these goals directly depends on availability of water, land, food and energy (Rasul 2014; Weitz et al. 2014; Rasul and Sharma 2015; GDI 2015).

The region should also promote infrastructural growth. Both state and non-state actors should work together for reducing the infrastructural gap and facilitating efficient and quality services to all. Countries in the South Asian region are characterized by huge infrastructural gap. Sustainable development goals on infrastructural development (SDG-9), ensuring access to drinking water and sanitation (SDG-6), provisions for electricity (SDG-7), and development in the information technology sector are crucial for the development of the region and achieving the 2030 agenda.

In May 2018, UNICEF noted that only about half of primary-aged children are now receiving education with minimum learning standards (UNICEF 2018). Achieving universal access to education is another priority agenda for achieving SDGs in the region. As enshrined in the SDGs, investment in health for ensuring universal health coverage (SDG-3) and providing quality education and vocational training to all (SDG-4) is essential to encourage the growing youth population in the region. Investment in quality education will ultimately result in resource generation and poverty reduction in the region. Previous studies have proven a significant connection between education and poverty (Khan et al. 2008; Govinda 2008; Danacica et al. 2010; Chaudhry and Rahman 2009;

Krueger and Maleckova 2003; Janjua and Kamal 2011; Adawo 2011). However, it is stressed that investment in quality and universal education through educational institutions will also help in reducing poverty (Chaudhry and Rahman 2009) and inequality (Danacica et al. 2010), reducing crime rate in society (Krueger and Maleckova 2003), decline in child labour and terrorism (Krueger and Maleckova 2003). Another group of researchers found that education helps in higher income generation (Durston and Nice Nashire 2001), economic growth is impacted by quality and quantity education provided (Adawo 2011), education helps in skill and productivity enhancement of poor households (Awan et al. 2011), empowering the poor is possible through investment in education (Jacovkis and Tarabini 2012). Studies have also proved that opportunities to enhance income are high among the educated (Jacovkis and Tarabini 2012); income of an individual can be enhanced with an additional year of schooling (Awan et al. 2011). Many other researchers have also expressed their concerns towards the contemporary education system that despite the fact that higher education institutions can play a role of effective communication channel in sustainable development (Lertpratchya et al. 2017), the teaching programmes in many universities and educational institutions, however, are not concerned about the environment (Schein 2015).

Research studies have also proven that only basic literacy will not help in poverty reduction (Raditloaneng 2004), hence thereby achieving sustainable development goals. In order to attain environmental sustainability, a more holistic curriculum needs to be adopted and future organizational leaders need to be trained for practices of sustainability leadership (Lertpratchya et al. 2017; Schein 2015; Stubbs and Cocklin 2008). Contribution of quality research is also an important aspect in achieving sustainability, whereas research so far has focused on profit maximization (Schein 2015; Stubbs and Cocklin 2008) rather than enhancing the quality of life and well-being. As a result, academics or the corporate literature failed to incorporate a cohesive theory of sustainability leadership (Schein 2015). Introduction of topics related to sustainable development is required to orient students towards world views and career links in sustainable development (Schein 2015; Stubbs and Cocklin 2008). Meanwhile, there is increased literature recommending inclusion of sustainability issues in the business school curriculum (Stubbs and Cocklin 2008). Educating students at their graduate school years towards sustainability concepts has proven to be effective as the ecological world view of the students contributes to their learning during school learning. Studies have also shown students' environmental concerns (Schein 2015).

Governments in South Asian countries are adopting right-based approach in order to provide universal education; however, these governments should also pay attention to higher education and research where interdisciplinary research and teaching curriculum on sustainability can be initiated. Incorporation of sustainability research and teaching contents will enable students to better integrate sustainability practices in their profession (Stubbs and Cocklin 2007). Research studies have proven the effectiveness of educational institutions in transformation of young individuals and incorporating lasting values among young people (Lertpratchya et al. 2017).

Ensuring social protection to the vulnerable population in the region is important to achieve well-being of all and leaving no one behind. Social protection is important, as it addresses income poverty and provides income, capital and assets to address the issues related to poverty and inequity. Social protection also serves as a means to ensure well-being, as it promotes welfare of the population. In addition, it helps the vulnerable and marginalized sections to avail their rights and entitlements. Social security refers to policies and programmes that enable a right-based approach to development. Greater attention on social protection all over the globe as well as in South Asia was necessitated by the process of globalization and its adverse effects on society (Sudarshan 2007; Samaratunge and Nyland 2006). While countries in the South Asian region are currently facing challenging issues related to social protection provisions, Sri Lanka continues to perform stronger than the rest of the countries in the region. Sri Lanka has a number of social protection provisions for the poor and vulnerable groups and together these programmes cover a third of the population, which is the highest among the countries in the South Asian region (World Bank 2007). Universal social protection and financial inclusion, especially for the poor and vulnerable population is an essential need in achieving sustainable development goals. Social protection and financial inclusion helps in poverty reduction (SDG-1) and in reducing inequality among different groups (SDG-10). Hence, both state and non-state actors in the region need to focus on developing the social protection programmes, including conditional cash transfer, employment guarantee and income support provisions.

Countries in the region need to focus on ensuring food security for all people. Ensuring food for all, irrespective of socio-economic condition, is one of the major focuses of the 2030 agenda. Achieving food security for all and zero hunger (SDG-2) is a key development challenge with high priority in the sustainable development goals. Hunger reduction efforts in the South Asian region is important, as the region is one of the largest hunger spots in the world; the 2018 Global Hunger Index reports a sharp increase in child wasting in South Asia since 2010 (Grebmer et al. 2018). Furthermore, the region accounts for nearly two-thirds of the global undernourished population, people in the region still have poor dietary habits with poor nutrition and lack of micronutrient contents (Dame and Nusser 2011; Headey and Hoddinott 2016; Jewitt and Baker 2007; Pingali 2012). As a result, malnutrition of various kinds remains a serious issue in the region (Gómez et al. 2013). According to a World Health Organization report, malnutrition of various kinds in the region is resulting in health complications such as obesity, diabetes and hypertension (WHO 2018). Another study shows that more than two billion people in the region suffer from nutrition deficiency (IFPRI 2016; Spielman and Pandya-Lorch 2009). Even though progress has been reported from East and Southeast Asia, performance of the South Asian region is not satisfactory, as people still face micronutrient deficiency (Beall et al. 2017). Studies have recommended addressing the triple burden as a policy priority in the region (Kulkarni et al. 2017; Meenakshi 2016; Shankar et al. 2017). South Asian countries need to envisage policies that ensure food and nutritional safety to the people. Policy action is much needed in order to eradicate extreme poverty and ensure better access to food and thereby reduce the inequality in the society. Ensuring food safety will also ultimately help in reducing anaemia and vitamin A deficiency.

The region also faces a high degree of vulnerability to climate change. Ensuring environmental protection and conservation of resources is an essential aspect for the sustainability in the region. The region needs to formulate policies related to environmental protection and strategies to ensure low carbon emission from the industries and economic business undertakings. The goals related to environmental protection can be achieved by focusing on renewable sources of energy, such as hydro, solar and wind, which are available abundantly in the region. Depending on these conventional energy sources will also help in reducing the carbon emission and help in the decarbonization strategy for South Asia. Researchers have proved sustainable business movements are the most important environmental movements (Schein 2015). Researchers have also shown the failure of the current environment policies (Allenby 2011; Chevalier et al. 1974; Roome 2001; Baumgartner and Korhonen 2010). Further, considering humans separate from and superior to nature is the major reason for environmental degradation (Stubbs and Cocklin 2008). People think they have ownership over nature, thus exploiting natural resources for their own benefit. This is also due to the fact that economic growth is measured on the basis of production and consumption of goods and services (Stubbs and Cocklin 2008). Most importantly, creation of wealth is the central focus of business organizations (Delios 2005) in the present context and thus environmental exploitation is at the maximum.

To conclude, sustainable development in the South Asian region needs the attention of policy makers and policy. Acknowledging the complexities of the policy problems, strategic thinking and addressing the issues at hand are the major priorities for achieving sustainability in the region. Despite its popularity and universal acceptance (Talberth et al. 2006), measuring economic progress based on gross domestic product (GDP) is criticized by many economists (Goodstein and Polasky 2014). Measuring unsustainable economic growth based on GDP may diminish the quality of life of people, as economic growth is their only objective and not the sustainable use of the environment. As recommended by the studies, economic production needs to ensure that increased welfare for the current generation is not incurred at the expense of the future generation (Goodstein and Polasky 2014; Talberth et al. 2006).

This part of the book makes an effort to document the civil society, NGO and third-sector contribution in the region for realizing the sustainable development goals. Chapters in this part mainly focus on the CSOs' role in environmental protection, women empowerment, social innovation social mobilization, tribal community right and elderly care in Bangladesh.

Chapter 31 explores the underlying cultural and cognitive differences which led Western and Eastern people to have very different concepts on environment and of their roles in protecting it. Then, it develops non-Western alternative strategies for popular participation in environmental protection law and policy consistent with non-western culture and philosophy. The chapter is based on secondary data collected through content analysis method from relevant literatures, research journals, books, news articles, websites and various published and unpublished materials. The outcome of the study would lead to better environmental results in reality. Chapter 32 outlines the gender balances in Bhutan and analyses women empowerment through political participation. The chapter discusses the efforts and the intervention programs and projects initiated by CSOs and NGOs and their critical role in narrowing the gender gaps. NGOs and civil society organizations are playing a considerable role in reducing the existing gender gaps and encouraging women's participation. The chapter recommends more intervention activities by NGOs and CSOs in meeting the practical and strategic needs of women.

Chapter 33 discusses the Forest Rights Act (2006) and community conflicts in the perspective of Sustainable Community Development. The chapter is based on a Case Study on Lodha Tribes of West Bengal in India. The authors in this chapter make an attempt to understand the specific knowledge system used by the tribal people to perceive and confer the ecological resources, determine the selection process to utilize the livelihood capitals to sustain their life style and to structure their well-being dimension in achieving sustainable community development. Chapter 34 illustrates the approaches and needs for ensuring safety of women in public spaces in the Indian context. The authors in this chapter discuss in detail the outlook of preventive, prohibition and redressal in ensuring the safety of women at the public spaces.

Chapter 35 examines the impact of the Kudankulam Nuclear Power Plant (KKNPP) on the livelihood, stress levels, awareness of risks on the people of Idindhakarai (Tamil Nadu, India). By doing so, the chapter takes an activist stand to highlight the ongoing struggle of people in

Idindhakarai for over 6 years and urges to close the plant and bring in the imperative need for Social Action among this fisher folk community.

Chapter 36 discusses the issues and challenges experienced by Bangladeshi pilgrims during their pilgrimage to Saudi Arabia. The author's takes a human rights perspective to discuss the issue as this event has some implications on the human rights and religious freedom of pilgrims. The chapter recommends possible solutions for addressing the existing problems in pilgrimage and *umrah* management.

References

Ackerman, B. P., Brown, E. D., & Izard, C. E. (2004). The relations between contextual risk, earned income and the school adjustment of children from economically disadvantaged families. *Developmental Psychology, 40*, 204–216. https://doi.org/10.1037/0012-1649.40.2.204.

Adawo, M. A. (2011). Has Education (Human Capital) Contributed to the Economic Growth of Nigeria? *Journal of Economics and International Finance, 1*(3), 46–58.

Allenby, B. (2011). Thoughts on Industrial Ecology, Emerging Technologies, and Sustainability Science. *Integrated Research System for Sustainability Science, 6*, 119–122.

Awan, M., Malik, N., Sarwar, H., & Waqas, M. (2011). Impact of Education on Poverty Reduction. *International Journal of Academic Research, 3*(1).

Baüer, P., Gérard, F., & Minster, J.-F. (2006). *Observing the Earth: An International Endeavour*. Paris: Académie de Sciences.

Baumgartner, R. J., & Korhonen, J. (2010). Strategic Thinking for Sustainable Development. *Sustainable Development, 18*(2), 71–75.

Beall, T., Massiot, E., Arsenault, J. E., Smith, M. R., & Hijmans, R. J. (2017). Global Trends in Dietary Micronutrient Supplies and Estimated Prevalence of Inadequate Intakes. *PLoS One, 12*, 20.

Benjamin, N., & Mbaye, A. A. (2012a). *The Informal Sector in Francophone Africa: Firm Size, Productivity and Institutions*. Washington, DC: World Bank.

Benjamin, N., & Mbaye, A. A. (2012b). Informality, Productivity, and Enforcement in West Africa: A Firm-Level Analysis. *Review of Development Economics.*

Blair, C. (2010). Stress and the Development of Self-Regulation in Context. *Child Development Perspectives, 4*(3), 181–188.

Carley, M., & Christie, I. (1993). *Managing Sustainable Development.* Minneapolis, USA: University of Minnesota Press.

Chaudhry, I. S., & Rahman, S. (2009). The Impact of Gender Inequality in Education on Rural Poverty in Pakistan: An Empirical Analysis. *European Journal of Economics Finance and Administrative Science, 15*, 174–188.

Chevalier, M., Bailey, L., & Burns, T. (1974). Toward a Framework for Large-Scale Problem Management. *Human Relations, 27*(1), 43–69.

Dame, J., & Nusser, M. (2011). Food Security in High Mountain Regions: Agricultural Production and the Impact of Food Subsidies in Ladakh, Northern India. *Food Security, 3*, 179–194.

Dănăcică, D., Belaşcu, L., & Llie, L. (2010). The Interactive Causality Between Higher Education and Economic Growth in Romania. *International Review of Business Research Paper, 4*(6), 491–500.

Delios, A. (2005). How can Organizations be Competitive but Dare to Care? *Academy of Management Perspectives, 24*(3), 25–36.

Durston, S. & Nice Nashira. (2001). Rethinking Poverty and Education: An Attempt by an Education Programme in Malawi to Have an Impact on Poverty. *Compare, 31*(1).

GDI German Development Institute. (2015). *Post 2015: Why Is the Water Energy Land Nexus Important for the Future Development Agenda?* DIE's Post-2015 Briefing Paper Series.

Gómez, M. I., Barrett, C. B., Raney, T., Pinstrup-Andersen, P., Meerman, J., Croppenstedt, A., Carisma, B., & Thompson, B. (2013). Post-green Revolution Food Systems and the Triple Burden of Malnutrition. *Food Policy, 42*, 129–138.

Goodstein, E.S., Polasky, S. (2014). Economics and the Environment. Seventh Edition.

Govinda, R. (2008). *Non-Formal Education and Poverty Alleviation, Analysis o Field Experiences from Asia*. Working Document, July 2008.

Hayles, C. S. (2015). A Case Study of Capacity Building: Embedding Sustainable Development Principles and Practices at the University of Wales, Trinity Saint David. Proceedings of the 5th International Conference on Building Resilience, Newcastle, New South Wales, July 2015.

Headey, D. D., & Hoddinott, J. (2016). Agriculture, Nutrition and the Green Revolution in Bangladesh. *Agricultural System, 149*, 122–131.

Hémous, D. (2016). The Dynamic Impact of Unilateral Environmental Policies. University of Zürich, Switzerland. *Journal of International Economics, 103*, 80–95.

Henley, A., Arabsheibani, R., & Carneiro, F. G. (2006). On Defining and Measuring the Informal Sector. IZA Discussion Papers 2473, Institute for the Study of Labor (IZA) and World Bank Policy Research Working Paper 3866, March 2006.

IFPRI. (2016). *Global Nutrition Report 2016, From Promise to Impact: Ending Malnutrition by 2030*. Washington, DC, USA: International Food Policy Research Institute.

Iqbal, M., & Amjad, R. (2010). Food Security in South Asia: Strategies and Programmes for Regional Collaboration. *Regional Integration and Economic Development in South Asia, 357*.

Jacovkis, J., & Tarabini, A. (2012). The Poverty Reduction Strategy Papers: An Analysis of a Hegemonic Link Between Education and Poverty. *International Journal of Educational Development, 32*, 507–516.

Janjua, P. Z., & Kamal, U. S. (2011). The Role of Education and Income in Poverty Alleviation: A Cross-Country Analysis. *The Lahore Journal of Economics, 16*(1), 143–172.

Jewitt, S., & Baker, K. (2007). The Green Revolution Re-assessed: Insider Perspectives on Agrarian Change in Bulandshahr District, Western Uttar Pradesh, India. *GEOFORUM, 38*(1), 73–89. https://doi.org/10.1016/j.geoforum.2006.06.002.

von Grebmer, K., Bernstein, J., Hammond, L., Patterson, F., Sonntag, A., Klaus, L., Fahlbusch, J., Towey, O., Foley, C., Gitter, S., Ekstrom, K., & Fritsche, H. (2018). *2018 Global Hunger Index: Forced Migration and Hunger*. Bonnand Dublin: Welthungerhilfe and Concern Worldwide. Retrieved from https://www.globalhungerindex.org/results/.

Khan, S. K., Nassr, M., & Hamdani, N. (2008). *Impact of Education and Health on Poverty in Pakistan A Critical Study*. Park Road, ChakShahzad Islamabad, Pakistan: COMSATS Institute of Information Technology.

Krueger, A. B., & Maleckova, J. (2003). Education, Poverty and Terrorism: Is There a Causal Connection? *Journal of Economic Perspectives, 4*(17), 119–144.

Kulkarni, V. S., Kulkarni, V. S., & Gaiha, R. (2017). Double Burden of Malnutrition: Reexamining the Coexistence of Under Nutrition and Overweight Among Women in India. *International Journal of Health Services, 47*, 108–133.

Lertpratchya, A. P., Besley, J. C., Zwickle, A., Takahashi, B., & Whitley, C. T. (2017). Assessing the Role of College as a Sustainability Communication Channel. *International Journal of Sustainability in Higher Education.*

Leventhal, T., & Brooks-Gunn, J. (2011). Changes in neighborhood poverty from 1990 to 2000 and youth's problem behaviors. *Developmental Psychology, 47*(6), 1680–1698. https://doi.org/10.1037/a0025314.

Meenakshi, J. V. (2016). Trends and Patterns in the Triple Burden of Malnutrition in India. *Agricultural Economics, 47*, 115–134.

National Aeronautics and Space Administration. (2006). *Applications of Earth Observations for Societal Benefit: Results of Project Grants Beginning in 2006.* NASA Earth Science, Applied Sciences Program.

Olivier, J.G.J., Shure, K.M., & Peters, J.A.H.W. (2017). Trends in Global CO2 and Total Greenhouse Gas Emissions: Summary of the 2017 Report. PBL Netherlands Environmental Assessment Agency.

Pingali, P. (2012). Green Revolution: Impacts, Limits, and the Path Ahead. *Proceedings of the National Academy of Science of the United States of America, 109*, 12302–12308.

Raditloaneng, N. (2004). *Post Literacy and Poverty Eradication: Implications for Adult Education. Adult Education and Poverty Reduction: A Global Priority*. Gaborone: Department of Adult Education, University of Botswana.

Raphael D. (2004). Introduction to the social determinants of health. *Social Determinants of Health: Canadian Perspectives*, (pp. 1–18). Toronto, ON: Canadian Scholars Press.

Rasul, G. (2014). Food, Water, and Energy Security in South Asia: Anexus Perspective from the HinduKush Himalayan Region. *Environment and Social Policy, 39*, 35–48.

Rasul, G., & Sharma, B. (2015). The Nexus Approach to Water Energy Food Security: Adoption for Adaptation to Climate Change. *Climate Policy*. http://dx.doi.org/0.1080/14693062.2015.1029865.

Roome, N. (2001). Conceptualizing and Studying the Contribution of Networks in Environmental Management and Sustainable Development. *Business Strategy and the Environment, 10*, 69–76.

Samaratunge, R., & Nyland, C. (2006). Globalisation and Social Protection Reforms in Sri Lanka. *South Asia: Journal of South Asian Studies, 29*(3), 415–434.

Schein, S. (2015). *A New Psychology for Sustainability Leadership*. Sheffield, UK: Greenleaf Publishing Ltd.

Shankar, B., Agrawal, S., Beaudreault, A. R., Avula, L., Martorell, R., Osendarp, S., Prabhakaran, D., & McLean, M. S. (2017). Dietary and Nutritional Change in India: Implications for Strategies, Policies, and Interventions. *Annals of the New York Academy of Science, 1395*, 49–59.

Sianipar, C. P. M., & Yudoko, G. (2012). Understanding Issue Dissemination and Arrival Patterns on Supply-Chain Using Network Analysis and Social Media. *Journal of Asia Pacific Business Innovation and Technology Management, 2*(3), 93–110.

Spielman, D. J., & Pandya-Lorch, R. (2009). *Millions Fed: Proven Successes in Agricultural Development*. Washington, DC, USA: International Food Policy Research Institute.

Steel, W. F., & Snodgrass, D. (2008). *Raising Productivity and Reducing Risks of Household Enterprises: Diagnostic Methodology Framework. World Bank Africa Region Analysis on the Informal Economy*. Washington, DC: World Bank.

Stubbs, W., & Cocklin, C. (2007). Teaching Sustainability to Business Students: Shifting Mindsets. *International Journal of Sustainability in Higher Education (IJSHE), 9*(3), 206–221.

Stubbs, W., & Cocklin, C. (2008). Teaching sustainability to business students: Shifting mindsets. *International Journal of Sustainability in Higher Education, 9*(3), 206–221.

Sudarshan, R. M. (2007). *Social Protection, the Informal Economy and Globalization*. Paper Prepared for the 'Globalisation and the State' Project of the National Foundation for India, New Delhi.

Sumner, A. (2012). Where Do the Poor Live? *World Development, 40*(5), 865–877.

Talberth, J., Cobb, C., & Slattery, N., (2006). The Genuine Progress Indicator: Executive Summary. Redefining Progress.

United Nations. (1987). Report of the World Commission on Environment and Development, General Assembly Resolution 42/187, 11 December 1987.

UNICEF. (2018). South Asia Leaders Meet in Nepal to Radically Improve Education in the Region. Retrieved from https://www.unicef.org/rosa/press-releases/south-asia-leaders-meet-nepal-radically-improve-education-region.

Weitz, N., Nilsson, M., & Davis, M. (2014). Anexus Approach to the Post 2015 Agenda: Formulating Integrated Water, Energy, and Food SDGs. *SAIS Review, 2*, 37–50.

WHO. (2018). *Global Health Observatory Data*. Geneva, CH: NCD Mortality and Morbidity. World Health Organisation.

World Bank. (2007). *Sri Lanka Poverty Assessment. Engender in Growth with Equity: Opportunities and Challenges*. Washington, DC: World Bank.

31

Popular Participation in Environmental Governance in Non-Western Societies: Procedure and Application

A. K. M. Mahmudul Haque

Introduction

Environment basically refers to the things and beings that make our surroundings and affect our ability to live on earth. Popular participation means general participation by the people in the society. In the actual sense of the term, mass participation in the environmental governance process is not welcome in the non-western societies in consequence of which non-western societies have become contaminated and polluted by the second-hand participants.[1] In most countries, the demand for environmental conservation percolates upward, from ordinary people to decision-makers. The people form environmental concern groups like Earth System Governance Project, International Union for Conservation of Nature (IUCN), World Nature Organization (WNO), Centre for

[1] Those who are ill-motivated over the world are called second-hand participants.

A. K. M. Mahmudul Haque (✉)
Department of Political Science, University of Rajshahi, Rajshahi, Bangladesh
e-mail: akmmahmudul@ru.ac.bd

© The Author(s) 2020

Md. N. Momen et al. (eds.), *Building Sustainable Communities*,
https://doi.org/10.1007/978-981-15-2393-9_31

Science and Environment (CSE), even the Green Party, and put the environment in the headlines and make it impossible for the State or major opinion-makers to ignore or forget the Earth.

In America, in the 1960s, it was the "highway revolts" (Mohl 2004) against building of new highways that helped to formulate the Environmental Protection Act 1970, creating the Environmental Protection Administration. It was ordinary people, sharing information, holding meetings and demonstrations, lobbying politicians and so on whose pressure produced green stickers,[2] saved the whales[3] and the baby fur seals,[4] got genetically modified food labeling requirements or bans (Bello 2013). International environmental NGO Greenpeace almost started a war between France and New Zealand over nuclear fallout from bomb tests in the Pacific Ocean in 1985 (Wilshire 2015).

However, in Bangladesh, environmental protection has been a "top-down" affair (Islam 2000). Donor countries pushed Bangladesh to act and the Government acted, mainly to keep the donations coming. Some say that the real purpose was merely to create an image of environmental protection and not to do anything for the environment. There was no groundswell of support or public demand. Environmental policy is not an issue in Bangladeshi elections, except sometimes at the local level. The majority of the people are apathetic, uninformed or even hostile to environmental protection, especially when it interferes with popular infrastructure construction (Mahmudul Haque 2015).

This chapter will first explore why Bangladeshis have less interest in protecting the environment than Europeans, Americans and Australians. Then the chapter will consider how an effective environmental protection policy can be constructed without such popular support as exists in other countries, in ways more consistent with Asian culture.

---

[2] *See e.g.* International Standards Organisation, ISO 14001 and ISO 14000 series.

[3] International Whaling Commission Schedule, para (e) (1982).

[4] At least from having their fur sold in the European Union, Commission Implementing Regulation (EU) 2015/1850.

Concept of Environmental Governance

The environment is what we have around us. The environment is composed of the natural, the artificial and the supernatural. The environment is a system. People interacting with the environment constitute a social system. A social system must be governed purposively or it will govern itself, without purpose and without regard to its effects on other systems in society. A social system may, by luck, be effective for those involved in it and not produce social problems, without governance. However, experience shows that the opposite is the more likely result. Everything we know about the environment today, especially in Bangladesh, shows that we have not been that lucky. Environmental governance is the purposeful government of the environment, for its own protection and for the protection of social systems with which the environment interacts.

The words "environmental governance" have become poignant, pithy and terse. The term has an intuitive appeal and most people think that they understand it when they hear it. Yet, when they begin to discuss it, most people arrive at the conclusion that environmental governance is a difficult concept which they barely understand at all. Environmental governance is a concept that coexists, and interacts, with the core concept of governance. If political science is the theory of governance, environmental governance is "applied governance". So the relationship of environmental governance and governance in general is complex and central to both issues. Without governance, environmental governance has no content. Without environmental governance, governance, along with the governed, will eventually cease to exist, along with the rest of the world.

Governance

Governance generally involves political, economic, theocratic and social issues: no less does environmental governance. The concept of "environmental governance" is a consolidation of the learning from the study of government and the learning from the study of the environment. In fact, environmental governance has appeared from recent pedantic endeavors to standardize the ambience governance rapport. Such endeavors are elaborately based on the presumption that the presence or absence of democratization or sound governance consequently affects the environment (Mugabe and Tumushabe 2011).

According to Mugabe, the difficulty in conceptualizing "environmental governance" lies precisely in the complexity of defining "governance" in general (Mugabe and Tumushabe 2011). Thus, before even trying to discuss environmental governance, we need to review what we know about governance in general. Then perhaps we can arrive at sense rather than continuing nonsense and confusion.

Governance is related to maintenance: of a thing living or non-living. Governance is the control of human action. Policies are formed by policymakers to control the actions of people. In the economic, social, environmental and political disciplines, the term "governance" has multiple interpretations (Harman 2008). Rhodes describes this term with six segregated uses: as marginal state, as consolidated governance, as the innovative public management, as accountable governance, as social cybernetic process and as organizing rapport-building procedure (Rhodes 1997). The World Bank defined governance as the exercise of political power to manage a nation's affairs (Hardallu 2011). The Institute for Global Environmental Studies explained governance as a complex set of values, norms, processes and institutions by which society manages its development and resolves conflicts (Institute for Global Environmental Strategies 2010). The United Nations Development Program (UNDP) explained governance as interrelated with sustainable human development. According to it, governance is:

> "the exercise of economic, political and administrative authority to manage a country's affairs at all levels. It comprises the mechanisms, process and institutions, through which citizens and groups articulate their interests, exercise their legal rights, meet their obligations and mediate their differences". (UNDP 1997)

UNDP denotes that the aim of governance should be to develop capacities promoting uplift, giving priority to the pauper, sustaining the environment and creating chances for employment for survival (UNDP 2011).

Similarly, Louis O. Dorvilier said that governance implements itself through the application of authority in the socio-economic and political domain as a way to actualize the matters of the country. It includes the institutions, procedures and apparatus through which citizens and groups promote their common factors, orientation, interests and rights: and also through which they meet their responsibilities and reconcile their differences (Dorvilier 2001).

United Nations Educational, Scientific and Cultural Organization in its *Capacity Building for Governance Monograph* opines that governance connotes more than government referring to a political process, encompassing the whole society and delivering the making of massive citizen agile uplift to the social contract binding them together duly (Quoted in Parvin 2007). According to United Nations Economic and Social Commission for Asia and the Pacific, governance refers to the process of decision-making and the method or device by which decisions are implemented or not implemented (United Nations Economic and Social Commission for Asia and the Pacific 2013).

According to Mugabe and Tumushabe, governance is beyond government. Governance is thought as an authority that enacts laws for an authority; government is the authority and its political administration and decision-making procedure. In the process of governance of a state, civil society and private sector must be planned to contribute to tolerate human amelioration. Different types of associations grouped in civil society also lead the static world to the proper implementation of the environmental governance. It ensures the orchestration of the environmental melioration with a view to enhancing the proper habitation of the mob in the planned environment. It may well be said that environmental governance is connoted by the enactment of laws for the welfare of the people and implemented by the authority concerned.

The institutions of governance in a state, civil society and private sector must be planned to contribute to tolerate human amelioration. Governance can effectuate all these things actualizing socio-political and economic factors for poverty minimization, job assurance, environmental protection and empowerment of women (Fig. 31.1).

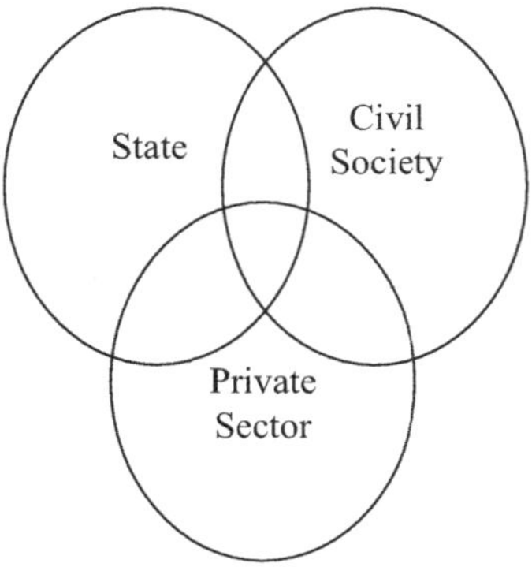

Governance includes the state, but transcends it by taking in the private sector and civil society. All three are critical for sustaining human development. The state creates a conducive political and legal environment. The private sector generates jobs and income, and civil society facilitates political and social interaction mobilizing groups to participate in economic, social and political activities. As each has weaknesses and strengths, a major objective of our support for good governance is to promote constructive interaction among all three.

Fig. 31.1 Good Governance reaches beyond the State. (Source: UNDP 1997 quoted in Lima 2002)

Civil society lying betwixt the individual and the state consists of individuals and groups working socio-politically and economically controlled by efficacious laws. The political system (the state) can actualize many things for the preservation of the environment, maintaining micro and macro-economic condition, ensuring public health and safety for all, standardizing resources for better livelihood. All these can be authorized by the state for the benefit of infrastructural development, uplift of law and order, security and socio-political and economic upkeep and uplift (UNDP 1997 quoted in Lima 2002).

As a consequence of post-Cold War governance crises, the United Nations Commission on Global Governance defined the term governance as:

> "the sum of the many ways through which individuals and institutions, public and private, manage their common affairs. It is a continuing process through which conflicting or diverse interests may be accommodated and cooperative action may be taken". (United Nations Commission on Global Governance 1995)

Among the actors involved in governance, the Government is one of them. Other actors lying in governance vary depending on the level of governance. It may well be that in the rural areas where intellectual landlords are included. Different institutions like NGOs, peasant associations, cooperatives, research institutes, religious leaders, defense institutes, political parties and other organizations assist in governance, but it is more complex in the urban areas.

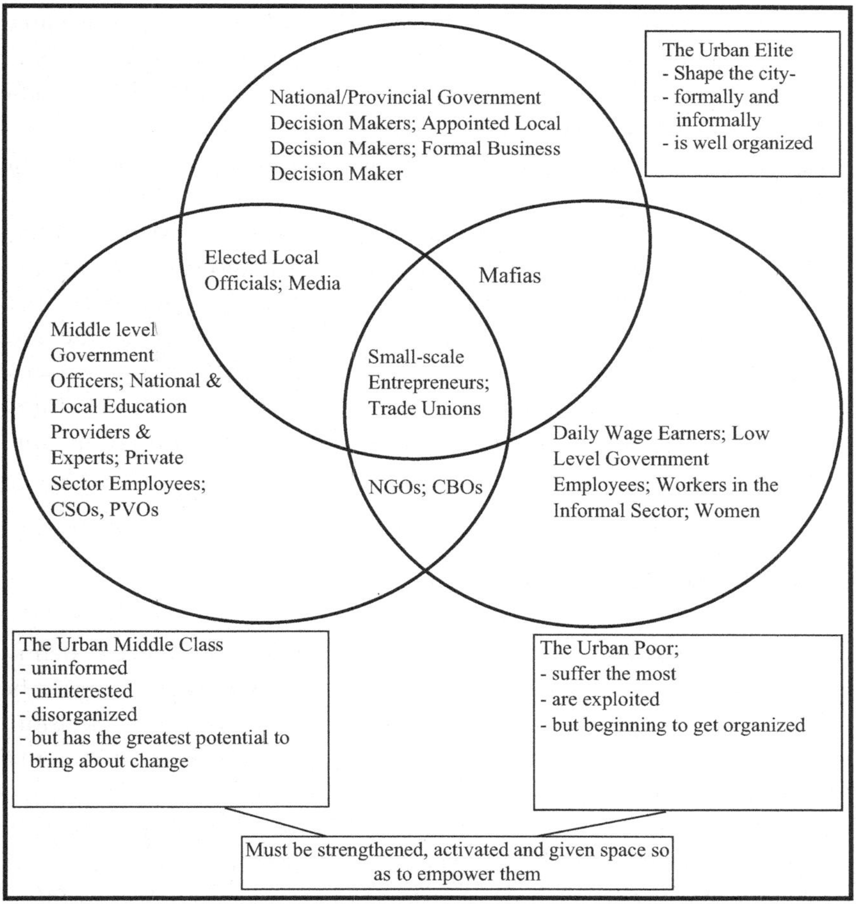

Fig. 31.2 Interconnection among Urban Actors in Governance. (Source: United Nations Commission on Global Governance 1995)

Figure 31.2 delineates interrelationship betwixt actors convolved in urban governance. In addition to the above actors, media, lobbyists, international donors, multi-national corporations, and so on at the national level may influence the decision-making procedure. Without the government and the military, all the actors are a part of civil society. In some countries like Bangladesh, crime syndicates also influence governmental mechanism.

At the same time, government structures are the devices through which decisions are made and actualized. Informal decision-making procedures at the national level exist as well. The organized crime syndicate influences decision-making in urban areas. Locally powerful families in some rural areas influence decision-making procedure. This type of decision leads to corruption (United Nations Commission on Global Governance 1995).

The above connotations indicate that governance, as a concept, recognizes that power lies inside and outside the formal authority and institutions of the state. In many cases, governance combines both central and local government, the private sector and civil society. Governance includes the institutions, processes and instruments through which a society collectively takes decisions and acts upon them. It recognizes that decisions are made taking the common interests of the citizens and groups into account.

Environmental Governance

From the discussion of governance, we can arrive at a more precise definition of environmental governance: Environmental governance means the interaction of social systems to take, and act upon, collective decisions which concern the interaction of the social and natural environments. Environmental governance is an important instrument, governance systems influence environmental outcomes. The term "environmental governance" is used to describe how decisions about the environment are made and who makes such decisions. It maintains structural and informal institutional arrangements for wealth and environmental decision-making process. It expands beyond the state to convolve the private sector and civil society basically. Ergo, it involves a level of institutions and other organizations together with traditions and taboos; they influence the way of power exercise by all these performances and ensure the accountability of governance in various related aspects of the state.

Asian Development Bank (ADB) defines "environmental governance" in the following way:

> "Environmental governance is the manner in which people exercise authority over nature. It concerns the actors—within each level of government, between elected and appointed officials, and among 'traditional' private and non-governmental bodies—and the power that they exercise to make decisions about the disposition of natural resources and benefits that flow from the environment". (ADB 2000)

According to Graham, Amos and Plumptre, governance, in the context of the environment, encompasses the nexus and interplay among state and other related entities. In this connection, it maintains all the procedures of the environment concerned (Graham et al. 2003). Environmental governance thus concerns legal and policy decisions to manage environmental issues; compliance with those policies in development management; and the participation of common people who are directly affected by the outcome of such decisions (Olowu 2007).

Mugabe and Tumushabe observed that the literature related to environmental issues failed to define the concept of environmental governance. According to them, environmental governance is a process of democratization in which everybody has a share. Participation of many actors in the static system can develop the concept of governance of environment. The political system in a broad sense or the static system at the global level and delegation of power managing and preserving the environmental governance must be combined cohesively and inherently. The juxtaposition of governance and the environment has a number of features (Mugabe and Tumushabe 2011).

- Firstly, the democratic process and the civil society are held responsible for environmental governance according to their domains;

- Secondly, the nexus between governance and environment prescribes that environmental management convolves political matters and procedures;
- Thirdly, the linkage denotes that environmental preservation is reciprocal work shared betwixt government and the civil society; and
- Last of all, environmental governance throws light on different yet connected roles of State, civil sector and participation of the people, democratization, civil society and the like.

An effective policy framework is important in developing an integrated environmental governance system. The frameworks will permit and inspire participation of all stakeholders: government, business and public. In actualizing these, they ensure sufficient and accountable spreading of data for all groups because of interaction.

The Environmental Governance Cycle (Fig. 31.3) recognizes the necessity for a gradual and continuous assessment for pollution and reduction initiatives. Through initiating this cycle, the central and local government will enable to display the necessity for heavier local steps according to demands.

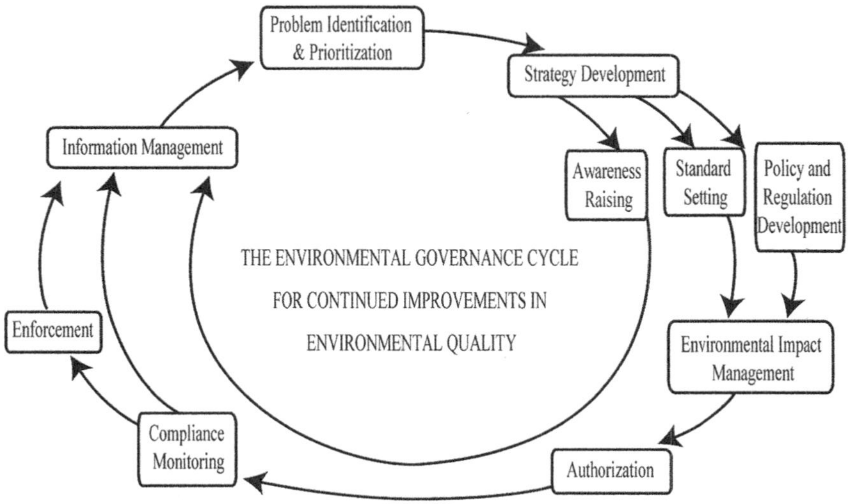

Fig. 31.3 The environmental governance cycle. (Source: Department of Environment and Tourism of South Africa 2007)

Part I

Environmental Policy Without Environmental Consciousness: The Bengali Concept of the Environment

Bengalis have a concept of the environment much like that of other tropical, non-western people, like Africans and South Americans, who, not so long ago, lived in nature in a jungle setting (Swan and Conrad 2014). They never looked after nature: nature looked after them. They have great faith in the power and perfection of nature and cannot believe that anything they do could damage it.

Western people think of themselves as custodians of the Earth and responsible for maintaining it.[5] Most Bangladeshis are Muslims and the Surat al-Hajj of Al-Quran (22:10) says:

> Do you not see that Allah has made subject to you whatever is in the heavens and whatever is in the earth and amply bestowed upon you His favors, [both] apparent and unapparent?

Of course, if you believe that nature is created "subject to you", you do not care so much about what happens to nature. You use it and enjoy, without any responsibility or care for it.

There are other cultural aspects to the difference in thinking too. Western people are teleological (Popkin 2013): they always ask "what is the purpose?" and believe that they can change the future from action today. Asians conceive of time as the ebb and flow of the sea, believe that they cannot change anything and consider that all they can do is to adjust as everything changes around them (Ames 2016). Of course, if you believe that you can change the future, you want to act today to save the Earth for your great grandchildren. Yet, if you believe that you can do nothing, the concept of trying to do something is quite silly.

[5] One of the most clear examples was the British Government's White Paper, "This Common Inheritance" in 1990, HMSO, Cmnd 1200, "We have a moral duty to look after our planet and to hand it on in good order to future generations".

The role of science is also an important source of the differences in thinking between Bengalis and western people about nature. Science is not a Bengali concept. Science is thoroughly western in its emphasis on the objective, on logical inference and on its attempt to understand processes and predict their outcomes (Siddique 2009). That is not to say that Bengalis do not study science but it is difficult for them: someone who gets "golden A+" (all subjects A+) in the Science stream in the Bangladesh Higher School Certificate is thought of as extremely intelligent, as few can do it. Indeed, a few years ago, almost all the Bangladesh Bank Governors (a highly prestigious position) were Physics graduates, not economists.

The point here is not about who gets better marks in Science but to emphasize that western people instinctively believe and trust scientists but Bengalis do not. If scientists say that the world is being destroyed by our actions, westerners feel that they must stop it urgently. Bengalis are unlikely to understand the scientists, less likely to believe them and may find their predictions of doom rather humorous. The scientists are not like them: they are different.

A Different Kinds of Politics

Western politics is fundamentally interest group politics (Cigler et al. 2015). Parties appeal to different interest groups by offering them the policies they want. In Government, the interest groups jostle for influence and negotiate among themselves to arrive at policy decisions. The pluralist model (Dahl 1961) predominates and no one can make a decision by himself.

Bangladesh politics is not like that. Policies are handed down by a charismatic Party leader. When there is a coalition Government, the policies may be the result of negotiations between the Prime Minister and her coalition partners. The leader's Party members follow blindly to support the policies, with their lives if necessary and the opposition Party members oppose root and branch. There is no negotiation and interest groups do not take a leadership role to get their preferred policy adopted. Rather, the Parties try to infiltrate, intimidate and puppetize the civil society organizations to get them to support their Party's policy choices.

Therefore, there is no obvious aperture for civil society organizations to use in influencing policy decisions in the direction of protecting the environment. There has never been a Green Party in Bangladesh, although there is now one in almost every western country. The one successful civil society organization on environmental issues, BELA (Bangladesh Environmental Lawyers Association), has not really tried to use the political process to get environmentally friendly policies. They use the courts to stop environmentally unfriendly actions and have been very successful in getting High Court injunctions to stop various acts of central and local government.

Foibles to Environmental Protection

The project is simply poignant, terse and pithy. But the matter is that our popular participation is not adequately conscious about the phenomenon. This unconsciousness leads to the incompletion of environmental governance. Moreover, unconsciousness prevents all from doing the work properly. Lack of planning is a great threat to environmental governance in non-western societies. Western societies actualize any work on the basis of scientific and long-term planning while non-western societies are averse to it. In consequence of it, in the air of the non-western societies, there is much plum bum/lead (PB). At the same time, we find much CO_2 and other destructive gases that suffocate our breathing system. Our water has been contaminated by industrial wastage. The popular participation in conservation and percolation seems to be nonchalant. Their involvement is also very imprudent. In consequence of it, we do not see perfect environmental governance in non-western societies. Popular participation is habituated to filthy environment for which they keep their shelter, atmosphere, even a thing poor and dirty. Due to habit, we see occupational hazard in non-western societies that creates severe problems of basic needs and rights in the region. Popular participation cannot realize occupational hazard unless and until they suffer from tuberculosis, dermatological problem, reproductive problem, respiratory problem and the like. The most crucial problem is environmental hazard. That is the resultant of nonchalance of popular participation in environmental governance.

In the process of application, popular participation is found inactive, irksome and derailed. The application and procedure of environmental governance is not properly enforced due to the ignorance of the popular sovereignty and the most important thing is that they are not willing to preserve their environment. We have delineated the process of environmental governance in non-western societies. But it depends on the eagerness and literacy of the popular participation. It does not have any long-term planning regarding the conservation of environment. We talk about coordination in matters of cleaning of the environment. But we do not visualize it. Indian Prime Minister is of the opinion that *swachhata* (cleanliness) is second to the development of our country. It is because it leads us to good health, sound environment, pollution-free society and good nourishment that result in our combined flourishment. People should believe in the protection of environment by being clean and cleaning up.[6]

Our popular participation believes in naturalism as a result of which they think that what is looted cannot be blotted. So, they do not try to change themselves in matters of environmental protection. River erosion is one of the problems of environment in this region. Hill side is a great threat to environmental conservation. If our popular participation wishes to clean up the hilly region, it cannot be possible easily. Here, many open fields are covered recklessly with the filthy and dirty things. Policy-making procedure is not suitable for our popular participation regarding environmental governance. Since non-western political system is not efficacious, political parties fail at making effective environmental policies. Only stable political parties can bring good policies regarding environmental governance. But it is a matter of great regret that we do not have stability in our political system that results in bizarre policies regarding it. Finance is a great problem to the conservation of environment. In the policy or structure of anything, finance is allocated equally. But owners of the structures do not pick finance up in their policy and they also do not

[6]Television Interview with Narendra Damodardas Modi by Sony TV, 19 September 2017.

finance environmental governance as a result of which environment becomes corrupted and bare. Industrial hazard comes out of their structure which pollutes the environment severely and appallingly. Non-western societies have overpopulation which cannot favor the concept truly. As a result, the environmental governance remains haggard and hazy. So we must remove and eradicate all these foibles from non-western societies in matters of environmental conservation and protection. It has been statistically proven that non-western societies do not have inspiration and aspiration for a sound environment for their sound habitation. They believe in industrial development without proper consideration of environment issues for which the most important phenomenon environmental governance is not properly done in this region. Since they believe in short-term policy, they do not accelerate the development of the environment from which they can get multifarious benefit. According to Huntington, we see the clash of civilization which includes flow of water, excavation, prevention, transformation and good coordination. These are not properly done due to the conflict and clashes, resulting in the uncleanliness of the society. Here we also mention illiteracy of the popular sovereignty together with policing a state instead of welfare state. He also mentions about group work for the development of the global development (Huntington 1990). Overall Huntington talks about lack of spontaneity among the non-western society regarding their development and the mise-en-scène. Last of all, we can say that there are many weaknesses and limitations in the conservation, we have to overcome all these for better environment through popular participation. It is deeply believed that environmental pollution is great treat to our body, mind and soul which can be easily eradicated and eliminated by the prescription we have jotted down here.

Some Environmental Movements

Yet things might be changing. BAPA (*Bangladesh Poribesh Andolon*) meaning "Bangladesh Environment Movement"), is a Bangladesh environmental voluntary society (which stresses that it is not an NGO and receives no contributions) that has achieved some major successes on environmental policy issues. They proceed through seminars, conferences, deputations, press conferences, publications and public rallies. On their website, they claim among their policy successes the banning of plastic bags, leaded petrol and two-stroke engine vehicles. The views and opinions of civil society and NGOs are now given more importance by the Government (Mollah 2007).

The formulation of NEMAP (National Environmental Management Action Plan), in 1995, was the first time that the Bangladesh Government actively involved the NGOs and civil society in policy development. The Government "proactively initiated a consultation process, in association with NGOs, journalists, academics, and other segments of civil society" (Ahmed and Roy 2015)".

The only NGO directly related to Environmental Impact Assessments (EIA) in Bangladesh is the National EIA Association of Bangladesh (NEAB) set up in late 1997, providing a platform for the EIA planners and others involved. NEAB has been working to ensure consciousness of EIA in all connected sectors, assisting in the development of EIA and some other related things in Bangladesh. A Memorandum of comprehension signed between NEAB and International Union for Conservation of Nature (IUCN), Bangladesh Country Office, in 1999 is working to multiply environment-assessment ability in the country. Bangladesh Environment Network is based in New York City and is composed of Bangladeshi emigrants. They have been mainly assisting in communication about the environment since 1998 and work mostly through BAPA.

Many electronic and print media have shown and written about environmental development and awareness. All these media have focused on the consciousness regarding environmental uplift, especially on June 5 (Environment Day).

The civil societies, democrats and NGO, have linkage with the national and international organization for strengthening environmental programs. They have forwarded and extended training, workshop, seminar, and so on to promote environmental issues. In this respect, different sister concerns of UNO played a crucial role.

Part II

A Non-Western Strategy for Environmental Protection

There are some truths which transcend culture and one of them is that the environment must be protected. People in some cultures may have less information and less concern about the environment, but that does not change the fact that their environment is endangered now too. In such cases, dying may be culturally sensitive but it can never be an acceptable result: at least because what happens to one culture on this little globe affects other cultures. Slash and burn agriculture was very much a part of Indonesian culture but the release of huge amounts of greenhouse gases from burning forests created a big hole in the ozone layer (Augustyn 2007), from which we have still not recovered.

So an alternative strategy must be found that must be culturally sensitive as well. The alternative strategy cannot depend on the sort of popular movements that have powered the environmental movements in the west but must obtain its legitimacy from the sorts of places that non-western leadership obtains it.

Now let us revise for a moment where non-western leadership gets its legitimacy:

- **Ascribed Status**—age, gender, family background, and so on.
- **Wealth**—in Asia, wealthy families are liked instinctively and often without reason, sometimes because they bribe everyone generously.
- **Association**—people who are associated with persons of high status and influence often obtain some vicarious power and influence from that association.
- **Required Status**—as in the Bangladesh local government legislation in 2009 (Local Government (Union Parishads) Act 2009 and Local Government (City Corporations) Act 2009), when the leaders require people to participate in a process, even when they would not have dared try, participation becomes acceptable and possible.

Using these basic concepts of what gives an Asian the right to participate in a process and to monitor the actions of his social superiors, we can begin to develop the idea of a participatory process of environmental protection policy (Bulkeley and Mol 2003) not based on environmental consciousness. Participants should be mostly older, male[7] and from famous or wealthy families. They should be associated with people of high status. The legislation should require the participation of other persons who do not meet these requirements, to give them their missing status. Inclusion of other persons without the required social status and without legislative mandates might cause the members to lack popular support and respect, so that they are ignored or laughed at in practice.

The picture emerging seems to be one of a kind of Environmental Congress: a group of wealthy, high-status people who are required to act on environmental issues. We can imagine the kinds of issues they would be suitable to act on, not being technical experts:

- auditing expenditures on environmental protection;
- hearing complaints against the enforcement or non-enforcement of environmental policy;
- developing new ideas for environmental policies;
- setting out goals and objectives for environmental policies;
- setting up meetings for community participation in developing environmental policies.

[7] It is true that there are women leaders in Asia: in Bangladesh, two women have been Prime Ministers for almost the whole time from 1990 to the present. Yet those women are only there because one is the daughter of the founder of the nation and one is the wife of a national hero who was also a very popular President: both assassinated. This is true in other nations of Asia too: Prime Minister Sirimavo Bandaranaike of Sri Lanka (daughter of a former Prime Minister); Prime Minister Aung San Suu Kyi of Myanmar (daughter of Gen Aung San, who liberated the nation from the Japanese in World War II); Prime Minister Indira Gandhi of India (daughter of Prime Minister Jawaharlal Nehru); Prime Minister Park Choong Hae of South Korea (daughter of President Park Choong Hee); Prime Minister Shinawatra of Thailand (sister of Prime Minister Thaksin Shinawatra). Yes, women can be leaders in Asia: but they need to draw their legitimacy from some popular male member of their family who was a politician or national hero. Similarly, women can be a part of local environmental councils but preferably the daughter or sister or wife of the popular local male MP or male local council Chairman.

Such a local Environmental Congress could inject communities with some interest and information about the environment. They could start discussions on needs and priorities. They could hold officials responsible (depending on the powers given to them in the legislation establishing them) or at least ask them questions about what they are doing and why. They could make plans and lay out goals and objectives for improving the environment in their local areas.

Obviously the establishment of such Environmental Congresses would have to be a "top-down" action. No one is rushing to create them. It is their creation in legislation that would give them the legitimacy they need to accomplish their goals. The legislation should set out areas of their authority and give a mandate to investigate official actions, make plans, organize popular meetings, audit expenditures, adjudicate complaints and to submit policy proposals directly to the Minister.

Britain started to involve individuals in environmental policy, but this was participation in implementation, not in policy development (Eden 1996). Mainly, these were campaigns to educate the public and ask them to carry out policy decisions like making their homes more energy efficient or reduce carbon emissions (Eden 1996).

Decisive Statement

While some environmental organizations have been created "from the bottom up" in recent years, there has not been a fundamental change in the ways most Bangladeshis think about the environment. There is still no widespread demand for environmental protection.

This situation could continue indefinitely; however, there are real consequences to doing so. In the west, strong civil society movements for environmental protection pressurize the State to show action and results. When this is absent, things can tend to settle into a "no noise" situation, where nothing is done and no one complains about it. In politics, "the squeaky wheels get the grease".

There have been many reports that the environmental laws and policies in Bangladesh and other developing countries are mainly for show and do not lead to any results (Faure 1995). This is the direct result of the absence of political and public pressure in favor of these issues.

This is not only a phenomenon of Bangladesh. This is a common reaction of ordinary people to environmental protection in many parts of the developing world. At one level, hungry people care about roads more than trees: yet it runs deeper. As discussed at the beginning of this chapter, the concept of the role of people in the environment is fundamentally, philosophically and culturally different in non-western societies. What has happened in Bangladesh (or, more accurately, the fact that little has happened) was therefore totally predictable and understandable. Leaving aside the broader issues about non-western societies, trying to copy western law to become "modern" is not the right way to do for the environment.

In Brazil, Jose Puppim de Oliveira found that the government authorities sometimes fail to materialize environmental protection policies generally because of their lack of political system. This happens due to inadequate wealth, problem of institutional capacity and the like. All these lead to weaken the significance of cooperation at the local sphere and phase (Oliveira 2009). The story sounds the same as Bangladesh. Gamman found a similar picture in his study of the West Indian islands, St Lucia, Barbados and St Kitts. He ascribed this to four major elements: national politics, behavior in the donor agency, the culture of decision making, and economic necessity. So he was one of the first to recognize the role of political culture in the failure of environmental protection in developing countries (Gamman 1990).

Bell and Russell begin their article by saying (Bell and Russell 2002):

> Most developing countries have long since established laws and formal governmental structures to address their serious environmental problems, but few have been successful in alleviating those problems.

Now we are arguing that we need to look more deeply, to culture, to understand why. That article said that the problem was caused by international finance organizations pushing market-based solutions on people who lacked the infrastructure and what we might call the "market-based culture" to carry them out. Yet it is so pervasive a problem that we believe it runs deeper than that.

If we care about the environment, then we need to care about what happens to it in the developing world, where many of the problems are. So far, there has been a lot of action in the western countries and not that much real change outside them. If this goes on, the environment will not be improved.

This chapter has put the argument that there is a fundamental difference in world view between the western and non-western societies about the environment. When we accept and understand it, we can understand why the global summits failed and why environmental protection, western-style, is not working in the non-western societies. So, we need to start thinking about ways in which we can be more effective in protecting the environment in the non-western world. This chapter has put forward one idea for trying to do that: the centrally sanctioned, locally based "Environmental Congress". It is certainly not the only possible way, yet let us begin to think in such a culturally sensitive way of new alternatives to get better results in protecting the environment in non-western societies. Hence, we can show two figures for the protection of environment where hierarchy will be a catalyst of popular participation of environmental governance in non-western societies in matters of procedure and application.

In the Fig. 31.4, we see that grass root level workers will maintain the environment through a process. The process will have different stages. At the bottom, workers will be engaged directly in environmental activities such as pollution control, waste management, forestation, sanitation and the like. It will be monitored by a coordination of the Ward Councilor. Ward Councilor will be supervised by the mayoral administration. It will be liable to the district level administration which includes District Council Chairman, Deputy Commissioner, law enforcing agencies and big guns[8] of the society. This administration will be checked and reviewed by the head of the government. Simultaneously, it will be justified by the head of the state. According to Gramsian theory, this hierarchy is followed and materialized all over the world. If we follow this hierarchy for the protection of environment, it will be the big success in course of time.

[8] Those who are included in civil society and sophisticated are known as big guns.

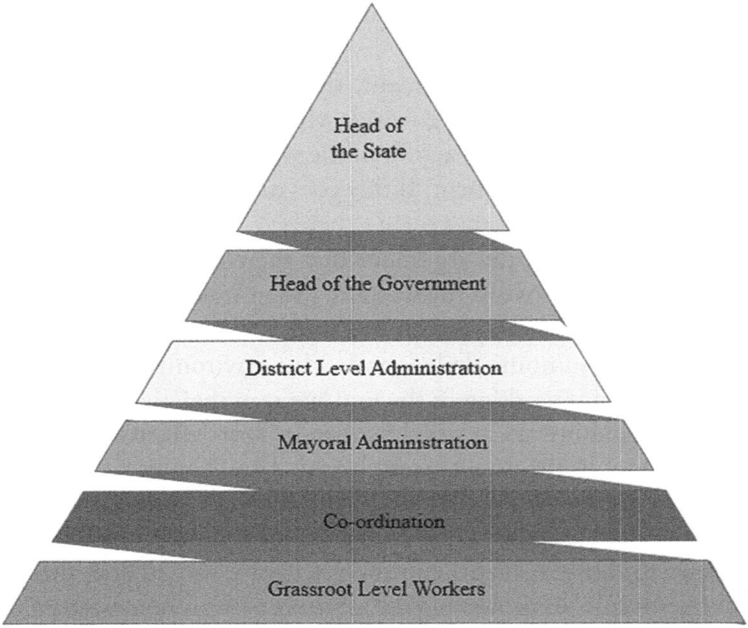

Fig. 31.4 Town/city congress to conservation. (Source: Developed by the researcher)

According to Fig. 31.5 as is stated in the Gramsian theory (book: Selection from the Prison Notebooks) we can say that rural level workers will do their environmental duties to the right direction like burning and removing dirty things, watering away the filthy things, covering dirt with soil and the like. They can also excavate different channels to let the filthy things go away. The whole work will be monitored by the Union Council Chairman. He/she will be liable and answerable to local government as per the rules of Local Government Division of the Ministry of Local Government and Rural Development (LGRD) and Cooperatives. It will be justified by the district level administration, which consists of different government and non-government organizations. Obviously, there will be coordination between urban government and rural government by ministerial level in matters of popular participation for environmental governance in non-western societies. At the end of the journey, the

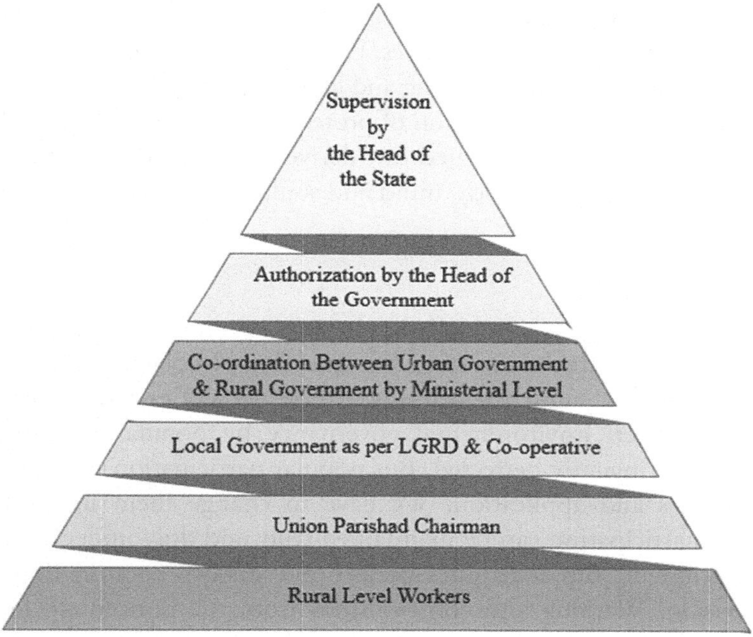

Fig. 31.5 Village/rural congress to conservation. (Source: Developed by the Researcher)

environmental protection and conservation will be authorized and reviewed by the head of the state government together with the head of the state.

World Health Organization (WHO) is of the opinion that health is the summation of proper physique, sound mental condition and environmental protection. First time, the WHO did not include environmental protection in sound health. Afterward, understanding the importance of environmental condition, it included it in overall sound health. There is no denying the fact that environmental protection must follow the code of conduct which has been delineated in the article entitled Figs. 31.1 and 31.2. Environmental Congress will play a vital role to clean up our society and to enjoy sound health. The process depicted by Gramsi and A. G. Frunk through satellite and metropolitan processes, the environment may be governed in toto and to the hilt. It is ineluctable that participation of popular sovereignty only can ensure environmental governance in non-western societies. Effectivity, enforcement, consciousness, motivation, dream, planning and long term-process must be treated as the catalyst for the application of protection of the environment from which all the people concerned can derive the best benefit which will ensure the perfection in body, mind and soul.

Conclusions

At the end of our discussion it can be stated that percolation process must be actualized truly and scientifically. If not, we will endanger ourselves sooner or later. Environmental governance by popular participation should be stringently perfected. But popular participation is not true to this process and application. We have to change them up properly. Popular participation can be maintained truly and duly under any form of government. But it requires perfect and smooth coordination with supervision. Working and consciousness from the bottom up level is essential. The success of this matter depends on overall cooperation and sense of hygiene. Since hygiene keeps us sound, we have to work unitedly in this connection. Environmental education is a catalyst in matters of popular participation in the protection of environment. Sincere working with technical knowledge is sine-qua-non for this phenomenon. The environmental Congress must finance the project properly. Along with it, monitoring, supervision, maintenance and all other necessary things must be accomplished. If the recommendation of the environmental Congress is not followed properly, the whole project will go in the darkness. Seminars, symposiums, and elaborate discussion must be continued to get good results. To fructify the project, all must be serious about it. If not, the concept with the recommendations will not be translated into reality.

References

ADB. (2000). *Environment Outlook 2001*. Manila: Asian Development Bank.

Ahmed, R., & Roy, C. (2015). An Integrated Approach To Environmental Management in Bangladesh. In R. Werlen (Ed.), *Global Sustainability, Cultural Perspectives and Challenges for Trans Disciplinary Integrated Research*. Cham: Springer International Publishing.

Al Hardallu, A. (2011). *Environmental Governance: Rio+10 Review Report*. The Environmentalists Society, EDGE for Consultancy and Research.

Ames, R. (2016). *East Asian Philosophy*. Routledge Encyclopedia of Philosophy, Informa UK Ltd.

Augustyn, H. (2007, July/August). A Burning Issue. *World Watch, 20*(4).

Bell, R. G., & Russell, C. (2002, Spring). *Environmental Policy for Developing Countries, 18*(3), Issues in Science and Technology.

Bello, W. (2013, October 29). 26 Countries Ban GMOs. *The Nation*.

Bulkeley, H., & Mol, P. J. (2003). Participation and Environmental Governance. *Environmental Issues, 2*, 143–154.

Cigler, A., Loomis, B., & Nownes, A. (2015). *Interest Group Politics*. Washington D.C.: CQ Press

Dahl, R. (1961). *Who Governs?* New Haven: Yale University Press.

Dorvilier, L. O. (2001). Challenges for Development and Social Stability Least Developed Countries (LDCs) Social Alert. Governance, Peace and Stability in LDCs. 3rd United Nation Conference on LDCs. May 9, 2001. Brussels, Belgium, Paragraph 3. Retrieved from http://www.socialalert.org/doc/PMAMay9-louis(en).htm.

Eden, S. (1996). Public Participation In Environmental Policy. *Public Understanding of Science, 183*(5).

Faure, M. (1995). Enforcement Issues For Environmental Legislation In Developing Countries, UNU/INTECH Working Paper No. 19, (March 1995). "Causes of Environmental Law Failures". Retrieved from www.ecovitality.org/badlaw.htm.

Gamman, J. (1990). *Environmental Policy Implementation in Developing Countries*.

Graham, J., Amos, B., & Plumptre, T. (2003). *Governance Principles for Protected Areas in the 21st Century*. Ottawa: Institute of Governance. Retrieved from http://www.environment.gov.u/soe/2006/publications/integrative/local government/references.html.

Harman, S. J. (2008). *The relationship between Good Governance and Environmental Compliance and Enforcement*. Retrieved from http://www.inece.org/conference/7/vol2/08_Harman.pdf.

Huntington, S. P. (1990). *The Clashes of Civilization*. Texus: Oriental Publication.

Institute for Global Environmental Strategies, Environment and Governance. (2010). Retrieved from http://www.iges.net/environment/governance/gov-1a.html.
Islam, N. (2000). Protecting Bangladesh's Environment. 88 Journal of Social Studies 34, at 36 *et seq.* Who Lamented the Unmet Need for a Broad-Based Social Movement for the Environment in Bangladesh That Never Existed.
Lima, I. B. (2002). Deforestation in Amazonia and Local Environmental Governance. Masters Thesis, International University of Japan.
Mahmudul Haque, A. K. M. (2015). Environmental Governance of Urban-Local Government in Bangladesh: Policies and Practices. PhD Dissertation, Institute of Bangladesh Studies, University of Rajshahi.
Mohl, R. (2004). Stop the Road. *Journal of Urban History, 30.*
Mollah, M. A. H. (2007). The Role of Civil Society in Good Governance. *Social Science Journal, 13*(45).
Mugabe, J., & Tumushabe, G. W. (2011). Environmental Governance: Conceptual and Emerging Issues. Retrieved from http://www.acts.or.ke/GE-Chapter1.pdf.
Oliveira, J. (2009). *Implementation of Environmental Policies in Developing Countries.*
Olowu, D. (2007). Environmental Governance Challenges in Kiribati: An Agenda for Legal and Policy Responses. Law, Environment and Development. Retrieved from http://www.Leadjournal.org/content/07259.pdf.
Parvin, S. (2007). Good Governance: A Theoretical Analysis. *The Journal of Bangladesh Public Administration, 12.*
Popkin, R. (2013). *The Columbia History of Western Philosophy.*
Rhodes, R. A. W. (1997). *Understanding Governance: Policy Networks, Governance, Reflexivity, and Accountability.* London: Open University Press. Quoted in Yoichiro Usui, (2000). Governance Legal Order, Social Integration: Reviewing New Governance Approaches in EU Studies. http://www.nuis.ac.jp/~usui/Usui_2000_Journal_of_NUIS.htm.
Siddique, M. N. A. (2009). Values Promoted Through Secondary Science Education In Bangladesh. Paper Presented at International History, Philosophy and Science Teaching Group Biennial Conference, Notre Dame University, June 24–28, 2009.
Swan, K., & Conrad, K. (2014). The Conflict Between Chinese Cultural and Environmental Values in Wildlife Consumption. Routledge Handbook of Environment and Society in Asia and Meinert, C., (2013) ed. Nature and Environment. Indeed, Oberlin College Offers a Paper Entitled "Nature and

the Environment In East Asian Culture". Retrieved from http://obiemaps.oberlin.edu/courses /nature-and-the-environment-in-east-asian-culture/.

UNDP. (1997). *Governance for Sustainable Human Development Program*. New York: United Nations Development Program.

UNDP. (2011). United Nations Development Program, Governance for Sustainable Human Development. Retrieved from http://magnet.undp.org/policy/chapter1.htm.

United Nations Commission on Global Governance. (1995). *Our Global Neighborhood*. New York: Oxford University Press. Retrieved from http://www.unescap.org/pdd/prs/ProjectActivities/Onging/gg/governance.asp.

United Nations Economic and Social Commission for Asia and Pacific. (2013). What Is Good Governance? Retrieved from http://www.unescap.org/pdd/prs/ProjectActivities/Ongoing/gg/governance.asp.

Wilshire, K. (2015, September 6). French Spy Who Sank Greenpeace Ship Apologises for Lethal Bombing. *The Guardian*.

32

Bridging Gender Gap in Bhutan: CSOs' Response to Gender Disparity

Pema Rinzin

Introduction

Bhutan is globally known for its philosophy of gross national happiness (GNH), a holistic development paradigm that puts peoples' wellbeing at the centre of development. In 2017, Bhutan had a total population of 7,27,145 individuals, of which 52.3% are males and 47.8% females (National Statistics Bureau [NSB] 2018). Rural population makes up 62.2% of the total population (NSB 2018) who are directly or indirectly dependent on agriculture and livestock-related activities for livelihood. Until the 1960s, Bhutan remained off the global stage owing to its self-imposed isolation, and gradually opened itself to the outside world. The fact that Bhutan had never been colonized in its entire recorded history and that Bhutan remained in isolation for most of its history enabled the Bhutanese to inherit and preserve much of its traditional value systems even to this day. Socio-cultural notion about the roles and position of

P. Rinzin (✉)
Department of Sustainable Development, College of Natural Resources, Royal University of Bhutan, Punakha, Bhutan

© The Author(s) 2020
Md. N. Momen et al. (eds.), *Building Sustainable Communities*,
https://doi.org/10.1007/978-981-15-2393-9_32

men and women in the society is one such traditional value system that has been passed down for generations.

Traditionally, social relationship between men and women in Bhutan is characterized by gender equality which is distinctive in South East Asian region where discrimination against women is quite common. In addition, gender equality is ensured by the national laws and policies of Bhutan that accrue equal rights and status to both men and women in all aspects. However, disparities exist especially in decision making, labour force participation and tertiary education amongst others. Recognizing the need to address such disparities, responses from various ministries of the RGoB,[1] commissions and civil society organizations (CSO) has been rapid (Verma and Ura 2015).

Civil societies' response to gender disparity in Bhutan is diverse in scope. Amongst other aspect of women empowerment that the civil society plays a role in, this exercise explores five significant areas such as education and skills development, entrepreneurial support, micro-finance for women, leadership capacity building, and awareness and advocacy. These five aspects play a strategic role in reducing gender gaps in Bhutan.

This exercise attempts to provide an analysis of synthesized information about the role of civil society and its significance in narrowing gender disparities in Bhutan. The author does so by aggregating scattered information from different organizations' reports and augmenting the discussions and analysis of these information with literature reviews. But first, the author explores the current state of gender gaps and also the traditional notions about men and women in Bhutanese society.

Socio-Cultural Notions About Women in Bhutan

The basic value systems in Bhutan are generally believed to have been influenced by Buddhism and pre-Buddhist ancient values (Crins 2004). But, several religious practices coexist in Bhutan, and as such, the value systems differ among communities of different cultures and religion.

[1] RGoB: Royal Government of Bhutan.

Thus the perception about men and women also differ between different regions and cultures. Generally in Bhutan, women are not allowed to enter the *Goenkhang*[2] due to their menstrual cycles (National Commission for Women and Children [NCWC] 2008). On the contrary, the core teachings in Buddhism do not discriminate individuals based on gender (NCWC 2008). Moreover female symbolism is an important aspect in Tantric Buddhism which "offers a variety of ideas about the role and philosophy of the female, both in terms of the place of women within its theocracy, and the esoteric meaning of female being" (Crins 2004).

According to Gross National Happiness Commission (GNHC[3]), socio-cultural perceptions see women as less capable and less confident compared to men in terms of "governance and interactions with external agencies" (2001). This notion attributes male to public sphere and female to private sphere (NCWC 2008), because women are considered physically weak and unsuitable for leadership role like that of a *Gup*[4] which is thought to be physically demanding (GNHC 2001). Such socio-cultural notion about women persists especially in rural areas where men performs most of the physically labour-intensive and necessary tasks, such as tilling land and carrying heavy loads. Yet, there is no sharply defined male or female domain in the Bhutanese society (Choden 1999, as cited in Crins 2004).

Division of Labour

Division of labour between men and women is not sharply defined in Bhutan. In general, both men and women supplemented and shared their work. But, certain differentiation in the kind of work between men and women exists based on their physical strength (GNHC 2001). Agricultural activities such as digging, planting, weeding and harvesting

---

[2] *Goenkhang*: a room, in Buddhist temples, reserved as the place for deities.

[3] Gross National Happiness Commission (GNHC): a govt. body, previously known as Planning Commission, that ensures policies and plans are formulated and implemented in line with GNH principles.

[4] *Gup*: Head of a local govt.

were mostly considered to be women's task while land preparation, ploughing and firewood collection were men's (ibid.). In addition to agricultural works, women were also tasked with cooking, cleaning house, washing clothes and caring for their children (ibid.), while the traditional role of tilling the land with oxen was regarded as a man's job (Priyadarshini 2014). This notion of differentiation in tasks might however evolve with the farm mechanizations, diversifying economic activities and education.

Policy Impetus Towards Gender Equality

Bhutan's national laws accord equal status and rights to both men and women. The Constitution of the Kingdom of Bhutan 2008, which provides an overarching foundation, ensures gender equality at all levels. It safeguards fundamental rights of all Bhutanese citizens irrespective of gender, religion or ethnicity, to voting, access to join public services, equal pay for work of equal value and property ownership without any discrimination. The philosophy of Gross National Happiness (GNH) guides all national policies and governmental actions in Bhutan. To ensure tangible manifestation of the central tenets of GNH, all proposed policies and projects are scrutinized using GNH Screening Tool.[5] The screening exercise provides a systematic appraisal of the potential effects of proposed policies and projects and it ensures compliance with GNH principles (Ura and Penjore 2008). Discrimination based on gender, ethnicity or religion is one of the indicators on the screening tool thus ensuring policy impetus towards promoting gender equality.

However, certain gender gaps still exists against women in the tertiary education, labour force participation and decision making.

[5] GNH Screening Tool: a systematic tool that ensures all development policies and projects of Bhutan are reviewed under the lens of GNH.

Gender Parity

In 2016, Bhutan ranked 121st from among 144 nations on the Global Gender Gap Index which used four indicators such as political participation, health, education and economic empowerment to assess gender parity (Leopold et al. 2016). The report highlighted a significant disparity against Bhutanese women on most of the sub-indices used to assess gender parity except for enrolment in primary and secondary education and healthy life expectancy where the gender parity is in favour of women. Disparity against women in three prominent areas such as tertiary education, labour force participation and decision making are discussed further.

Gender Gaps in Education

Bhutan has achieved gender parity with regard to school enrolment in the primary and secondary level education. In 2016, female to male ratio for enrolment in primary and secondary education was 1.02 and 1.14, respectively (Leopold et al. 2016). However, this ratio decreases for women in the tertiary-level education. In 2018, female enrolment in the tertiary education both within and outside the country on scholarships and self-funding was lower compared to male (Ministry of Education [MoE] 2018). Proportionately, tertiary-level educational attainment by female was also found to be lower compared to male (NSB 2017). Financial barriers, family problems and lack of support are some reasons that were found to be constraining female students' progression to higher-level education (J-F. et al. 2015) and also the parental notion, especially in rural areas, that girls needed less education than boys (GNHC 2001). Whatsoever, gender gaps in education is an important issue, as it is a major impediment in achieving gender equality in Bhutan (Wangmo 2004).

Gender Gaps in Labour Force Participation

A significant gender gap exists in terms of labour force participation in Bhutan. In 2017, Bhutan's working-age population was estimated at 506,611 persons, of which female comprised 51.9% (NSB 2017). However, the labour force participation rate for female at 49.6% was considerably lower than males which stood at 73.6% (ibid.). Factors that impede women's participation in the labour force are the domestic role of women as mothers and caregivers, and lack of skills, education and financial barriers. These are some aspects that the civil societies in Bhutan have recognized as crucial for women empowerment as discussed later.

Women's representation in the civil service is another aspect of gender disparity that solicits emphasis. In 2017, female constituted 36.43% of the 28,070 civil servants in the country (Royal Civil Service Commission [RCSC] 2017). In 2014, women's share of civil service was lower compared to men at all levels of position and this gap was particularly marked at the executive level in which women comprised 6.5% and men 93.4% (RCSC 2013, as cited in NCWC 2014). Women's highest share in the civil service was in the supervisory and support category constituting 37.7% (ibid.). This could be due to women's lack of higher education. On the other hand, the pattern could also suggest societal notions that undermine women's leadership capabilities.

In 2014, the unemployment rate for women was higher than men at all levels of education and this gap was even wider for those with tertiary education (Asian Development Bank [ADB] 2014). This difference was particularly striking in urban areas, where 6.6% unemployment rate for women was significantly higher than their male counterparts, 1.4% (NSB 2017). The overall national unemployment rate in 2017 was 2.4%, of which female comprised 2.9% and male 2.1% (NSB 2018). The aggregation of these patterns suggests that women could be facing certain discrimination in the labour market (ADB 2014). Women comprising higher proportion of unemployment rate could also be due to lack of required skills and education, which is one aspect of CSOs' engagement that seeks to empower women.

Gender Gaps in Decision Making

The form of governance in Bhutan peacefully transitioned to Democratic Constitutional Monarchy with the signing of the Constitution of the Kingdom of Bhutan 2008. This transition to democracy allowed for a fair participation of both men and women in the electoral process. Despite equal opportunity being guaranteed by law, women representation in the electoral politics was lower compared to men. Bhutan's parity score for political empowerment index on the Global Gender Gap Index 2016 was 0.06 indicating a very high disparity against Bhutanese women, the perfect parity score being 1 (Leopold et al. 2016). Women's representation in the parliament[6] was significantly lower than men in all the three parliamentary terms as shown in Tables 32.1 and 32.2.

In the 2008 parliamentary elections, 16 women candidates contested and 8 were elected; 4 for National Council (NC) and 4 for National Assembly (NA). Two females were nominated as Eminent Members[7] in the NC by his majesty the king. The significantly lower female representation in the parliament signposts that the decision-making power could be skewed towards men. Female representation was not only lower in the parliamentary sitting, but it was also significantly lower in the number of women candidates who contested for parliamentary elections[8] compared to men as shown in Table 32.2.

Table 32.1 Male and female representation in the Parliament of Bhutan

Year	Male	Female	Total
2008	62	10	72
2013	66	6	72

Source: ECB (2013, as cited in Chuki 2015)

[6] Parliament of Bhutan: Composed of His Majesty The King of Bhutan, National Council and National Assembly. Parliament of Bhutan convenes at least twice a year and the term is for five years.

[7] Eminent Member: Five eminent persons nominated by the King of Bhutan as members of parliament in the National Council.

[8] Parliamentary Election: Held every five years. For NA, two rounds of elections occur, first (primary) round selects two highest voted parties, and second (general) round elects members from 47 constituencies. The party with highest elected member forms the ruling govt. For NC, 20 non-partisan members are elected from each *Dzongkhags* (districts) in a single round of election.

Table 32.2 Male and female composition in the Parliamentary Elections 2018

		National Council			National Assembly			
Year	Gender	Contested	Elected	Eminent members	Contested (primary round)	Contested (general round)	Elected	Total elected
2018	Male	121	18	3	170	84	40	58
	Female	6	2	2	18	10	7	9

Source: Election Commission of Bhutan (ECB 2018a, b)

In the 2018 NA elections, 18 women candidates contested in the primary round and 10 in the general round, of which 7 were elected (ECB 2018a). For the 2018 NC elections, 6 women candidates contested, of which 2 were elected (ECB 2018b). Some factors that constrained women from participating in the political and public sphere were women's traditional roles as housewives, caregivers and even income earners in some cases (NCWC 2008).

In 2018 parliamentary elections, female voters who actually cast their votes comprised a majority although marginally; 50.8% for NA elections and 51.13% for NC elections (ECB 2018a, b). Despite women voters comprising marginally major population of the voters who actually cast their votes, the significantly lesser number of women candidates who contested for the elections did not get elected. This pattern could confirm not only societal notions and attitudes against women but also women's notions about their role in the public sphere and decision making. This also confirms the findings by GNHC (2001) that women are viewed as less capable for responsibilities of governance and leadership. The traditional notion that undermines women's leadership capabilities still exists in Bhutanese society.

Civil Society and Women Empowerment

The formalized Civil Society Organization, although a recent development in Bhutan, carries out a wide range of activities to fill the gaps that are not addressed in the government's overall plans. Functioning under the overall governance of Civil Society Organization Authority (CSOA), Bhutan Network for the Empowerment of Women (BNEW), Bhutan Association of Women Entrepreneurs (BAOWE) and Respect, Educate, Nurture and Empower Women (RENEW) are some of the CSOs particularly focused on reducing gender gaps in Bhutan. The following sections explore five prominent aspects of women empowerment where the CSOs in Bhutan have intervened.

Education and Skill Development

Before the 1950s when monastic education was the only form of formal education in Bhutan, families preferred to send only boys, which accrued men significant advantage in the religious, politics and socio-economic aspects of society (Sinha 2009). Girls on the other hand were retained back home where they are needed and also because of the traditional notion that considered daughters vulnerable to be sent away from home (Wangmo 2004). Today, in the modern education system as well, gaps between men and women manifests particularly in tertiary-level education.

Women's lack of education and skills especially in rural areas accentuates the traditional view that considers women inferior to men. Lack of education and skills are the main factor that makes women less employable, which consequently incapacitates their earning capabilities (GNHC 2001). Providing education and skills for women is one strategic area where CSOs intervened to enable women to venture out of the domestic bondage. For example, the Livelihood Skills Training Program organized by RENEW in 2017, that trained 367 women on tailoring, fabric printing, dyeing, embroidery and ginger candy making enabled women to not only acquire skills but also enabled them to showcase and sell their products through the RENEW's Souvenir Shop (RENEW 2017). This is just one example of the skills development intervention initiated by the civil society. Education and trainings specifically tailored to suit the needs of women play a critical role in economic and social empowerment of women. Ensuring women's access to education and skills development programmes increases their feeling of self-worth, and even develop confidence and bargaining power in the household and societal decision-making process. It is therefore imperative that the CSOs identified education and skills for women as strategic measure for empowering women in Bhutan.

Women Entrepreneurship

Considering the increasing youth unemployment rate and the potential of entrepreneurship to gainfully employ self and others, the Bhutanese government is progressively encouraging youths to venture into entrepreneurship. This is another crucial aspect of civil societies' engagement towards empowering women in Bhutan.

Women entrepreneurship in Bhutan is an important untapped potential for poverty reduction and economic growth. BAOWE works towards promoting women entrepreneurs at the grassroots level to contribute towards poverty reduction, self-reliance and empowerment. Advancing Economic Opportunity for Women and Girls[9] was one such project initiated by BAOWE, which is aimed at positively changing livelihood practices of women and girls through promotion of groups, especially in deprived rural areas where large segment of the population are small and marginal farmers (BAOWE 2017). This intervention is significant considering that more than 60% of the total Bhutanese population resides in rural areas and are dependent on agriculture and livestock for livelihoods. Specifically to facilitate women's social empowerment in group-based enterprises that are facilitated by BAOWE, more than 90% women membership is encouraged (BAOWE 2017).

Women who venture into entrepreneurship not only become economically independent, but it also enhances their social interactions and decision-making capacity at the household and community level (Sharma et al. 2012). The targeted economic and social development interventions by CSOs that particularly focuses on women empowerment not

[9] Advancing Economic Opportunity for Women and Girls: Project initiated by BAOWE which facilitates formation of self-help groups (SHG) and farmer's cooperatives to enable production and marketing of products as an income generating basis.

only supplements governmental works but also fills the gap that government cannot address in its overall plans. However, for some potential women entrepreneurs, lack of access to credit facilities is one main factor that incapacitates their ability to venture into entrepreneurship.

Microfinance Services for Women and Girls

The financial institutions' requirement for collateral and high interest rates particularly hinders individuals from availing credit services, which incapacitates aspiring entrepreneurs from being able to start-up businesses. Today, the Bhutanese government has introduced affordable and accessible micro-financing schemes specifically targeted to benefit aspiring entrepreneurs. In addition, CSOs also play an important role in this aspect specifically towards empowering women.

Through micro-financing schemes, CSOs in Bhutan aim to improve the living status of women and their families. RENEW's Micro-Finance Project (MFP) initiated in 2012 and BAOWE-Pelzing launched in 2017 are two such schemes. MFP consists of micro loans, savings opportunities, credit facilities and insurance services and as of December 2017, MFP benefited 11,543 members (RENEW 2017). BAOWE-Pelzing is a micro finance institution that supports infrastructure and skills development especially for single mothers, general women and disadvantaged youths to help them grow their business (BAOWE 2017). These interventions are geared towards addressing problems of access to finance, youth unemployment and gender disparity.

The accessibility to affordable micro-finance for women plays a significant role in empowering women in Bhutan. Lack of financial capital was one main issue faced by aspiring entrepreneurs. However, with CSOs micro financing schemes especially designed for women and girls, it offers them the opportunity to shed their domestic shackles and venture into income generating tasks. Earlier studies concluded that when women have access to loans, they not only become able to contribute economically to their household but also bring about positive change in the household decision making (Kabeer 2001). The study also suggested that women's access to loans could result in a long term reduction in domestic violence as well (ibid.). As such, the civil societies' intervention with micro-financing schemes for women entrepreneurs is another strategic measure to addressing gender disparity in Bhutan.

Leadership Capacity Building

Despite having equal access and opportunity to compete for leadership roles, women's composition in leadership category is significantly lower compared to men. Cultural and stereotypical notions against women pertaining to women's leadership capabilities were common barriers that impede women's progression to responsibilities of leadership. The notion that men are more capable leaders for decision making at higher level resonates with women as well (GNHC 2001). This view constrains gender parity in leadership and governance.

Enhancing women's leadership capacity is another aspect where the CSO's intervention is crucial in Bhutan. By conducting workshops and conferences for aspiring women candidates for electoral politics, BNEW equips women with the knowledge and skills needed to equally participate in a male dominated politics. Civil societies engage elected women leaders, from local government and parliament in a dialogue to advocate for women's issues and also to prepare them for male dominated work environment. When women are provided such platform to come together and discuss on issues of significance especially pertaining to women, a common understanding of significantly critical issues can be developed for advocacy. Such activities can also morally inspire women to come forward and partake in electoral processes and to address women's issues.

Awareness and Advocacy

One important factor contributing to gender gaps was the stereotypical mindset against women. A study has concluded that women in Bhutan believe men to be superior and better equipped to comprehend and participate in matters of governance (GNHC 2001). This view, especially among women, clearly explains the comparatively lesser number of women coming forward as candidates for local government and parliamentary elections. Awareness, advocacy and education are therefore crucial in changing traditional mindset and encouraging women to enter public sphere.

CSOs play a crucial role in raising awareness on gender stereotypes in Bhutan. By providing networking platforms for women, BNEW encourages women of different background to come together and engage in dialogue for promoting women empowerment. Such platform offers women from diverse background a platform to share their stories of discrimination, violence, challenge, struggles and of successes to inspire each other and to advocate for more gender equality in the society. CSOs also advocate for gender sensitive policies to address issues of gender disparity. Transforming societal mindset and attitudes towards women's leadership role is an important aspect of women empowerment strategy that the civil societies in Bhutan are engaged in.

Conclusion

Bhutan's national policies and laws are gender neutral and do not discriminate individuals based on any differences. However, gender disparities exist against women in tertiary education, labour force participation and decision making. The amalgamation of patterns emerging out of gaps in employment, labour force participation and leadership suggests that women in Bhutan still face challenges stemming from gender stereotypes. The significantly lower number of women participating in the electoral process and representation in the leadership positions signposts that the decision making could be skewed towards men. This pattern suggests that the traditional notion which attributes men to the public sphere and women to the domestic sphere and the notion that considers women as less capable still exists in Bhutan.

The civil societies' recognition of gender disparity was rapid and their response covers a wide range of interventions. Amongst many other responses from the civil societies towards empowering women in Bhutan, this exercise highlighted five prominent areas such as education and skills enhancement, entrepreneurship development, micro-financing for women, leadership capacity development and awareness and advocacy. By intervening in these five strategic areas, civil societies are playing a significant role in challenging and overcoming gender stereotypes in Bhutan.

References

Asian Development Bank. (2014). *Bhutan: Gender Equality Diagnostic of Selected Sectors*. Mandaluyong City, Philippines: O Books Sun & Moon. Retrieved December 24, 2018, from www.adb.org.

Bhutan Association of Women Entrepreneurs. (2017). Advancing Economic Opportunity for Women and Girls. Retrieved December 19, 2018, from www.baowe.org.

Chuki, S. (2015). *Women in Politics in Democratic Transition: The Case of Bhutan*. PhD Thesis. Queensland University of Technology. Retrieved December 19, 2018, from https://eprints.qut.edu.au/85085/1/Sonam_Chuki_Thesis.pdf.

Crins, R. (2004). Religion and Gender Values in a Changing World. In K. Ura & S. Kinga (Eds.), *The Spider and the Piglet—Proceedings of the First International Seminar on Bhutan Studies* (pp. 581–596). Thimphu, Bhutan: Centre for Bhutan Studies.

Election Commission of Bhutan. (2018a). Press Release: Declaration of Results of the General Elections of the Third National Assembly 2018. Retrieved December 29, 2018, from http://www.ecb.bt/?p=6292.

Election Commission of Bhutan. (2018b). Press Release: Declaration of the Results for the National Council Elections 2018. Retrieved December 29, 2018, from http://www.ecb.bt/?p=5903.

GNHC (Gross National Happiness Commission). (2001). Gender Pilot Study. Bhutan: GNHC. Retrieved December 21, 2018, from http://www.gnhc.gov.bt/publications/rep/gpsr.htm.

J-F., Gyamtsho, S., Swabey, K., & Pullen, D. (2015). Capturing the Voices Drop-Outs in Bhutan. In *The Australian Association for Research in Education (AARE) Conference* (pp. 1–11). Retrieved December 16, 2018, from https://www.researchgate.net/publication/296294812_CAPTURING_THE_VOICES_OF_SCHOOL_DROP-OUTS_IN_BHUTAN.

Kabeer, N. (2001). Conflicts Over Credit: Re-Evaluating the Empowerment Potential of Loans to Women in Rural Bangladesh. *World Development, 29*(1), 63–84. https://doi.org/10.1016/S0305-750X(00)00081-4.

Leopold, T. A., Ratcheva, V., & Zahidi, S. (2016). *The Global Gender Gap Report 2016*. World Economic Forum. Retrieved from http://www3.weforum.org/docs/GGGR16/WEF_Global_Gender_Gap_Report_2016.pdf.

Ministry of Education. (2018). *State of Tertiary Education in Bhutan 2018*. Thimphu, Bhutan. Retrieved from http://www.dahe.gov.bt/images/pdf/HEPD/State-of-tertiary-education_2018.pdf.

National Commission for Women and Children. (2008). Study of Gender Stereotypes and Women's Political Participation (Women in Governance). Retrieved December 16, 2018, from http://www.ncwc.gov.bt/publications/Study%20of%20Gender%20Stereotypes%20and%20Womens%20Political%20Participation%20(Women%20in%20Governance).pdf.

National Commission for Women and Children. (2014). Gender at a Glance. Retrieved December 18, 2018, from http://www.ncwc.gov.bt/publications/Gender%20at%20a%20Glance.pdf.

National Statistics Bureau. (2017). *Bhutan Living Standards Survey Report 2017*. Thimphu, Bhutan: National Statistics Bureau of Bhutan. Retrieved December 16, 2018, from http://www.nsb.gov.bt/publication/files/pub2yo10667rb.pdf.

National Statistics Bureau. (2018). *Population and Housing Census of Bhutan 2017*. National Statistical Bureau of Bhutan. Retrieved December 15, 2018, from http://www.nsb.gov.bt/publication/files/PHCB2017_national.pdf.

Priyadarshini, V. (2014). *Women in Bhutan: Exploring Their Socio Cultural Status in the Late 20th Century*. Proceedings of the Indian History Congress, 75, 920–927. Indian History Congress. Retrieved December 15, 2018, from http://www.jstor.org/stable/44158477.

RENEW. (2017). RENEW Annual Report. Retrieved December 19, 2018, from http://renew.org.bt/publications/RENEW-Annual-Report2017.pdf.

Royal Civil Service Commission. (2017). Civil Service Statistics 2017. Retrieved December 18, 2018, from http://www.rcsc.gov.bt/wp-content/uploads/2018/05/RCSC-Statistcs-2017-final.pdf.

Sharma, A., Dua, S., & Hatwal, V. (2012). Micro Enterprise Development and Rural Women Entrepreneurship: Way for Economic Empowerment. *Artha Prabandh: A Journal of Economics and Management, 1*(6), 114–127.

Sinha, C. (2009). Effects of Education and ICT Use on Gender Relations in Bhutan. *Information Technologies & International Development, 5*(3), 21. Retrieved from https://itidjournal.org/index.php/itid/article/view/379/175.

Ura, K., & Penjore, D. (2008). *GNH Policy and Project Selection Tools*. Center for Bhutan Studies. Retrieved December 21, 2018, from http://www.gross-nationalhappiness.com/docs/GNH/PPT/Pol&Pro_Scr_Tools.ppt.

Verma, R., & Ura, K. (2015). Gender Differences in Gross National Happiness in Bhutan: Abridged Analysis of GNH Surveys. In K. Ura & D. Penjore (Eds.), *GNH: From Philosophy to Praxis-Proceedings of the Sixth International Conference on Gross National Happiness* (pp. 196–248). Center for Bhutan Studies. Retrieved December 25, 2018, from http://www.grossnationalhappiness.com/Conference%20Proceedings/6th%20GNH%20Conference/6th%20GNH%20Proceedings%20DP%2006.01.2018.pdf#page=202.

Wangmo, S. (2004). Country Case: Bhutan. In S. Wangmo, S. Violina, & M. M. Haque (Eds.), *Trend and Status of Gender Perspectives in the ICT Sector: Case Studies in Asia-Pacific Countries* (pp. 26–66). ITU (International Telecommunications Union). Retrieved from http://citeseerx.ist.psu.edu/viewdoc/download?doi=10.1.1.118.7888&rep=rep1&type=pdf.

33

Forest Rights Act 2006, Customary Laws, and Sustainable Community Development: Study on Lodha Tribe of West Bengal, India

Koustab Majumdar and Dipankar Chatterjee

Background

Since the pre-independence period, various decisive legislation and policies have adopted in India so far to foster forest and forest resources conservation agenda. However, espousing the Forest Rights Act (FRA) 2006 has been one of the historic legislations. The inception of FRA was to respond towards undue harvest and misuse by the potential beneficiaries of the forest resource and sustainable growth of the forest, which is conducive to forest resource conservation and livelihood security (Bhullar 2008). The FRA is also known as the Scheduled Tribes and Other Traditional Forest Dwellers (Recognition of Forest Rights) Act 2006, which aims at recognizing and entrusting the forest rights and occupation to scheduled tribes and other traditional forest dwellers who have

K. Majumdar (✉) • D. Chatterjee
Ramakrishna Mission Vivekananda Educational and Research Institute, Ranchi, Jharkhand, India

© The Author(s) 2020
Md. N. Momen et al. (eds.), *Building Sustainable Communities*,
https://doi.org/10.1007/978-981-15-2393-9_33

been residing in forest land for generations; but the rights could not be recognized and recorded (MoTA 2014).

The high dependency for surviving and customary rights over forest resources of tribal communities indeed confer an opportunity to make a unique reciprocal relationship with nature; hence the co-existence of forest and tribal or indigenous people can be reckoned as an integral part of ecological sustainability (Bhullar 2008) and indigenous identity of tribal people. The tribal people have authoritarian compliance and institutionalized rules regarding cultural practices, regulated uses of natural resources as they are well cognizant in using and conserving forest resources (Gadgil and Berkes 1991; Singh et al. 1996) without hampering the environmental ecosystem and biodiversity (Stevens 1997). The tribal people who are the predominant consumer of forest resources (Sarangi 2006) use traditional knowledge (Beltran 2000; Furze et al. 1996) through which they perceive the ecological resources. The ecological perception of available resources determines the livelihood and structures the wellbeing of the tribal community that ensures sustainable community development.

The discourse of sustainability has had considerable attention in the field of social sciences, development studies in general, and tribal studies in particular. The notion of sustainable community development can be linked with the concept of sustainable development as defined in the Brundtland Report:

> Sustainable development is development that meets the needs of the present without compromising the ability of future generations to meet their own needs. (WCED 1987)

The above conception focuses on the resource conservation that considers the present need and intergenerational equity. The discourse of natural resource depletion has been included as an imperative aspect in sustainable development debate (Hembd and Silberstein 2011). Community development involves the local community people, especially the tribal people due to their close juxtaposition with nature (Yu 2018); therefore, it is closely linked with sustainable development (Szitar 2014; Kapera 2018). Community development ensures the continuous development of a community in social, economic, and environmental aspects. Rogers and Ryan (2001) mentioned some guiding principles which must be considered as the criteria of recognizing a sustainable community:

- firstly, the human needs should be satisfied by utilizing the natural resources without undermining the nature's ability;
- secondly, ensuring the wellbeing of the community members and also offer and encourage the tolerance, participation, creativity, and safety;
- thirdly, empowering the members with shared responsibilities, access to knowledge and decision-making power that affect them; and
- lastly, an institutional collaboration that is environmentally sound, economically viable, socially responsible, and investing the community members in various dimensions.

The rapid population growth, expanded human settlement, and territory have gradually been susceptible to societal mainstream and public interference (Kemf 1993). The public interference has created a state of competition, and the tribal people have lost their land, culture, social structures, and traditional knowledge as a consequence (Colchester 1997; Feeney 1993; Ghimire and Pimbert 1997). Furthermore, the public interference regarding imposed decisions on restricted use of forest has steered to the unsustainable harvest by the tribal people who were earlier managed by tribal people themselves. The enactment of FRA has been articulated as a commendable move in correcting the historical injustice regarding over-exploitation of forest and environmental resources (Bhullar 2008) by the forest dwellers. Notwithstanding, various conflicts have transpired due to external influences that invigorate the case of livelihood loss of the tribal and other forest dwellers who are significantly dependent on forest resources (Saravanan 2018). The sudden diversification from forest dependency to other occupations (agriculture, wage labour) has created various conflicts.

Moreover, it became more prominent after the enactment of FRA. Livelihood loss, inter-community disagreement on forest resource access, and negligence of customary laws have emanated as the consequence of the implementation of forest rights act; therefore, countersigning the sustainable community development through executing FRA has been coming out as a vexing issue in the territory of Lodha community.

In this perspective, this present study explores the critical issues (i.e. customary laws and inter-community conflicts) concerning to FRA.

Following this introductory section, this study proceeds as follows. The second section presents the historical profile of the Lodha tribal community, that is, transition from criminal tribe to the Particularly Vulnerable Tribal Group (PVTG). The third section highlights the research methods, including the study area, sampling, data collection, and analysis. The fourth section presents the results of this study, such as conflict issues (disagreement with FRA-2006), various customary laws of the Lodha tribal community. The fourth section presents an in-depth discussion of the obtained results. Moreover, it proposes a sustainable community development (SCD) model, and the last section concludes and reiterates the significant findings and policy suggestions of this study.

The People: Lodha Tribal Community

The Lodha tribe of West Bengal is considered as Particularly Vulnerable Tribal Group (PVTG). The Lodha tribes are highly concentrated in *Purba* (East) Medinipur, *Paschim* (West) Medinipur, and Jhargram district of West Bengal (Devi 1983). Lodha community people have spread over Odisha, Bihar, and Chhattisgarh states in India. Earlier, the Lodha people were recognized as 'Criminal Tribe' during the British rule and till the revocation of the Criminal Tribes Act, 1952. Hunting and gathering were the primary sources of livelihood of the Lodha tribal community members, and they were highly dependent on forest resources and lived in the dense forest. In 1960–1961, Lodha tribe was regarded as the Primitive Tribal Group by Dheber Commission due to negative population growth, hunting and gathering livelihood practice, and pre-agricultural stage of living. In 2006, the Government of India designated the Lodha tribe as one of the Particularly Vulnerable Tribal Groups (PVTG) in India (Panda 2015; Panda and Guha 2015). However, during the post-independence period, the government has emphasized to mainstream Lodha people by restructuring their livelihood. Consequently, they have recently shifted in settled agriculture and even have engaged in agricultural and various non-agricultural occupations (Bhowmick 1994). The community people can speak both their traditional language and Bengali language. Despite practicing agriculture-based livelihood, the Lodha tribal community members are still highly reliant on forest and forest resources. Cultural hunting[1] of animals of dense forest, collecting non-timber forest products, and worship of nature are highly appreciated cultural practices in Lodha.

Materials and Methods

This present empirical study was based on the qualitative approach of research. Qualitative research method accentuates social reality through the constrictive lens that helps in recognizing the holistic aspect of a particular phenomenon (Merriam 1998). Furthermore, qualitative research imparts the ground reality of the research problem in a better way (Caldas-Coulthard 1987). The Lodha tribal community from Paschim Medinipur (Muchiberiya village) and Jhargram (Govindapur and Lohomaliya village) district of West Bengal was purposively selected.

This present study involved a total of 90 key respondents (30 respondents from each village) age 25 years or higher as the sample respondents Data was collected using two tools, such as in-depth interviews and focused group discussions (FGDs). Data were collected in two phases (first phase in May 2018 to August 2018 and the second phase in October 2018 to November 2018).

The respondents were interviewed through the open-ended questionnaire to decipher the emerging issues related to FRA and its implementation process. The in-depth interview was based on open-ended questions and taken for 30 to 45 minutes (average). Each interview was taken in the local language (Bengali) and recorded with the verbal consent of the interviewers. It was also ensured to all the interviewers that their anonymity their identity will be maintained. Three focused group discussions were also conducted in selected villages to explore the impact of FRA on community life.

The interview transcripts were translated into English and analysed through the qualitative approach; that is, thematic analysis. This study used the thematic analysis as a data analysis method to recognize the critical theme of data set and explore the meaning of the responses given by the interviewers. Thematic analysis was used in analysing the data. Thematic analysis is a crucial method that analyses the meaning of the data set and also helps in constructing the themes of the data (Neuendorf 2019). The mentioned quotes were adopted from transcripts that reflect the emerged theme of the data.

Results

Emerging Conflicts on Forest

The emergence of conflicts can be traced back from the historical perspective of the enactment of forest policy that envisioned to conserve the forest and environmental resources and recognizing the rights of traditional forest dwellers. In 2008, the Forest Rights Act came into effect in the studied areas. The non-functioning Forest Protection Committee[2] (FPC) at the village level was reformed based on revised guidelines to protect the forest from anthropogenic activities and prepare a list of forest-dependent households with the help of *gram-sabha* in order to facilitate three acres of land to conduct agricultural practices. The new guidelines of FRA further sanctioned to facilitate some arrangements such as the demarcation of forest land, sustainable harvest practice (or restricted access) of forest resources, rights to access the forest and forest resources, and, most importantly, an opportunity to claim against different injustice related to the forest.

The forest dependent livelihood has gradually been translated into agriculture and other non-agricultural occupations as a consequence of the forced restriction towards the access of forest. The prime cause of conflicts or disagreement with FRA was due to the livelihood diversification (forest-based economy to agriculture and non-agricultural occupations). However, this diversification failed to provide adequate input supply, extension, agricultural training, marketing services, and work opportunities in non-farm sector. It revolves around conflict with government policy directly. Two community members stated this issue:

> Earlier, my grandfather was exclusively dependent on the forest. But my father got three acre of land from the state government for cultivation, which is not enough for conducting commercial agriculture. Therefore, we have to depend on forest resources. The forest resources are the supplementary source of livelihood to us. The restriction in accessing forest has become an obstruction to us. So that we have been suffering a lot. #R33

> Our forefathers were solely dependent on the forest. They did not confront much struggle for livelihood, because there were no external interferences. They were free and could do whatever they wanted. There was no scarcity of food. Now we are doing agriculture. We are strongly connected with nature and forests. We do not have registered land. We are dependent on little leased land. The external interference and rapid destruction of large trees have led to food scarcity of food and shelter, and consequently, elephant attack to our storeroom and agricultural field. Hence, we confront a large amount of economic loss in cultivation that is kept for self-consumption only. Furthermore, there are many restrictions in accessing forest. Now tell me what should we do, and where should we go to ensuring my two-time meals? Only one option we have, i.e., go outside to work as wage labour. #R17

The people of Lodha community have been residing the forest and its adjacent areas since long, and their exclusive dependency had been on the forest resources (non-timber forest products, roots, fruits, mushrooms including hunted meat of wild animals). During the post-independence period, the community people adopted different agriculture and non-agricultural occupations with the help of government interventions. Despite the adoption of these kinds of occupations, forest still has been a great source of livelihood support to them. The forest can be considered as an integral part of the socio-cultural life of Lodha community. Livelihood diversifications and possession of little agricultural land[3] directed to rejection of permanent land and forest rights application by the forest department. Failure in securing such rights (land and forest) also engendered the conflict with the forest department.

On the contrary, the community members also reported about the inter-community forest conflict. The people from other communities sometimes access the forest resources illicitly. Furthermore, it was a prominent issue that a lot of tribal families claimed application to get registered land and forest access rights; unfortunately, the applications got rejected. One respondents said:

> I do not have registered land. I cultivate two acres of leased land. I am highly dependent on forest resources because the income from agriculture is not sufficient enough. During 2013 the forest department recorded landless households. I was also recorded. But, they rejected my claim. And I did not have any registered land to date. Furthermore, due restriction on forest resource access I can not rely on the forest resources only. #R52

The non-tribal people tampered predominantly to the forest areas to cut the mature trees illegitimately. It breaches the customary law of the community one hand and generates inter-community conflict on the other hand. There were intra-community level conflicts reported by the community members. Sometimes, the community member themselves violate the customary laws or imposed restrictions; subsequently, the members confront respective punishments over breaching custom rules. Furthermore, it was emerged from the interview that rejection of filed applications regarding land and forest rights was a prominent issue.

Customary Laws on Forest Resource Conservation

Customary law, which is passed through one generation to another, is a set of knowledge that generated continuously 'in the course of long history' through the human experiences, behaviour, and interactions between nature and human beings. It is a kind of teaching and practices that regulate the relationship with the natural environment with the man. Customary laws have an enormous spiritual significance and are accepted by the entire community people for the sake of maintaining a balanced relationship with the natural environment and community development (CIRUM 2012). There were distinct customs and rules governed among the Lodha community members. The customary laws were made and supported by the community people. The penalty in terms of money or other means of punishment is charged in case of breaking the customary laws by any of the community members. There was a particular cause in making such customary laws. The core customary laws are based on the belief system, values, and social norms. These custom laws make nature more sustainable and help in managing resources. One community member stated:

> This forest law ensured us on registered land and rights to access the forest resources. It is our forest, and we have been maintaining it very well. We have been here for more than three generations. However, the forest right has been taken away from us. We have our law to protect it. We can take care of our forest. #R4

The community people believe that elder community members are the pioneers of the society, so that the elder community members are the principal customary laws maker. There were different customary laws within the community to use forest resources. Community members stated about the sentiment, rules of customary law, which can be considered as cultural practice. Two elderly members mentioned:

> We usually do not cut any tree. We have to take permission from FPC in case if we want to do so. Sometimes community members do not follow the community rules. Then they are punished. There have been so many cases where the community members themselves breach the law that creates create conflict among us. #R43

> We are the tribal people ... we cannot live without forest, because the forest resources provide the major life support to us ... we need to ask for permission from FPC for having commercial timbers; otherwise, we will be punished. We worship and love the forest as it gives us so many things; therefore, we should too take care of the forest. #R62

The forest resources contribute three-dimensional benefits (food, commercial consumption, and medicinal values) to the Lodha community. There were several customary laws associated with a community that regulate the sustainable harvest of forest resources and forest resource conservation. Some of the customary laws are as follows:

1. The community members cannot access the resources of the dense forest[4] that is demarcated by the forest department.
2. Unlimited dry leaves and non-timber produces of the forest areas can be collected for fuel purposes.
3. The timber of some selected species (mango, jackfruit, and other fruit trees) of large trees can be taken for home consumption (making house, fencing the boundary of house, etc.) only after permission granted by FPC. Those breaching the law will be charged Rs. 500/- as a penalty.
4. The timber of a high-valued tree (Sal) can be taken only during a financial emergency (extra-ordinary debt, marriage ceremony, death, etc.). It is subjected to further verification and sanctioned by FPC.
5. The particular family (who consumed the timber of high-valued tree) will have to plant two same trees and also will be responsible for managing it.
6. The community members cannot hunt the animals of the dense forest. Furthermore, they only can hunt local forest animals during cultural or religious festivals.

FPC is the fundamental responsible institution that verifies, scrutinizes, regulates, and determines the status of law. FPC also has the power to include, change, or remove any particular rules from their custom guidelines. The community people admitted that the customary laws are meant to regulate the members from unsustainable harvest and conserve the forest resources.

Discussions

The Lodha community people were the traditional forest dwellers. Since the post-independence period, they have gradually been adopting agricultural occupations due to imposed restrictions over forest resource use. The community people have been conducting agriculture on leased land or working as a hired agricultural labourer and daily wage labourer. This transformed occupation has not been enough to ensure livelihood security. Historically the socio-cultural system has been closely associated with forest. The community members still highly dependent on forest resources not because of livelihood, but with their emotion too. They acknowledge the forest resources as the common property resources of the community, which has been maintaining efficiently by community.

However, the implementation process of FRA in the studied area has emerged as a quite vexing issue. The tribal households reported their names at the FPC and *gram-sabha* to get the registered land for the sake of agricultural practice and vested forest rights. Nevertheless, most of the claims got rejected for unknown reasons. The community members cannot access the forest areas to a sufficient extent due to imposed restrictions and not having vested forest rights. Moreover, it has emerged from the study that the members of *gram-sabhas* and FPC were neither sufficiently trained nor fully aware of the appropriate process in filling reported claims. Notably, majority of the tribal community people do not know the vested rights. However, they were well-known about restricted access to forest resources.

It has emerged that most of the community members have less than two acres of leased land in which they carry out agricultural activities. Being smallholder agrarian community, the Lodha tribal community practices agriculture in leased land and works as agricultural or daily wage labourer. On the contrary, imposed restriction on forest destructs livelihood of the tribal community members.

There were many inter- and intra-level community conflicts with the implementation of FRA as the executed FRA failed to provide the rights of land and forest that have been well maintained by the tribal people themselves. Moreover, the community as a social institution has customary laws for its community members to conserve and management of natural resources, but the interference of non-tribal communities in accessing forest resources provoked many inter-community level conflicts as a consequence.

Therefore, it can be argued that the FRA 2006 failed to provide the land and forest rights to the Lodha tribal community on the one hand and created conflicts with the community by providing the imposed restricted actions in using forest resources on the other hand. Consequently, this circumstance has been leading towards the state of unsustainability in livelihood and community development as well. In this context, this present study suggests a comprehensive model of sustainable community development by promoting synchronization between customary law and imposed forest policy, that is, FRA 2006.

Figure 33.1 represents a sustainable community development (SCD) model. The customary law, which encompasses the community is based on a belief system and community feelings. On the contrary, the government institution, that is, forest department as an external public domain imposes different forest-related laws, orders (land and forest rights) to regulate the unsustainable harvest of forest resources and ensures the livelihood security to the forest-dependent people through bestowing land or forest rights. The government institution should consider the community-specific customary laws in policy-making approaches, such as environmental decisions (resource use, restriction in using forest resources, forest resource management, and tracking the external issues related to injustice regarding forest resource use) and vesting the rights (land and forest

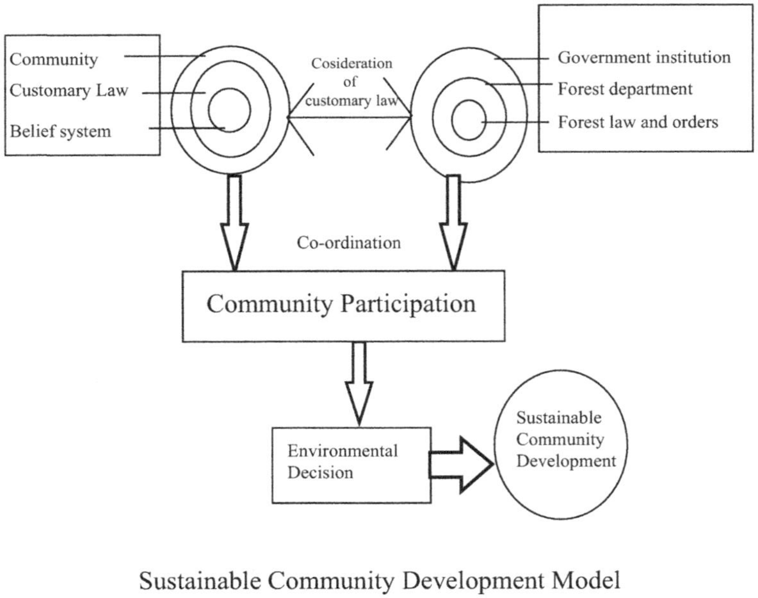

Fig. 33.1 Sustainable Community Development (SCD) framework in perspective of FRA implementation

rights). This consideration will construct a synchronized relationship between two institutions (community and government department). The community participation in environmental decision-making process will reinforce the environmental planning and natural resource management that will further satisfy the conditions of SCD and sustainable development at the larger process.

Conclusions

This study explores the issue of conflicts or disagreement regarding the implementation of FRA 2006 in the tribal territory with particular reference to Lodha community of West Bengal, India. The smallholder Lodha community still has a significant dependency on forest for livelihood. The culture of Lodha community has highly been associated with forest. Based on the qualitative research approach, it reveals that FRA 2006 failed to provide both land and forest rights to the Lodha community. Conflicts regarding community rights, land and forest rights, including the imposed restriction on forest resource access, have emerged as a prominent issue in FRA implementation process. The community members, including the *gram-sabha* members, were less aware of the suitable process in reporting the claims of land and forest rights. Although the community members have been associated with the forest for more than three generations, but many reported claims were rejected for unknown reasons. Furthermore, the community members cannot access the forest resources to a sufficient extent as they did not get the legal rights over forest resource.

The ultimate aim of the FRA 2006 was to provide the land and forest rights to the tribal peoples that will enable the people to have justice regarding the rights to access the forest and its management to conserve the forest resources. However, the ground reality of the studied area (Lodha tribal community) contends the poor implementation that remains as a conflict issue.

Based on policy gaps and statement given by the community people, we would like to recommend the following points to be considered in achieving sustainable community development:

1. Forest dependency or resource use is a cultural practice among the tribal people. We need to consider this fact and reconsider the rules and regulations mentioned in FRA-2006.
2. The *gram-sabha* members, including the FPC, should be trained properly in documenting the list of potential beneficiaries of tribal households.
3. Reconsidering the implementation process of providing the land rights and forest rights because in many cases, the forest-dependent tribal families have been excluded.
4. The rules and guidelines should be focused on community-specific needs.
5. Active community participation should be ensured in forest resource conservation—the environmental decision-making process in considering the customary laws of the community.

Notes

1. Traditionally 'hunting' was one of the fundamental occupations of Lodha community. Since the imposed restricted forest access, the community members only hunt some selected animals live in forest during the local festivals only. The cultural hunting which has been the customary law of Lodha tribal community is a method of forest protection.
2. However, limited access over forest resources had been imposed on the people during 1990's by the forest officers (Range officers).
3. Most of Lodha households in the studied areas did not have permanent land (i.e. land ownership). They cultivate on the leased land that is taken for five years of contract from the other non-tribal people. Very few families in the study had 3 acres of registered land, which was provided by state government.
4. According to the customary laws of Lodha tribal community, forest areas up to 6 kilometres (approximately) from the habitat is regarded as local forest areas and beyond that is considered as dense forest.

References

Beltran, J. (Ed.). (2000). *Indigenous and Traditional Peoples and Protected Areas: Principles, Guidelines and Case Studies* (Vol. 4). IUCN-The World Conservation Union.

Bhowmick, P. K. (1994). *The Lodhas of West Bengal*. Kolkata: Institute of Social Research and Applied Anthropology (ISRAA).

Bhullar, L. (2008). The Indian Forest Rights Act 2006: A Critical Appraisal. *Law, Environment and Development Journal, 4*(1), 396–408.

Caldas-Coulthard, C. R. (1987). Reporting Speech in Narrative Written Texts. *Discussing Discourse, Discourse Analysis Monographs, 14*, 149–167.

CIRUM (Culture Identity and Resources Use Management). (2012). *Customary Law in Forest Resources Use and Management: A Case Study Among the Dzao and Thai People in North-West Vietnam*. Thailand: AIPP Printing Press, Ltd.

Colchester, M. (1997). Salvaging Nature: Indigenous Peoples and Protected Areas. In K. B. Ghimire & M. P. Pimbert (Eds.), *Social Change and Conservation: Environmental Politics and Impacts of National Parks and Protected Areas*. London: Earthscan Publications.

Devi, M. (1983). Lodhas of West Bengal. *Economic and Political Weekly, 18*(22), 947–949.

Feeney, T. (1993). The Impact of a European Community Project on Peasant Families in Uganda. *Oxfam Brief, 6*, 1–7.

Furze, B., De Lacy, T., & Birckhead, J. (1996). *Culture, Conservation and Biodiversity: The Social Dimension of Linking Local Level Development and Conservation Through Protected Areas*. Chichester, UK: John Wiley and Sons Ltd..

Gadgil, M., & Berkes, F. (1991). Traditional Resource Management Systems. *Resource Management and Optimization, 18*, 127–141.

Ghimire, K. B., & Pimbert, M. P. (Eds.). (1997). *Social Change and Conservation: Environmental Politics and Impacts of National Parks and Protected Areas*. London, UK: Earthscan Publications Limited.

Hembd, J., & Silberstein, J. (2011). Sustainable Communities: Sustainability and Community Development. In W. Robinson Jr. & G. P. Green (Eds.), *Introduction to Community Development: Theory, Practice, and Service-Learning* (pp. 261–278). Los Angeles: Sage.

Kapera, I. (2018). Sustainable Tourism Development Efforts by Local Governments in Poland. *Sustainable Cities and Society, 40*, 581–588.

Kemf, E. (Ed.). (1993). *Law of the Mother: Protecting Indigenous Peoples in Protected Areas*. San Francisco: Sierra Club Books.

Merriam, S. B. (1998). *Qualitative Research and Case Study Application in Education*. San Francisco: Jossey-Bass.

MoTA (Ministry of Tribal Affairs). (2014). *Forest Rights Act 2006 Act, Rules and Guidelines*. Retrieved from https://tribal.nic.in/FRA/data/FRARulesBook.pdf.

Neuendorf, K. A. (2019). Content Analysis and Thematic Analysis. In P. Brough (Ed.), *Research Methods for Applied Psychologists: Design, Analysis and Reporting* (pp. 211–223). New York: Routledge.

Panda, S. (2015). The Development of Primitive Tribal Groups: A Case of Lodha and Birhor of West Bengal. *World Wide Journal of Multidisciplinary Research and Development, 1*(6), 28–35.

Panda, S., & Guha, A. (2015). *'Criminal Tribe' to 'Primitive Tribal Group' and the Role of Welfare State: The Case of Lodhas in West Bengal, India*. NOVA Science Publishers, UK: Nova Science Publishers.

Rogers, M., & Ryan, R. (2001). The Triple Bottom Line for Sustainable Community Development. *Local Environment, 6*(3), 279–289. https://doi.org/10.1080/13549830120073275.

Sarangi, T. K. (2006). Contextualising the Conflict Between Livelihood and Conservation: Implementation of FRA 2006 in Odisha. Retrieved from http://www.nird.org.in/nird_docs/srsc/srsc230217-20.pdf.

Saravanan, V. (2018). Forests and Tribals, 1950–2000. In V. Saravanan (Ed.), *Environmental History and Tribals Modern India* (pp. 87–127). Singapore: Palgrave Macmillan. https://doi.org/10.1007/978-981-10-8052-4_4.

Singh, G. S., Sexena, K. G., Rao, K. S., & Ram, S. C. (1996). Traditional Knowledge and Threat of Its Extinction in Chhankinal Watershed in North-Western Himalaya. *Man in India, 76*(1), 1–17.

Szitar, M. A. (2014). Learning about Sustainable Community Development. *Procedia-Social and Behavioural Science, 116*, 3462–3466.

WCED (World Commission on Environment and Development). (1987). *Our Common Future*. Oxford, UK: Oxford University Press.

Yu, C. Y. (2018). An Application of Sustainable Development in Indigenous People's Revival: The History of an Indigenous Tribe's Struggle in Taiwan. *Sustainability, 10*(9), 3259. https://doi.org/10.3390/su10093259.

34

Ensuring Safety for Women at Public Space: Need and Approaches

M. Nagaraj Naik

Introduction

A public place is a societal space that is commonly open and available to individuals. Streets, open squares, parks, shorelines, government buildings, offices and libraries that are accessible to use by the public are typically considered public place or space, though they tend to have restricted areas and greater limits upon use. The elements of safety in the public space have a profound effect on the three notions of safety such as comfort, belonging and commitment (Tovi Fenster 2005). The South Asian countries being a patriarchal society, public spaces are male-centric with expanding populace thickness, a differing blend of individuals in urban areas, developing imbalances and absence of opportunities to offending youth, urban crime, are all these on the ascent. Violence against women is existing at different dimensions, starting with separation during childbirth, propagated through separation in education, nourishment, work, compensation and direct and indirect differentiation of

M. Nagaraj Naik (✉)
Department of Social Work, Bangalore University, Bengaluru, India

© The Author(s) 2020

Md. N. Momen et al. (eds.), *Building Sustainable Communities*,
https://doi.org/10.1007/978-981-15-2393-9_34

sexual animosity. Violence and crimes against women are increasing as part of this process (Jagori and UN Women 2011). Violence against women, particularly in the community, is underneath and underreported, and for that reason, it is hard to evaluate the degree of the incident.

Women's safety involves policies and measures that happen before the violence has occurred to put off perpetration or maltreatment. This may occur by improving awareness and frames of mind that relate to the domestic or sexual harassment, adherence to societal standards, nonviolence and male sexual privilege. Global data shows that where laws are in place against domestic violence, its prevalence is lower and people also think they have to obey the law (UNDP 2010). In such cases, more legislations have to be brought in changing attitudes and behaviour towards women safety at the public spaces which are imperative. The treatment of women in the public sphere, particularly about sexual harassment, has recently got importance because of media inclusion and worldwide recognition of the abuse and exploitation of women (Manjoo and McRaith 2011; Neupane and Chesney-Lind 2014). To prevent violence, women's and young girls' cooperation must be developed; associations between neighbourhood, network associations, local authority and the government must be pursued, including decision making; and community participation and the leadership forms must be promoted. The preventive measure includes planned long-term, wide-ranging initiatives that address the risk and defensive factors related to perpetration, harassment and spectator behaviour (Curtis 2014).

The Need for Achieving Women Safety in South Asia

South Asia is behind in terms of gender equality, the endeavours of individual nations have not yet been liberal, profound, continued or sufficiently genuine to undermine the different types of discrimination that persist. The ample interest has not been given to the economic, political and legal level that could improve women in many areas. Women's constant poor representation in political, economic and legal institutions across the section has shaped a deficit in power and voice, which thus enables inequalities to go unchallenged.

Developing a safe environment in public, domestic and working places should be the priority of the government, private organization, corporate companies, non-governmental organizations (NGOs), institutions and even civil society, it is a collective effort. They need to consider women's safety as a key element in the public space and that could set to encourage the psycho-social, behavioural and cultural changes among the men towards the women. The need or importance of protecting women's rights is seen in the UN's Sustainable Development Goals (SDGs), mainly those on Gender Equality (SDG 5) and Decent Work and Economic Growth (SDG 8). The South Asian countries are developing countries and have no proper safety for women. Out of nine nations in South Asia, (Bhutan, Pakistan, India, Sri Lanka and Bangladesh) have existing legislation against sexual harassment, whereas Sri Lanka, India, Nepal and Bangladesh have laws to forbid Domestic Violence (UN Women 2018). Worldwide data demonstrates that where laws are set up against violence, its predominance is lower and hence more legislation has to be brought towards women's safety.

While deliberating on violence against women, few incidences are worth understanding in terms of extent, nature, type, intensity and consequences of harassment means.

Firstly, the gang rape of a medical undergraduate in Delhi on 16 December 2012, by six persons under intoxication in a private bus, in the presence of her friend (a male), during busy hours (9.30 pm), in the capital city, whose chief minister was a woman, and with a woman as the leader of the decision party, had shaken the nation (Sharma 2015).

Secondly, in 2018, a US-based journalist accused MJ Akbar, Minister of State External Affairs, India, for sexual harassment. Majlie de Puy Kamp has informed that she was sexually harassed by Akbar in 2007 when she was an 18-year-old intern. He grabbed her right under shoulders, on (her) arms, pulled towards him and kissed her forcefully (India Today 2018).

Thirdly, previous Miss Universe and film heroine Sushmita Sen claimed that she was explicitly bugged by the advertising head of Coca-Cola India, Shripad Nadkarni, and her endorsement contract was unexpectedly ended because she opposed his advances. Coca-Cola India paid Rs 1.45 crore to get her quiet over a supposed inappropriate behaviour case (Bhatia 2003).

Women Safety: The Current Scenario in South Asia

A large number of countries in the South Asian region have no laws on women's safety in public spaces. Even where domestic violence laws exist, legislation is not effectively implemented and the statutory compliance concerned at the organizations and institutions level they are not maintained. The maltreatment of women in the public is about sexual violence has very recently got importance as a result of media inclusion and global acknowledgement of the abuse and exploitation of women (Manjoo and McRaith 2011; Neupane and Chesney-Lind 2014).

Women Rehabilitation Center in Nepal demonstrated that 1563 cases against women violence were registered from April 2012 to 2013. Aside from that, it likewise called attention to that somewhere in the range of 5000–12,000 young women and girls aged 10–20 are trafficked every year.

A research conducted by development organization BRAC, shows around 94% of women commuting in public transport in Bangladesh have experienced sexual harassment in verbal, physical and other forms. Males belonging to the age group of 41–60 years are the main perpetrators and responsible for 66% of such incidents. The research likewise mentioned components, including misusage of laws, the exorbitant crowd in transports and fragile or no checking (e.g. absence of closed-circuit cameras) as the real causes behind the sexual harassment on streets and open transport, particularly in buses (Kabir 2018).

Another study conducted by ActionAid in 2017 based on five indicators—violence against women, legal assistance to fight against violence, women-friendliness in the budget, women-friendly urban cities, safety in public transports—revealed a total of 54.7% of women living in urban areas face violence in Bangladesh. Further study also showed 57% of women allege that their complaints are not taken seriously by the law enforcers, 65% of women believe that the law enforcement authority blames the exploited instead of the accused, 49% women feel risky in open transports, and 48% feel the same while benefiting public services including health services.

Pakistan the world's fifth most crowded nation and the second biggest South Asian nation (UN Women). An examination was directed in Karachi announced that a large proportion of women are exposed to physical savagery that has genuine physical and emotional outcomes (Fikree and Bhatti 1999). Honour killing is one more type of familial savagery against women in Pakistan. The act of *karo kari* is known to happen in numerous places in the country (HRCP 2000; Felix 2007). The whole situation obviously mirrors the brutality against women, an enormous public health and social problem in Pakistan, which has never been properly reacted and managed by the Pakistan government (Bettencourt 2000).

In such a situation, Pakistan has opted many international commitments towards gender equality and women's human rights—the Universal Declaration of Human Rights, Beijing Platform for Action, the Convention on the Elimination of all forms of Discrimination against Women, and the Sustainable Development Goals. Despite these commitments, Pakistan's ranking for gender equality remains one of the lowest in the world. With less than 12 years to achieve the 2030 agenda for Sustainable Development and in ensuring that underprivileged women and girls are not left behind, under its new five-year plan (2018–2022), Pakistan is committed to the following:

1. An empowering condition is set up to translate, monitor and report on the execution of gender equality and women's empowerment commitments, to worldwide standards, approaches and norms.
2. Gender-responsive plans, strategies and frameworks of administration are set up with foundations being increasingly available to and conveying for women and girls.
3. To create an atmosphere where women benefit from decent work, income security and socio-economic development

4. A safe environment where women and young girls are free from all types of harassment in private and public spaces, where survivors can get quality and essential services.

In Sri Lanka, a study on sexual harassment on public transport was commissioned by UNFPA in 2015. This study had a sample of 2500 women between 15 and 25 years of age from all 25 districts who commute on public buses and trains. Data were collected using both qualitative and quantitative (focus group discussion and key informant interviews) methods. Women who were using public transport about 90% of the women reported having undergone sexual harassment in buses and trains. Only 4% sought help from the police and 82% of women had seen others being subjected to harassment but not spoken up or reacted in consequences. About 44% of the women who were harassed said that their personal life had been affected by it.

In 2015, in Bhutan, Domestic Violence Prevention Rules and Regulation 2015 was formulated. It defines the functions and roles of the National Commission for Women and Children (NCWC) as the appropriate authority of the law, with all of those protection officers, adjudicators, central and local governments, police, social welfare officials, and other related institution and stakeholders. The criminal law that was sanctioned in 2004 supplements the Rape Law of 1996, which got more illumination of the meaning of rape and serious discipline for brutality. The Bhutanese government is attempting to address brutality against women. Notwithstanding correcting laws, it is dealing with a progressively compelling approach to anticipate brutality through awareness to policemen, lawyers, doctors, educators, the media and government officials. Through these actions, it attempts to improve the safety of women and children.

Afghanistan has encountered over four decades of conflict and has a profoundly man-centric gender regime, dependent on moderate cultural practices and traditionalist Islamic interpretation. Women have no power, harassment against women and young girls is endemic, and unsafe conventional practices are common. The respect killings, early and constrained marriage of young women (Central Statistics Organization 2016), bride price (walwar or sherbaha), baad (the giving of young women to a disputant gathering to settle an obligation or struggle) and badal (women and girls exchanged between families for marriage) (Gibbs et al. 2018).

India's Experience in Ensuring Women Safety

Sexual harassment at the workplace or public space is as old as the historical backdrop of humanity; its awareness of the public has received attention recently. Indeed, until the decision of *Vishaka vs. State of Rajasthan (1997)* case by the Hon'ble Supreme Court of India, there were no official rules to deal with respect to sexual harassment. The case identifies with a sexual assault of a social worker in a village of Rajasthan. For this situation, the court opined that inappropriate behaviour at work spot adds up to breach of individual rights ensured under Articles 14 (equality before law), 15 (prohibition of discrimination on the ground of sex), 19 (right to practice freely any profession, trade or occupation), 42 (provision for humane conditions of work), and the citizen's duties under Article 51A to renounce practices derogatory to the dignity of women. This case gave guidelines in the year 1997, however it took 15 years for the execution of the Sexual Harassment of Women at Workplace (Prevention, Prohibition and Redressal) Act.

Concerning to harassment at public spaces, Jagori, a nongovernmental organization led an investigation on women's safety, published a significant report featuring the different types of inappropriate behaviour (from verbal, visual, and physical to glimmering, rape or assault) that ladies in urban communities face in open spaces, for example, markets, transport terminals and roadsides. The report discovered open transport as the most common public space where most inappropriate behaviour happens. This finding is especially disturbing, given that the Nirbhaya assault episode occurred on open transport a year later of this report. From many points of view, this report fills as an essential report looking at the sexual harassment in open spaces (Jagori and UN Women 2011).

The Nirbhaya gang rape case, wherein injured girl was on 16 December 2012 beaten, assaulted and tormented in private transport, caused broad challenge all over India with good media coverage. In this case, an Enquiry

Commission was set up and J.S. Verma Committee was constituted to make recommendations to improve the safety of women and young girls. It was the hand of public and media activism that helped and provoked the court to shape several rules in the Vishaka case to shield ladies from harassment (Mampilly 2018). This occurrence which shook the whole Country set off the quick necessity of law. Thus, the 2013 Ordinance was declared. This was followed by the constitution of the "Sexual Harassment of Women at Workplace (Prevention, Prohibition and Redressal) Act, 2013." The bill got the consent from the President of India in April 2013 and it came into effect from December 2013.

The Sexual Harassment of Women at Workplace (Prevention, Prohibition and Redressal) Act, 2013, denotes statutory obligations to the employer or other responsible persons in the workplace or institution as follows:

1. Recognize sexual harassment as a serious crime.
2. Recognize the accountability of the factory, workplace or company premises to prevent and deal with sexual harassment.
3. The workplace includes mines, factories, shops, plantations, commercial establishments, hospitals, educational institutions, sports facilities, places where woman employee travels on duty, charitable organizations, and residence place or house.
4. The law recommends constituting two types of committees:
    (a) Internal Complaints Committee (ICC) to be established at the workplace by the employer.
    (b) Local Complaints Committee (LCC) to be established at the district level by the government, where smaller organizations having less than ten employees (District Officer designates one officer in every taluka/block/ward to receive the complaint from an aggrieved woman) may complain and sought help from the committee.
5. All women who draw salary, compensation, get an honorarium or work voluntarily in the government, private sector or unorganized subdivision come under the domain of these rules.

6. The committee must include NGOs/individuals familiar with the issue of sexual harassment.
7. The complaints procedure must be time-bound and confidentiality must be maintained.
8. To formulate an anti-sexual harassment policy in the working premises.

This act for the Indian women safety concern has brought new hope among working women in the organized and unorganized sectors, and government and private, hospitals, institutions and the places where women directly or indirectly are connected to work. The act compulsorily constitutes the Internal Complaint Committee (ICC) and Local Complaint Committee (LCC) to handle the matter related to women's safety or harassment at work.

Approaches Towards Ensuring Women Safety

There are many types of harassment that a victim can undergo, the impact and consequences of violence/harassment differ from individual to individual and its duration, severity and the offensive actions, but yet people usually react to harassment by showing certain psychological, behavioural, emotional and mood symptoms. Harassment can have a demoralizing effect on everyone within range of it, and it often negatively impacts on individual and organizational productivity. Disregard for harassment and discrimination doesn't influence it to leave. Rather, the culprit may see the quiet an inferred acknowledgement of his or her conduct. The conduct will probably proceed and may even increase, aggravating its impending impacts. Hence to combat harassment at public spaces a multidisciplinary approach has to be adopted in the community. Accordingly, there have been several approaches that are used to counter harassment under preventive, prohibition and redressal stages as follows:

Preventive Method

1. Smart city initiatives: A smart city enables its women social security, gender-inclusive, urban plan safeguards, women's right; provides decent jobs; and puts an end to discrimination and violence at the workplace. Besides, engaging women in the smart city planning and design of these cities will help in addressing their needs. Women's Safety Audits (WSA) can help to build smart communities and surveillance by putting a CCTV camera.
2. Code of conduct: As like in certain industries have their code of conduct, the code of conduct should extend and implement to all type of working organizations, workplace, school, dwelling, apartment and public places with respect to the relationship of men and women.
3. Sexual orientation refinement: Gender sensitization by guardians and teachers is required with respect to the sensitivities, limits and social interaction in a man–women relationship.
4. Quality education and employment prospects for youths.
5. Public awareness approaches: Techniques such as sensitivity training, cultural and diversity awareness training, campaigns, harassment simulations, role-plays, behaviour modelling, and team building programmes that give actual knowledge in dealing with difficult situations. These empirical preparation strategies are most suitable for changing individuals' mentalities, conduct and inspirations.
6. Anti-harassment policies: Effective policies and laws, combined with against harassment preparing for all staff, will help with counteracting provocation and encourage people who are being bothered to come forward and guarantee that the issue is tended to rapidly and viable. Hostile to harassment strategies should likewise put forward an itemized instrument by which people can make objections when inappropriate behaviour happens.
7. Policing approaches: In sexual harassment, proof and data about the offense are collected basically from three sources: the person in question; the suspect; and the crime scene (counting different observers). Officers ought to altogether examine every one of the three sources and some other accessible wellspring of evidence. In any case, cases can be effectively indicted regardless of whether the evidence uncovers

that proof from each of the three sources. In reacting to rape grievances, each exertion must be made to alleviate unfortunate potential sentiments of humiliation and embarrassment.

8. Central and state governments must adopt measures, including legislation, to ensure that private employers also observe and recognize sexual harassment as a serious offense. They should formulate an anti-sexual harassment policy at their organizational level and it should encompass prevention, prohibition and redressal of complaint mechanism.

Prohibition Method

1. The law generally disseminates charter or resolution or declaration for the prohibition of sexual harassment at all places and proposes to advance sexual orientation at public spaces and expels fundamental factors that contribute towards an unfriendly workplace against women.
2. If the accused found guilty, the authority may proceed disciplinary action like a written apology, warning, demotion, withholding of increments and promotion, termination from service, referring to a counselling session or carrying out community service.
3. If accused found guilty, imprisonment for a term which may extend to one year, or with fine, or with both may be imposed.

Redressal Method

1. Redressal is the aftermath of the harassment, every possible measure has to be taken to prevent harassment in the beginning stage only. Redressal of harassment differs from nation to nation as indicated by the type of legitimate protection accessible to victims of harassment. Depending on the severity of sexual harassment, complaints and findings of the investigator, redress actions may be extended.

2. The other purpose of redressal is, as far as possible, to put accused in the previous place if he would have not been involved in the harassment or if he found not a perpetrator.
3. Every organization or institutions has to adopt mechanisms to redress the complaints.

Conclusion

In South Asia, unbending cultures and prejudiced attitudes towards women have led to the harassment of women, from individual to community. The harassment has to be controlled at the preventive level rather than at the later stage. The poor legislation encourages abusers to commit offensive acts in opposition to women without any repercussions. To eliminate harassment against women in South Asia, certain fundamental areas should be strengthened such as an inclusion of family, community, organizations, institutions, state and civil society in sensitizing women issues, changing social and individual attitudes towards women, implementing policies and legislation for equity, sustaining the political will to achieve significant results, and following the best practice of other countries.

References

54.7% Urban Women Victims of Violence in Bangladesh: Action Aid Study. (2017, November 30). Retrieved from https://www.thedailystar.net/country/547-urban-women-victims-violence-bangladesh-study-1498573.

Bettencourt, A. (2000). Violence Against Women in Pakistan. Human Rights Advocacy Clinic; Litigation Report Spring. Retrieved from www.du.edu/intl/humanrights/violencepkstn.pdf.

Bhatia, G. (2003, December 22). Sensex Sniffs At Coke. *Outlook Magazine*. Retrieved from https://www.outlookindia.com/magazine/story/sensex-sniffs-at-coke/222436.

Central Statistics Organization. (2016). Afghanistan Living Conditions Survey 2013–14. National Risk and Vulnerability Assessment. In. Kabul. Google Scholar.

Curtis, M. J. (2014). Engaging Communities in Sexual Violence Prevention: A Guidebook for Individuals and Organizations Engaging in Collaborative Prevention Work. Texas Association Against Sexual Assault. Retrieved from http://taasa.org/wp-content/uploads/2014/10/Engaging-Communities-in-Sexual-Violence-Prevention.pdf.

Felix, Q. (2007). Honour Killing and "Karo Kari" in Pakistan. [Online]. Retrieved May 30, 2007, from http://www.asianews.it/view.php?l=en &art=1187.

Fenster, T. (2005). The Right to the Gendered City: Different Formations of Belonging in Everyday Life. *Journal of Gender Studies, 14*(3), 217–231.

Fikree, F. F., & Bhatti, L. I. (1999). Domestic Violence and Health of Pakistani Women. *International Journal of Gynaecology and Obstetrics, 65*, 195–201.

Gibbs, A., Corboz, J., Shafiq, M., Marofi, F., Mecagni, A., Mann, C., et al. (2018). An Individually Randomized Controlled Trial to Determine the Effectiveness of the Women for Women International Programme in Reducing Intimate Partner Violence and Strengthening Livelihoods Amongst Women in Afghanistan: Trial Design, Methods and Baseline Findings. *BMC Public Health, 18*(1), 164.

Human Rights Commission of Pakistan (HRCP). 2000). State of Human Rights in 1999. Lahore.

India Today. (2018, October 14). MJ Akbar Grabbed Me and Forced His Tongue into My Mouth: CNN Journalist. Retrieved from https://www.indiatoday.in/india/story/metoo-mj-akbar-cnn-journalist-1367090-2018-10-13.

Jagori & UN Women. (2011). *Safe Cities Free of Violence Against Women and Girls Initiative: Report of the Baseline Survey Delhi 2010* (p. 64). New Delhi, India: Jagori.

Kabir, R. (2018, March 7). *Study: 94% Women Victims of Sexual Harassment in Public Transport.* Retrieved from https://www.dhakatribune.com/bangladesh/crime/2018/03/07/study-94-women-victims-sexual-harassment-public-transport.

Mampilly, A. (2018). The Road from Codification to Implementation: A Critical View of Vishaka Guidelines to the Present Law. *The Penstand Journal, 1*(3). Retrieved from https://www.thepenstandjournal.in/2018/06/the-road-from-codification-to.html.

Manjoo, R., & McRaith, C. (2011). Gender-Based Violence and Justice in Conflict and Post-conflict Areas. *Cornell International Law Journal, 44*, 11–31.

NCRB (National Crime Records Bureau). (2016). *Crimes in India 2016-Statistics*. National Crime Records Bureau, Ministry of Home Affairs. Retrieved December 22, 2018, from http://ncrb.gov.in/StatPublications/CII/CII2016/pdfs/NEWPDFs/Crime%20in%20India%20-%202016%20Complete%20PDF%20291117.pdf.

Neupane, G., & Chesney-Lind, M. (2014). Violence Against Women on Public Transport in Nepal: Sexual Harassment and the Spatial Expression of Male Privilege. *International Journal of Comparative and Applied Criminal Justice, 38*, 23–38.

Sharma, I. (2015). Violence Against Women: Where Are the Solutions? *Indian Journal of Psychiatry, 57*(2), 131–139. https://doi.org/10.4103/0019-5545.158133.

United Nations Development Programme (UNDP). (2010). Power, Voice and Rights: A Turning Point for Gender Equality in Asia and the Pacific. Retrieved from https://www.undp.org/content/undp/en/home/librarypage/hdr/asia_and_the_pacifichumandevelopmentreport2010.html

35

We Are Stressed of Nuclear Power Plant: 2014 and 2017

Ghouse Basha Ahameed Mustafa
and Prince Annadurai

Introduction

The development in technology and scientific inventions with regard to the nuclear power plant in India is fast expanding. Some of these so-called innovative inventions and development from human beings often have caused detrimental effects to the environment. This can be considered to be equal to "harming the God's creation by the creation itself." In this globalist world, the constant thirst for urban development has become an unfair dominant source of pressure on the more deprived communities. One such example is the construction of nuclear power plants in villages, which are known to be highly dangerous for the environment and human beings (Malathi et al. 2008). This raises a significant

G. B. Ahameed Mustafa (✉) • P. Annadurai
The Department of Social Work, Madras Christian College & Ashokam, Chennai & Madurai, India

© The Author(s) 2020

Md. N. Momen et al. (eds.), *Building Sustainable Communities*,
https://doi.org/10.1007/978-981-15-2393-9_35

question, "what is the real situation among the people living nearby a nuclear power plant?"

The state and central governments were actively involved in the process of developing a nuclear power plant in Kudankulam, despite the active protests from the community against this project. Both the governments illustrate a complete lack of empathy and care for those affected, including any concerns for the future generation (Udayakumar, 2012). As a result of the power plant, the nearby village people's fishing has been terribly affected. They have been subject to immense amount of psychological stress since 2011 and, till this day, are still protesting against the power plant as a collective group of people. This study will discuss the emotional impact of the people who have for countless years protested against the development of the nuclear power plant.

J. Prabhakaran's (2013) book *Anu Ulai Arivom* (translating to "Atomic Reactor Knowledge") has provided extensive information about the actively running nuclear power plant. Currently, about 440 nuclear power plants are running all over the world. Worldwide, roughly 35 nuclear reactors had been operating for less than 10 years, 51 nuclear reactors between 11 and 20 years, 206 reactors had been running for 21 to 30 years, 135 reactors were operating worldwide from 31 to 40 years and finally 13 reactors had been operating for more than 40 years. About 129 nuclear reactors had shut down worldwide in the last decade (Table 35.1).

Table 35.1 Different countries' desire against the nuclear power plant: pre- and post-Fukushima incident

Countries	Before Fukushima (%)	After Fukushima (%)
Germany	67	79
Ukraine	52	71
Brazil	55	69
France	52	67
Russia	52	62
Japan	28	58
China	28	58
England	36	48
India	19	39

Majority of the people from various entices oppose nuclear power after the Fukushima incident, on average of 55%

Mental Health Report of the Chernobyl Nuclear Disaster

Harold M. Ginzburg (1993) in his book *The Psychological Consequences of the Chernobyl Accident—Findings from the International Atomic Energy Agency Study* had conducted an analysis of the symptoms experienced by the patients affected, their clinical examinations and the magnitude of anxiety and stress levels as a result of the Chernobyl accident. To summarise, the levels of psychological stress were recorded to be unequal to the biological importance of the radioactive contamination. Significant levels of stress based on sleep disturbance and other forms of distress were identified using standardised checklists and questionnaires, and interestingly, high levels of alcohol consumption patterns were also recorded in the control villages.

Radiophobia, a Soviet psychiatric term commonly used to describe the fear of radioactive material, was discovered to be of less significance or correlation when it came to applying the concept to a clinical situation. Subjects from the Chernobyl accident had no concept of the sinister effects of radiation and the different kinds of physical and mental illnesses that are associated with acute and chronic exposure to low levels of radiation. These subjects were uninformed about the levels of contamination, and taking this into consideration, it is completely understandable the high levels of distress, anxiety and concerns they would have. Answers given by the villagers to questions concerning relocation strongly suggested that those living in the contaminated villages wanted to be relocated. Over 70% of the inhabitants of contaminated villages preferred to move, in comparison to over 80% of the inhabitants of control villages who were not willing to relocate.

The importance of human life and health does not only concern the physical part but also the mental side. Mental health and physical health always influence and complement each other. One would not attain complete wellbeing without having both physical and mental fitness. According to Samuel Butler in his book *Stress, Trauma, Anxiety, Fears and Psychosomatic Disorders*, human beings tend to be in a constant state of anticipation about what is to happen in the future. More importantly, we tend to dwell on the negative consequences of our actions and choices. We constantly think about what others would comment on our future, allowing it to push us into further never-ending cycles of stress and negativity. Also, we often tend to misconceive the reasons behind other people's actions. On a more positive note however, such prospective thinking and stringent planning for the future may help us achieve our goals easier and, thus, reduce the stress levels. Despite this, pessimistic predictions tend to lead us in a negative and dark path where we may end up feeling down constantly (Butler 2004).

For some peculiar and unknown reasons, we are always blind to the more positive and practical of handling and reacting to a situation. The help is there if one needs it. The information regarding awareness about any issue is available and easily attainable. The ways to deal with a complexity are also there. However, we still choose to ignore such ways of dealing with difficulties. With technology, we have become too focused on the technical developments of a society yet choose to omit this side of development, which is undeniably more significant. These psychological theories could be applied to the Kudankulam Nuclear Power Plant (KKNPP) situation and the inhabitants in Idindhakarai village, and this forms the basis of this study. How aware of the seriousness are the people and the law makers? What kind of actions has been taken to deal with such issues? Or has more importance been given to technological developments?

Methodology

Idindhakarai village is located about 4.3 km away from Kudankulam, in Tirunelveli (district of the Southern Indian state of Tamil Nadu). The nuclear power plant in Kudankulam can be found about 1.5 km from the main village of Kudankulam and 2.3 km from the main village of Idindhakarai. The construction work started in March 2002. Delays in construction occurred for about 11 years because of anti-nuclear protests by the locals and people's movement against nuclear energy. In 2013, the plant was ready to start its power production, but another two years of delay occurred as a result of technical faults in the power plants. In 2015, the plant started successfully and now the plant is producing some electricity for several Indian states.

The people living in Idindhakarai village belong to a fishing community. Most of them belong to the same caste (backward caste), and the main religion is Christianity. There is a Roman Catholic Church, where the people would gather for meetings whenever the struggle committee calls for. Response was collect from 20 samples; there the universe was about 10,000 people in the village as per the source given by Dr S.P. Udayakumar, member of the struggle committee (Udayakumar, 2013). The research design used in this study was descriptive in nature. The study aimed to show how the livelihood of people of Idindhakarai village is affected because of the nuclear power plant, the awareness among the people of Idindhakarai village about the nuclear power plant and its hazards, the reasons for the stress among the people, the level of stress prevalence among the people due to KKNPP and reasons for cooperation (or the lack of) among the people in pushing toward social action.

Fundamentally based on the social learning theory of American psychologist Bandura, keeping social action as the main goal and utilising purposive sampling technique, the following criteria were used to select the sample. The respondents must have lived and be living in the Idindhakarai village at least for the past 20 years. The respondents' age should be above 20 years, as this age group is more likely to have more experience of participating in the protests against KKNPP. The sample size taken for the research was 40 respondents (20 males and 20 females). Though the researcher wanted to have more respondents, because of the tension that prevailed in the community at the time of data collection, he was asked to leave the community at the earliest by the struggle committee during the data collection period in January 2014. To record and analyse the updated situation, the researcher went to the village in August 2017, met the villagers and did a qualitative study with six respondents. During this period, the researcher faced a police enquiry for visiting the village and was unfortunately asked to leave the premises at once.

The researcher used a structured interview style to collect the primary data from the Idindhakarai village people. A stress scale was used to measure the stress experienced by the people, known as Perceived Stress Scale (PSS) developed by Sheldon Cohen. Statistical Product and Service Solutions (SPSS) software was used for the statistical analysis of the data collected in 2014. To analyse the information collected in 2017, the researcher used open-ended general questions from the 2014 interview questionnaire.

Common Response Regarding KKNPP in 2014

Every respondent faced challenges in their lives due to the nuclear power plant. An example is the quality of fishing being compromised due to the waste material/discharge being expelled from the nuclear plant. This is not only a huge concern for the wellbeing and economic stability for the families of this community but also illustrates a frightening environmental concern. In the study by TRANSCEND Media Service, S.P. Udayakumar explained about the transport and evacuation of the community people during any disasters, and he suggested that "more than 1 million people live within the 30 km radius of the KKNPP which far exceeds the Atomic Energy Regulatory Board stipulations. It is quite impossible to evacuate this many people quickly and efficiently in case of a nuclear disaster at Kudankulam." The researcher found that the respondents were facing problems related to the transport in and around the village because of the power plant. The respondents specified that the government has stopped transport services flowing in and out of the village. The government bus facility, being that of the cheapest transport service, was not readily available for the people. In this case, how would the villagers be evacuated during times of disaster? This situation also revealed the importance of transportation for civilians, which is an essential need for their daily life. And from personal experience of the researcher visiting the village several times, it was evident that the roads were not suitable for vehicle transport. Proper road works had not been implemented to solve this problem, and this had been used as an excuse for the government to halt all public transport services for the village.

The villagers were knowledgeable about the alternative sources to generate electric power, such as fossil fuel, biomass energy, geothermal energy, wind energy, tide energy, solar energy and water/hydro energy, and this was highly encouraging in terms of the level of awareness and education that prevailed among them. Being aware of this kind of information allowed them to make informed decisions about what was best for themselves and their village. In addition to this, all of the respondents were very much aware of the nuclear disasters and accidents that had occurred, as they were educated by the struggle committee and also through other sources such as all kinds of multimedia. The respondents were aware of other examples of nuclear disasters which occurred in Three Mile Island in 1979, Chernobyl in 1986 and Fukushima in 2011.

Being very emotionally attached to the village, the respondents did not want to be displaced. This was the reason behind them standing up for their rights and for their land. During the interviews, the researcher came across this particular phrase from the respondents stating that "This is our mother land (Idindhakarai) not India, we do not care about India but we care about Tamil Nadu." They were not at all worried about the government which functions only to cause more inconvenience to its people; however, they all were extremely worried about their fellow beings and nearby states.

An additional concern was the fear of displacement from their native village. The whole community was concerned and was worried about the future generation. Most of the respondents' thought comprised the following: "What will be the impact on my village and for my children if something like Fukushima happens here?" They were worried not for themselves but more so for their children and the future. A community member explains that, "We never care whether we die or live, we have already lived our lives. But what about our children? They need to live, don't they?"

Not only have the respondents of the study faced police arrest, but the researcher found out that the entire village have faced this situation, everyone including children. Cases have been filed on almost every single being in that village, for protesting against KKNPP. The researcher also found that the government has not provided any proper awareness about the safety measures on disaster management or emergency secure at all, till this day. And the government had not shown any concern for the people living in the village. In the beginning the government officials held a meeting for the village heads who were living around the power plant and had asked them to convey a message to the people that, "If they hear an emergency alarm sound from the plant they should immediately walk into any house which is open at that time, close the doors and cover their mouth with wet cloth." Did this sound like a plausible safety measure? Everyone in the village was genuinely angry about this suggested precautionary measure that was encouraged to follow during times of emergency, and this drove them to demand their rights even more strongly.

Common Response Regarding KKNPP in 2017

The researcher visited Idindhakarai after three years in 2017, and still there were no proper roads constructed; however the frequency of the public transport had increased, which was somewhat a positive news. The villagers themselves hired vehicles for their transport and for emergency purposes. Generally most of the respondents implied that they have started praying to god to protect them and their state people from this nuclear plant and that hopefully it should never face any disaster or accident. Some of them said that, "We are getting used to most of the things that were happening here, still the struggle is happening but nobody is ready to listen to us. Everyone were busy with Jallikattu Protest and Hydrocarbon Protest. But we strongly believe that all these protests were happening only because of us. We started to fight for our rights and this is still continuing now all over Tamil Nadu in different forms via different requests." Few said that, "We tried our best every time and we were pushed back by the government of Tamil Nadu, therefore we get stronger every time and now we are waiting to show this world how dangerous these Nuclear Power Plants are."

(****Quoted are statements translated from Tamil (primary language) as the respondent stated.****)

Results and Findings

Livelihood Affected in Idindhakarai Village Due to KKNPP: 2014

Mostly the youth and adults were the respondents for this study. The data was collected from them, in terms of the availability of the respondents in the village. A vast majority (94%) of the respondents' livelihood is solely based on fishing. The police have restricted the transportation to this village; because of this restriction, their livelihood was seriously affected. One fifth of the respondents said that they use motor bikes and shared auto for the transportation. Very few mentioned the rare availability of private buses and public buses for their transportation. In case of any emergency situations, the village people were also using sea for transport. There is a fear that the nuclear waste from KKNPP is mixing with seawater and spoiling the seafood. Consequently, these people have to sail far into the sea in order to find edible fishes, and this is a big challenge for them.

Due to the boundaries drawn around KKNPP limiting their fishing area, 45% of the respondents acknowledged the fact that there has been a decline in the availability of fishes from the places where they have worked for years. The rest said that they were not able to guess the climate so they were very much capable of making their assumption on sea. Some of them put their lives at great risk trying to fish over long periods of time looking for a reasonable amount of catch to sell, getting caught in hazardous weather in the sea. This is an extremely unfortunate and risky situation for the villagers. Because of KKNPP, 35% of the respondents were forced in such situations to change their livelihood which they were not used to in the past, which further highlights the complexity of the issue.

Some of the village people also lost their properties (land) which were located around the power plant construction. The government has taken these lands by giving false promises that they will pay the actual market cost, but they did not end up fulfilling their promise. These lands were agricultural lands in which they have been using to maintain an alternative livelihood such as farming. Loss of livelihood, loss of land, loss of rights and loss of respect. The female respondents had stated that hopefully as time goes by, they will get used to everything and things will start to settle. They seemed both dejected and hopeful at the same time. This was the state of Idindhakarai people in 2014.

Livelihood Affected in Idindhakarai Village Due to KKNPP: 2017

The unfortunate state of the villagers has not changed much since 2014. "It's been few years and after doing so many years of struggle no government body has responded to our demands nor any political parties especially those who begged us for votes. No one consider us as humans. Instead they have treated us as prisoners and terrorists." This was a statement provided by one of the respondents from 2017. As recorded in 2014, 94% of the respondents' livelihood was fishing. But now a respondent stated,

> We are still in fishing but not like how we were doing before 10 years. We have motor boats of course, we need fuel for that and we can go to deep sea for fishing, this is possible only for those who have motor boats. But for those don't have, they were fishing near the sea in kattumaram. After KKNPP they were not able to find fishes near the shore in their usual distance, and they don't have money to buy motor boats. They tried to apply for government schemes, but it's been 4 years and still nobody has got any reply. I am one among them, I left the village along with my family to Tootukurin and am doing fishing with my uncle. His family is here, I came to meet them. We just don't need this government, we don't need this development, let us live in peace please.

Another respondent said that,

> What we do in the sea with our boats when we don't find fishes in the sea? Should we catch the black material from KKNPP that floats on the sea water where the fishes were floating once? Will anyone buy and eat that? Government and banks are willing to give funds for catching those black material but not fishes. The banks will add more interest, which we will never be able to pay and one fine day we will end up killing ourselves. This is what the government is doing.

It was true that the people of Idindhakarai were migrating, and some were settling with the situations, as female respondents said in 2014 that they will be able to cope up with the situations and settle down as days go.

Awareness Among the People of Idindhakarai About Nuclear Power, Power Plant and Its Effects: 2014

The people are knowledgeable and well trained by the struggle committee. Most of the people know about the nuclear reaction as a heat energy (52%) and chemical reaction (38%). Other than this, they are aware of the nuclear power plant disasters and accidents (90%) such as Three Mile Island (TMI) in 1979 and Fukushima in 2011, as mentioned before.

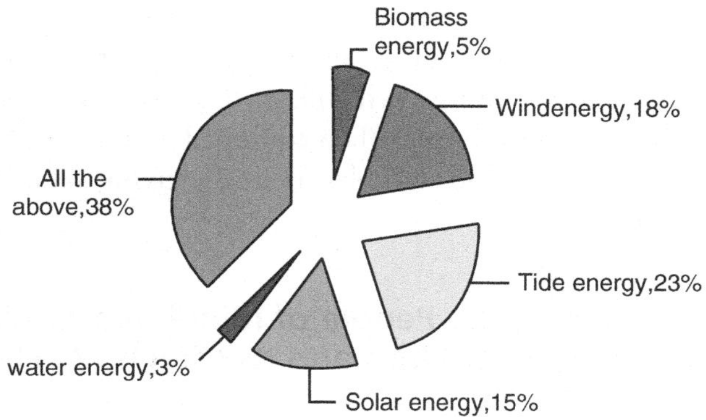

Fig. 35.1 The pie chart shows the findings of the respondents' knowledge on power generation sources

From Fig. 35.1 it is clear that majority of the respondents are very much aware and knowledgeable about the alternate sources. Around half of the respondents were able to point out all the known possible effects to the environment after a nuclear disaster or accident such as air pollution, seawater and seafood pollution, soil pollution, spoilage of drinking water and continuous radiation of the radioactive elements for years. It is very important that everyone knows the effects.

About 45% of the respondents mentioned the human being will suddenly die and will disappear like ash; 40% of the respondents were able to list down all the possible effects on the human being if there is a nuclear disaster or accident such as effects on food chain, alteration in DNA, spacing cancerous cells, loss of hair, bleeding from various parts of the body and sudden death. About 13% of the respondents said that bleeding from body parts and 2% of the respondents said that alteration in DNA will be present.

These results demonstrate that people from Idindhakarai village have a great knowledge on hazards of KKNPP than those who live outside the village. The reason why we try to know more and to gain extensive knowledge is because people are struggling for their life, culture, values and their practices with which they have lived all these years and are simply wanting to live a great life within their homeland. Knowing what nuclear power is and the history of nuclear disasters and accidents forced this community to live in a constant state of panic. Despite this, it is still better for the villagers to be aware of all the facts in order to make better informed decisions regarding their village, and also their hard-fought protests should not go to waste. When equipped with the right information and facts, justice should stand by their side giving value to all those years of protest.

Awareness Among the People of Idindhakarai About Nuclear Power, Power Plant and Its Effects: 2017

"As a result of all the effort put on us by the struggle committee, we came to know each and every thing about the Nuclear and its effect. But now we are living our life with a disaster nearby. We wake up every day looking at it. Most of us pray to it not to explode at any cost. Do you think we are ready to die here with this danger next to us?" "Though we have other forms of technology to produce power, why use this nuclear? Where is the educational system? Why our youngsters were not able to bring innovative ideas to produce power instead of this nuclear. We don't need this government nor its educational system." This is the common voice of the villagers.

More or less, the people of Idindhakarai village wanted the government to respond to their demands, and they were completely against the government (both the central and the state—Indian and Tamil Nadu government). Meanwhile, they were questioning the young generation about their education which has no use for the people of their country. They wanted the young generation to be innovative and discover something which will cause no harm for the nature. We all need to consider this valuable point, when it comes to educating the future generation.

Social Action Aptitudes from the Struggling Committee: 2014

Each individual in the village works democratically, and they all actively participate in all programmes. About 92.5% of the respondents said that they were committed to the struggle committee for the future planning and implementation. About 60% of the respondents said that they are able to adapt to the various situations imposed on them. A vast majority of the respondents said that they meet with the struggle committee and act according to the announcement. Simultaneously, they sit together, spend time and discuss things that have happened till date and plan for the future along with the struggle committee. The struggle committee works transparently with all the village people, so that everyone knows what is happening and that is the way of building trust within the community (Gabriel Paul 2006).

When it comes to improving health through community organisation and community building under the social action theory, uniform empowerment occurs when the community works together to strengthen their identity. Therefore from this research, it can be evidently seen that most of the village people are committed and accept each other since they work together, and this in return leads to strengthening them as a strong community (Minkler 2008).

Social Action Aptitudes from the Struggling Committee: 2017

"Struggle committee was organised by us, we decided to struggle for our people's life and for the future generation. We called S.P. Udhaya Kumar sir to lead us, because we had no knowledge about anything so we need a leader and he is our leader. Now, even without him we will be able to lead our struggle," was a respondent's statement. Another said, "We are waiting for the right time, with the right government in its position, then we will again raise our voice against KKNPP. When we raise again it won't be just Idindhakarai people but the whole Tamil Nadu or the whole India against Nuclear Power Plant" (Kumar 2012).

A protest or a struggle is not started from one particular individual, and that individual cannot be forced into leadership. Instead a protest or a struggle is a collection of individuals' decision after discussions within and with others to decide on the purpose whereby these individuals come together to stand for their rights with or without the aid of a leader, but preferably with a leader. This is the main findings observed from KKNPP struggle. Right now the protests might seem to have quietened down, but the drive within people of Idindhakarai has not. They are trying to give time for themselves to restore their energy for the future struggles. They are in need of more support from people who are living in Tamil Nadu as well as from other states of India. At the same time, they are very proud of themselves for such an endless struggle nearly 30 years now (1988 to 2017).

Reason for Stress Among the Idindhakarai Village People: 2014

Most of the respondents feared that they would not be able to survive in other places because they didn't know any other work than fishing, and fishing is their profession by birth and they didn't want to change that at any cost. They were willing to develop their fishing methods using technologies but never change their profession completely. Their worse fear was them being forced to leave their profession and their village, relocating to another areas starting from the beginning. On the other hand, each one of the villagers worried about their future generation; 23% of the respondents felt that this panic and stressful situation will keep continuing. About 47% of them felt that it would be extremely difficult to develop themselves without proper livelihood. Majority of the respondents were worried about their social status which might get stuck if they are forced to leave their own land and live like a refugee in some other places. On the other hand, about 30% of the respondents were reasoning that one day they will be facing a situation where they will have to migrate from their own land. And at least 22.5% of the respondents were thinking that stress will continue for everyone seeing this power plant, if they had continued living just behind it.

Some feared for police arrest (especially the 25% of the villagers who were caught and beaten by the police, most of them being middle-aged civilians), while most of them did not have the fear of getting arrested (75%). There were more reasons for stress which have not been included, as each individual of this village seems to be emotionally depressed and was facing stressful life for years.

Reason for Stress Among the Idindhakarai Village People: 2017

"After what we faced during the struggle, we have become more resilient and stronger. However after continuous police arrest, the struggle committee's strength was tried to be weakened by the government via different means. The government used the police and the border security force to get into the village and take over the situation. But unfortunately they were not able to, they were completely avoided by us (Idindhakarai villagers). Right now very few of them are here for the government's record sake. Some of our villagers here are informers of the police and CBI, we know that too, we know who they are but still we accept them because we believe they will change one day," a respondent said. "Right now the CBI and the police might have got information about you (the researcher) who is now inside the village and talking to people," another respondent said. These responses highlight just how closely every single civilian in the village is being watched, which shows the complete lack of privacy for an individual.

Prevalence Rates of Stress Among the People Living in Idindhakarai Village Because of KKNPP: 2014

The Perceived Stress Scale shows that majority (80%) of the respondents were experiencing high stress, 12.5% of the respondents were experiencing very high stress and 7.5% of the respondents were under the moderate stress level, as shown in Fig. 35.2.

As shown in Fig. 35.3, the elderly people (50 years and above) of the community experienced high stress level compared to the younger generation. The stress level was very high among female respondents (20%) than the male respondents (5%). More stress was experienced by the respondents whose monthly income was more than 8000 rupees per month (88%). About 45% of the respondents were facing very high stress because there was no freedom on sea which was one of the challenges while fishing. Those respondents who had awareness on environmental issues had experienced very high stress of 78%. In contrast, 40% of the

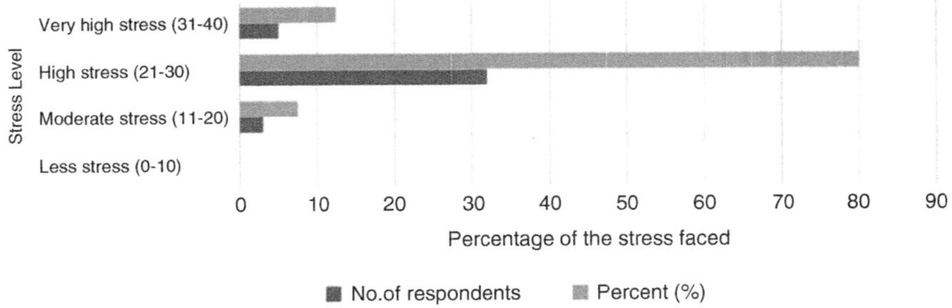

Fig. 35.2 Findings of the stress level among the people of Idindhakarai. (*** It is a measure of the degree to which situations in one's life are appraised as stressful. Items were designed to tap how unpredictable, uncontrollable and overloaded respondents find in their lives. ***)

Variables and Stress Level / **Hypothetical Analysis on stress variables**

Gender	Male		Female		Stress level %
Stress level	No. R	%	No. R	%	%
Moderate stress (11-20)	2	10	1	5	7.5
High stress (21-30)	17	85	15	75	80
Very high stress (31-40)	1	5	4	20	12.5

Very high stress was experienced by female respondents than male respondents. Whereas high stress is prevailed among male respondents.

Challenges faced in the sea	No freedom on sea		Change in fishing pattern		Can't guess the climate		Stress level %
Stress level	No. R	%	No. R	%	No. R	%	%
Moderate stress (11-20)	1	5.6	2	11.1	0	0	7.5
High stress (21-30)	14	77.8	14	77.8	4	100	80
Very high stress (31-40)	3	16.7	2	11.1	0	0	12.5

Null Hypotheses – H_0: There is no significant association between stress level and challenges faced in sea.
Alternative Hypotheses – H_1: There is a significant association between stress level and challenges faced in sea.
Level of significance: 5%
Test Statistic: x^2= 1.704 for 7 cells (77.8%),
Expected count = <5. Minimum expected count = 0.30
Table value = 0.202 Inference:
This table value is less than Pearson Chi-Square value. Therefore there is a significant association between stress level and the challenges faced by the respondents in the sea.

Respondents who have faced no freedom on sea experienced high stress than others. Change in fishing pattern affected their livelihood. This could be the reason for high stress. Therefore, this community people were facing very stress because of no freedom in the sea which is because of Kudankulam Nuclear Power Plant.

Age in years	20 – 30		31 – 40		41 – 50		50 <		Stress level %
Stress level	No. R	%	No. R	%	No. R	%	No. R	%	%
Moderate stress (11-20)	0	0	1	9.1	1	14.3	1	7.7	7.5
High stress (21-30)	6	66.7	10	90.9	5	71.4	11	84.6	80
Very high stress (31-40)	3	33.3	0	0	1	14.3	1	7.7	12.5

Null Hypotheses – H_0: There is no significant difference between stress level and age of the respondents.
Alternative Hypotheses – H_1: There is significant difference between stress level and age of the respondents.
Test Statistic: t^2= -0.609, for up to moderate stress level t^2= -0.709, for very high stress level Inference: Not assuming the null hypotheses where 95% confidence interval of the difference is nearly 1. Therefore there is significant difference between stress level and the age of the respondents.

This table shows that a majority 84.6% of the respondents who are above the age of 50 years experienced high stress. Invariably all respondents in the age group experienced high stress.

Monthly income in rupees	>5000		5000-7000		8000-10000		<10000		Stress level %
Stress level	No. R	%	No. R	%	No. R	%	No. R	%	%
Moderate stress (11-20)	0	0	0	0	1	5.9	2	11.1	7.5
High stress (21-30)	4	100	1	100	14	82.4	13	72.2	80
Very high stress (31-40)	0	0	0	0	2	11.8	3	16.7	12.5

The respondents who earn more than ten thousand rupees per month experiencing high stress which is maximum of 72.2% and minimum of 11.1% of the respondents were experiencing moderate. The respondents who earn eight to ten thousand rupees per month were under high stress with the maximum of 82.4% and minimum of 5.9% of the respondents were under moderate stress. From this cross tabulation, the people who were earning more than eight thousand rupees per month were facing very high stress and it has to be considered that even people with good earnings are also under stressful life.

Opinion on the environment	Air pollution		Sea water and food pollution		Continuous radioactive emission		All the other		Stress level %
Stress level	No. R	%	No. R	%	No. R	%	No. R	%	%
Moderate stress (11-20)	0	0	2	12.5	0	0	1	5.6	7.5
High stress (21-30)	3	75	13	81.2	2	100	14	77.8	80
Very high stress (31-40)	1	25	1	6.2	0	0	3	16.7	12.5

This tabulation reveals that the respondents who had more awareness on the effects of the environment, if a nuclear disaster happens in future experiences very high stress. Environmental effects such as air pollution, sea water and sea food pollution, continuous radioactive emission, drinking water pollution and soil pollution making the respondents to live with fear. Stress level increases based on awareness level.

Fig. 35.3 Findings of the stress with other variables: a statistical data analysis respondents said that they are afraid of their ability to survive in the migrated place in future.

Roughly 75% of the respondents have faced police arrest which had an impact among them that leads to a high level of stress. On average, the stress level of the general population has been ascertained to be a score of 7.6/10. Invariably, all respondents in the age group experienced high stress level according to the analysis. Not assuming the null hypotheses where 95% confidence interval of the difference was nearly 1. This shows that there was a significant difference between the stress level and the age of the respondents. This community people were facing very high stress because of the complete lack of freedom in the sea which was because of KKNPP, and there was a significant association between the stress level and the challenges faced by the respondents in the sea.

Prevalence Rates of Stress Among the People Living in Idindhakarai Village Because of KKNPP: 2017

One of the respondents said, "Supreme Court, High Court, Police or politician, none of them can ever frighten us. We are all answerable to god, no other than him." No scale was used to measure the stress in 2017, but from the discussion and interviews, it is possible to assume that the village people were getting used to the situation and cope with the stress they face every day through prayer meeting in the church to save them and prayers toward the power plant not to explode at any point. Moreover, women in the village were set their mind that this situation will get used to them, as it was a finding in 2014. Hence, one could conclude that they must be coping with the stress slightly better than the men of the village.

As the respondents stated that most of the youngsters had migrated to other places in India and abroad, it is possible to conclude that the stress level among the youngsters is still lower compared with the older adults as it was measured in 2014. Youngsters are much easier to migrate and they are more flexible; hence looking for more opportunities outside the village is easier for them.

Stress is part of everybody's life, which varies with different situations in which an individual is managing the stress. This is possible only by tackling the reasons that cause stress. Once the basic and the day-to-day needs of the people are met, these people will come out of such stressful situations. When the researcher questioned the respondents, they were clear in their answers and they had no doubts in their statements. Apart from this, the researcher felt like an alien when staying in Idindhakarai, even though he was a person from the state of Tamil Nadu itself.

Suggestions to Resolve Conflict

The government policies have no supportive measures in place to provide updated technologies which they can use to ameliorate fishing. They also have no proper safety measures to be conducted during emergency. Incentives, loans and grants should be provided to the people to develop their livelihood, and the government should not encourage migration of the people from their own homeland. The fishing boundaries between the seashore and the power plant construction have been fixed, making catching fishes become an impossible task near the power plant. Now there is no alternative plan for the fishermen community given by the government. After these many years, there have been no disaster management teams set up in the villages, which have to be set and pre-prepared for the safety purposes in all the villages that are closer to the nuclear power plant within the radius of 20 km.

The government should provide proper compensation to the respondents for having taken their land toward KKNPP. The government should provide basic facilities such as hospitals, drinking water, education and so on at an accessible distance, and it should have to respond to the people's basic needs. A government is nothing but "for the people, by the people and to the people." So government officials should work along with the people not only with the multinational companies, industries, factories and people living in the cities. Human rights, environmental protections and acts should be followed by the government (Dike 2000).

For example (taken from the Indian Constitution), following these acts could be a good starting point.

Articles on Rights:
Articles 19 to 22—Right to Freedom
Articles 38, 39, 39A, 40, 47, 48A, 51 and 51 A—Directive Principles of State Policy
Laws and Rules on Environment:

Water (Prevention and Control of Pollution) Act, 1974
Water (Prevention and Control of Pollution) Cess Act, 1977
The Air (Prevention and Control of Pollution) Act, 1981
The Environment (Protection) Act, 1986
The Hazardous Wastes Part II (Management and Handling) Rules, 1989

Conclusion

Concentrating on the development of the country lies not only in the hands of industries, companies, factories, IT sectors and so on. The development of any nations also depends on the villages, agriculture, fishing and animal husbandry. It is inevitable that it is these sources that provide us food and keep us alive. Therefore it is our duty to take care of them. Better plans on transport facilities, hospitals and safety measures in emergency situations, opportunities for loans and grants should be given to the people to aid their sustenance (Dike, 2000). Even after 968 days of protest, there is no change in the government policies and no proper response from the government (WNA Information Papers, 2013). This has been observed even now in 2017 and 2020, ongoing for more than a decade. This, in return, is causing an immense amount of distress for the Idindhakarai villagers, forcing nearly 36% of the population to migrate to different parts of India in search of a living. This is an actual shame on state and central governments (Tamil Nadu and India).

Every individual deserves to live a peaceful life. Does the world need the kind of development that affects the environment and violates human rights? All in all, this whole research is a critical comparative case study of a struggle for heading social action and highlights some key issues that not only every social worker but every human being needs to be aware of.

Appendices

Tools used for data collection

Stress faced by the people of Idindhakarai due to KKNPP

PERSONAL DETAILS

1) Name

2) Age

1. 20 – 30 years

2. 30 – 40 years

3. 40 – 50 years

4. 50 < years

3) Sex

1. Male

2. Female

3. Transgender

4) Marital Status

1. Married

2. Not married

3. Divorce

4. Widow

5) Educational Qualification

1. Illiterate

2. Non matric

3. Matric

4. Intermediate

5. Under graduate

6. Post graduate

6) Religion

1. Hindu
2. Muslim
3. Christian
4. Other, specify

7) Caste

1. BC
2. MBC
3. SC
4. ST
5. OC

8) Duration of living in this place

1. 1 – 20 years
2. 20 – 30 years
3. Native place

9) Type of House

1. Hut
2. Concrete
3. Tilled
4. Others, specify

B. To study how livelihood of the people of Idindhakarai village is affected because of KKNPP

10) What are your livelihood capability options?

1. Agriculture
2. Animal husbandry
3. Fishing
4. Construction labourer

5. Household industry
6. Others, specify

11) How much can you earn per month?
1. Less than 6000 rupees
2. 6000 – 8000 rupees
3. 8000 – 10000 rupees
4. More than 10000 rupees

12) Is that money is enough for you and your family?
1. Enough
2. Not enough

13) Apart from your income from livelihood do you have any other source of income?
1. I have some source
2. I do not have any sources

14) What are the sources of transport?
1. Bike
2. Bi-cycle
3. Share auto
4. Bus
5. Others, specify

15) How much do you spend for transport per month?
1. 100 -300 rupees
2. 300- 600 rupees
3. 600- 900 rupees
4. 900 & more rupees

16) Do you have saving accounts?

1. Have savings accounts
2. Do not have savings accounts

17) Did you try to buy any property?

1. Tried to buy and bought
2. Tried to buy but could not
3. Didn't try

18) Have you taken any steps initiated to develop your social status?

1. Taken steps
2. Not taken steps

19) Do you get any other livelihood options that you get to develop your life?

1. I get other livelihood options
2. I don't get any other livelihood options

20) What are the problems are you facing now towards your livelihood because of Kudankulam Nuclear Power Plant?

1. Change in livelihood options
2. Change in natural resources
3. Can't earn enough money
4. No Changes

21) Are you facing any challenges in your capability because of Kudankulam Nuclear Power Plant?

1. Facing challenges
2. Not facing challenges

If yes, what are the challenges are you facing?

1. Don't have freedom on our sea
2. Change in fishing pattern
3. Can't guess the climate
4. Others, specify

22) Is there any problems regarding your properties because of Kudankulam Nuclear Power Plant?

1. There are problems
2. There are no problems

23) Do you have any problems related to transport because of Kudankulam Nuclear Power Plant?

1. There are problems with transport
2. No problems with transport

If yes, what are those problems?

1. No proper roads
2. Government stopped local transportation
3. Roads are blocked for certain circumference
4. Others, specify

To understand the awareness among the people of Idindhakarai about the nuclear energy, nuclear power plant and its hazards

24) What do you know about nuclear?

1. It is a heat energy
2. It is a chemical reaction
3. Do not know
4. Others, specify

25) What do you know about power plant?

1. It is a big construction
2. It will generate electric current
3. Do not know
4. Others, specify

26) Do you know the other sources to generate power?

1. Yes, I know other sources
2. No, I don't know

If yes, what are they?

1. Fossil fuel
2. Biomass energy
3. Geothermal energy
4. Wind energy
5. Tide energy
6. Solar energy
7. Water energy
8. All the above

27) What are the sources of information, from whom you got?

1. Media and publications
2. Struggle committee
3. Community
4. Education
5. All the above

28) Are you aware of nuclear disasters and nuclear accidents?

1. Aware of nuclear disasters and accidents
2. Not aware of nuclear disasters and accidents

29) Are you aware of the nuclear hazards?

1. Aware of nuclear hazards
2. Not aware of nuclear hazards

30) What will happen to the environment if there is a nuclear disaster or accident?

1. Air pollution
2. Seawater and seafood pollution
3. Soil pollution
4. Drinking water will be affected

5. Continuous radioactive all over the place

6. All the above

31) Because of Nuclear disaster what will happen to the human beings?

1. Effects on food chain
2. Alteration in DNA
3. Spacing cancerous cells
4. Loss of hair
5. Bleeding from mouth and other parts of the body
6. Sudden death
7. All the above

32) What are the changes that have to face in future if there is a nuclear disaster?

1. Neonatal death
2. Birth defects in off springs
3. Reduction of lifespan
4. Skin and other organic cancers
5. All the above

33) Are you aware of Fukushima Nuclear disaster?

1. Aware of Fukushima
2. Not aware of Fukushima

To understand the reason for stress among Idindhakarai village people

34) Are you happy with your present livelihood?

1. Happy with the present livelihood
2. Not happy with the present livelihood

35) Are you emotionally attached with this village?

1. Emotionally attached
2. Not emotionally attached

If yes, what makes you emotionally attached?

1. This is your native place, culture
2. Your property is here
3. Your livelihood is here
4. A sense of belongingness of your spouse

36) Do you fear displacement?

1. Fear displacement
2. Don't fear displacement

If yes, why do you fear displacement?

1. Can't survey in other places
2. Won't have livelihood option
3. Can't stay freely, independently
4. Others, specify

37) Are you worried about your future generation?

1. Worried about the future generation
2. Not worried about the future generation

If yes,

What are the reasons to cause this worry?

1. Stress will continue
2. No development in social status
3. Migrate from their land
4. Others, specify

What do you think that can be the solution?

1. Don't need this power plant
2. Need life safety protection
3. Need incentives
4. Others, specify

38) Do you feel safe enough in this village?

1. Safe enough

2. Not safe

39) Do the government give you enough awareness about the safety measures during emergency?

1. Government have given enough awareness

2. No, government have not given any awareness

40) Whom do you get the awareness?

1. Media and publications

2. Community

3. Struggle committee

4. Education

5. All the above

41) Do you have the fear of getting arrested by the police?

1. Yes, I fear for getting arrest

2. No, I am not fear of getting arrest

42) Have you ever faced such situations?

1. Yes, I had

2. No, I don't

43) How many times did you face such situations?

1. 1 – 5

2. 5 – 10

3. 10 & more

44) Do you feel that issues relating to KNPP are creating stress in your family?

1. Creating stress

2. Not creating any stress

3. Not able to decide

To study the reasons for the cooperation among Idindhakarai village people in order to implement social action

45) Are you working democratically towards the issue?

1. Yes I am democratic
2. No I am not democratic

46) Do you actively participate in all the programmes regarding this issue?

1. I, participate
2. I don't participate

47) Do you feel that you all committed to the struggle committee?

1. We are committed
2. No we are not committed

48) Do you agree that each of you accept each other in this community?

1. I agree
2. I don't agree

49) What are the ways you communicate with the community?

1. Through meetings
2. By visiting homes
3. Others, specify

50) What are the ways in which you communicate with the struggle committee?

1. Meeting in live
2. Through mobile
3. Visit as a group
4. Others, specify

51) Are you ready to adapt to the situations?

1. Adaptable
2. Not adaptable

52) When do the community share information or plan for the protest?

1. Day
2. Noon
3. Evening
4. Night
5. According to the announcement from the struggle committee

53) Where do you gather to discuss?

1. Sea shore
2. In front of church
3. Struggle committee's place
4. Others, specify

54) How long does the meeting last?

1. 1 - 2 hrs.
2. 2 - 3 hrs.
3. 3 - 4 hrs.
4. 4 hrs & more

55) How often do you meet for discussions?

1. Once in a day
2. Twice in a week
3. Weekly once
4. According to the announcement from the struggle committee

To study the level of stress prevalence among the people, due to Nuclear power plant

The questions in this scale focuses primarily on your feelings and thoughts, experienced during the past month. Regarding each situation/case, you will enquired about how you felt or thought and asked to circle the respective answers.

Never =0, Almost never = 1, Sometimes =2, Fairly often= 3, Very often =4

1. As a result of KKNPP, how frequently did you feel sad, due to the happenings of unexpected events?

   0 1 2 3 4

2. As a result of KKNPP, how frequently did you feel like you were not in control of the situation?

   0 1 2 3 4

3. As a result of KKNPP, how frequently were you anxious and stressed?

   0 1 2 3 4

4. As a result of KKNPP, how frequently did you feel like you had lost the confident to handle daily problems and challenges in life?

   0 1 2 3 4

5. As a result of KKNPP, how frequently have you felt like problems were getting out of control?

   0 1 2 3 4

6. As a result of KKNPP, how frequently did you feel like you were unable to cope with the daily pressures of life?

   0 1 2 3 4

7. As a result of KKNPP, how frequently are you able to control irritations and difficult situations in your own life?

   0 1 2 3 4

8. As a result of KKNPP, how frequently have you felt like you were in control of things in your life?

   0 1 2 3 4

9. As a result of KKNPP, how frequently did you feel like frustrated due to things getting out of your control?

   0 1 2 3 4

10. As a result of KKNPP, how frequently did you feel overwhelmed with difficulties and felt like you were not able to overcome them?

    0 1 2 3 4

Informed Consent Form

This form was used to provide information to the research participant or and to attain the respondent's informed consent and permission.

Researcher: Ghouse Basha. A.M. BSc, MSW, BL

Title of Project: Stress faced by the people of Idindhakarai village about Kudankulam Nuclear Power Plant (KKNPP)

I am kindly requesting for your voluntary participation in my research study. Please browse through following information, regarding the project and if you are keen to participate, kindly sign in the appropriate place as indicated.

Objective of project:
To study and analyse the people of Idindhakarai village undergoing psychological stress due to Kudankulam Nuclear Power Plant.

If you participate, you will be asked to:
Share about your livelihood experience, awareness about the nuclear, nuclear power plant and its hazards. Then about your stress level towards the nuclear power plant and their factors. And finally the factors that binds you all strongly to keep going for a social action.

Time required for participation: Minimum 30 minutes and maximum 45 minutes

Potential Risks of Study: I may contact you in future to collect more details for this study and you need to help me.

Benefits: The central, state government will understand how Nuclear power plants are causing stress to the human beings, how confidentiality will be maintained.

I promise you that I won't discuss about your information with anyone except the professionals related to this study and also I will not mention your personal details anywhere in this study. If you have any questions about this study, feel free to ask about this study.

Voluntary Participation in project:

As stated above, subjects included in this study are volunteers. Deciding to not participate at any stage of the study will not have any negative consequences. However, if you do decide to take part, you have all the rights to stop participating or answering any questions which you would not to answer, at any time of the study.

I am consenting that I have read and understood the information stated above and I give my full consent to participate in this study, by signing this form.

Date / Signed on:

Name of the Participant:

Contact details: Signature of the participant

References

Butler, S. (2004). *Stress, Trauma, Anxiety, Fears and Psychosomatic Disorder* (pp. 35–40). Retrieved from http://www.psychologicalselfhelp.org/Chapter5.pdf.

Dike, V. (2000). *Leadership, Politics and Social Change, Nigeria and the Struggle for Survival.* Sacramento, CA: African Economic Analysis (pp. 3–5). Retrieved from http://www.afbis.com/analysis/leadership.htm.

Gabriel Paul, W. (2006). *A Study on the Impact of Ceasefire Among the Students of Jaffna University.* Chennai: Department of Social Work, Madras Christian College.

Ginzburg, H. M. (1993). The Psychological Consequences of the Chernobyl Accident—Findings from the International Atomic Energy Agency Study. *Public Health Reports, 108*(2), 184–192.

Kumar, D. (2012). Case Study on Kudankulam Nuclear Power Plant. *International Journal of Management Research*, p. 169. Retrieved from http://eprints.manipal.edu/139604/1/article%203.pdf.

Malathi, J., Selvasekarapandian, S., Brahmanandhan, G. M., Khanna, D., Meenakshisundaram, V., Santhanam, S., et al. (2008). Thoron Levels in the Dwellings of High Background Radiation Area Located Around Kudankulam Nuclear Power Plant. *Atmospheric Environment, 42*(21), 5494–5498. https://doi.org/10.1016/j.atmosenv.2008.02.039.

Minkler, M. W. (2008). *Improving Health through Community Organization and Community Building*. San Francisco: John Wiley & Sons.

Prabhakaran, J. (2013). Anu Ulai Arivom (First Edition, March 2013 ed.). (S. J. P. Kalingam, Ed., & R. Ravichandran, Trans.) Penny Cuic, Madurai, Tamil Nadu, India.

Udayakumar, S. P. (2012). People's Movement Against Nuclear Energy (PMANE) National Alliance of Anti-nuclear Movements (NAAM). Retrieved from http://www.dianuke.org/thirteen-reasons-against-the-koodankulam-nuclear-power-project.

Udayakumar, S. P. (2013, January 28). Thirteen Reasons Why We Do Not Want the Koodankulam Nuclear Power Project. Activism, Asia & The Pacific, Energy, Kudankulam Anti-Nuclear Satyagraha India (pp. 1–2.) Retrieved from http://www.transcend.org/tms/?p.

WNA Information Papers. (2013, March). Current and Future Generation Plans for New Reactors Worldwide. Retrieved from http://www.world-nuclear.org/info/Current-and-Future-Generation/Plans-For-New-Reactors-Worldwide.

36

Hajj and *Umrah* Management in Bangladesh: Issues, Challenges and Human Rights Perspective

Isahaque Ali, Azlinda Azman,
Paramjit Singh Jamir Singh, Mahathir Yahaya,
Zulkarnain A. Hatta, and Muhammud Shariful Islam

Introduction

Hajj is the pilgrimage to the sanctified places of Mecca and Medina in the Arabic month of *Dhulhijah* and *Umrah* means intentional trips to divine sites and visits to the sanctuaries of holy personages (Kessler 1992). The act of pilgrimage is the fifth among the five pillars[1] of Islam. *Hajj* is

I. Ali • A. Azman (✉) • P. S. Jamir Singh • M. Yahaya
Social Work Programme, School of Social Sciences, Universiti Sains Malaysia, Penang, Malaysia
e-mail: azlindaa@usm.my

Z. A. Hatta
Faculty of Social Science, Arts & Humanities, Lincoln University College, Petaling Jaya, Malaysia

M. S. Islam
Department of Social Work, University of Rajshahi, Rajshahi, Bangladesh

© The Author(s) 2020

Md. N. Momen et al. (eds.), *Building Sustainable Communities*,
https://doi.org/10.1007/978-981-15-2393-9_36

obligatory at least once in a lifetime for every Muslim who is economically solvent to travel to Mecca and Medina. A total of 126,000 pilgrims from Bangladesh performed the pilgrimage in the year of 2018 (Arab News 2018). Out of an estimated 126,000 Bangladeshi Muslims performing pilgrimage, 120,000 went through private pilgrimage agencies and only 6000 went under government arrangement (Bdnews24.com 2018). In Bangladesh, pilgrimage operations are conducted both under government and private schemes (The Independent 2015a). The figure shows that the number of pilgrims is larger in the private arrangement than in the government system; however, mismanagement and agonies reported by the pilgrims were mostly against the private scheme.

Bangladeshi pilgrims have also indicated that there has been increased security restrictions on the issues of pilgrimage and *umrah* visas by the Saudi government, human trafficking, the government's monopoly on pilgrimage airfare, higher pilgrimage fees, flight delays, poor accommodation as well as transport issues.

[1]There are five pillars in Islam: declaring one's complete faith that Allah (God) is the only Supreme Being and Muhammad (SAW) is the messenger of Allah; performing five prayers a day; donating 2.5% of annual income through *zakat* (a charity tax to help the needy); fasting (which includes no eating, drinking, nor intimacy) during the daytime in Ramadan; and making a pilgrimage to Mecca (Muhammad's birthplace) at least once in a person's life if one is able.

In short, mismanagement by private as well as government pilgrimage agencies have compounded the problems of many aspiring pilgrims in Bangladesh. The plight of Bangladeshi pilgrims is pathetic right from the planning stage and the intensity of problems keeps increasing with their journey to their destination. In India and Pakistan, pilgrimage are professionally managed as compared to Sri Lanka and Bangladesh, where the control of private agency and broker are much prevalent (Bianchi 2017, p. 13). A sustainable management system should be developed to perform pilgrimage but is very difficult to execute in Bangladesh (Getz 2009; Henderson 2011). This chapter sets out to discuss the problems and difficulties being faced and the violation of religious freedom and human rights of Bangladeshi pilgrims and suggests practical solutions for mitigation of sufferings.

Methodology

This chapter depended on the review of secondary data and used different secondary sources for analyzing the data and literature on *hajj* and *umrah* management. The limitation of this chapter perhaps, is the limited source of data. Evaluating the significance of the issue, the existing data and literature published on *hajj* and *umrah* management issues in Bangladesh is very scarce. However, according to Saunders et al. (2011), secondary data can be used to generate new knowledge, new hypotheses or that reduces the burden placed on respondents by negating the need to recruit further subjects, which allows the wider use of data from rare or inaccessible respondents. The secondary sources included scholarly journals, articles, government documents, daily newspapers and websites have been used as the data for the article.

Issues and Challenges for Pilgrims in Bangladesh

This section deliberates the concerns and challenges faced by the Bangladeshi pilgrims in their own country and Saudi Arabia. The concerns and challenges are as follows:

Allegation for Excessive Pilgrimage Fees

The Cabinet of Bangladesh has approved the National Pilgrimage and *Umrah* Policy and Pilgrimage Packages, 2018, which settled on a point that there will be two packages under the government management in which Bangladesh Taka (BDT) 3,97,929 (US$4974.16) was fixed under Package-1 as a fee, including *Qurbani* (sacrificial ritual) (The Independent 2018). In Package-2, BDT 3,31,359 (US$4142) was fixed as a fee, excluding Qurbani fee and it was BDT 3, 19,355 (US$3991.94) the previous year in 2017. The fee is much higher in comparison to other countries such as Malaysia, India and Nepal (The Independent 2018).

In comparison to Malaysian practice, the cost for pilgrimage in 2018 is RM18,890 (US$ 4551.80) per person in Malaysia. For a pilgrim performing their pilgrimage for the first time, the cost is only RM 9980 (US$2404) per person, where the Pilgrimage Fund Board Malaysia (*Tabung Haji*) will subsidize another RM 8910 (US$2147) and total subsidy provided was Malaysian Ringgit (RM) 4 million (US$975,140) (Salim 2018). But in the case of Bangladesh, there is no such subsidy from the government. Moreover, the majority of the private pilgrimage agencies are reluctant to follow government policy in Bangladesh. In recent years, with the introduction of middlemen, prospective pilgrims have become more vulnerable to exploitation (The Observer 2016). Similarly, the sacrificial ritual fee under the package includes a premium; however, sometimes these agencies demand more charge for that (The Independent 2015a).

To aggravate matters, the majority of Bangladeshi pilgrimage agencies appoint agents, sub-agents and brokers to help them meet their quotas. In the yearly pilgrimage process scene '*Imams* (religious leaders)' have achieved a robust position between the pilgrims and owner of private pilgrimage agencies (Siddiqi 2014). Indeed, pilgrimage agencies have become completely reliant on the so-called *Imams* of several local mosques of the country for business and make profit from the pilgrims in Bangladesh. These brokers have become so dominant that they demand the price of the pilgrimage package and obtain secure benefits for themselves from these pilgrimage agencies. The attendance of a large number of private pilgrimage agencies in the market makes the job of bargaining by *Imams* easy because innocent pilgrims trust them (Ummah Link 2014). A big number of private pilgrimage agencies are likely to be motivated by the *Imams* to attend private packages. It has become a business and they are in the marketplace to generate revenue by cheating pilgrims in Bangladesh (Ummah Link 2014). There are allegations against many licensed agencies for charging large amounts of money from the pilgrims of private agencies pilgrimage package. It is also apparent that private agencies not only charge undue amounts but do not provide appropriate services to the pilgrims that has been promised. Such allegations are also under inquiry against 119 private pilgrims agencies in Bangladesh (Siddiqi 2014).

Human Trafficking and Visa Issues

Nearly 60,000 pilgrims from Bangladesh perform *umrah* every year (Arab News 2015). Each year, many pilgrims do not return to their countries and hence become illegal immigrants in Saudi Arabia for earning of remittances. A total of 11,485 pilgrims did not return in the year 2015, as reported by the Saudi Authorities (The Independent 2015b). There were numerous allegations regarding human trafficking against 104 agencies, including the president of the Private Agencies Association of Bangladesh (HAAB) (The Independent 2015a). A large number of youth pilgrims traveling to Saudi Arabia to escape from poverty and unemployment in home country agree to put their life in danger by illegally staying and pursuing a position there. They are likely to engage in a host activities such as sleeping on the streets, scrounging for odd jobs, risky behaviors, and trying to keep a step ahead of the law (Bianchi 2017, p. 3). The Saudi authorities have stopped issuing visas to Bangladeshi citizens from March 2015, claiming illegal human trafficking through some corrupt agencies (Karim 2015). *Umrah* visas are issued only for a period of 14 to 28 days and do not require any police verification or ministry certification. It leaves scope for unscrupulous local and *umrah* agencies to take advantage of the human trafficking in the name of pilgrimage and *umrah* (Karim 2015).

Around 21,000 pilgrims failed to get visa for the pilgrimage in the year of 2015 (The Independent 2015a) and 6000 pilgrims failed to collect visa due to cheating of unscrupulous agents in 2018 (Bdexpress24news 2018). Besides, the suspension of visa for agency representatives called *Monazzem* (pilgrimage manager), has been creating more complications for the management of pilgrimage in Bangladesh. In earlier periods, *Monazzem* received multiple entry visas from the Saudi administration and therefore could often travel on behalf of pilgrimage agencies from Bangladesh and Saudi Arabia to carry out their executive tasks. Nevertheless, in 2015, the Saudi delegation refused such multiple entry visas to *Monazzem* amongst rumours that they were given to state-managed pilgrims and to civic official and employees (The Observer 2016). According to the Saudi government, "The Royal Embassy of Saudi Arabia in Bangladesh presents its compliments to the Ministry of Foreign Affairs Bangladesh, in respect of issuance of *umrah* visa for Bangladeshi pilgrims for this year in 2016, has the honour to convey that the authorities concerned in the Kingdom of Saudi Arabia have informed that *umrah* visa has 'been allocated to Bangladeshi nationals through Saudi *umrah* agencies and companies applying for that" (The Independent 2015a).

Monopoly over Air Fare by Government

Pilgrims have to pay an increasing figure every year for air ticket, food and accommodation because of the fixing of the airfare by the government of Bangladesh (The Observer 2016). Additionally, by limiting pilgrimage operations to only two airlines (Bangladesh and Saudi Airlines) and fixing pilgrimage charges at high prices, the government is striking higher rate on the pilgrims by refuting the benefit of market competition in Bangladesh. It is evident that the participation of additional airlines into the pilgrimage process would result in more capacity, better, efficient and effective of the pilgrimage operations in Bangladesh (The Observer 2016).

Moreover, if government stops setting the fare and leaves it to the airlines, the pilgrims are likely to get cheaper air fare in Bangladesh. The HAAB filed a writ petition against the government decision on the monopoly of airfare in Bangladesh. The pilgrims and organizers have been urging and complaining about the growing problem; the government needs to agree to the operation a third carrier to offer competitive prices for airfare. In response to the petition, The High Court termed the government decision as illegal and allowed the pilgrims to choose from a wider option of carriers. However, the government ignored the ruling of the High Court (The Observer 2016).

In the year 2018, the government fixed US$1550 as the airfare for pilgrimage (Ministry of Religious Affairs 2018), quoting an example from India where airfare for pilgrimage is subsidized by the government. That is why in 2018, pilgrims who performed Pilgrimage Committee paid only Rupees 20,000 (US$286.64) as airfare in India (The Times of India 2018). India is one of the most important non-Muslim countries that provide of publicly subsidized *hajj* (Bianchi 2017, p. 13). Another example is Nepal, where the Qatar Airways carried pilgrims at a fare of only US$821 in 2018 (The Times of India 2018). In the year of 2018, the government of Bangladesh fixed US$1550 air fare for pilgrimage (Ministry of Religious Affairs 2018). This is higher in comparison to pilgrimage fare from neighbouring countries such as India and Nepal.

Accommodation and Transport Issues

There is practice of falsehood while booking the accommodation for the pilgrims in and around Mecca for pilgrimage and *umrah*. A group called 'Building Selection Team' selects accommodation for pilgrims in different categories according to their pilgrimage scheme. Usually photographs of beautiful and attractive buildings with quality services are advertised at the time of booking and the price is charged accordingly. But when the pilgrims reach Mecca, they often find that there has been a breach of the agreement. The accommodation offered is of poor quality and service, for example, without air-conditioners, lifts, availability of water and clean toilets. Bangladeshi pilgrims have complained that the units where they were made to stay were far away from the Ka'bah (sanctum sanctorum) of Mecca (The Observer 2016).

The cost of the accommodation is determined by its distance from the Ka'bah—the closer one is to it, the higher will be cost. There are cheaper houses in hilly areas around Mecca's Ring Road but only strong and young people can manage to live there. Yet elderly pilgrims were placed around these hills. Usually, pilgrims pay the total amount in advance for their entire journey, which includes air-conditioned travel by vehicles from Jeddah to Mecca and all other places to perform pilgrimage rituals, but they are often compelled to travel in old and worn-out buses with no proper doors, windows and air-conditioners. These buses literally burn in the desert heat with the mercury rising as high as 65° Celsius (The Independent 2015a).

Lack of Proper Management and Flight Delay

In Bangladesh, there are two categories of pilgrimage: public and private. The government makes all arrangements for the pilgrims traveling under the public scheme. But ironically, arrangements for the majority of pilgrims (more than 99,000) are left in hands of private pilgrimage agencies (Siddiqi 2014). According to the Policy and Pilgrimage Packages 2016, a total of 113,868 devotees were able to go to Saudi Arabia in 2016. Among this number, only 5000 pilgrims were to go under government management, while 108,868 under private management (The Observer 2016). The government of Bangladesh lacks the capacity to manage this large number of pilgrims in Bangladesh and once they are in Saudi Arabia.

The absence of appropriate administration for pilgrimage has created a few problems for pilgrims in Bangladesh. The majority of the pilgrims are heedless of the on-board security checks and the insufficient number of flights causes flight delays in Dhaka and Jeddah Airport (Dhaka Tribune 2014). Sometimes, the pilgrims have to wait for several hours at the airport due to inordinate delay in boarding a flight and collecting their baggage that adds to the sufferings of the pilgrims. In addition, pilgrims are forced to stay much longer in Saudi Arabia due to higher fare and trouble in getting seats due to the restriction of pilgrimage to only two airlines. A large number of pilgrims have to stay for about 50 to 60 days instead of normal stay of about 22 days. Pilgrims have to spend additional money for food and other basic needs during this unnecessary and longer stay (Siddiqi 2014). The absence of a pivotal action taken by the pertinent law enforcement authorities to address this issue causes anger and frustration on the part of pilgrims in Bangladesh.

Issues of Police Clearance Certificates

Along with the administrative problems, a new rule affecting Bangladeshi pilgrims is the demand for a police clearance certificate for pilgrims of age less than 50 years for the prevention of human trafficking in Bangladesh. To prevent human trafficking, the police clearance certificate is essential for pilgrims to collect visa (Ummah Link 2014). Moreover, the police harasses and forces most of pilgrims to pay bribe in the name of issuing police clearance certificate. People have opposed this decision arguing that as every pilgrim gets a passport after police verification and clearance certificate in Bangladesh, this certificate is unnecessary (The Observer 2016).

Human Rights and the Constitution of Bangladesh

Human rights are rights inherent to all human beings, regardless of sex, race, ethnicity, nationality, religion, language, or any other status. Human rights include the freedom from slavery and torture, right to life and liberty, the right to work and education, freedom of opinion and expression, and many more. Every citizen can enjoy these rights without any discrimination (United Nations 1948). The constitution of Bangladesh includes some noble provisions from Universal Declarations of Human Rights (UDHR) of 1948, the Declaration on the Granting of Independence to Colonial Countries and Peoples of 1960 (DGICCP), the International Covenant on Economic, Social and Cultural Rights of 1966 (ICESCR), the International Covenant on Civil and Political Rights of 1966 (ICCPR) and the Declaration on the Right to Development of 1986 (DRD) and is also a signatory of these organizations (Mondol and Mondol 2017). The Constitution of Bangladesh, Article 27, mentions "all citizens are equal before law and are entitled to equal protection of law"; Article 28 (1) states, "the State shall not discriminate against any citizen on grounds only of religion, race, caste, sex or place of birth" (MLJPA 2014, p. 16).

Moreover, Article 31 states, "to enjoy the protection of the law, and to be treated in accordance with law, and only in accordance with law, is the inalienable right of every citizen, wherever he may be, and of every other person for the time being within Bangladesh, and in particular no action detrimental to the life, liberty, body, reputation or property of any person shall be taken except in accordance with law" (MLJPA 2014, p. 17). Article 10 determines "a socialist economic system shall be established with a view to ensuring the attainment of a just and egalitarian society, free from the exploitation of man by man" (MLJPA 2014, p. 12). According to Article 12, the principle of secularism shall be realized by the elimination of (b) The granting by the State of political status in favour of any religion; (d) Any discrimination against, or persecution of, persons practicing a particular religion (MLJPA 2014, p. 12). And finally, Article 41(1), mentions, (a) Every citizen has the right to profess, practice or propagate any religion (MLJPA 2014, p. 20). As a citizen, it is the obligation of the Government of Bangladesh to promote and protect human rights of pilgrims in terms of law, exploitation, and harassments (Mondol and Mondol 2017).

Discussion

Undoubtedly, the Bangladeshi pilgrimage agency makes exaggerated promises and raises pilgrim expectations, both verbally and in terms of advertising, for setting a relatively high cost against low standards. It is unfortunate that these package suppliers are benefiting from helpless pilgrims including the elderly, sick and disabled and charging them high and unjustified prices for the pilgrimage and *umrah* packages. The latter is mostly a major problem for the uneducated, poor and vulnerable pilgrims. It is the crucial accusations that these agencies claim on extra charges for visa fees, airfare and performing rituals in Mecca and Medina in Saudi Arabia (The Independent 2015a). Accountability for pilgrimage management belongs to the Ministry of Pilgrimage, Saudi Arab, directed by the Supreme Hajj Committee. The ministry of Hajj, Saudi Arabia, mentioned that profitable and non-profitable interest should be balanced and to ensure that "religious duties are not transformed into a money-making product" (Ministry of Hajj 2010). The negligence of Ministry of Religious Affairs Bangladesh provides a great chance for operating fraud travel agencies, without fear of the grip of the law (The DAWN 2017). Article 7 of the UDHR and Article 27 of the Constitution of Bangladesh mention all citizens are equal before the law and are entitled without any discrimination to the equal protection from the law. It is the statutory duty of the Ministry to protect the citizens from any exploitation; however, they fail to protect human rights of pilgrims in Bangladesh.

The pilgrimage to Mecca and Medina is one of the five pillars of Islam and it is mandatory for Muslims who are financially capable and physically able to undertake this sacred journey once in life. Performing pilgrimage is not obligatory for a Muslim who cannot afford it. However, in Bangladesh, a large number of people who lack financial affordability and are physically unable to withstand the hardship, ignore all odds to make the pilgrimage because these pilgrims are honoured in their family and community; what has been reckoned as an incentive to perform the pilgrimage (Siddiqi 2014). Pilgrimage has a religious implication among Muslims; they believe that through pilgrimage they can come to closer to their Creator (God) (Timothy and Iverson 2006). Keeping this divine faith and innocent view, many pilgrims are likely to have spent their life-savings to fulfil their desires and wishes to perform their religious obligation (McLoughlin 2013). Article 14 of the constitution of Bangladesh mentions that it is a fundamental responsibility of the state to protect the workers and backward sections of the people from all forms of exploitation. However, the poor pilgrims are always cheated by fraud travel agencies, which hampers the religious freedom for pilgrims in Bangladesh. Public awareness should be created by the Ministry of Religious Affairs through print and electronic media in village-, Thana- (sub-district) and district-levels, indicating the approved and authorized Pilgrimage Agencies listed by the Ministry with a fixed quota (Siddiqi 2014; The DAWN 2017). The Pilgrimage Agencies will be liable to extend a guarantee to the pilgrims that the indicated details of expenses and facilities will be provided to them during transit from Bangladesh to Saudi Arabia, including food, lodging, transport and other facilities (The DAWN 2017).

There is a lot of corruption in the management of pilgrimage and this extends to various agencies, both in the government and private sectors. Allegations of irregularities and bribery have been raised against high officials of the Ministry of Religious Affairs (MORA), Bangladesh, about the out-of-the-way allocation of pilgrimage quota for the pilgrimage agencies in Bangladesh. A quota mechanism has been started by the Saudi administration that indicates that the official control is strictly exercised to determine pilgrim entrance into Saudi Arabia (Henderson 2011).

However, the quota has reduced the number of pilgrims and pilgrimage has moved to *umrah* where travels are controlled by weaker regulated private segment travel agencies in Bangladesh (Bianchi 2017, p. 3).

Moreover, there is very limited or no control by the government of Bangladesh on the pilgrimage agency and service packages industry. The Pilgrimage Agency allegedly deceived a large number of innocent pilgrims and made a lot of money. Sometimes, the owners of travel agencies disappear and pilgrims are unable to travel and their money is not refunded; this indicates the violation of the human rights and freedom of expression and worship of pilgrims in Bangladesh (The DAWN 2017). Senior officials from Ministry of Religious Affairs, Bangladesh, including the HAAB leaders, are involved in those corruptions in Bangladesh, which is a violation of human rights and country's laws (The Independent 2015a). According to the constitution of Bangladesh, the victims whose rights have been violated might file a trial in the court against the fraud pilgrimage agency. It is the constitutional and statutory obligation of police, authorities and relevant ministries for the prevention of such crime within the country, but they fail.

The MORA has been wasting time, manpower and money by conducting less public pilgrimages and leaving the rest at the mercy of private agencies. A large number of government officials are amassing benefits annually during pilgrimage in the name of organizing public pilgrimages (Siddiqi 2014). In fact, through their illegal actions, they are helping the private agency owners by taking bribes and other illegal benefits. Moreover, without proper verification, they are giving licences to pilgrimage and *umrah* agencies that are not eligible for getting such permission (The Independent 2015a). In Bangladesh, travel agents are operating with political connections (Bianchi 2017). These kind of policies violate freedom of religion and worship of the Bangladeshi pilgrims. Furthermore, some agencies whose licences were cancelled earlier for their alleged involvement in human trafficking and fraudulence are getting new licences by giving illegal benefits to concerned official authorities. According to the Saudi administration, application for pilgrim's visa must be collected by a licensed agency in their home country and Ministry of Religion encourages pre-departure training with details of pilgrims (Henderson 2011). However, new travel agents and other private agencies with no experience in pilgrimage operations have entered this sector and are responsible for such malpractice in the entire pilgrimage system, causing agony to the pilgrims in Bangladesh (McLoughlin 2013). Article 21(1) of the constitution of Bangladesh highlights that it is the prime duty of every citizen and public servant to observe the Constitution and the laws, to maintain discipline, to perform public duties and to protect public property (MLJPA 2014, p. 14). However, a majority of the public servants of Bangladesh are not following their obligation according to the country's constitution and laws that protect the religious freedom of pilgrims in Bangladesh.

The skilled human resource in pilgrimage management has no alternatives, since, the entire management includes range of tasks, for example, registration and selection of pilgrims, renting of lodging in Mecca and Medina, issuance of security clearance certificate, collection of visa, plan of air tickets, medical check-up, immigration, pre-departure food and lodging of pilgrims in Bangladeshi and Saudi Arabia (Ummah Link 2014). The MORA has totally failed to monitor the activities of the private pilgrimage agencies due to severe shortage of skilled and experienced persons in MORA and pilgrimage mission that is responsible for arranging pilgrimage for most of the pilgrims in Bangladesh (Siddiqi 2014). The government of Saudi Arabia asks for the careful screening and processing of pilgrims by the Bangladeshi officials, which indicates that they respect the human rights and freedom of religion for pilgrims in Bangladesh (Mason 1993).

Additionally, private pilgrimage agencies provide poor food, lodging and other basic needs by the assistance of *Imams* (Ummah Link 2014). The lodging offered to Bangladeshi pilgrims is also located at a distance from the two holy mosques because it is cheaper. The number of pilgrims placed in a room is more that it can accommodate (The Guardian 2016). It is exploitation of the pilgrims by the *Imams* in the name of religion. The *Imams* brainwash the pilgrims in such a way that the latter bears all the sufferings as a part of pilgrimage, with the expectation of reward from Allah against facing these difficulties. They are told that during pilgrimage if they express any kind of displeasure about food, lodging, transport and other kinds of worldly luxuries, their pilgrimage will not be gratifying or even may be unacceptable to the creator (God) (Siddiqi 2014).

There is an indication that a huge number of pilgrims for pilgrimage or *umrah* experience a lot of problems due to the carelessness and inexperience of pilgrimage and *umrah* agencies in Bangladesh. A majority of pilgrims have been dissatisfied and upset by the insufficient services, lack of guidance and misbehaviour of their tour managers; altogether these poor services and mismanagement during pilgrimage and *umrah* have made the experience of the majority of pilgrims a hostile one during last few years (The Observer 2016). Human rights in Islam stem from two foundational principles: dignity and equality. Dignity is a fundamental right of every human being merely by virtue of his or her humanity. As God states in the Quran, "We have honored the children of Adam and carried them by land and sea; We have provided good sustenance for them and favored them specially above many of those We have created" (17:70). Additionally, Article 1 of the UDHR and Article 11 of the Constitution of Bangladesh say, all human beings are born free and equal in dignity and rights. It indicates that it is an obligation to respect the worth and dignity of a person regardless of age, religion, race and status of life. However, the people from pilgrimage agency always mistreat pilgrims, which is violation of human rights and religious freedom in Bangladesh.

Currently, it is a considerable public and political concern of trafficking of human beings and human trafficking is a more profitable global criminal business that claims millions of victims. Human trafficking represents one of the most serious human rights problem in the contemporary world, especially in the developing countries (Davidson 2010, p. 244). In reality, for a long time, Bangladesh pilgrimage agencies operated more as a medium of human trafficking of young men who desired to go abroad as illicit workers. Eventually, the government of Saudi Arabia pressurized the Bangladesh government to restrict the travel agents that aid human trafficking (Bianchi 2017; Financial Express 2013; Dhaka Tribune 2013). Therefore, the unscrupulous pilgrimage agencies are earning lot a money by human trafficking in the name of pilgrimage and *umrah* in Bangladesh. However, the government of Bangladesh failed to take necessary action against these agencies in Bangladesh. After observing such malpractices in the system, the Saudi authorities expressed their dissatisfaction over the Bangladeshi pilgrimage management system. The

Saudi authority may stop or restrict pilgrimage visa in the future, if the fraudulent agents remain unpunished through legal procedures (Karim 2015).

According to legal experts, no instance of punishment for human trafficking activities has yet been reported in Bangladesh since the enactment of two separate laws named Human Trafficking Deterrence and Suppression Act-2012 and Overseas Employment and Migration Act-2013 for combating human trafficking and labour trafficking respectively (Karim 2015), which speaks volume about the serious lack of effective law in this regard. However, as the host nation, it is very important and significant on the part of the Saudi Arabia administration that they should inspire all pertinent governments throughout the world to take the pilgrimage and *umrah* seriously to assist and defend their future pilgrims from suffering, and to avoid the loss of valuable human lives. According to the present government rules of Saudi Arabia, the state will be accountable of taking full take care of pilgrims in Mecca and Medina. The state should provide safety and care to the pilgrims to assist them to complete pilgrimage, *umrah* with amity (Ministry of Hajj 2010).

The MORA, Bangladesh, should show zero-tolerance policy for cracking down on the responsible persons for this increasing dishonesty and corrupt practices in the pilgrimage and *umrah* service provider agencies. This will ensure the gratification and satisfaction of Bangladeshi pilgrims in accomplishing their religious duty with peace harmony. However, Bangladesh as a country, has so far failed to develop a smooth and well-organized annual pilgrimage operation for its citizens to perform pilgrimage that reflects the protection of human rights and freedom of worship (Siddiqi 2014).

Policy Recommendations

The study suggested few measures to improve the present pilgrimage and *umrah* management system in Bangladesh. These include scrutinizing passports by a committee to identify genuine pilgrims, holding back pilgrims below 50 years of age, dropping the names of those pilgrims who have traveled for pilgrimage more than once in the past five years, distributing the names of all the pilgrims among the agencies and creating a fresh database. The policy makers of Bangladesh may follow the Malaysian policy as a model which does not permit a person for registration as a pilgrim if they have performed pilgrimage in the last five years (Tabung Haji 2016). The pilgrimage database should be linked to the immigration department, especially at airports and other strategic locations, and be easily accessible to the concerned personnel. The data in the pilgrimage database would be very helpful in tracing and identifying the illegal migrants once they are caught.

The government of Bangladesh could learn from the Pilgrimage Fund Board Malaysia called *Tabung Haji*. Malaysia can be taken as one of the outstanding and excellent examples in managing pilgrimage as compared to the other Islamic countries. The *Tabung Haji* is a body that is responsible for pilgrimage and *umrah* management in Malaysia. *Tabung Haji's* principal objective is to assist a smooth and truthful service for pilgrims, especially the rural and urban poor, who had previously been ragged off by dishonest brokers and travel operators (Parker 2010).

Tabung Haji is a nationwide and local institution that would be a mutual savings institutions and not a profit-oriented organization where limited stockholders (very often *Imams* and so-called religious leaders) are the profit recipients. Pilgrims are main beneficiaries, in the case of adopting *Tabung Haji* model. The senior management and the Board of Trustees are selected by the government in Malaysia (Parker 2010). *Tabung Haji* is not only an investment fund and a savings but also an educational organization thus preparing a would-be pilgrim on the climate, culture, attire and health care on the religious rituals of pilgrimage. This would create healthier, informed future pilgrims, which might add to the smoother functioning of pilgrims in the holy cities of Mecca and Medina (Parker 2010).

Furthermore, the rule of law should be prevailed in the case of any allegation of fraud and irregularities committed by pilgrimage and *umrah* agencies and culprits should be penalized, and deposit money should be confiscated and their licences as operators should be cancelled. Without proper verification, pilgrimage and *umrah* agencies should not be provided licences which are not eligible for getting such permission as per set requirements. The licences should be given properly after scrutinizing all the submitted documents. During this scrutiny process, those agencies whose previous activities were found unsatisfactory should be blacklisted and terminated from the list of applicant agencies. When pilgrims make complaint for malpractice against any agency, the matter should be immediately investigated by the concerned law-enforcement agencies and if the complaint is valid, such agencies should be banned for further intake of pilgrims. The freedom of religious belief and religious exercise shall be guaranteed by the present Constitution in Bangladesh. The state should be better able to protect citizens by national laws with international laws in order to prevent the violation of religious freedom against human rights (Mason 1993). Pilgrims not only travel within the country but also travel to other countries to visit sacred places. The government of Bangladesh should ensure the smooth travel of pilgrims to sacred places as acts of devotion prescribed by their religion or belief—whether inside or outside their own country.(Mason 1993).

MORA should undertake the operation and strict supervision of the agencies through constituting a Management Committee. The established Management Committee later can undertake the role in monitoring the service agencies in their activities directing towards the welfare of the pilgrims in Bangladesh. The Committee may start a scrutiny of the existing licenced agencies as a first step towards bringing discipline in the pilgrimage management in Bangladesh. The licence of those agencies which have violated terms and conditions of their appointment can be cancelled. Moreover, strict action may be taken for the punishment of those agencies that have not complied to the conditions of their contracts made with the government and for their negligence in giving due services to the pilgrims for which the agencies have taken money from the pilgrims. The provision could be made for punishment of the agencies by way of penalty of their security money, cancellation of their licence or imposition of fine; in a case of grave allegation, necessary step may be taken to punish the agencies through the judiciary for dishonesty.

Finally, the fear of Creator (God) (man is answerable to Him for all his activities) and spirituality (divine pleasure) should be preached among the pilgrimage management personnel that all the pilgrims are the guests of Creator (God). Hence, they should fear Creator (God) before deceiving the pilgrims. Muslim society is based on purity of feeling, love, sincerity towards every other Muslim and fulfilment of promises to every member of society. Cheating and deception are alien characters in contrast to the noble character of a true Muslim. Islam has declared dishonesty and fraud as a terrible sin and a source of shame to the one guilty of pledging them, both in this world and the Doomsday. The Prophet (PBUH) did not simply condemn them by excluding them from the Muslim in this world, he also proclaimed that on the Day of Judgement, every traitor would be raised carrying the flag of his betrayal. A caller will cry out from the vast arena of Judgement, pointing to him, drawing attention to him: "Every traitor will have a banner on the Day of Resurrection and it will be said: This is the betrayer of so-and-so" (*Saheeh Al-Bukhari*). Therefore, a Muslim who has true Islamic sensitivities must avoid deceit, cheating, treachery and lying, no matter what benefits or profits such activities might bring for him/her under the pretext of pilgrimage services in Bangladesh.

Conclusion

It has become crucial that the government of Bangladesh should not act as a mere spectator of this continuing terrible disorder surrounding pilgrimage management, rather it must take appropriate and stern actions against the unscrupulous and corrupt government officials and private pilgrimage agencies that handle pilgrimage affairs. The government of Bangladesh should give serious attention for collaboration with Malaysia, which have a practical and effective model for pilgrimage and *umrah* management. The government of Bangladesh could develop its own model for pilgrimage and *umrah* management by reviewing the model of other countries. There should not be any shyness to adopt a good model from other countries. Good lessons should be learned from previous experiences and training programs should be made obligatory for these prospective pilgrims for effectively performing *hajj* and *umrah*, and the rituals must be in compliance with the defined Islamic ways (*syariat*). With regard to human trafficking, the Saudi Arabian authorities have banned 67 Bangladeshi agencies from sending pilgrims in the year of 2015 for their alleged involvement in human trafficking, fraudulent activities, irregularities and mismanagement during the year 2014. This has badly damaged the image of the country. The time has come to end this abuse and MORA can do it by executing the pilgrimage policy for the advantage of all the pilgrims, not just within a minor section of public scheme in Bangladesh. MORA should be well organized with truthful, devoted and trained manpower in order to facilitate the guests of Creator (God) from Bangladesh to perform their religious obligations in peace and sanity during the holy pilgrimage. Last but not least, the government of Bangladesh should initiate research on that issue with the collaboration of experts and experienced researchers in order to find the proper solution for pilgrimage and *umrah* management in Bangladesh. Finally, according to the holy text of the Quran, Muslims travel to Saudi Arabia to sanctify the Allah. Therefore, it is the duty and responsibility of the Saudi administration to provide hospitality and take care of pilgrims who, under Islamic law, enjoy the human rights (Fahim and Dooty 2014).

Conflicts of Interest The author declares no conflict of interest.

References

Arab News. (2015). Bangladeshi Umrah Visa Ban Lifted as Corrupt Travel Agencies Fixed. Retrieved from http://www.arabnews.com/saudi-arabia/news/854656.

Arab News. (2018). 126,000 Pilgrims from Bangladesh Will Perform Hajj 2018. Retrieved August 20, 2018, from http://www.arabnews.com/node/1357746/saudi-arabia.

Bdexpress24news. (2018). *About Six Thousand Pilgrims Hajj Is Uncertain.* Retrieved August 23, 2018, from https://bdexpress24news.com/4886/about-six-thousand-pilgrims-hajj-is-uncertain/.

Bdnews24.com. (2018). Biman Bangladesh Cancels Two Hajj Flights for Shortage of Pilgrims. Retrieved August 20, 2018, from https://bdnews24.com/bangladesh/2018/07/27/biman-bangladesh-cancels-two-hajj-flights-for-shortage-of-pilgrims.

Bianchi, R. R. (2017). Reimagining the Hajj. *Social Sciences, 6*(2), 1–36.

Dhaka Tribune. (2013). Riyadh Blacklists Thirty-One Hajj Agencies for Trafficking. Retrieved September 4, 2018, from http://archive.dhakatribune.com/bangladesh/2013/apr/30/riyadh-blacklists-31-hajj-agenciestrafficking.

Dhaka Tribune. (2014). Returning Pilgrims Held Up by Delayed Flights. Retrieved from http://www.dhakatribune.com/bangladesh/2014/oct/13/returning%2D%2Dpilgrims-held-delayed-flights.

Fahim, S. T., & Dooty, E. N. (2014). Islamic Tourism: In the Perspective of Bangladesh. *Global Journal of Management and Business Research, XIV*(I), I–VII.

Financial Express. (2013). Private Hajj Operators Suffer Blow over Human Trafficking Claim. Retrieved September 4, 2018, from http://print.thefinancialexpress-bd.com/old/more.php?page=detail_news&date=2013-06-22&newsid=173927.

Getz, D. (2009). Policy for Sustainable and Responsible Festivals and Events: Institutionalisation of a New Paradigm. *Journal of Policy Research in Tourism, Leisure and Events, 1*(1), 61–78.

Henderson, J. C. (2011). Religious Tourism and Its Management: The Hajj in Saudi Arabia. *International Journal of Tourism Research, 13*(6), 541–552.

Davidson, J. O. (2010). New Slavery, Old Binaries: Human Trafficking and the Borders of 'Freedom'. *Global Networks, 10*(2), 244–261.

Karim, M. (2015). Ministry Yet to Take Actions against Errant Agencies. Retrieved from http://www.observerbd.com/2015/06/04/92333.php.

Kessler, C. S. (1992). Pilgrims' Progress: The Travellers of Islam. *Annals of Tourism Research, 19*(1), 147–153.

Mason, P. W. (1993). Pilgrimage to Religious Shrines: An Essential Element in the Human Right to Freedom of Thought, Conscience, and Religion. *Case Western Reserve Journal of International, 25*(3), 619–653.

McLoughlin, S. (2013). Organising Hajj-Going in Britain: A Preliminary Report. Retrieved from http://www.arts.leeds.ac.uk/crp/current-research/hajj/.

Ministry of Hajj. (2010). Responsibilities of the Ministry of Hajj. Retrieved September 2, 2018, from http://www.hajinformation.com.

Ministry of Religious Affairs. (2018). Hajj Package 2018. Retrieved August 20, 2018, from http://www.hajj.gov.bd/wp-content/uploads/2018/02/haj-package-2018.pdf.

Mondol, M. S., & Mondol, R. (2017). Constitution of Bangladesh and Human Rights: A Comparative Study. *IOSR Journal of Humanities and Social Science (IOSR-JHSS), 22*(11), 09–18.

MLJPA. (2014). Bangladesh's constitution of 1972, reinstated in 1986, with amendments through 2014. Retrieved July 10, 2019, from http://extwprlegs1.fao.org/docs/pdf/bgd117108E.pdf.

Parker, M. (2010). Tabung Haji, a Truly Shari'ah-Based Institution. Retrieved from http://www.arabnews.com/economy/islamicfinance/article190862.ece.

Salim, Z. (2018). The Cost of Performing the Haj Remains at RM9,980 per Pilgrim. Retrieved August 20, 2018, from https://www.nst.com.my/news/nation/2018/06/385901/cost-performing-haj-remains-rm9980-pilgrim.

Saunders, M. N., Saunders, M., Lewis, P., & Thornhill, A. (2011). *Research Methods for Business Student, 5/e*. Pearson Education India.

Siddiqi, R. (2014). Most Bangladeshi Hajj Pilgrims Are Exposed to Exploitation. Retrieved from http://www.greenwatchbd.com/most-bdeath-hajj-pilgrims-are-exposed-to-exploitation/.

Tabung Haji. (2016). Services and Operators. Retrieved from http://www.tabunghaji.gov.my/.

The DAWN. (2017). Supreme Court Regrets Deceiving Of Haj Pilgrims. Retrieved August 20, 2018, from https://www.dawn.com/news/1317464.

The Guardian. (2016). Scandal of the Hajj Pilgrims Who Are Cheated by Devious Tour Operators. Retrieved August 20, 2018, from https://www.theguardian.com/money/2016/oct/08/scandal-hajj-pilgrims-cheated-devious-tour-operators.

The Independent. (2015a). Aggrieved Agencies Seek PM's Intervention to Sort Out Anomalies. Retrieved from http://www.theindependentbd.com/printversion/details/7226.

The Independent. (2015b). 95 Agencies Punished for Human Trafficking. Retrieved from http://www.theindependentbd.com/printversion/details/23556.

The Observer. (2016). Cabinet Okays Policy. Retrieved from http://www.observerbd.com/2016/01/12/130654.php.

The Times of India. (2018). Haj 2018: Relief for Pilgrims as Airfares Slashed. Retrieved August 20, 2018, from https://timesofindia.indiatimes.com/india/haj-2018-relief-for-pilgrims-as-airfares-slashed/articleshow/63098995.cms.

Timothy, D., & Iverson, T. (2006). Tourism and Islam: Consideration of Culture and Duty. In D. Timothy & D. Olsen (Eds.), *Tourism, Religion and Spiritual Journeys*. London: Routledge.

Ummah Link. (2014). Airfare Monopoly Hurdles Bangladesh Hajj. Retried from http://www.ummahlink.org/airfare-monopoly-hurdles-bangladeshi-hajj/.

United Nation. (1948). Universal declaration of human rights. Retrieved July 10, 2019, from https://www.un.org/en/universal-declaration-human-rights/.

37

Conclusion: The Modern Welfare States–Civil Society Organization and the Welfare Challenges—Global and Local Debates

Md. Nurul Momen, Rajendra Baikady, Cheng Sheng-Li, and M. Basavaraj

The point of departure for this volume is that globalization and neoliberalism have resulted in tremendous challenges to the governments of the South Asian countries. Despite development and economic progress, countries in the region continue to suffer from enormous social problems and challenges. The region is the home for 'poverty', one of the most distressful and disturbing conditions that poses serious challenges to the development of the region as well as of the world (Mohanty et al. 2015).

Md. N. Momen (✉)
Department of Public Administration, University of Rajshahi, Rajshahi, Bangladesh

R. Baikady
Paul Baerwald School of Social Work and Social Welfare, Hebrew University of Jerusalem, Jerusalem, Israel

Department of Social Work, University of Johannesburg, Johannesburg, South Africa

© The Author(s) 2020

Md. N. Momen et al. (eds.), *Building Sustainable Communities*,
https://doi.org/10.1007/978-981-15-2393-9_37

The range of chapters covered in this volume from India, Pakistan, Nepal, Bangladesh, Bhutan and Sri Lanka are divided into seven parts in which a wide range of issues are discussed, including understanding and explaining poverty, women empowerment, refugee crisis and organic farming community, a broad range of civil society organizations (CSOs) working for the welfare of marginalized groups in society, the governance process, and the political transformation, democratic practices and the challenges faced by South Asian civil society. All of these chapters deal extensively with the role of the third sector and focus on the welfare of the population. However, there are debates and discussions about the strategic concerns pertaining to civil society's role both as a direct social welfare provider and as a watchdog of the democratic welfare of state provisions in the South Asian region.

The international community has pledged to reduce poverty and hunger by half. Scholars argue that the issues related to poverty in the South Asian society and their solutions still remain an important agenda in global development debates (Handley et al. 2009). However, the poverty agenda in South Asia is still critical to global poverty reduction, and the region accommodates more than 33 percent of poor people in the world (World Bank 2016), and almost 15 percent of the population were living under the poverty line in 2013 (Ortiz and Roser 2015). The United Nations General Assembly (2015) acknowledged that poverty continues to be the most vital challenge faced by the world, whereas alleviating poverty in all its forms and dimensions by 2030 is a prerequisite for sustainable development (World Bank 2016). In the South Asian region, the incidence of poverty has become a matter of national and regional concern.

Throughout the book, we have tried to explore the strengths and shortcomings of CSOs, and their role in addressing the challenges and gaps

---

C. Sheng-Li
Department of Social Work, Shandong University, Shandong, China
e-mail: chengsl@sdu.edu.cn

M. Basavaraj
Department of Economic Studies and Planning, Central University of Karnataka, Kadganchi, India

faced by diverse marginalized populations in the South Asian society. It is worth noting that 36 specially commissioned chapters in this volume, each representing a South Asian country, shed light on the issues that are most experienced by the South Asian society. Authors in this volume have explored the role of civil society in the welfare of different marginalized large sections of people across the South Asian countries. Studies have already documented the civil society's contributions in social welfare, such as health services, education, social and other services, and recognized it as a critical contributor to promoting a healthy democracy (Waters et al. 2003; Peters et al. 2004; Casey 2004; Frumkin 2002; Phillips 2003; Putnam 2000). Overall, the volume is an exploration of the relation of civil society and the state, women's empowerment, women's rights, child welfare, globalization and its impact on the CSOs, sustainable development, building sustainable communities in the South Asian region and the challenges faced by CSOs in conducting their welfare activities. These discussions are important in the South Asian context, as the region continues to remain trapped in poverty and is struggling to tackle issues related to poverty, while the rest of the world is busy celebrating the technological advancement and accumulating the wealth gained from globalization.

Globalization–Neoliberalism and Welfare Challenges

The terms globalization and neoliberalism are widely used both in academics and in political debates, yet the impact and meaning of the concepts of globalization (Gordon 1987; Mittelman 1996; Amin 1997) and neoliberalism (Schmidt and Thatcher 2013) are highly contested. Neoliberalism has become a catchword among the social sciences to denote the socioeconomic and political changes that have taken place in industrial societies since the 1970s (Crouch 2011; Thelen 2014; Brown 2015; Davies 2017a). In an argument, Gordon 1987; Mittelman (1996) contest the direction, involvement and impact of the word globalization. Similarly, Amin (1997) argues that the rising literature on globalization is confusing, and leaving people unclear about the concept and implication of globalization. Some other scholars have also questioned the emergence of globalization (Yeates 2001), which has created differences of opinion and provoked arguments. Scholars have also expressed their concern that it is difficult to find an academic definition of globalization which is holistic and inclusive of all its aspects (Scholte 2005). However, one definition associates globalization with a "process of compaction and intensification of connections and relationships of the economic, political, military, cultural, ideological, and between human societies" (Robertson 1992; Scholte 2000). The most commonly used definition of globalization is "the process of increasing interdependence among the nations" (Chase-Dunn et al. 2000; Guillén 2001; Meyer 2017; Rugman and Verbeke 2004; Verbeke et al. 2018). However, the discussion here is about the impacts of globalization on the provision of welfare. We have tried to explore the facts and figures as to how globalization affects the life of poor and marginalized sections in the less developed economies.

As explained in the earlier definition, the process of globalization leads to international trade and financial liberalization, which may result in exploitation of the previously protected sectors, and their workers (Buckley and Ghauri 2004). Further, globalization leads to low wage or loss of employment opportunities for the lesser skilled and unskilled workers, resulting in increasing wage inequality (Acemoglu et al. 2016; Autor et al. 2013). While examining the globalization and welfare nexus, there are two hypothetical explanations: (a) the race-to-bottom hypothesis holds that people may enjoy reduced corporate tax rates and reduction on tax rates on interest incomes as a provision of the national government, but these reductions are only made with the intention of attracting more foreign investment (e.g., Sinn 1997, 2003); (b) the compensation hypothesis puts forward an optimistic view of globalization (Cameron 1978; Rodrik 1998), in which the government increases the social expenditure in order to compensate for the uncertainty and risks; however, available evidence shows that this hypothesis appears to be more applicable in terms of tax rates and public spending (Cameron 1978; Dreher et al. 2008b; Potrafke 2009; Walter 2010; Meinhard and Potrafke 2012; Gaston and Rajaguru 2013a, b; Herwartz and Theilen 2014; Gozgor and Ranjan 2017; Yay and Aksoy 2018; Gründler and Köllner

2018—for surveys see Schulze and Ursprung 1999; Ursprung 2008; Potrafke 2015). Further, globalization-induced effects on welfare spending also differs across different welfare state regimes, such as social democratic, conservative and Mediterranean welfare state regimes and liberal welfare state regimes (Yay and Aksoy 2018).

The negative effect of globalization is expected to be stronger on the developing welfare states when compared to the developed welfare states (Rudra 2002; Wibbels 2006). The reason behind this adverse effect on developing countries is the lack of manifold institutions at hand to compensate for the risks and threats of globalization while developed countries could manage these situations with advanced institutional settings. Further, the weak labor union and its bargaining power in developing countries make these countries more vulnerable to reduced social expenditure by the government in the course of globalization, whereas a strong trade union and labor unity are key to a collective voice in developed countries that prevent the government from decreasing social expenditure (Rudra 2002). The workers in the developed economies are skilled, and they have managed to consolidate their collective interests, whereas workers in developing countries are mostly unskilled, and have weak trade unions as a result (Rudra 2002). It is stressed that weak trade unions prevent workers from getting the benefits of globalization. In addition, the government's capabilities of borrowing from capital markets to spend on social affairs are comparatively less than in industrialized countries (Wibbels 2006). As evidence to this, Latin American countries have procyclical fiscal policies that support social affairs spending during an emergency situation. Further, increase in public expenditure has not been equal around the world; for instance, social expenditure tends to increase among the high-income countries in contrast to decrees among the low-income countries in the East (Leibrecht et al. 2011; Onaran et al. 2012; Onaran and Boesch 2014). Schuknecht and Zemanek (2018) describe this trend of increasing social expenditure and its impact as 'social dominance'. However, there is heterogeneity among the OECD countries with respect to the globalization effect, despite a drastic increase in social expenditure among these countries in recent years.

Globalization effects with respect to welfare provisions are more pronounced in Asian countries rather than in other developed countries.

This is because many Asian countries do not have the financial resources and the cultural inclination to support the large welfare state. It is important to note that in the East Asian countries social assistance is often provided either by the family or by the firm. In fact, this is the reason behind developed non-OECD countries being more marked with effects of globalization on social expenditure than the less developed OECD countries. Further, in the high-income countries income dominates the citizen's views on government spending for social protection (Lim and Burgoon 2018). Evidence to this is the fact that citizens exposed to economic globalization tend to support welfare spending more than citizens who are not exposed in Japan and Singapore (Lim and Burgoon 2018).

The neoliberal economic policy paradigm emerged, and got prominence in the West in the late 1970s (Hall 1993; Fourcade-Gourinchas and Babb 2002), and proposed the governance of societies based on increased competition and free markets and criticized the dominant state control over the market (Davies 2017b). However, this discourse soon lost its legitimacy (Hall 1993; Crouch 2011). The neoliberal reform proposed an economic policy to bring the Western capitalist societies back to a state of steady growth, low inflation and higher level of company profit (Boltanski and Chiapello 2005; Evans and Sewell 2013; Jenson and Levi 2013). Similar to globalization, neoliberal policies were also based on incentives for individuals (Foucault 2008; Connell and Dados 2014). In addition, neoliberalism advocated for punishing those who did not pay their taxes, according to the rules. As the neoliberal policies spread in the society, the competition and market mechanisms became dominant. Thus, the principles of neoliberalism started reflecting in modified business strategies (Dobbin and Jung 2010), health care, educational reforms (Le Galès and Scott 2008 cited in Beckert 2020), housing, infrastructure, education privatization (Crouch 2011), wealthy friendly tax reforms (Bartels 2004; Harrington 2016; Palan 2017), liberalization of financial market (Krippner 2011; Streeck 2014; Fligstein and Goldstein 2015), welfare reforms (Kronauer and Linne 2005 cited in Beckert 2020), cultural reorientation (Evans and Sewell 2013) and dependency on environmental and social standards (Bartley 2018).

Undoubtedly, neoliberalism had a positive impact on the societies in the low- and middle-income countries. The global market concept with the value chain also resulted in availability of inexpensive commodities and industrial profit in the developed countries. In addition, investors, owners of the financial capital and the highly elite class involved in the market economy got benefits and profits. However, similar to globalization, the neoliberal economic policy lost its legitimacy and led to social inequality (Hall and Lamont 2013; Blyth 2013; Piketty 2014), and also detached from the society's quest for justice and stability (Streeck 2014) by mainly focusing on economic gains and profit-oriented markets. As a result of profit rootedness, higher education for the middle class became expensive and led to the burden of education loan on some and denial of access to education to many. Further, lower salaries forced people to become debtors (Krippner 2017; Gonzalez 2019). Globalization and neoliberalism promoted the global integration of value chains and production, which resulted in job loss for many people who were not compensated for their loss even though they were directly associated with the effects of globalization and the consequences of the adoption of the neoliberal policy (Hochschild 2016). Globalization has also been linked to the degradation of the environment and the violation of labor rights (Bartley et al. 2015). The criticisms of these moves have been recorded by research studies across the social science discipline focusing on income inequality and wealth (Piketty 2014; Atkinson 2015; Korom 2018).

Need for Governmental–Nongovernmental Partnership in Building Just Societies

Civil society organizations are the organizations working outside the state and are made up of individuals with common interests coming together to achieve a common goal. Further, Newton (2001, p. 206) stated that voluntary associations are crosscutting ties between the societies, and acting as the social networks in building societies together. These associations bridge the gap between the people and social groups by creating social bonds. These organizations contribute in different social arenas, such as education, human services, social welfare, health care and elderly care among others. The modern NGOs are considered one of the major collective actors in the socioeconomic development process; particularly in developing countries they play a very important role in bringing about social transformation. Health care, education and welfare services for people are increasingly provided by NGOs, which are also ensuring social welfare for the people (Jianxiu 2006). Further, NGOs in developing countries also provide services to disadvantaged people suffering from poverty, unemployment, malnutrition and natural disasters (Islam and Morgan 2012a). NGOs' contribution in social capital development is also evident (Islam and Morgan 2012a, b; Abom 2004; Islam 2014a, b), and further helps in developing social networks, social trust and community empowerment among vulnerable people (Islam and Morgan 2012a, b; Abom 2004; Islam 2014a, b). NGOs are also active in leading the popular peoples' movement (Islam and Morgan 2012a) and advocating for peoples' rights. CSOs also contribute to community development through resource management. Further, they encourage a participatory approach by involving local communities. Evidence exists that participation of the community helps in the empowerment of that community in order to increase control over their quality of life (Islam and Mahjabeen 2003; Jenkins 2001), and helps in the redistribution of natural resources, which contributes to improving social justice outcomes (Leuenberger and Wakin 2007). Meeting the needs of the poor through a participatory approach also helps in achieving a socially just society in the region (see Davoudi 2000; Sandercock 1997). Social movements are another aspect of CSOs that ensures the functioning and protection of social, civil and political rights and activities and governance through regular campaigning. The recent inclusion to this social movement campaigning is social justice in urban planning and practice, as the cities and urban settlements have started developing rapidly.

Despite their high impact and commitments for social good, CSOs are being watched very closely by the state. Evidence in the literature shows that a number of studies have examined the accountability practices of different NGOs (see, for example, Goddard and Assad 2006; O'Dwyer and Boomsma 2015; O'Dwyer and Unerman 2010, 2007). Scandals continue to plague the NGOs and most of them are questioned and monitored for their transparency in the public domain (see, for example, Humanitarian Accountability Partnership 2014; Chowdhury 2017;

Ebrahim 2010). NGOs have contributed to the public domain in different ways, such as democratization of states (Clark 1991), facilitating development alternatives (Drabek 1987), developing innovative approaches (Newaz 2003), creating a vehicle for popular participation (Farrington and Bebbington 1993), providing advocates for the poor (Newaz 2003), and implementing cost-effective and cost-efficient means (Bebbington 2004).

In conclusion, many chapters in this book have addressed the valuable contributions by CSOs in the welfare of the South Asian society on the one hand, and the challenges and the perpetual uncertainty of CSOs in many South Asian countries on the other hand. It is also important to note that despite the innovations, technological advancement and economic boom across the globe, societies, in general, continue to face innumerable challenges related to poverty, climate change, violence and unrest. The fruits of globalization and neoliberalism have had not only positive impact, but also negative impact resulting in societal and economic problems in the countries of the South. Social inequality occurs when resources are allocated and distributed unevenly in many societies, and many countries in the South continue to face this in the worst forms and dimensions (UNDP 2017). As a result, an estimated 650 million people still live in extreme poverty as evidence to the income gap between developing and developed countries (Johnson 2018). The overall finding in this volume shows that NGOs' role in social welfare initiatives is largely successful and effectively contributes to people's development in South Asian countries. This was also evident in previous studies (Kwok 2003; Newton 2001; Buckland 1998; Mondal 2000; Islam and Morgan 2012a, b; Abom 2004; Islam 2015; Islam et al. 2013). NGOs are the manifestation of social activities, and usually the expression of grassroots initiatives. These organizations draw attention to social problems that are otherwise not considered by either politicians or producers. These non-governmental initiatives address and attempt to solve societal challenges and issues, and provide indispensable support to the marginalized sections in a society. Despite all these valuable contributions by NGOs, they are largely neglected and not supported in many South Asian countries. In the backdrop of issues related to governmental provisions related to social welfare and their failure in reaching out, there is a need for new research and further evidence. Finally, in the contemporary South Asian society social welfare is a matter of right rather than of need. Hence, the social welfare provisions need to be utilized to bridge the gap between the poor and the rich (Kwok 2003), and to aim to build a sustainable just society.

References

Abom, B. (2004). Social Capital, NGOs, and Development: A Guatemalan Case Study. *Development in Practice, 14*(3), 342–353.

Acemoglu, D., Autor, D., Dorn, D., Hanson, G. H., & Price, B. (2016). Import Competition and the Great US Employment Sag of the 2000s. *Journal of Labor Economics, 34*(S1), S141–S198.

Amin, A. (1997). 'Placing Globalization. 'Theory, Culture & Society, 14 (2), 123–137

Atkinson, A. (2015). *Inequality What Can Be Done?* Cambridge: Harvard University Press.

Autor, D. H., Dorn, D., & Hanson, G. H. (2013). The China Syndrome: Local Labor Market Effects of Import Competition in the United States. *The American Economic Review, 103*(6), 2121–2168.

Bebbington, A. (2004). NGOs and uneven development: geographies of development intervention. Progress in Human Geography, 28(6), 725–747.

Bartels, L. M. (2004). Unenlightened Self-Interest: The Strange Appeal of Estate Tax Repeal. *The American prospect magazine*, A17–A19.

Bartley, T. (2018). *Rules without Rights. Land, Labor, and Private Authority in the Global Economy*. Oxford: Oxford University Press.

Bartley, T., et al. (2015). *Looking behind the Label: Global Industries and the Conscientious Consumer*. Bloomington, IN: Indiana University Press.

Beckert, J. (2020). The Exhausted Futures of Neoliberalism: From Promissory Legitimacy to Social Anomy. *Journal of Cultural Economy, 13*(3), 318–330. https://doi.org/10.1080/17530350.2019.1574867.

Blyth, M. (2013). *Austerity. The History of a Dangerous Idea*. Oxford: Oxford University Press.

Boltanski, L., & Chiapello, È. (2005). *The New Spirit of Capitalism*. London: Verso.

Brown, W. (2015). *Undoing the Demos: Neoliberalism's Stealth Revolution*. New York: Zone Books.

Buckland, J. (1998). Social Capital and Sustainability of NGO Intermediated Development Projects in Bangladesh. *Community Development Journal, 33*(3), 236–248.

Buckley, P. J., & Ghauri, P. N. (2004). Globalisation, Economic Geography and the Strategy of Multinational Enterprises. *Journal of International Business Studies, 35*(2), 81–98.

Cameron, D. (1978). The Expansion of the Public Economy: A Comparative Analysis. *American Political Science Review, 72*(4), 1243–1261.

Casey, J. (2004). Third Sector Participation in the Policy Process: A Framework for Comparative Analysis. *Policy and Politics., 32*(2), 239–256.

Chase-Dunn, C., Kawano, Y., & Brewer, B. D. (2000). Trade Globalization since 1795: Waves of Integration in the World-System. *American Sociological Review, 65*(1), 77–95.

Chowdhury, R. (2017). The Rana Plaza Disaster and the Complicit Behavior of Elite NGOs. *Organization, 24*(6), 938–949.

Clark, John (1991). Democratizing development. London: Earthscan

Connell, R., & Dados, N. (2014). Where in the World Does Neoliberalism Come From? The Market Agenda in Southern Perspective. *Theory and Society, 43*, 117–138. https://doi.org/10.1007/s11186-014-9212-9.

Crouch, C. (2011). *The Strange Non-death of Neoliberalism*. Malden: Polity Press.

Davies, W. (2017a). *The Limits of Neoliberalism. Authority, Sovereignty and the Logic of Competition*. Los Angeles: Sage.

Davies, W. (2017b). *Moral Economies of the Future – The Utopian Impulse of Sustainable prosperity*. Working Paper No 5. CUSP Centre for the Understanding of Sustainable Prosperity.

Davoudi, S. (2000). Sustainability: A New Vision for British Planning System. *Planning Perspectives, 15*, 123–126.

Dobbin, F., & Jung, J. (2010). The Misapplication of Mr. Michael Jensen: How Agency Theory Brought Down the Economy and Why It Might Again. In M. Lounsbury & P. M. Hirsch (Eds.), *Markets on Trial: The Economic Sociology of the U.S. Financial Crisis* (pp. 29–64). Bingley: Emerald Group Publishing Limited.

Drabek, A. G. (1987). Development alternative: the challenges for NGOs—an overview of the issues. World Development, 15(Supplement), ix–xv.

Dreher, A., Sturm, J.-E., & Ursprung, H. W. (2008b). The Impact of Globalisation on the Composition of Government Expenditures: Evidence from Panel Data. *Public Choice, 134*(3-4), 263–292.

Ebrahim, A. (2010). The Many Faces of Nonprofit Accountability. In D. O. Renz (Ed.), *The Jossey-Bass Handbook of Nonprofit Leadership and Management* (3rd ed., pp. 110–121). San Francisco: Jossey-Bass.

Evans, P., & Sewell, W., Jr. (2013). Neoliberalism. Policy Regimes, International Regimes, and Social Effects. In P. Hall & M. Lamont (Eds.), *Social Resilience in the Neoliberal Era* (pp. 35–68). Cambridge: Cambridge University Press.

Farrington, J. and Bebbington, A. (1993). Reluctant partners? Non-governmental organizations, the state and sustainable agricultural development. London: Routledge.

Fligstein, N., & Goldstein, A. (2015). The Emergence of a Finance Culture in American Households 1889–2007. *Socio-Economic Review, 13*, 533–579. https://doi.org/10.1093/ser/mwu035.

Foucault, M. (2008). *The Birth of Biopolitics: Lectures at the Collège de France, 1978–1979: Lectures at the College de France, 1978–1979*. Houdmills: Palgrave.

Fourcade-Gourinchas, M., & Babb, S. L. (2002). The Rebirth of the Liberal Creed: Paths to Neoliberalism in Four Countries. *American Journal of Sociology, 108*, 533–579. https://doi.org/10.1086/367922.

Frumkin, P. (2002). *On Being Nonprofit: A Conceptual and Policy Primer*. Cambridge, MA: Harvard University Press.

Gaston, N., & Rajaguru, G. (2013a). International Migration and the Welfare State Revisited. *European Journal of Political Economy, 29*(1), 90–101.

Gaston, N., & Rajaguru, G. (2013b). International Migration and the Welfare State: Asian Perspectives. *Journal of the Asia Pacific Economy, 18*(2), 271–289.

Goddard, A., & Assad, M. J. (2006). Assad Accounting and Navigating Legitimacy in Tanzanian NGOs Accounting. *Auditing & Accountability Journal, 19*(3), 377–404.

Gordon, D. M. (1987). 'The Global Economy: New Edifice or Crumbling Foundations', New Left Review 168: 24–64

Gozgor, G., & Ranjan, P. (2017). Globalisation, Inequality, and Redistribution: Theory and Evidence. *World Economy, 40*(12), 2704–2751.

Gründler, K., & Köllner, S. (2018). *Culture, Diversity and the Welfare State*. Working Paper.

Guillén, M. F. (2001). Is globalization Civilizing, Destructive or Feeble? A Critique of Five Key Debates in the Social Science Literature. *Annual Review of Sociology, 27*(1), 235–260.

Hall, P. A. (1993). Policy Paradigms, Social Learning, and the State: The Case of Economic Policymaking in Britain. *Comparative Politics, 25*, 275–296. https://doi.org/10.2307/422246.

Hall, P., & Lamont, M. (2013). Introduction. Social Resilience in the Neoliberal Era. In P. Hall & M. Lamont (Eds.), *Social Resilience in the Neoliberal Era* (pp. 1–31). Cambridge: Cambridge University Press.

Handley, Geoff, Kate Higgins, Bhavna Sharma, Kate Bird, and Diana Cammack. "Poverty and Poverty Reduction in sub-Saharan Africa: An Overview of the Issues." Working paper 299. London: Overseas Development Institute, January 2009. www.odi.org.uk/sites/odi.org.uk/files/odi-assets/publications-opinionfiles/860.pdf

Harrington, B. (2016). *Capital without Borders. Wealth Managers and the One Percent.* Cambridge, MA: Harvard University Press.

Herwartz, H., & Theilen, B. (2014). Partisan Influence on Social Spending under Market Integration, Fiscal Pressure and Institutional Change. *European Journal of Political Economy, 34*, 409–424.

Hochschild, A. (2016). *Strangers in Their Own Land: Anger and Mourning on the American Right.* New York: The New Press.

Humanitarian Accountability Partnership [HAP]. (2014). HAP Standard. Retrieved September 8, 2014, from http://www.hapinternational.org/what-we-do/hap-standard.aspx.

Islam, M. R. (2014a). NGOs' Role for Social Capital and Community Empowerment in Community Development: Experience from Bangladesh. *Asian Social Work and Policy Review, 8*(3), 261–274.

Islam, M. R. (2014b). Improving Development Ownership among the Vulnerable People: Challenges of NGOs' Community Empowerment Projects in Bangladesh. *Asian Social Work and Policy Review, 8*(3), 193–209.

Islam, M. R. (2015). Non-governmental Organizations and Community Development in Bangladesh. *International Social Work.* https://doi.org/10.1177/0020872815574133.

Islam, N., & Mahjabeen, Z. (2003). The Role of Civil Society Organizations in Urban Development in Dhaka City. *Oriental Geographer, 47*(2), 13–47.

Islam, M. R., & Morgan, W. J. (2012a). Non-governmental Organizations in Bangladesh: Their Contribution to Social Capital Development and Community Empowerment. *Community Development Journal, 47*(3), 369–385.

Islam, M. R., & Morgan, W. J. (2012b). Agents of Community Empowerment? The Possibilities and Limitations of Non-governmental Organizations in Bangladesh. *Journal of Community Positive Practices, 12*(4), 703–725.

Islam, M. R., Siti Hajar, A. B., & Haris, A. W. (2013). Local Knowledge in the Lips of Globalization: Uncertainties of Community Participation in NGO Activities. *Revista De Cercetare Si Interventie Sociala, 43*, 7–23.

Jenkins, P. (2001). Relationships between the State and Civil Society and Their Importance for Sustainable Development. In M. Carley, P. Jenkins, & H. Smith (Eds.), *Urban Development and Civil Society: The Role of Communities in Sustainable Cities*. London: Earthscan Publications.

Jenson, J., & Levi, R. (2013). Narratives and Regimes of Social and Human Rights. In P. Hall & M. Lamont (Eds.), *Neoliberalism. Policy Regimes, International Regimes, and Social Effects* (pp. 69–98). Cambridge: Cambridge University Press.

Jianxiu, G. (2006). The Role of NGOs in Social Welfare. *Canadian Social Science, 2*(4), 15–18.

Johnson, P. (2018). *Global Philanthropy Report* (1st ed.). Cambridge: Harvard University's John F. Kennedy School of Government.

Korom, P. (2018). Inherited Advantage. Comparing Gifts and Bequests Receiving and Non-receiving Households across the Distribution of Household Wealth in 11 European Countries. *European Sociological Review, 34*, 79–91. https://doi.org/10.1093/esr/jcx084.

Krippner, G. R. (2011). *Capitalizing on Crisis: The Political Origins of the Rise of Finance*. Cambridge, MA: Harvard University Press.

Krippner, G. R. (2017). Democracy of Credit: Ownership and the Politics of Credit Access in Late-Twentieth Century America. *American Journal of Sociology, 123*, 1–47. https://doi.org/10.1086/692274.

Kwok, J. (2003). Social Welfare, Social Capital and Social Work: Personal Reflection of a Hong Kong Social Worker. Retrieved April 23, 2014, from http://www.lwb.gov.hk/download/services/events/040411_wel_forum/16_joseph_kwok_paper.pdf.

Leibrecht, M., Klien, M., & Onaran, O. (2011). Globalisation, Welfare Regimes and Social Protection Expenditures in Western and Eastern European Countries. *Public Choice, 148*(3-4), 569–594.

Leuenberger, D. Z., & Wakin, M. (2007). Sustainable Development in Public Administration Planning: An Exploration of Social Justice, Equity, and Citizen Inclusion. *Administrative Theory & Praxis, 29*(3), 394–411. https://doi.org/10.1080/10841806.2007.11029601.

Lim, S., & Burgoon, B. (2018). Globalisation and Support for Unemployment Spending in Asia: Do Asian Citizens Want to Embed Liberalism? *Socio-Economic Review*, 0-0, 1–35. https://doi.org/10.1093/ser/mwy032.
Meinhard, S., & Potrafke, N. (2012). The Globalisation-Welfare State Nexus Reconsidered. *Review of International Economics, 20*(2), 271–287.
Meyer, K. E. (2017). International Business in an Era of Anti-globalization. *Multinational Business Review, 25*(2), 78–90.
Mittelman, J. H. (1996). 'How Does Globalization Really Work?', in J. H. Mittelman (ed.) Globalization: Critical Reflections (pp. 229–41). London: Lynne Reinner
Mohanty, P., Chandran, A., & Gantait, A. (2015). Is Tourism Driven Poverty Alleviation 'Rhetoric' or 'Pragmatic Ideal'? Scanning the Academic Discourses. *South Asian Journal for Tourism & Heritage, 9*(2), 51–61.
Mondal, A. H. (2000). Social Capital Formation: The Role of NGO Rural Development Programs in Bangladesh. *Policy Sciences, 33*(3 & 4), 459–475.
Newton, K. (2001). Trust, Social Capital, Civil Society, and Democracy. *International Political Science Review, 22*, 201–214.
Newaz, W. (2003). Impact of micro-credit programs of two local NGOs on rural women's lives in Bangladesh, academic dissertation. Finland: The Faculty of Economics and Administration, University of Tampere.
O'Dwyer, B., & Boomsma, R. (2015). The Co-construction of NGO Accountability: Aligning Imposed and Felt Accountability in NGO-Funder Accountability Relationships Accounting. *Auditing & Accountability Journal, 28*(1), 36–68.
O'Dwyer, B., & Unerman, J. (2007). From Functional to Social Accountability: Transforming the Accountability Relationship between Funders and Non-governmental Development Organisations. *Accounting Auditing & Accountability Journal, 20*(3), 446–471.
O'Dwyer, B., & Unerman, J. (2010). Enhancing the Role of Accountability in Promoting the Rights of Beneficiaries of Development. *NGOs Accounting and Business Research, 40*(5), 451–471.
Onaran, O., Boesch, V., & Leibrecht, M. (2012). How Does Globalisation Affect the Implicit Tax Rates on Labor Income, Capital Income, and Consumption in the European Union? *Economic Inquiry, 50*(4), 880–904.
Ortiz, E., & Roser, M. (2015). *World Poverty*. Our World in Data. Retrieved https://ourworldindata.org/world-poverty/.
Palan, R. (2017). Futurity, Offshore, and the International Political Economy of Crime. In J. Beckert & M. Dewey (Eds.), *The Architecture of Illegal Markets.*

Towards an Economic Sociology of Illegality in the Economy (pp. 108–120). Oxford: Oxford University Press.

Peters, D. H., Mirchandani, G. G., & Hansen, P. M. (2004). Strategies for Engaging the Private Sector in Sexual and Reproductive Health: How Effective Are They? *Health Policy and Planning, 19*(Suppl 1), i5–21.

Phillips, S. D. (2003). Voluntary Sector—Government Relationships in Transition: Learning from International Experience. In K. Brock & K. G. Banting (Eds.), *The Nonprofit Sector in Interesting Times: Case Studies in a Changing Sector* (pp. 17–71). Montreal and Kingston: McGill-Queen's University Press.

Piketty, T. (2014). *Capital in the Twenty-First Century*. Cambridge: Harvard University Press.

Potrafke, N. (2009). Did globalisation Restrict Partisan Politics? An Empirical Evaluation of Social Expenditures in a Panel of OECD Countries. *Public Choice, 140*(1-2), 105–124.

Potrafke, N. (2015). The Evidence on Globalisation. *World Economy, 38*(3), 509–552.

Putnam, R. D. (2000). *Bowling Alone: The Collapse and Revival of American Community*. New York: Simon & Schuster.

Rodrik, D. (1998). Why Do More Open Economies Have Bigger Governments? *Journal of Political Economy, 106*(5), 997–1032.

Rudra, N. (2002). Globalisation and the Decline of the Welfare State in Less Developed Countries. *International Organization, 56*(2), 411–445.

Rugman, A. M., & Verbeke, A. (2004). A Perspective on Regional and Global Strategies of Multinational Enterprises. *Journal of International Business Studies, 35*(1), 3–18.

Robertson, R. (1992). Globalization: Social Theory and Global Culture. London: Sage

Sandercock, L. (1997). The Planner Tamed: Preparing Planner for the Twenty First Century. *Australian Planner, 34*(2), 90–95.

Scholte, J.A. (2005) Globalization: A Critical Introduction, second edn. Basingstoke: Palgrave Macmillan.

Scholte, J.A. (2000). Globalisation. A Critical Introduction. Basingstoke: Palgrave.

Schmidt, V., & Thatcher, M. (Eds.). (2013). *Resilient Liberalism in Europe's Political Economy*. Cambridge: Cambridge University Press.

Schuknecht, L., & Zemanek, H. (2018). *Social Dominance*. CESifo Working Paper No. 6894.

Schulze, G. G., & Ursprung, H. W. (1999). Globalisation of the Economy and the Nation State. *World Economy, 22*(3), 295–352.
Sinn, H.-W. (1997). The Selection Principle and Market Failure in Systems Competition. *Journal of Public Economics, 66*(2), 247–274.
Sinn, H.-W. (2003). *The New Systems Competition*. Oxford: Blackwell.
Streeck, W. (2014). *Buying Time: The Delayed Crisis of Democratic Capitalism* (P. Camiller, Trans.). London: Verso.
Thelen, K. (Ed.). (2014). *Varieties of Liberalization and the New Politics of Social Solidarity*. Cambridge: Cambridge University Press.
UNDP (Ed.). (2017). *UNDP Strategic Plan 2018–2021 Special Session 2017* (1st ed., pp. 1–15). New York, NY: UNDP.
UNGA. (2015). Transforming Our World: The 2030 Agenda for Sustainable Development. A/RES/70/1, 21 October.
Ursprung, H. W. (2008). Globalisation and the Welfare State. In S. N. Durlauf & L. E. Blume (Eds.), *The New Palgrave Dictionary of Economics* (2nd ed.). Köln: Palgrave Macmillan.
Verbeke, A., Coeurderoy, R., & Matt, T. (2018). The Future of International Business Research on Corporate Globalization that Never Was…. *Journal of International Business Studies, 49*(9), 1101–1112.
Walter, S. (2010). Globalisation and the Welfare State: Testing the Micro foundations of the Compensation Hypothesis. *International Studies Quarterly, 54*(2), 403–425.
Waters, H., Hatt, L., & Peters, D. (2003). Working with the Private Sector for Child Health. *Health Policy Plan, 18*, 127–137.
Wibbels, E. (2006). Dependency Revisited: International Markets, Business Cycles, and Social Spending in the Developing World. *International Organization, 60*(2), 433–468.
World Bank. (2016). Poverty Overview. *Worldbank.Org*. Retrieved from http://www.worldbank.org/en/topic/poverty/overview.
Yay, G. G., & Aksoy, P. (2018). Globalisation and the Welfare State. *Quality and Quantity, 52*(2), 1015–1040.
Yeates, N. (2001). Globalization and Social Policy. London: Sage.

Index[1]

[1] Note: Page numbers followed by 'n' refer to notes.

© The Author(s) 2020
Md. N. Momen et al. (eds.), *Building Sustainable Communities*,
https://doi.org/10.1007/978-981-15-2393-9

R

The manufacturer's authorised representative in the EU is Springer Nature Customer Service Centre GmbH, Europaplatz 3, 69115 Heidelberg, Germany. If you have any concerns regarding our products, please contact ProductSafety@springernature.com

Printed and bound by CPI Group (UK) Ltd, Croydon, CR0 4YY
30/09/2026
02225460-0012